Ecological Studies, Vol. 118

Analysis and Synthesis

Edited by

O.L. Lange, Würzburg, FRG
H.A. Mooney, Stanford, USA

Ecological Studies

Volumes published since 1989 are listed at the end of this book.

Springer
New York
Berlin
Heidelberg
Barcelona
Budapest
Hong Kong
London
Milan
Paris
Santa Clara
Singapore
Tokyo

Susan Fox Robert A. Mickler
Editors

Impact of Air Pollutants on Southern Pine Forests

With 88 Illustrations, 25 in color

Springer

Susan Fox
USDA Forest Service
Southern Research Station
Raleigh, NC 27606
USA

Robert A. Mickler
ManTech Environmental Technology, Inc.
Raleigh, NC 27606
USA

Library of Congress Cataloging-in-Publication Data

Impact of air pollutants on southern pine forests / Susan Fox and Robert A. Mickler, editors.
p. cm.—(Ecological studies ; vol. 118)
Includes bibliographical references (p.) and index.
ISBN 0-387-94382-X (hardcover)
1. Pine—Effect of air pollution on—Southern States. 2. Pine—Wounds and injuries—Southern States. 3. Forest ecology—Southern States. 4. Forest health—Southern States. I. Fox, Susan (Susan A.) II. Mickler, Robert A. III. Series: Ecological studies ; v. 118.
SB745.I48 1995
634.9'751619—dc20 95-16799

Printed on acid-free paper.

Production coordinated by Chernow Editorial Services, Inc., and managed by Karen Phillips; manufacturing supervised by Jacqui Ashri.
Typeset by Best-set Typesetter Ltd., Hong Kong.
Printed and bound by Braun-Brumfield, Inc., Ann Arbor, MI.
Printed in the United States of America.

9 8 7 6 5 4 3 2 1

ISBN 0-387-94382-X-Springer-Verlag New York Berlin Heidelberg

Preface

The public's attitude toward air pollution in the United States evolved substantially during the 1960s. One of the results of the nation's emerging environmental ethic was the creation of the U.S. Environmental Protection Agency (EPA) in December of 1970. Prior to this time, research was focused on the impacts of air pollution on human health and welfare and was largely conducted by several federal research agencies, which included the Department of Health, Education, and Welfare; the National Oceanic and Atmospheric Administration; and the U.S. Department of Agriculture. After the creation of the EPA, much of this work was consolidated in one regulatory agency, which resulted in periodic evaluations of the various effects of atmospheric pollution on human health, materials, agriculture, and forest ecosystems.

At the same time that environmental interest was growing in the United States, concern increased in the European scientific community and public over the ecological impacts of acidic deposition. As the magnitude of the damage to European lakes and streams and the widespread decline in Norway spruce and silver fir was reported, concern that similar problems were occurring in the United States increased substantially. This concern was heightened by press reports of high elevation spruce–fir forest declines in the Adirondack and Appalachian Mountains and the decline and death of sugar maples in the northeastern United States and Canada. During the early 1980s, it was not uncommon to read stories from respected news organizations quoting scientists as suggesting that our lakes and streams would soon be too acidic to support existing aquatic ecosystems and that much of the forests in the eastern United States were exhibiting symptoms of substantial decline and death.

During the early 1980s, acid deposition emerged as the dominant theme for air pollution research in the U.S. government. The National Acid Precipitation Assessment Program (NAPAP) was created by congressional action in 1979 to develop and implement an "acid rain" research strategy. NAPAP was a collaborative effort among several federal agencies. It was organized around a series of work groups responsible for the effects on various aspects of the acid deposition problem. NAPAP was authorized for 10 years and scheduled for a culminating assessment in 1990. Because of the interagency nature of the program, significant compromises and accommodations had to be made during the initial stages of research funding and implementation. One consequence of difficulties encountered in implementing such a large national research program was the five-year delay in starting the Forest Response Program (FRP). The FRP was a joint interagency research program initiated in 1985 between the USDA Forest Service and the U.S. EPA. The FRP was organized into four regional research cooperatives to address three principal policy questions: (1) Is there a significant problem of forest damage in North America that might be caused by acid deposition alone or in combination with other pollutants; (2) What is the causal relationship between air pollutants and forest damage; and (3) On a stand or regional basis, what is the dynamic relationship between air pollutants and forest damage?

As is often the case with federally funded research, events overtook the time frame of the Forest Response Program. The relatively slow start of the program combined with the scientific challenge of testing hypotheses on long-lived perennial trees within five years proved a major challenge to the scientific community. At the time of the enactment of the Clean Air Act Amendments, only the first policy question could be addressed in any significant way. The scientific results that would address the remaining two policy questions were barely underway before the enactment of the Clean Air Act Amendments.

In spite of the inability of scientific inquiry to meet political schedules, much important work has been accomplished to elucidate the relative impacts of acidic deposition and other air pollutants on forest ecosystems. This is particularly true in the southern United States, where researchers in the Southern Commercial Forest Research Cooperative (SCFRC) focused their attention on a reported growth decline in southern pine forests. The work reported in this book, while too late to impact the Clean Air Act amendments directly, will be of substantial use as the nation reevaluates its policies toward regulating and managing its ecological resources. While political considerations will often overshadow scientific data in the public policy-making arena, scientific data must continue to be the basis for making policy decisions. Continuing our efforts to improve our understanding of air pollution effects on forest ecosystems is essential, independent of the political time tables, which will change much more rapidly than our ability to understand the complex interactions that occur in our forest's biological systems.

Gerard Hertel
USDA Forest Service
Program Manager
Forest Response Program

Roger Blair
US Environmental Protection Agency
Deputy Program Manager
Forest Response Program

Acknowledgments

This book is the culmination of five years of research planning and implementation. Many individuals were responsible for the start of this work, and many others carried the work to its conclusion. We acknowledge the initial research planning efforts of Ellis Cowling, Donald Marx, and James Woodman, whose vision for the SCFRC began this work. Ann Bartuska served as the first SCFRC Program Manger and was assisted by Kimberly Joyner in the planning and implementation of the research program. Al Lucier and Charles Webb provided helpful insight from forest industry. Many research administrators from the USDA Forest Service and the U.S. Environmental Protection Agency helped define the forest management and regulatory goals of their agencies into the policy and scientific questions that focused and guided our research. We would like to thank especially Jerry Sesco, William Sommers, Richard Smythe, Gerard Hertel, Courtney Riordan, Gary Foley, and Roger Blair.

Many scientists and their technicians contributed a tremendous amount of time and effort to accomplish the research goals of the Southern Commercial Forest Research Cooperative. Their conscientious and dedicated work has advanced our knowledge of forest science. It is with our thanks that this book is dedicated to them.

The following individuals kindly agreed to review chapters of this book: Ann Bartuska, Margi Bohm, Stanley Buol, Noel Cost, Ivan Fernandez, Robert Goldstein, Walter Heck, Thomas Hennessey, Gerard Hertel,

Judson Isebrands, Keith Jensen, David Karnosky, H. Joseph Kleiss, John Lawrence, Peter Lorio, Margaret Miller–Weeks, Volker Mohen, Ram Oren, Gregory Reams, David Read, Patrick Reid, Wayne Robarge, Hugo Rogers, John Seiler, John Skelly, Boyd Strain, Susan Spruill, and Paul Van Deusen. We thank Art Johnson, Dale Johnson, and Ellis Cowling for reviewing the first draft of this book in its entirety. The thoughtful comments from all the reviewers increased the scientific quality of this book and are gratefully acknowledged by the authors and editors.

Special thanks go to Larry Royer for data analysis and geographic information system maps of pine species used in Chapter 1; Ralph Baumgardner, Doug Shadwick, and Luther Smith for providing many of the air quality and atmospheric deposition maps used in Chapter 3; and Tracy Huneycutt for graphic art support and the design of the book cover illustration.

Support for the Southern Commercial Forest Research Cooperative research discussed in this book was provided by the USDA Forest Service, the U.S. Environmental Protection Agency, and the National Council for Air and Stream Improvement. The support of the Southern Research Station of the USDA Forest Service was essential in the preparation and production of this book.

This book has not been subject to policy review by the USDA Forest Service and the U.S. Environmental Protection Agency and should not be construed to represent the policies of either agency.

Contents

Contributors

Eric R. Allen	Department of Environmental Engineering Science, University of Florida, Gainesville, FL 32611, USA
Paul Berrang	USDA Forest Service, Dry Branch, GA 31020, USA
Arthur H. Chappelka	School of Forestry, Auburn University, Auburn, AL 36849, USA
Michael L. Cline	Woodlot Alternatives, Inc., Topsham, ME 04086, USA
Phillip M. Dougherty	USDA Forest Service, Research Triangle Park, NC 27709, USA
Richard B. Flagler	Natural Resource Ecology Laboratory, Colorado State University, Fort Collins, CO 80523, USA

Susan Fox — USDA Forest Service, Southern Research Station, Raleigh, NC 27606, USA

Henry L. Gholtz — Department of Forestry, University of Florida, Gainesville, FL 32611, USA

Fred P. Hain — Department of Entomology, North Carolina State University, Raleigh, NC 27695, USA

Allen S. Heagle — USDA Agricultural Research Service, Raleigh, NC 27606, USA

John D. Hodges — Department of Forestry, Mississippi State University, Mississippi State, MS 39762, USA

Jon D. Johnson — Department of Forestry, University of Florida, Gainesville, FL 32611, USA

Daniel Markewitz — School of the Environment, Duke University, Durham, NC 27706, USA

Donald H. Marx — Mycorr Tech, Inc., Frogmore, SC 29920, USA

James S. Meadows — USDA Forest Service, Stoneville, MS 38776, USA

Robert A. Mickler — ManTech Environmental Technology, Inc., Raleigh, NC 27606, USA

Daniel D. Richter — School of the Environment, Duke University, Durham, NC 27706, USA

Thomas W. Sasek — Department of Biology, Northeast Louisiana University, Monroe, LA 71209, USA

Philip Schoeneberger — Natural Resource Conservation Service, Lincoln, NE 68508, USA

Robert O. Teskey — School of Forestry Resources, University of Georgia, Athens, GA 30603, USA

Introduction: The Southern Commercial Forest Research Cooperative

Susan Fox

Coordinated research programs are a relatively new approach to organizing and conducting research in the forestry community. Disciplines such as engineering, aeronautics, medicine, and others have cultivated a programmatic approach to addressing broad-scale, complicated topics, but such an approach has been the exception in forestry. Instead, the approach has emphasized funding individual research projects, organized under topical areas. This has been the primary method of addressing research. Much more funding has been allocated to understanding the function of the parts and relating them to the whole or understanding one aspect of the whole, such as hydrology, physiology, or biochemistry.

In 1980, Congress demonstrated interest in a new, comprehensive approach to science by mandating the National Acid Precipitation Assessment Program (NAPAP, 1985). NAPAP was established as a broad, interdisciplinary research program focused on providing information to decision makers to evaluate the Clean Air Act. NAPAP's target, defined in the Acid Rain Act, was to determine whether current emissions of sulfur and nitrogen in the atmosphere were harmful to people, materials, crops, lakes, streams, and forests. It established the framework for conducting an evaluation of acid rain impacts on people and resources.

The forestry component of NAPAP was funded with approximately $55 million. More than providing an adequate funding basis, the NAPAP organization forced federal agencies with very different mandates to work together. University and industry scientists collaborated because the scope of the research was so complex and broad. The collaborative nature of the research was particularly beneficial in addressing acid rain impacts on forest resources. These long-lived, complex communities of trees and other organisms required the expertise of scientists from many disciplines.

NAPAP provided funding for a five-year research program on the effects of acid rain on forests. This book describes how this was accomplished for southern commercial forests, one component of the NAPAP forestry research effort. It discusses the research agenda for this important resource and outlines the major findings of the Southern Commercial Forest Research Cooperative.

Development of a Coordinated National Research Program

The Clean Air Act was first passed in 1967. It signaled a new level of awareness about air pollution problems. But more importantly, the Clean Air Act highlighted the limitations of our current knowledge about the potential hazards air pollution presented to human health, vegetation, water resources, and materials. It authorized funding for research to evaluate air pollution impacts, with the primary focus on sulfur dioxide effects on vegetation around point sources such as coal-fired power plants and metal smelters (Shriner, 1990).

In 1980, new legislation—the Acid Precipitation Act—accelerated research on air pollution effects. The Acid Precipitation Act established a task force of federal agencies to develop and implement a comprehensive national program on acid rain impacts. The purpose of this program was to increase our understanding of the effects of acidic deposition in a manner and time frame that would aid decision makers (NAPAP, 1985). The task force named this new Program the National Acid Precipitation Assessment Program (NAPAP, 1985). NAPAP was conceived as a multidisciplinary, large-scale effort supported by 12 federal departments and agencies and four national laboratories (NAPAP, 1985).

Prior to the passage of the Acid Precipitation Act, several federal agencies were conducting acid rain research individually and with limited communication and coordination. One of NAPAP's unique contributions to air pollution research was its role as a catalyst in bringing federal agencies to work together. Acid deposition impacts on forest ecosystems and is such a complex issue that understanding current impacts and potential future threats necessitated cooperative research of chemists, meteorologists, biologists, economists, physiologists, and scientists from many other disciplines to exchange ideas and share findings. Research, monitoring, and assessment were the three primary methods of studying acid precipitation effects in this program. A significant portion of the funding (70%) was from ongoing research and monitoring or redirected from other research activities.

Air Pollution Research on Forests

The first years of NAPAP were devoted to the study of sulfur and nitrogen impacts on agricultural crops, materials and cultural resources, and aquatic ecosystems. A comprehensive component that addressed impacts on forests, called the Forest Response Program (FRP), was initiated in the mid-1980s. Public opinion galvanized support for a forestry component of NAPAP. The public was already sensitized to the potentially devastating effects of acid rain on forests in former Eastern block countries and, to a lesser extreme but still extensive, forests in western Europe.

The FRP was funded mainly by the U.S. Department of Agriculture Forest Service (USFS), the U.S. Environmental Protection Agency (U.S. EPA); with important contributions by the National Council of the Paper Industry for Air and Stream Improvement, and the Electric Power Research Institute. The goal of the FRP was to evaluate acid rain impacts on forests in the United States (Carey et al., 1986). The objectives were stated in three broad policy questions: (1) Is there a significant problem of forest damage in the United States that might be caused by acidic deposition, alone or in combination with other pollutants? (2) By what mechanisms does acid deposition, alone or in combination with other pollutants, contribute to forest damage in the United States? and (3) On a regional basis, what is the dynamic relationship between air pollutants and forest damage? These three policy questions were further defined as scientific questions around which research cooperatives developed their agendas. The policy questions and their accompanying scientific questions are listed in Table 1.

Monitoring

The FRP supported two monitoring efforts: one to monitor the quality of the air and to determine regional pollution concentrations/deposition, called the Atmospheric Exposure Cooperative; and one to monitor the health of forests, called the National Vegetation Survey.

The FRP established the National Vegetation Survey (NVS) to address the first policy question (see Table 1): to define if there is a problem with forest conditions that is being caused by air pollution. The basic approach of the NVS was epidemiological, characterized by broad surveys of forest conditions and a search for patterns of conditions that may relate to patterns of pollution distribution or other environmental factors. The

Table 1. Policy and Scientific Questions in the Forest Response Program

Policy	1.0	Is there a significant problem of forest damage in the United States that might be caused by acidic deposition, alone or in combination with other pollutants?
Scientific	1.1	Are changes in condition greater than can be attributed to typical trends and levels of natural variability?
	1.2	What spatial patterns exist in forest condition, and how do these patterns relate to spatial patterns of pollutant exposure?
Policy	2.0	What is the causal relationship between acidic deposition, alone or in combination with other pollutants, and forest damage in the United States?
Scientific	2.1	What is the effect of sulfur and/or nitrogen pollutants on forests through the mechanisms of (1) direct toxicity of roots, mycorrhizae, soil microbial populations by mobilized metals in acidified soil water; (2) nitrogen toxicity to mycorrhizae; (3) increased leaching of soil nutrients resulting in reduced nutrient availability?
	2.2	What is the effect of sulfur and/or nitrogen pollutants, alone or in combination with oxidants, on forests by the mechanism of increased leaching of foliar nutrients?
	2.3	What is the effect of sulfur and/or nitrogen pollutants, alone or in combination with oxidants, on forests by the mechanism of altering photosynthesis, respiration, and carbon allocation patterns with possible induction of water and/or nutrient stress?
	2.4	What is the effect of nitrogen compounds, possibly in combination with oxidants, on forests by the mechanism of delayed cold hardening or early break in dormancy resulting in increased winter injury?
	2.5	What is the effect of sulfur and/or nitrogen pollutants, alone or in combination with oxidants, on forests by the mechanism of disrupting reproduction or regeneration?
	2.6	What is the effect of sulfur and/or nitrogen pollutants, alone or in combination with oxidants, on forests by the mechanism of altered susceptibility to insects and pathogens?
Policy	3.0	What is the dose/response relationship between acidic deposition, alone or in combination with other pollutants, and forest damage in the United States?
Scientific	3.1	What is the dynamic response of forests to specified levels of acidic deposition, alone or in combination with other pollutants?

approach was inferential, utilizing existing information about forest conditions together with air quality data on pollution emissions and concentrations to determine where potential problems might occur (Carey et al., 1986). Two particular forest health issues were given special attention by the NVS: the decline of red spruce at high elevations in the Appalachians and southern pine growth decline.

The Atmospheric Exposure Cooperative (AEC) was formed to provide atmospheric monitoring data and technical information, such as gaseous pollutant concentrations within a forest canopy, to researchers and administrators in the FRP. Its activities included instrumentation and operation of sites to collect atmospheric data and to compile data collected outside the FRP that were useful for evaluating air quality, pollutant deposition and concentrations, and emissions of pollutants and their transport (Saint, 1987).

Research on Cause and Effect

Research addressing policy questions 2 and 3 was focused at four forest types: Appalachian spruce-fir forests, eastern hardwood forests, western conifer forests, and southern commercial forests. There were a number of informal selection criteria used in the choice of these four forest types: evidence of a potential air pollution–related health problem, elevated air pollution levels, existing information on species from which to build a research program, commercial importance of the forests, and geographic distribution (so information could be gained to help evaluate how widespread air pollution impacts might be nationwide).

The low pH, high acidity of rainfall in the northeastern Appalachian mountains was suspected to play a role in the decline of red spruce. This was an important factor in the establishment of the Spruce-Fir Forest Research Cooperative. Sugar maple trees in Canada and the northeastern United States were showing evidence of damage and decline, and there was concern that this might be related to air pollution. Due to the extensive location of industry in the Northeast, the United States was particularly concerned about export of pollution to Canada. The Eastern Hardwoods Forest Research Cooperative was established to assess the role of pollution in sugar maple and other hardwood species. In the southern United States, USDA Forest Service reports suggested air pollution as a potential causal factor in a decline in the growth rate of natural southern pine forests; this played a major role in the FRP's establishment of the Southern Commercial Forest Research Cooperative.

Rationale for a Southern Commercial Forest Research Cooperative

The reported decline of growth in southern forests was of tremendous interest. Forestry is big business in the southern United States. Almost half of the timber harvested in the United States comes from the South, and a very small percentage of this is from federal lands. Most is harvested off of small private land holdings (e.g., less than 100 hectares) or industry lands. These forests are a major source of employment for people in all

12 southern states. In 1985, forest surveys for three southern states showed a 19% drop in growth rates for some southern pines. Scientists were concerned and forest industry officials were alarmed. Industry is a major landowner in the region and buys even more timber than it owns off of private lands for sawlogs and pulp production. A reduction in growth rates from these forests was going to have a big impact on the forestry industry and, in turn, on regional employment and the national lumber supply if the growth reductions were actually occurring.

This started a big debate. While all landowners, public and private, rely on the periodic surveys of timber conducted by the Forest Service, there were also other data available from industry that, although not as extensive, did not indicate a growth reduction.

Were Southern Pines Declining?

Preliminary analysis by the Forest Service of the data suggested that there were a number of potential causes for the reductions, but no one cause was singled out as likely (Sheffield et al., 1985). It was suggested that such widespread reductions were not due to the suite of logical factors that could have resulted in such a decrease, including stand aging, increasing stand density, hardwood competition, drought, prescribed fire, and the fact that many of the stands had been planted on marginally productive agricultural sites.

The furor raised over this report has not yet died down. Intense debate has accompanied every mention of the growth decline since it was first reported in 1985. One reason for the intense debate is that the report raised questions about methods used by the Forest Service to collect and analyze inventory data. Actually, these questions had been brought up numerous times by both Forest Service personnel and industry users of the data, but the report revitalized and refocused the debate.

Additional model-based analyses demonstrated a reduction in growth rates of southern pines (Bechtold et al., 1991; Ruark et al., 1991; Zahner et al., 1989). These studies were evaluated in a study of regional growth trends (Van Deusen, 1992), which confirmed a general decrease in basal area growth starting in the mid-1970s. However, it also demonstrates that stand dynamics is a likely explanation for the slowing of growth (e.g., following a period of rapid growth prior to the mid-1970s, the stand canopy closed and competition led to a slowing of basal area growth; see Van Deusen, 1992).

Does this mean that pollution is affecting regional growth of pine stands in the South? No. Does it mean that pollution is not a factor in regional growth of southern pines? No. What the debate about southern pine growth decline did make clear was that current forest survey data are capable of defining regional timber volume, as they were designed to do, and further that the data from forest surveys can be used to identify changes in growth rates over time. What the survey data cannot show is why growth rates change. Certainly, analysis of forest survey data may suggest whether stand dynamics are involved, including stand age, density, and competition. But the surveys do not collect data, nor are there methods to permit an evaluation of other factors that can influence stand growth. Pollution is one of these factors, but pollution interacts with drought, temperature, insects and diseases, nutrient availability, and other site-specific characteristics that are capable of having a major impact on stand growth. There are no methods available even now, a decade after the initial debate on southern pine growth rates began, to determine the role of any stressor if it operates below the acute level, as a chronic stress.

The mid-1980s discussion on southern pine growth decline was one of several factors in the establishment of the Southern Commercial Forest Research Cooperative (SCFRC). It helped identify that forest survey methods need to be developed not only to document timber volume but also to indicate what is causing a change in volume.

Research Agenda for the SCFRC

The first planning for an SCFRC was initiated in 1984. The debate over regional growth reductions was used to justify funding the program, but there were other equally compelling reasons cited. The importance of the South as the major supplier of timber together with the rapid population increases and accompanying pollution was even more compelling to the U.S. Forest Service and the U.S. EPA in the establishment of the SCFRC (Marx et al., 1985). Another equally important factor in the establishment of the SCFRC was the tremendous amount of research that had been conducted for decades in the South. Of all the regions in the United States, forests in the southern region have one of the greatest legacies of research. Much had been done in the South to investigate and understand how trees function, ways to improve growth, how rapidly trees grow, and what damages trees and slows growth. Much was know about southern yellow pines, morphology, silviculture, physiology, and biology, which provided an excellent basis to develop a research program.

Preliminary efforts involved the evaluation of current information about southern commercial forests. These efforts were led by the Forest Service and EPA and involved scientists and administrators from federal agencies, private organizations, and regional universities. After careful scrutiny of the literature, the SCFRC formulated plans for a multiyear, multidisciplinary research program. The objective of the SCFRC was to determine if, when, where, and to what extent airborne toxic gases, nutrient elements, toxic metals, acidic substances, and growth-altering organic chemicals were causing or had the potential to cause beneficial and/or detrimental changes in the health, productivity, and stability of commercial forests in the southern United States. The SCFRC focused on the FRP's policy questions 2 and 3, which addressed the causal relationship between pollution and forest damage and the relationship between exposure and response to pollution. To formulate researchable hypotheses regarding the policy questions of the FRP, the questions were restated as scientific questions. Table 1 provides a list of the scientific questions for each policy question.

The primary focus of the SCFRC was on the scientific question (question 2.3 in Table 1) concerning the effect of sulfur and nitrogen pollutants, alone and in combination with ozone, on photosynthesis, respiration, and carbon allocation patterns. The secondary focus was on the question of pollution impact on foliar nutrient leaching (question 2.2). The question of sulfur and nitrogen deposition impacts on southern forest soils comprised the third focus, but only to a very limited extent based on the analysis of existing data collected from some long-term monitoring plots.

Species Selection

The SCFRC decided to work with three pine species: loblolly pine (*Pinus taeda L.*), slash pine (*Pinus elliottii* Engelm. var. *elliottii*), and shortleaf pine (*Pinus echinata* Mill). Loblolly pine was chosen because of its prevalence in southern commercial forests and its economic importance. With shortleaf pine, it comprises the major commercial forest ecosystem (24% of the commercial acreage), and it represents the keystone of the southern pine products industry. Loblolly pine is the most dominant single species of the region, occupying nearly 16% of the total acreage. Equally important in its selection was the availability of information on this species. It is likely that more is known about this species—its silvicultural characteristics, its wood properties, and its utilization—than any other species in the world. The availability of such extensive information on this species provides a strong basis from which experimental results can be extrapolated and interpreted.

Slash pine is the second most important pine species in southern forestry, occupying around 7% of forested acreage. By 1965, its range had been extended by artificial regeneration programs begun in the late 1930s well beyond its natural range. Slash pine became widely used in southern forestry based on its rapid growth, early culmination of mean annual increment, and favorable resistance to major insects and diseases.

Shortleaf pine has the widest range of any southern pine. In 1965, it was found in 22 states (although at one time, its range was even wider than today). It grows in humid areas; however, with regard to temperature and moisture, it is the least exacting of the southern pine. The wide distribution of shortleaf pine is partially due to its ability to grow on a great variety of soils.

Pollutants

The NAPAP was established to determine the impacts of acid rain and related pollutants on people and resources. Therefore, the SCFRC studied the effects of acid rain on trees but not acid rain alone. The scientific community was adamant about the need to study acid rain in conjuction with ozone. Ozone is the major pollutant of concern in the southern United States. It occurs in concentrations sufficient to cause visible injury to vegetation in many parts of the eastern United States (Garner et al., 1989). It is a particular concern in the South because rapid population increases resulted in dramatic increases in automobiles and industry, both of which release hydrocarbons and nitrogen oxides, precursors of ozone. In addition, the regional climate of the South is favorable to the conversion of these precursors into ozone (i.e., there are numerous sunny days). Furthermore, the typical growing season has many days characterized by stagnant air masses trapped over the central South, which results in a buildup of ozone in the growing season.

Methods

The SCFRC used a number of state-of-the-art research approaches to evaluate air pollution impacts (see Table 2). Controlled field studies were the most widely used approach, and almost all employed open-top chambers (Heagle et al., 1973). Controlled field research provided a good balance of realistic field conditions with sufficient control for determining air pollution impacts. Another controlled field approach used was branch chambers. Whereas the open-top chamber could only be used to test young, seedling material, branch chambers permitted testing of mature trees under relatively controlled conditions. Controlled laboratory and greenhouse studies were also employed. Both of these approaches were used with seedling material and mainly to screen large numbers of genotypes to pollution. Screening of genotypes was used to estimate the amplitude of species response and to select sensitive and nonsensitive genotypes for use in multiyear controlled field studies.

Several SCFRC studies used the uncontrolled field study approach. Similar to monitoring projects in the NVS, the uncontrolled field studies characterized field responses to ambient conditions. Climate and pollution concentrations were monitored closely, as were various biological response variables, determined by the goals of the project. A final approach used was literature reviews. The purpose of the literature reviews was to synthesize existing information and to evaluate whether, based on current knowledge, additional research was warranted.

Table 2 shows all projects funded by the FRP that supported the goals and objectives of the SCFRC. It identifies thc principal scientist, provides a brief title, characterizes the research approach, and indicates the funding

Table 2. FRP Projects Funded in Support of the Southern Commercial Forest Research Cooperative

Scientific Question	Lead Scientist	Title (Abbreviated)	Research Approach	Funded By*
1.1, 1.2	E. Allen	Monitoring air quality	Monitoring	AEC
1.1, 1.2	J.P. Bennett	Air pollution impacts on eastern white pine	Uncontrolled field	NVS
1.1, 1.2	T.R. Dell	Statistical methods for evaluation of pollution impacts on forests	Methods development	NVS
1.1, 1.2	T.R. Dell	Southern forest growth trends	Data analysis	NVS
1.1, 1.2	J.P. McClure	Testing hypotheses of southern pine growth decline	Methods development	NVS
1.1, 1.2	J.R. Saucier	Survey of eastern forest condition	Monitoring	NVS
1.2	D.H. Marx	Long-term monitoring pilot study	Monitoring	NVS
2.1	D. Binkley	Impacts of sulfate and nitrogen deposition on southern soils	Uncontrolled field	SCFRC
2.1, 2.2, 2.3	D. Brown	Development of an air pollution testing center	Controlled lab	SCFRC
2.1	M. Cline	Nitrogen deposition impacts on mycorrhizae	Literature review	SCFRC
2.1, 2.2, 2.3	E.B. Cowling	Assessment of pollutants impacts on forests	Literature review	SCFRC
2.1, 2.2	H. Gholz	Development of intensive forest study site (Florida)	Uncontrolled field	SCFRC
2.1, 2.3, 2.5, 2.6	J.D. Hodges	Effects of natural and airborne chemical stresses on growth and development of individual trees and forests	Literature review	SCFRC
2.1, 2.3	S. McLaughlin	Acid rain and ozone stress on loblolly pine	Controlled lab and controlled field	SCFRC
2.1, 2.3	R.A. Reinert	Acid rain and ozone stress on loblolly pine	Controlled lab	SCFRC
2.3	F. Fong	Acid rain and ozone stress on loblolly pine	Controlled lab	SCFRC
2.1, 2.3	L. Kress	Acid rain and ozone stress on loblolly	Controlled field	SCFRC
2.3	C. Richardson	Acid rain and ozone stress on loblolly physiology	Controlled field	SCFRC
2.3	R. Flagler	Acid rain and ozone impacts on shortleaf pine	Controlled lab	SCFRC
2.3	J. Johnson	Acid rain and ozone impacts on slash pine	Controlled field	SCFRC
2.3	A. Chappelka	Acid rain and ozone stress on loblolly pine	Controlled field	SCFRC
2.3	J. Reardon	Acid rain and ozone stress on shortleaf pine	Controlled field	SCFRC
2.1	B.L. Wong	Mycorrhizal accumulation of trace metals	Controlled lab	SCFRC
2.3	S.V. Kossuth	Use of clonal trees for acid deposition studies	Controlled field	SCFRC
2.3	G.E. Taylor	Air pollutant exposure of mature forest trees	Controlled field	SCFRC
2.3	R. Teskey	Ozone impacts on mature loblolly pine trees	Controlled field	SCFRC
2.5, 1.1	N. Christensen	Patterns of production and mortality of seedlings and trees		SCFRC

* AEC = Atmospheric Exposure Cooperative, NVS = National Vegetation Survey, SCFRC = Southern Commercial Forest Research Cooperative.

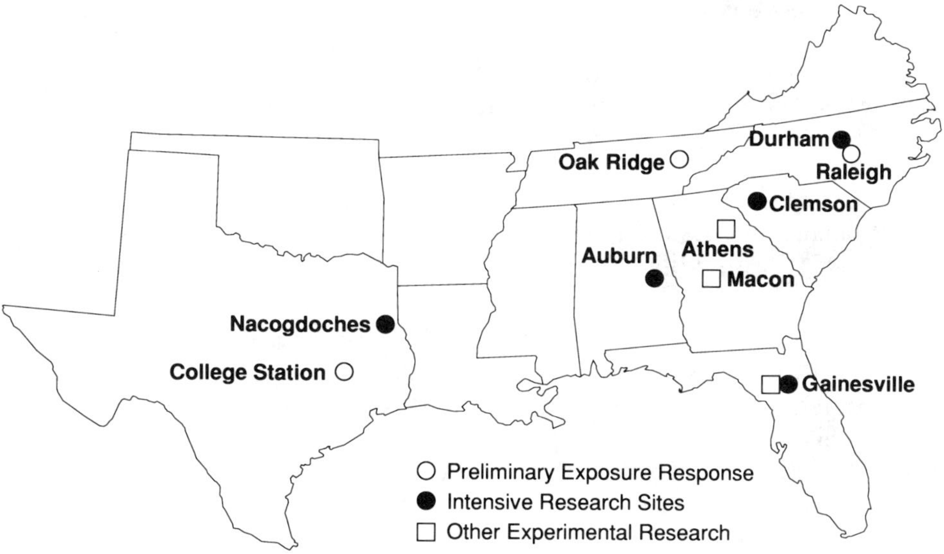

Figure 1. Location of research in the Southern Commercial Forest Research Cooperative.

source. The map in Fig. 1 shows the location of research funded by the SCFRC.

Screening Trials

An essential first step in the SCFRC was the screening of various species and genotypes within these species to acid deposition and ozone. The purpose of these screening studies was to determine the range of response to various levels of ozone: charcoal filtered, ambient, and several levels above ambient (to simulate potential future concentrations if no restrictions on the emissions of ozone precursors were mandated). Based on the screening studies, genotypes (from very ozone sensitive to very unsensitive) would be selected for use in more intensive controlled exposure studies. The species to be evaluated included the major southern pine and associated hardwood species. In the end, funding restrictions and time limitations resulted in only loblolly pine genotypes being tested or screened for sensitivity to various levels of ozone.

Three laboratories were selected to run the screening trials, and 100 genotypes of loblolly pine were chosen for testing. It was important to characterize the variation in response of the species by evaluating a subsample of the total population of genotypes. The evaluation included quantifying the effects of ozone on physiology, growth, and development. In 1986, 100 genotypes were selected based on availability of single tree seed collections in loblolly pine improvement programs. A thorough search for this seed revealed the availability of nearly 700 collections. Simple criteria were used to select the seed. Only one family per county per state was to be selected, and the collection had to have at least 1000 seeds. Following these criteria, 209 collections were identified and covered, evenly, about 90% of the geographic range of the species. Because space in testing facilities was limited and could only accommodate 100 families for testing in 1986, a further selection was made based on geographic distribution.

Intensive Research Sites

The SCFRC established intensive research sites to conduct controlled exposure studies on loblolly, slash, and shortleaf pine. These were field research facilities using open-top chambers to study physiological and growth responses of pines. Open-top chambers provided the means to evaluate tree response under relatively controlled conditions while growing trees in the more realistic harvested sites. They were a balance between reality and control.

Multiple sites were selected to allow for an evaluation of site impacts on species response. Five intensive sites were selected (1) to provide information with which to develop empirical models to estimate the effects of rain acidity and ozone dose on pine growth responses; (2) to test the interactions of acid rain and ozone impacts on pine growth and physiology; (3) to evaluate mechanisms of the effects of acid rain and ozone; and (4) to provide information that could serve as a basis for a regional assessment of the impacts of these pollutants on southern commercial pine forests.

The intensive sites were designed to help identify causes and effects of pollutants on trees. Processes such as photosynthesis, carbon allocation, respiration, nutrient mobilization, translocation, and nutrient leaching could be studied. Exposure response results from these studies were designed to evaluate the effects of acid precipitation with and without ozone on selected families of these important pine species.

Intensive site location was given careful consideration. The first site selected was in the Piedmont of North Carolina at Duke University. It was selected because it was representative of many central Piedmont sites, where a large percentage of southern commercial forests are grown, and because of its location in proximity to SCFRC headquarters. An additional four sites were selected to run smaller scale but otherwise comparable open-top chamber studies. Site distribution was important in the selection of the additional four sites. They were to be distributed across the region and located in the major physiographic regions of the South. Each has a different climate and soils. One criterion in the selection was the availability of research records and historical information on tree growth, stand mortality, burning, logging, soils, stand management, etc. The availability of meteorological and air quality data at or near the site was also a consideration. Another factor was the potential for mature tree research in adjacent stands.

The four selected sites, in addition to the Duke University Forest in North Carolina, were the Auburn University Forest in Alabama, Austin Cary Forest in Florida, the Clemson Experimental Forest in South Carolina, and the Stephen F. Austin Experimental Forest in Texas. Figure 1 shows the location of the intensive sites.

Mature Tree Research

Branch chambers were used in one study to determine the effects of ozone on mature trees. Branch chamber technology has unique advantages and disadvantages. Its advantages include the ability to test controlled exposure and response on mature trees, cost effectively, for multiple years. The major disadvantage is that branches may not be autonomous. So if ozone causes damage to a branch, then nutrients may be imported from the stem, roots, or other branches to repair damage. One study was supported by the SCFRC to determine if branches are autonomous with respect to carbon import and export, and they were found to be autonomous. However, additional research is needed. Nonetheless, other technology to study mature trees, such as whole tree chambers or stand-level chambers, was either cost prohibitive or not available when the SCFRC was established. Branch chambers provided the first opportunity to compare the response of seedlings to that of mature trees. (The SCFRC initiated a study to develop a whole tree chamber for mature trees. Some progress was made, but when funding levels were reduced halfway through the program, the study had to be cut.)

Synthesis and Integration of Knowledge

Four critical reviews, which synthesized existing information, were funded early in the program to assist in collecting existing knowledge and evaluating other research needs. In particular, the following reviews were conducted to address several key questions that arose during the establishment of the SCFRC:

1. An evaluation of the role of ozone, acid deposition, and other airborne pollutants on forests in the eastern United States
2. An evaluation of the role of acid deposition of forest soils
3. A review of existing information on the impacts of nitrogen deposition on mycorrhizae in southeastern commercial forests
4. A thorough review of the literature on natural and pollutant stress impacts on the growth and development of trees and forests.

These synthesis and integration studies provided valuable information on topics for which there were existing data in the scientific literature. Several of the projects formed the basis for chapters in this book.

A Retrospective

The SCFRC provided an organized framework for conducting research on a broad, complex subject. Given the importance of forests in the southeastern United States and the prevalence of ozone and acid rain, it was essential to evaluate the potential damage that these pollutants could have on forest health, function, and productivity. The SCFRC was a comprehensive program of research on this important topic. Given the limited time span in which it had to operate (five years) and the limited funding available, it established a research agenda and drew on the expertise of some of the best scientists in the region in accomplishing its agenda. But its biggest legacy was that it set the stage for interagency cooperative research.

Results from the program were used in evaluating the adequacy of air pollution legislation. Results were also used to help define the research agenda for several federal research programs that followed the SCFRC. Moreover, results have been used by the U.S. EPA in developing a research basis for recommending a secondary standard for ozone.

In addition to advancing our knowledge about pollutant impacts on trees and improving existing environmental legislation, the SCFRC was the catalyst for debate about forest monitoring and survey data. The ensuing evaluation of survey data and monitoring in the region demonstrated that information was inadequate to assist in determining whether forest health and productivity were being impacted by the host of biotic and abiotic stressors to which forests in the South are subjected. This led to the establishment of several nationwide monitoring and assessment activities. In addition, it resulted in an evaluation of Forest Service inventory and analysis methods and a push to standardize these across the United States.

Research from the SCFRC demonstrated that ozone has a detrimental effect on trees and suggested that forests in the South are also being impacted by ozone. Preliminary SCFRC research indicated that long-term sulfur deposition on soils may result in leaching of already limited nutrients. Even though nitrogen deposited in the region will augment this limited nutrient, the nutrients leached from sulfate deposition have the potential to limit tree growth.

Findings from the SCFRC clearly show that some genotypes are very susceptible to injury from current levels of ozone and that if ozone continues to increase, as is expected, it will substantially increase damage to trees in the South. Given the importance of southern forests in local, regional, and national economies, it is important that regional ozone levels be reduced. The restriction of logging in the northwestern United States will increase the South's role in providing timber. It is therefore important that research be conducted that will continue an evaluation of regional impacts of pollutants on forest health and productivity.

References

Bechtold WA, Ruark GA, Lloyd FT (1991) Changing stand structure and regional growth reductions in Georgia's natural pine stands. Forest Science 37:703–717.

Carey A, Janetos AC, Blair R (1986) *Responses of Forests to Atmospheric Deposition: National Research Plan for the Forest Response Program*. US Environmental Protection Agency and USDA Forest Service.

Engstrom A, Backstrand G, Stenram GH (eds) (1971) *Air Pollution across National Boundaries: The Impact on the Environment of Sulfur in Air and Precipitation*. Report No. 93. Ministry for Foreign Affairs/Ministry for Agriculture, Stockholm, Sweden.

Garner JHB, Pagano T, Cowling EB (1989) *An Evaluation of the Role of Ozone, Acid Deposition, and Other Airborne Pollutants in the Forests of Eastern North America*. USDA Forest Service General Technical Report SE-59.

Heagle AS, Philbeck RB, Heck WW (1973) An open-top chamber to assess the impact of air pollution on plants. J Environ Qual 2:365–368.

Marx DH, Cowling EB, Woodman JW (1985) (unpublished) *Effects of Air-borne Chemicals on Southern Commercial Forests: A Scientific Research Plan and Budget for the Southern Commercial Forest Research Cooperative*. USDA Forest Service, Raleigh, NC.

National Acid Precipitation Assessment Program (1985) Annual Report, 1985. NAPAP, Washington, DC.

Ruark GA, Thomas CE, Bechtold WA, May DM (1991) Growth reductions in naturally regenerated pine stands in Alabama and Georgia. Southern J Applied Forestry 15:73–79.

Saint C (ed) (1987) *Forest Response Program Progress Report 1987*. US Environmental Protection Agency and USDA Forest Service.

Sheffield RM, Cost ND, Bechtold WA, McClure JP (1985) *Pine Growth Reductions in the Southeast*. USDA Forest Service Resource Bulletin SE-83.

Shriner DS, Heck WW, McLaughlin SB, Johnson DW, Joslin JD, Peterson CE (1990) Response of Vegetation to Atmospheric Deposition and Air Pollution. In: Irving PM (ed) *Acidic Deposition: State of Science and Technology*. Vol. 3. Terrestrial, materials, health and visibility effects. National Acid Precipitation Assessment Program, Washington, DC.

Van Deusen PC (1992) Growth trends and stand dynamics in natural loblolly pine in the southeastern United States. Canadian J Forest Res 22:660–666.

Zahner R, Saucier JR, Myers RK (1989) Tree-ring model interprets growth decline in natural stands of loblolly pine in the southeastern United States. Canadian J Forest Res 19:612–621.

1. Characterization of the Condition of Southern Commercial Forests

1. Southern Pine Forests of North America

Robert A. Mickler

The South's pine forests, which were first glimpsed by English colonists at Jamestown, Virginia, in 1607, Charleston, South Carolina, in 1608, and Savannah, Georgia, in 1733, were part of the South's old-growth timber resource that was harvested by the early 1900s. The second growth, which naturally regenerated following the cutting of the primeval woodlands along with stands planted as a result of the United States' newly developing forestry policies and programs, became the South's "second forest." Second-growth pine forests supplied the wood for the first southern pine mill built in Texas in 1939, which utilized a new process to extract pine resin from pulp for the production of paper. From the 1930s to the 1960s, this forest was utilized by the rapidly growing pulp and paper industry and for other wood-using industries. By the 1960s, this forest had been almost entirely cut. Professional foresters and forest industry personnel call today's naturally regenerated and managed stands of southern pines and hardwoods the South's "third forest." This forest is being harvested today and is projected to sustain commercial harvesting into the next century.

Southern pine forests are subject to numerous physical and chemical stresses. Results from soil modeling suggests that the fertility of some southern soils may be at risk in the long term due to the combination of short rotations and acidic deposition. Ozone continues to occur at levels that have been shown to cause foliar injury, physiological alternations, and reductions in growth of several southern pine species in controlled exposure studies. Recent scientific debate over future climatic change has raised concern about the impacts of drought, high temperature, increased carbon dioxide, and nutrition on forest health and productivity. The interaction of natural and human disturbances along with changes in the chemical and physical environment have the potential to alter natural and managed forest ecosystems across the steep environmental gradients of the southern United States. The purpose of this chapter is to survey and describe the southern pine resource of the South's third forest. Subsequent chapters will describe the impact of chemical and physical stresses on this resource.

Forest Land Area

The southern region of the United States is composed of 13 states, which occupy 216,318,000 hectares (ha) from the Atlantic Coast west to Texas and Oklahoma. The region represents approximately 24% of the land area of the United States. In 1992, forest land—land at least 10% stocked by forest trees of any size—covered 85,730,000 ha or about 40% of the total land area in the region (Table 1.1). Timberland or forest land that is in production or capable of producing a commercial crop and has not been withdrawn from timber utilization by statute or administrative regulation comprised 80,660,000 ha. This represents a 1% increase in timberland area reported between 1987 and the subsequent forest survey report in 1992. Southern forests remain a vitally important resource for the region and the nation, encompassing 41% of the timberland for the entire United States.

Most of the 80,660,000 ha of timberland in the South are in nonindustrial private and forestry industry ownership. In 1992, 90% of southern timberland was owned by private citizens, farmers, and the forest industry (Table 1.2) Nonindustrial private ownership includes large numbers of small parcels and a smaller number of large tracts. This ownership group is found near urbanizing areas, dispersed among cultivated lands, and in remote areas. Changing ownerships and differing management objectives affect the availability of the resource for the production and harvest of timber for market. Forest industry timberland is managed primarily as a resource for wood products manufacturing facilities. The locations of the holdings correspond to the areas of the most highly productive forest land. Thus, the contribution of forest industry timberland as a source of wood raw materials exceeds the proportional area in the South. The southern United States encompasses 55% of all industry-owned timberland. The remaining 8,297,000 ha of timberland in the South is found in National Forests and lands held by other public agencies. The USDA Forest Service administers the largest portion of these lands.

Table 1.1. Total Land and Forest Area (Thousands of Hectares) in the United States

Region, Subregion, and State	Total Land Area[a]	Total Forest Land[b]
North total:	167,338	68,186
Rocky Mountains total:	300,372	56,548
Pacific Coast total:	231,901	87,666
South:		
Southeast:		
Florida	13,985	6,697
Georgia	15,001	9,768
North Carolina	12,618	7,802
South Carolina	7,799	4,960
Virginia	10,256	6,418
Total	59,660	35,645
South Central:		
Alabama	13,144	8,893
Arkansas	13,488	7,229
Kentucky	10,291	5,145
Louisiana	11,284	5,611
Mississippi	12,151	6,880
Oklahoma	17,788	3,051
Tennessee	10,676	5,509
Texas	67,837	7,767
Total	156,659	50,085
South total:	216,318	85,730
United States total:	915,928	298,130

Note: Data may not add to totals because of rounding.
[a]*Source*: U.S. Department of Commerce, *1990 Decennial Census*.
[b]*Source*: USDA Forest Service, 1992, *Forest Resources of the United States*.

Forest Type and Distribution

The southern United States has been divided into 10 geomorphic regions encompassing a diversity of climates, soil characteristics, and topography (Fenneman, 1983; Thornbury, 1965). This diversity leads to a corresponding diversity of tree species and forest types, including southern yellow pines, mixed pine-hardwoods, upland and bottomland hardwoods, and tropical hardwoods in Florida. The Society of American Foresters Type Group (Eyre, 1980) is primarily used in describing southern forests. This chapter emphasizes Eyre's classification scheme, which aggregrates groupings or associations of southern pine tree species and identifies species that occur in pure stands.

Table 1.2. Timberland Area (Thousands of Hectares) in the United States by Ownership Group, Region, Subregion, and State, 1992

Region, Subregion, and State	All Ownerships	Ownership Group[a]			
		National Forest	Other Public	Forest Industry	Nonindustrial Private
North total:	63,861	3,863	8,402	6,555	45,041
Rocky Mountains total:	25,346	14,732	2,423	1,181	7,010
Pacific Coast total:	28,256	10,991	4,507	4,983	7,774
South:					
Southeast:					
Florida	6,064[b]	401[b]	584[b]	1,930[b]	3,148[b]
Georgia	9,563	304	362	2,019	6,878
North Carolina	7,572	438	351	911	5,871
South Carolina	4,929	234	241	1,063	3,391
Virginia	6,189	585	205	653	4,745
Total	34,316	1,962	1,744	6,577	24,034
South Central:					
Alabama	8,880	249	225	1,941	6,465
Arkansas	7,051	946	321	1,775	4,008
Kentucky	5,002	255	133	83	4,531
Louisiana	5,607	230	301	1,593	3,483
Mississippi	6,876	463	292	1,322	4,799
Oklahoma	2,478	99	140	436	1,803
Tennessee	5,372	229	386	454	4,304
Texas	5,078	244	79	1,613	3,142
Total	46,344	2,714	1,877	9,217	32,536
South total:	80,660	4,676	3,621	15,793	56,570
United States total:	198,123	34,262	18,953	28,513	116,394

Note: Data may not add to totals because of rounding.
[a]*Source*: USDA Forest Service, 1992, *Forest Resources of the United States*.
[b]*Source*: The 1992 data have not been updated since 1987.

Forest Composition

Forest inventories conducted by the USDA Forest Service focus on forests that are available for harvesting, because of their commercial value and society's economic dependence on the timber resource as well as the need for information on timber supply to meet the nation's needs. In the United States, 6% of all forest land and 9% of southern forests are reserved from timber harvesting in National Parks, wilderness areas, designated scenic areas, and historic sites. Subsequent discussions of species distribution, timber volume, growth, and mortality are limited to unreserved forest land.

The timberland in the southern United States contains an estimated 7780 million cubic meters of timber. The South has seen an increase of 2.4% in total volume from 1987 to 1992. Southern pines and other softwoods account for 38% of the timber volume in the South and 22% of the softwood volume in the nation (Table 1.3). For softwoods, loblolly and shortleaf pine are 64% of the total growing stock volume on timberland in the South, longleaf and slash pines account for 16%, other yellow pines make up 10%, eastern white pine is less than 2%, and other softwoods comprise 8% of the remaining growing stock volume (Table 1.4). Almost 97% of the combined total net volume of southern yellow and eastern white pine is found in the South. From 1987 to 1992, softwood volume decreased by 2.5%. This is a reversal of an increasing volume trend since 1952 and can be attributed to increasing mortality and removals exceeding growth.

Loblolly pine is the most abundant of the southern pine species. It occurs in natural stands in the Coastal Plain and Piedmont (see Chapter 2), where it has frequently seeded in on idle cropland and pasture. Loblolly pine has also been planted extensively throughout the region. Shortleaf pine, in pure young stands and more commonly in association with loblolly pine, becomes an increasingly important component of stands at higher elevations, especially in the Cumberland Plateau, the southern Appalachians, and the Ouchita Mountains of western Arkansas and southeastern Oklahoma.

Longleaf pine was historically the predominant species on the lower Coastal Plain. Intensive logging, conversion of longleaf sites to loblolly and slash pine plantations, and absence of wildfires have greatly reduced the area of longleaf pine. The longleaf pine–slash pine association is now the dominant forest type primarily in southeastern Georgia and northern Florida.

Pond pine, slash pine, South Florida slash pine, Virginia pine, and sand pine occur in pure stands or comprise a majority of stocking in their forest type groups. These type groups can be found on the Piedmont and Coastal Plain from Virginia to Florida. Sand pine and Virginia pine are pioneer species that establish on disturbed and droughty soils. Pond pine

Table 1.3. Net Volme of Timber (Million Cubic Meters) on Timberland in the United States by class of Species Group, Region, Subregion, and State, 1992

Region, Subregion, and State	All Timber[a]		
	Total	Softwoods	Hardwoods
North total:	6,600	1,565	5,030
Rocky Mountains total:	3,518	3,182	336
Pacific Coast Total[b]:	6,375	5,666	709
South:			
Southeast:			
Florida	476	268	208
Georgia	925	446	480
North Carolina	983	357	625
South Carolina	513	225	287
Virginia	804	193	610
Total	3,700	1,490	2,211
South Central:			
Alabama	702	321	381
Arkansas	603	231	372
Kentucky	474	36	438
Louisiana	591	289	302
Mississippi	648	269	380
Oklahoma	115	30	85
Tennesse	521	86	436
Texas	424	228	196
Total	4,079	1,489	2,590
South total:	7,780	2,978	4,801
United States total:	24,269	13,391	10,877

Note: Data may not add to totals because of rounding.
[a]*Source*: USDA Forest Service, 1992, *Forest Resources of the United States*.
[b] Estimates of hardwood volume are not available for most National Forests in Oregon and Washington.

is most readily identified with its ability to establish and grow under soil conditions that eliminate other pine species. Slash pine is a subclimax species that is replaced by various hardwoods, depending on soil moisture.

Across much of the South, pine stands represent a transitional stage in natural succession to hardwood forest types. Mixed pine–hardwood stands are typically composed of 50% or more oak and other hardwoods and 25 to 50% pine. Common pine–hardwood associations include longleaf pine–scrub oak, shortleaf pine–oak, virginia pine–oak, loblolly pine–hardwood, and slash pine–hardwood.

Net annual growth, removals, and mortality are common measures of productivity of the timber resource. Net annual growth is a measure of annual timber volume growth less the volume lost to mortality. Removals are net volume of the growing stock trees removed from the inventory during a specified year by harvesting, cultural operations such as timber

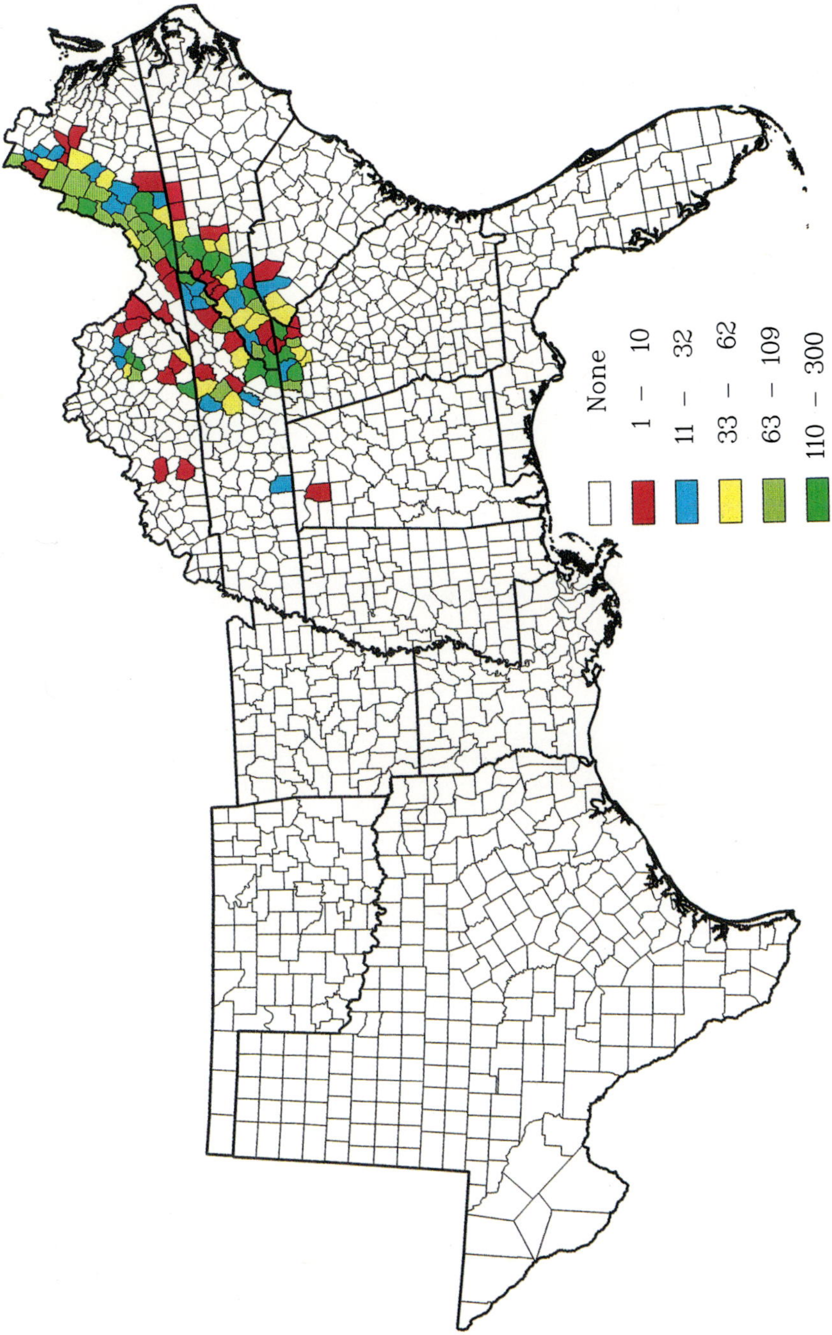

Figure 1.3. Eastern white pine trees per hectare of forest land by county, 1994. Note that data for counties in central and western Texas and Oklahoma are not included in the figure. Forest inventory surveys are not conducted in these areas due to the small amount of forest land. (*Source:* USDA Forest Service, Southern Research Station, Forest Inventory and Analysis.)

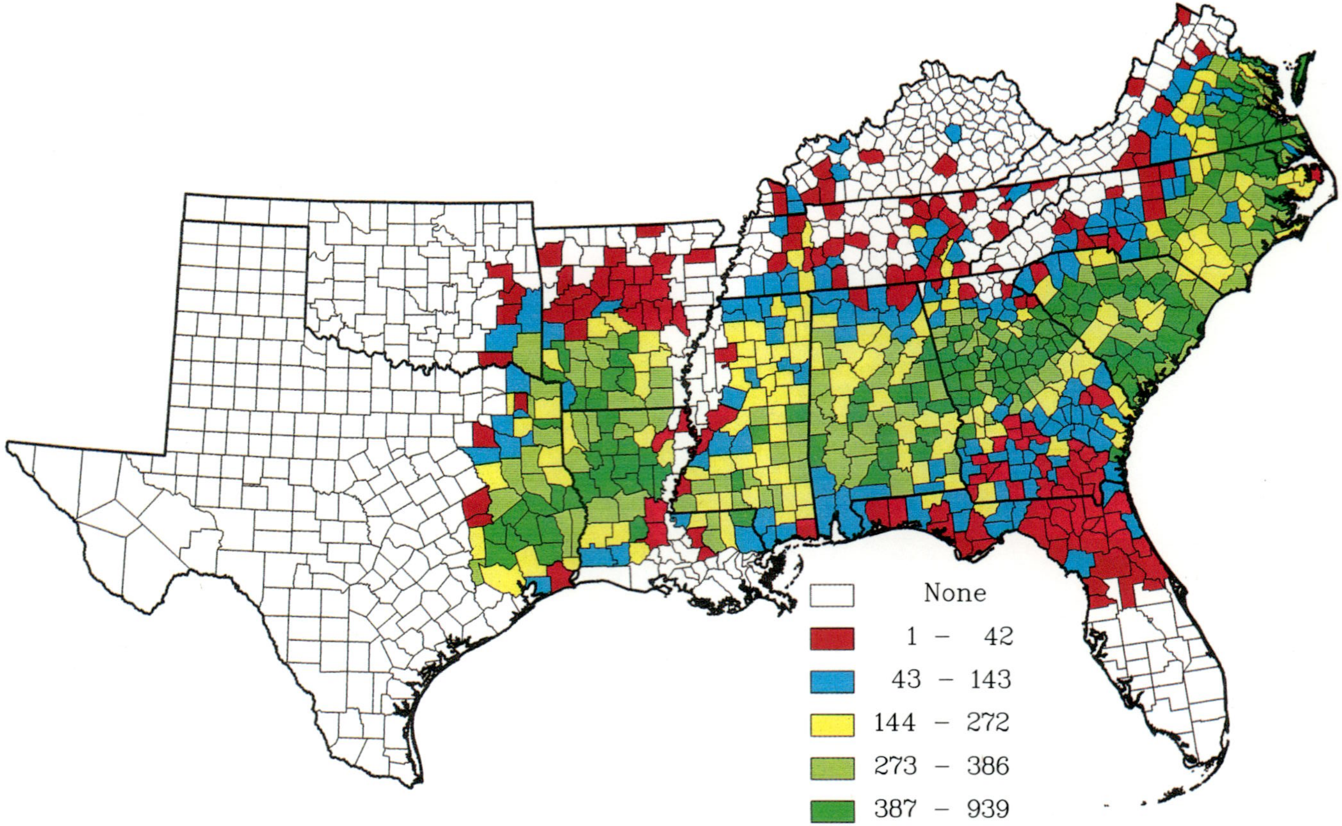

Figure 1.4. Loblolly pine trees per hectare of forest land by county, 1994. Note that data for counties in central and western Texas and Oklahoma are not included in the figure. Forest inventory surveys are not conducted in these areas due to the small amount of forest land. (*Source*: USDA Forest Service, Southern Research Station, Forest Inventory and Analysis.)

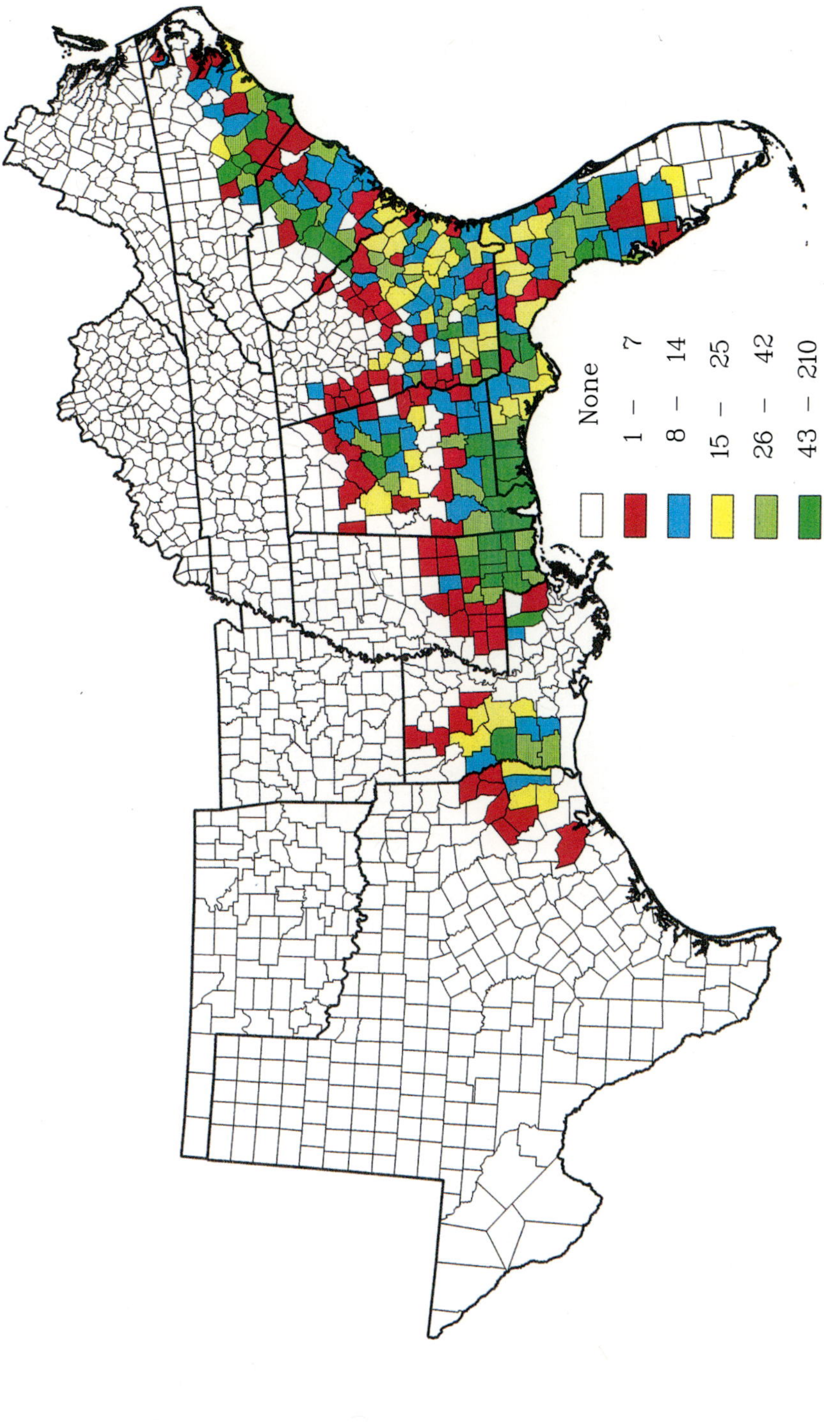

Figure 1.5. Longleaf pine trees per hectare of forest land by county, 1994. Note that data for counties in central and western Texas and Oklahoma are not included in the figure. Forest inventory surveys are not conducted in these areas due to the small amount of forest land. (*Source:* USDA Forest Service, Southern Research Station, Forest Inventory and Analysis.)

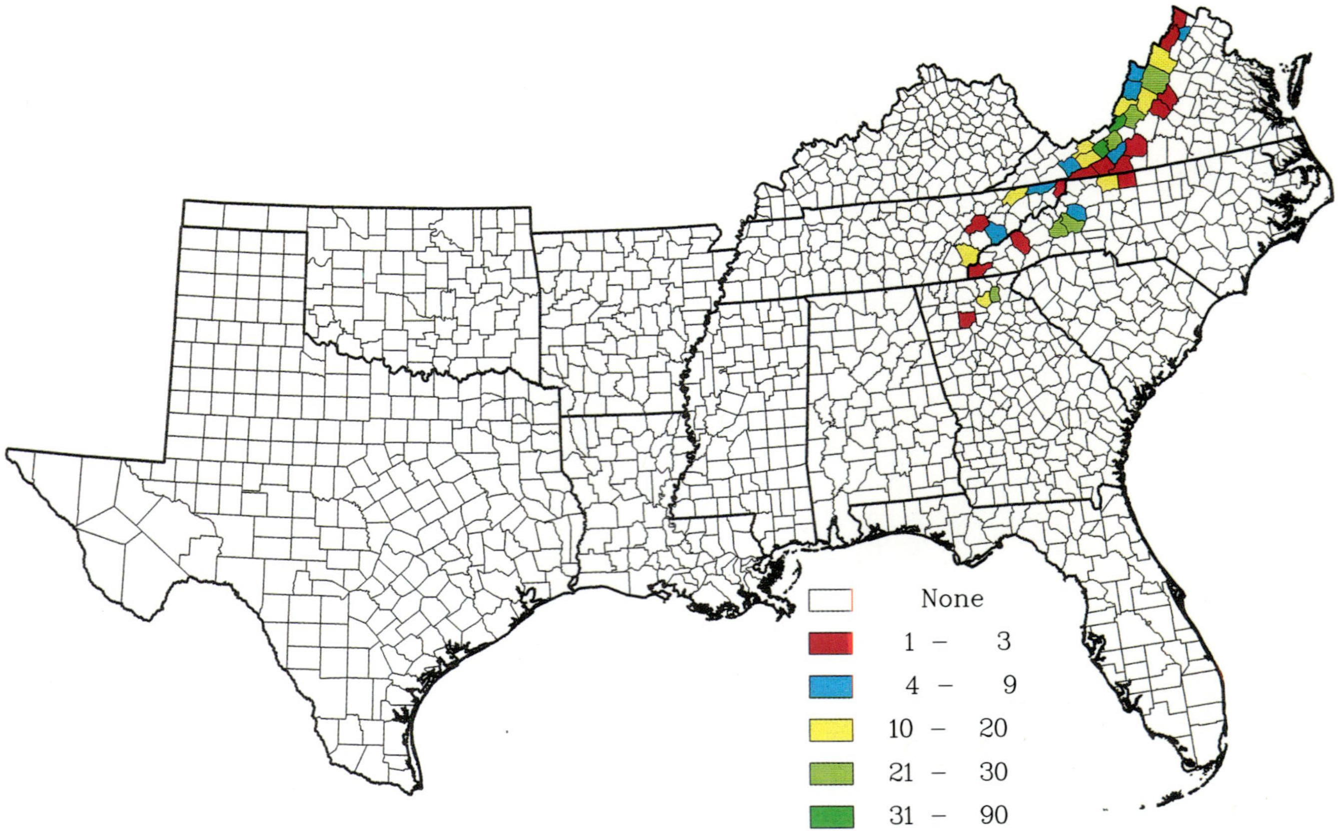

Figure 1.6. Table Mountain pine tress per hectare of forest land by county, 1994. Note that data for counties in central and western Texas and Oklahoma are not included in the figure. Forest inventory surveys are not conducted in these areas due to the small amount of forest land. (*Source*: USDA Forest Service, Southern Research Station, Forest Inventory and Analysis.)

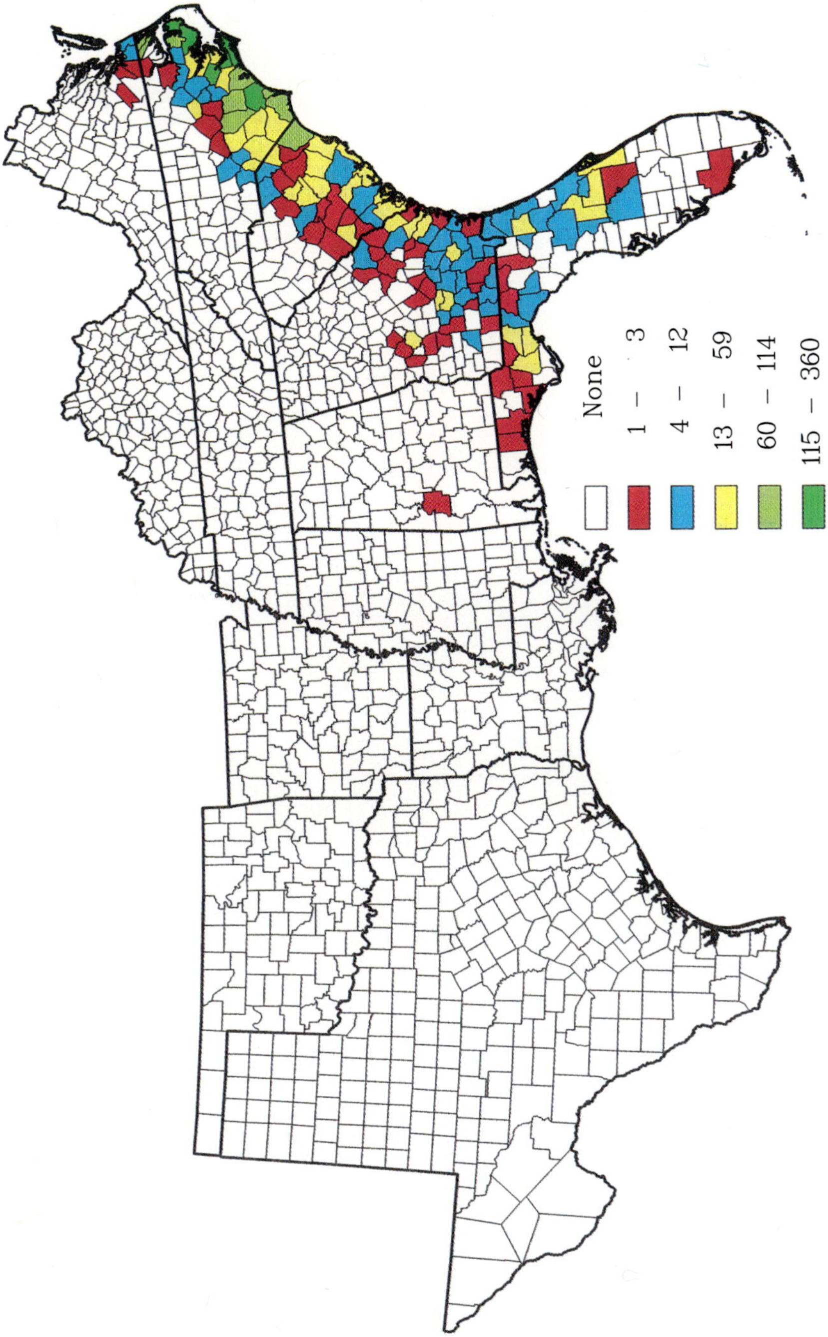

Figure 1.7. Pond pine trees per hectare of forest land by county, 1994. Note that data for counties in central and western Texas and Oklahoma are not included in the figure. Forest inventory surveys are not conducted in these areas due to the small amount of forest land. (*Source:* USDA Forest Service, Southern Research Station, Forest Inventory and Analysis.)

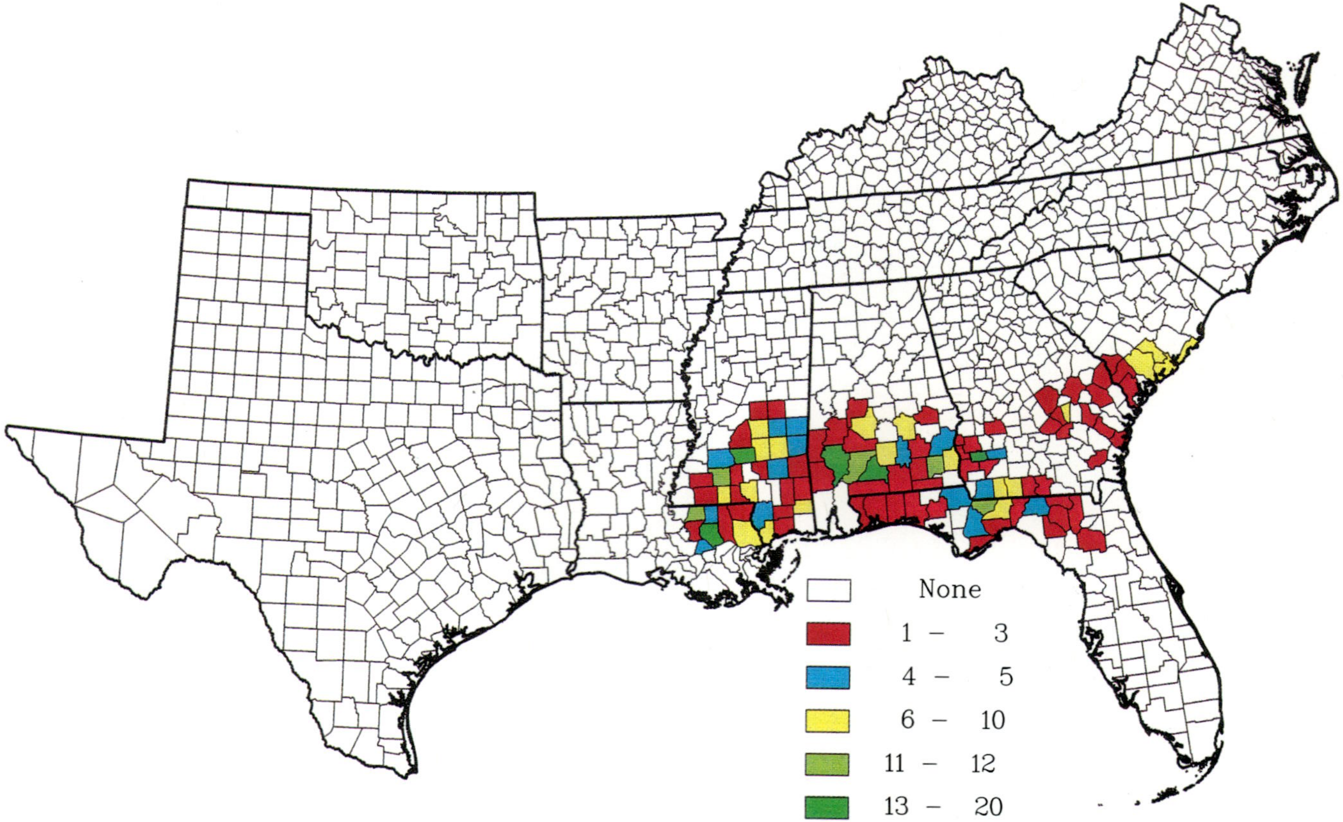

Figure 1.8. Spruce pine tress per hectare of forest land by county, 1994. Note that data for counties in central and western Texas and Oklahoma are not included in the figure. Forest inventory surveys are not conducted in these areas due to the small amount of forest land. (*Source*: USDA Forest Service, Southern Research Station, Forest Inventory and Analysis.)

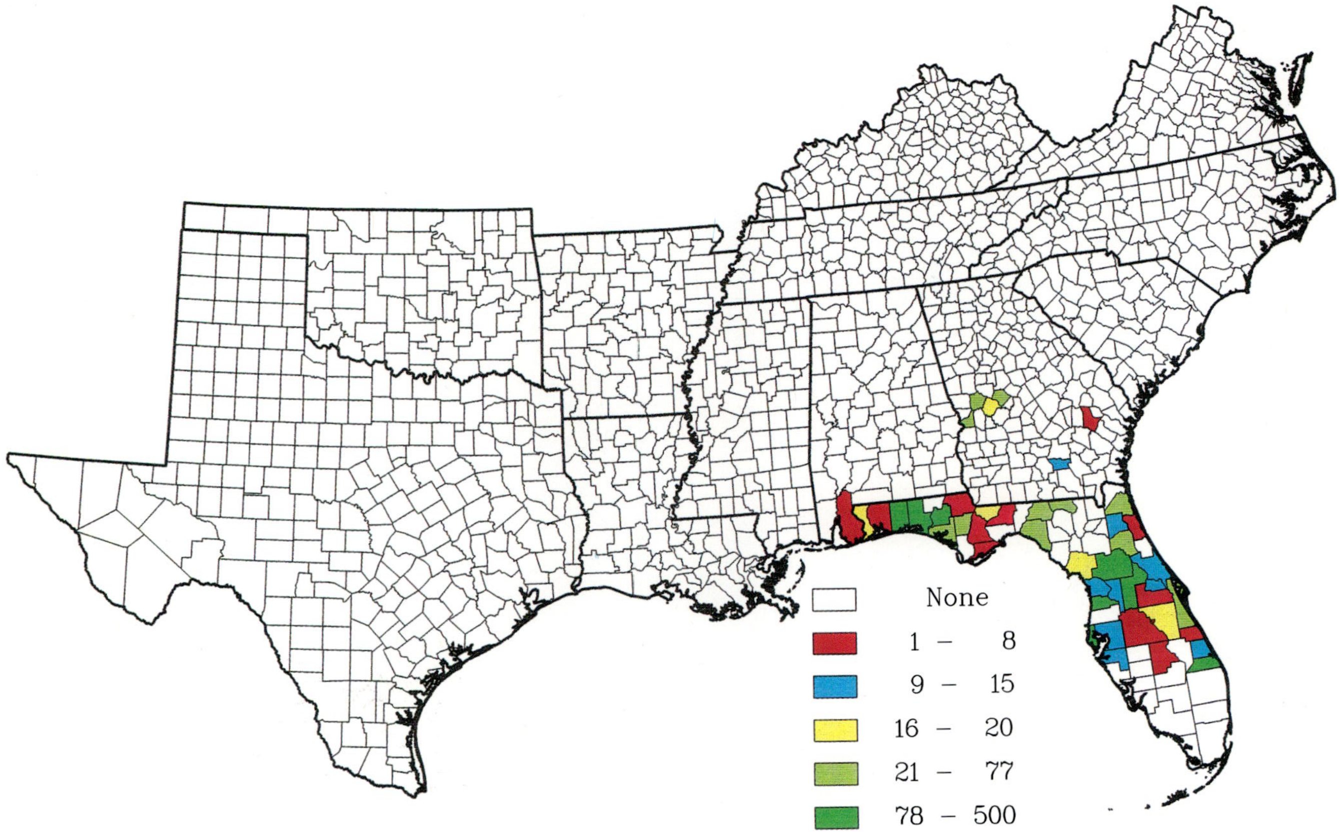

Figure 1.9. Sand pine trees per hectare of forest land by county, 1994. Note that data for counties in central and western Texas and Oklahoma are not included in the figure. Forest inventory surveys are not conducted in these areas due to the small amount of forest land. (*Source:* USDA Forest Service, Southern Research Station, Forest Inventory and Analysis.)

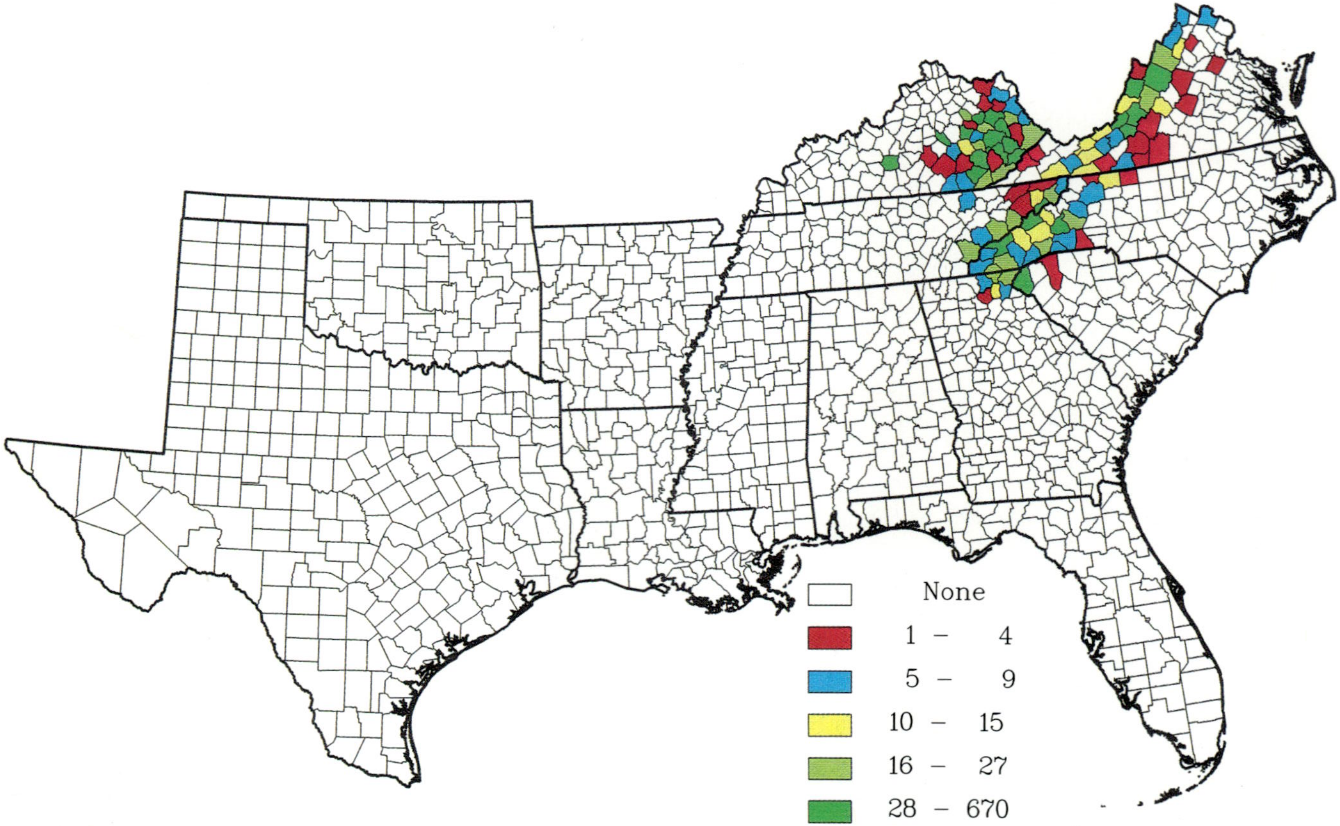

Figure 1.10. Pitch pine trees per hectare of forest land by county, 1994. Note that data for counties in central and western Texas and Oklahoma are not included in the figure. Forest inventory surveys are not conducted in these areas due to the small amount of forest land. (*Source*: USDA Forest Service, Southern Research Station, Forest Inventory and Analysis.)

Table 1.4. Net Volume (Million Cubic Meters) of Softwood Growing Stock on Timberland in the Eastern United States by Species, Region, and Subregion, 1992

Region and Subregion	Longleaf and Slash Pines	Loblolly and Shortleaf Pines	Other Yellow Pines	White and Red Pines
North:				
Northeast	0	19	43	238
North Central	0	20	9	137
Total	0	39	52	375
Rocky Mountains total:	0	0	0	0
Pacific Coast total:	0	0	0	0
South:				
Southeast	327	739	205	46[b]
South Central	135	1130	80	8[b]
Total	462	1869	285	54[b]
United States total:	462	1908	337	429

Note: Data may not add to totals because of rounding.
[a]*Source*: USDA Forest Service, 1992, *Forest Resources of the United States*.
[b] Volume estimates include only eastern white pine native range in South. No red pine native ranges extend into the South.

stand improvements, or land clearing. Mortality includes all losses of growing stock from natural causes leading to tree death. In 1991, the South accounted for more than 45% of all timber growth and 43% of all softwood growth for the nation (Table 1.5). Removals of growing stock on timberland in the South exceeded growth by 50,506 thousand cubic meters.

Projections of net annual growth and timber removals show that growth is estimated to equal removals to the end of the century and growth will exceed removals through 2040 (Fig. 1.1). Historical data shown in Fig. 1.2 were derived from the USDA Forest Service's Forest Inventory and Analysis units (Powell et al., 1993). Adams and Haynes (1991) have argued that shifting age class structure may lie at the root of the decline in southern softwood growth beginning in 1976. In addition, since the summary statistics are for all timberland owners, they are subject to changes in total and between owner timberland areas.

Projections developed for the 1993 Resource Planning Act (RPA) Timber Assessment Update (Haynes et al., 1995) are based on an age class inventory projection stystem Aggregate Timberland Assessment System (ATLAS). ATLAS is the standard USDA Forest Service inventory projection system used for private timberlands (Mills and Kincard, 1992). The projections reflect contained shifts in both area and the age class distribution. The effects of these later changes are illustrated for one ownership class consisting of private lands that are not owned by forest industry or farmers (Fig. 1.2). Given the projected harvest, Haynes et al.

Table 1.5. Net Annual Growth, Removals, and Mortality (Thousand Cubic Meters) of Softwood Growing Stock on Timberland in the United States by Region, Subregion, and State, 1991

Region, Subregion, and State	Sofwoods[a]		
	Growth	Removals	Mortality
North total:	34,357	20,989	12,649
Rocky Mountains total:	56,169	22,541	17,097
Pacific Coast total:	103,987	101,856	24,716
South:			
Southeast:			
Florida	13,797[b]	14,823	1,747[b]
Georgia	23,148	27,340	4,206
North Carolina	16,693	15,033	3,294
South Carolina	7,191	14,310	6,803
Virginia	6,480	6,639	2,038
Total	67,308	78,145	18,087
South Central:			
Alabama	18,618	20,240	3,008
Arkansas	10,952	13,998	1,221
Kentucky	722	221	369
Louisiana	14,983	16,604	2,195
Mississippi	14,428	17,483	2,649
Oklahoma	1,307	1,382	85
Tennessee	2,767	2,537	929
Texas	13,262	14,047	1,653
Total	77,039	86,512	12,110
South total:	144,347	164,656	30,197
United States total:	338,860	310,043	84,659

Note: Data may not add to totals because of rounding.
[a]*Source*: USDA Forest Service, 1992, *Forest Resources of the United States*.
[b]*Source*: The 1992 data have not been updated sine 1987 assessment.

(1995) expect that the average age of stands will decrease with proportionally more inventory in younger and faster growing stands. This is especially the case during the next two decades, when the authors project a transition from existing stands of mostly natural origin to the stands that will make up the South's "fourth forest," which will contain a higher proportion of managed hectares.

The Genus *Pinus*

Pinus is the largest genus of conifers (approximately 100 species) and the most widely distributed and economically important genus of trees in the Northern Hemisphere (Critchfield and Little, 1966; Mirov, 1967). The

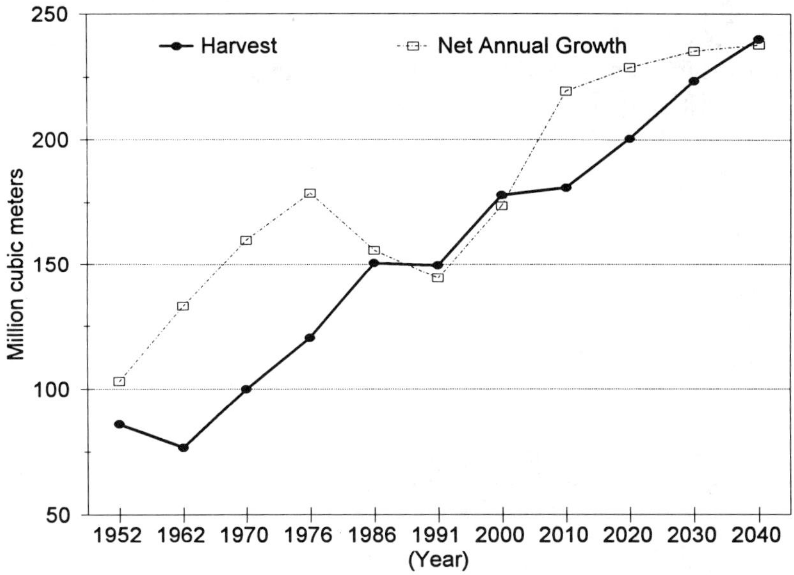

Figure 1.1. Softwood net annual growth and harvest in the southern United States for all ownerships, 1952–1991, with projections to 2040. (*Source*: Richard A. Haynes, USDA Forest Service, Pacific Northwest Research Station.)

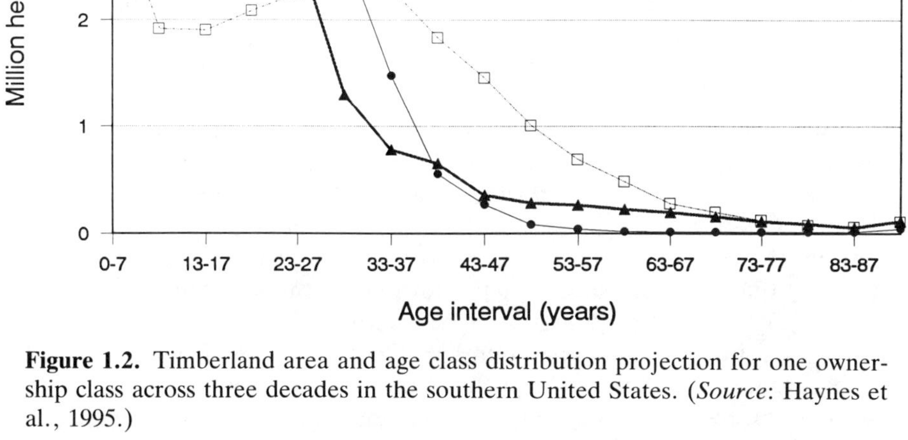

Figure 1.2. Timberland area and age class distribution projection for one ownership class across three decades in the southern United States. (*Source*: Haynes et al., 1995.)

genus ranges from arctic and subarctic areas of North America and Eurasia to subtropical and tropical areas of Central America and Asia. Several species reach to 12°N latitude in the highlands of Nicaragua and Vietnam. The largest number of species (approximately 65) are native to North America and are concentrated in the southern United States, California, and Mexico.

The classification within *Pinus* has been reviewed by Little and Critchfield (1969), Mirov (1967), and Price (1989). Subdivision of the genus dates to the 1838 classification of Loudon, which divided the genus into three sections and 15 groups corresponding to subsection based on number of needles per fascicle. Subsequent classification by Koehne in 1893 divided the genus into two main groups, based on the number of vascular bundles (one versus two) in the needle, *Haploxylon* and *Diploxylon* [= subg. *Strobus* (D. Don) Lemmon and subg. *Pinus*, respectively].

The 11 species of *Pinus* native to the southern United States are classified into two subgenera recognized by Little and Critchfield (1969). The subgenus *Strobus* [subg. *Haploxylon* (Koehne)Rehder], which includes the white pines (subsect. *Strobus*), is characterized by leaves with a single fibrovascular bundle, five needles per fascicle with a deciduous sheath at the base, and a nondecurrent subtending bract. The subgenus *Pinus* [subg. *Diploxylon* (Koehne)Rehder, subg. *Eupitys* (Spach)Rehder] includes the yellow pines (subsect. *Australes* Loudon and subsect. *Contortae* Little and Critchfield) and is characterized by two or three leaves per fascicle, two fibrovascular bundles per needle, a persistent fascicle basal sheath, and a decurrent subtending bract. This subgenus includes the remaining 10 species native to the area.

Southern Pine Species and Distribution

The distribution of the 11 southern pine species is shown in Figs. 1.3 through 1.13 for the southern United States. The maps show the distribution and estimated number of trees (2.54 cm diameter and larger) per hectare of forest land for individual counties. The maps are based on forest inventory data collected through the joint efforts of the Forest Inventory and Analysis (FIA) Units at two USDA Forest Service Experiment Stations: the Southern Research Station and the Northeastern Forest Experiment Station. The data were collected on FIA permanent ground-sample plots during the most recent forest inventories of individual states between 1987 and 1993. The data have not been projected to a common point in time. A companion map showing hectares of forest land per county (Fig. 1.14) is presented to account for the wide differences in the amount of forest land among counties. The maps differ from traditional timber volume estimates, which are typically based on poletimber and sawtimber trees as found in timber volume distribution maps (Beltz et al.,

1992) and forest resource tables (Powell et al., 1993) used to summarize forest inventory data.

Eastern White Pine

The only white pine native to the southern United States is eastern white pine, *Pinus strobus* (L.) Small, also called white pine and northern white pine. In Europe, where it is planted extensively, it is known as Weymuth pine. A related taxon from southern Mexico and Guatemala has been described as *P. strobus* var. *chiapensis* Martinez, but quantitative analysis by Andresen (1964, 1966) placed it in a distinct species, *P. chiapensis* (Matinez) Andresen. Eastern white pine is one of the most valuable tree species in eastern North America, and precolonial virgin stands across its range are estimated at 3.4 billion m^3 of lumber (Burns and Honkala, 1990). By the late 1800s, these stands were cut for the lightweight, straight-grained, light-brown wood for use in ship building, bridge building, and house building.

Eastern white pine has a larger range than any of the hapoxylon pines in the New World. It occurs broadly in the northeastern United States and the Great Lake states and in southern Canada from extreme southeastern Manitoba to Newfoundland. In the South, it occurs at up to 1500 m altitude in the Appalachian Mountain regions of Tennessee, the Carolinas, and northern Georgia (Fig. 1.3). It is also native to western Kentucky, western Tennessee, and Delaware (Little, 1971). Disjunct populations are found in central North Carolina (Chatham and Lee counties) and central Tennessee (Cheatham County).

The eastern white pine species is found in areas with a cool, moist climate where July temperatures average between 18° and 23°C. Annual average rainfall in the Southeast is 2030 mm, which is about 1 to 1.5 times the evaporation from shaded free-water surfaces (Fowells, 1965).

In the southern portion of its range, eastern white pine is found on the soil orders Ultisols and Inceptisols. In the central Appalachian Mountains, the Great Groups Dystrochrepts, Fragiodults, and Hapludults are typical of the weathered acid shale and sandstone residual and alluvial soils occupied by eastern white pine on lower slopes and stream terraces. These soils are well drained and have a coarse loamy to fine loamy texture. Eastern white pine occupies a unique ecological niche in the southern Appalachian northerly aspects, in coves, and on stream terraces.

Loblolly Pine

Pinus taeda L., loblolly pine, oldfield pine, North Carolina pine, Arkansas pine, is the most important commercial forest species in the southern United States. Loblolly pine was first observed by English explorers who settled on the South's Coastal Plain. The pines they encountered growing in murky appearing soil were called loblolly, which derived from the English dialect meaning "a thick gruel" (Walker, 1991).

More is known about loblolly pine's growth and physiology, silvical characteristics, genetics, culture and management, and wood properties and utilization than any other tree species. The species responds well to silvicultural treatments on a broad range of sites and can be managed as even-aged or uneven-aged stands. Inherent biological traits of the species predetermine its ecology and management. It is a light-demanding species, especially during juvenile growth, and has difficulty competing with more tolerant hardwoods in mixed stands and during establishment of artificially regenerated pure stands. Loblolly pine is responsive to spacing control, control of hardwood competition, fertilization, soil water management, and genetic selection. All of these cultural treatments, singly and in combination, are currently employed in its management throughout its range.

In the southern United States, loblolly pine occurs on the Coastal Plain and Piedmont from Virginia south to central Florida and west to southern Arkansas, eastern Texas, and southeastern Oklahoma (Fig. 1.4). These areas include the Atlantic Plain, the Piedmont Plateau, and the southern extremities of the Cumberland Plateau, the Highland Rim, and the Valley and Ridge Provinces of the Appalachian Highlands. Small populations are found in the lower Atlantic Plain, the sandhills of North and South Carolina, and localized areas of southeastern Georgia and northern Florida. The range of loblolly pine extends north from Virginia along the Coastal Plain and Piedmont to southern New Jersey (Little, 1971).

Loblolly pine is native to areas with humid, warm-temperate climate characterized by long, hot summers and mild winters. Average annual rainfall ranges from 1020 to 1520 mm, and mean annual temperatures vary from 13° to 24°C. Temperature extremes can range from an average monthly temperature of 27°C to in excess of 38°C during July, to a low average monthly temperature of 4° to 16°C in January with occasional drops to −23°C. Limits to northern extension of the range is associated with low winter temperatures accompanied by ice and snow damage and low spring temperatures during flowering. Limits to western extension of loblolly pine in Texas and Oklahoma are probably associated with low growing-season precipitation (Fowells, 1965).

Natural stands of loblolly pine are found predominantly on the Soil Order Ultisol and to a lesser extent on small areas of Entisols and Spodosols. Within these Orders, loblolly pine grows on a wide range of soils, ranging from the flat, poorly drained Suborder Aquults and Aquods of the Coastal Atlantic Plain to the dry, sandy, and humid Psamments, Udults, and Udalfs of the Atlantic Plain, Piedmont, and upland Provinces (Buol et al., 1989). Loblolly pine grows best on moderately acid soils with imperfect to poor surface drainage, a thick medium-textured surface layer, and a fine-textured subsoil (Burns and Honkala, 1990).

Longleaf Pine

Pinus palustris Miller (*P. australis* Michx. f., longleaf pine, longstraw pine, Georgia pine, yellow pine) once existed in pure stands with a park-like appearance that occupied over 24 million ha. Longleaf pine has been heavily exploited as a premier timber species and for gum and rosin naval stores. It has been estimated that by 1930, only 10% of its original timber volume remained (Snyder et al., 1977). The lack of planned regeneration, a reduction in fire occurrence, and conversion of longleaf pine range to other commercially important pine species had resulted in a reduction of longleaf pine forest by 1985 to less than 1.6 million ha. Recent interest in regeneration of longleaf pine from the implementation of an ecosystem management forestry policy may reverse this trend in the future.

Longleaf pine is one of the few pine species whose seeds germinate soon after dispersal and contact with mineral soil and in which the seedlings go through an unusual "grass-like" stage lasting for two or more years. It is highly resistant to fire, except in the earliest seedling stage, and is dependent on fire to reduce or eliminate competition. Growth of longleaf pine stands is dependent on duration of the grass stage. Release from the grass stage and onset of rapid height growth results in a broad range of size classes within the developing stands.

Longleaf pine is found on a wide variety of site conditions ranging from well-drained sandy soils near sea level (Entisols) and wet, poorly drained flatwoods (Spodosols) on the lower Coastal Plain to as high as 600 m altitude in the foothills of the adjacent Piedmont, from southeastern Virginia to south-central Florida and west to eastern Texas (Little, 1971) (Fig. 1.5).

The natural range of longleaf pine can be characterized as warm, wet temperate with hot summers and mild winters. Annual mean temperatures range from 16° to 23°C. Annual precipitation throughout the range varies from 1090 to 1750 mm. The lowest rainfall in its range occurs in the Carolinas and Texas, and the highest rainfall occurs along the Gulf Coast.

Coastal Plain soils in the longleaf pine range are derived from marine sediments and range from deep, coarse, excessively drained sands to poorly drained clays. Pedogenic processes result in surface horizons that can be characterized as sandy, acid, low in organic matter, and relatively infertile. At higher altitudes in the Mountain Province, soils are derived largely from felsic parent material consisting of granite, quartzite, and schist; whereas Ridge and Valley Province soils are derived predominantly from sandstone, shale, limestone, and dolomite (Hodgkins, 1965).

Table Mountain Pine

Pinus pugens Lamb., Table Mountain pine, prickly pine, or hickory pine, is a small, rugged tree of the Appalachian Mountains. The species is of no commercial value but provides valuable wildlife cover and watershed protection on high-altitude exposed ridges and outcrops. The species, most important ecological function is to provide vegetative stabilization to soil and minimize erosion from shale barrens and other rugged topographic features found in its native range (Hack and Goodlett, 1960). Table Mountain pine is often found in pure stands, and individual trees are characteristically small, poor in form, and limby. The tree's tough, hickory-like branches are often laden with numerous cones and armed with stout, hooked spines that persist on branches for many years.

Table Mountain pine is unevenly distributed on xeric sites in the Appalachian Mountains of the southern United States from northern Virginia south to extreme northeastern Georgia and Tennessee (Fig. 1.6). The northern range extends along the Applachian Mountains from Virginia into southeastern Pennsylvania and locally in New Jersey, Delaware, and the District of Columbia (Little, 1971). The range includes the crest and eastern escarpment of the prominent Blue Ridge Front and altitudes up to 1760 m on southwest-facing aspects in the Great Smoky Mountains. It has also been found growing on an island in the Susquehanna River in Pennsylvania and near bogs in North Carolina and Virginia (Zobel, 1970).

Climate within the range of Table Mountain pine varies greatly with latitude and altitude. Mean annual precipitation ranges from 760 mm in the northern portion of the range to 2030 mm in the Great Smoky Mountains. Average July temperatures range from 15° to 27°C throughout the range, and average January temperatures range from −7° to −1°C in Pennsylvania to 2° to 5° in the Great Smoky Mountains.

Table Mountain pine is found growing on gravelly and rocky substrate soils in the Order Inceptisols, commonly on associated Lithic Dystrochrepts and Typic Dystreochrepts. Half of all stands on sites with bedrock contact are located within 50 cm of the soil surface. The soil profile typically contains 24 to 59% stone (Zobel, 1969).

Pond Pine

Pinus serotina Michx. f. [*P. rigida* Miller var. *serotina* (Michx. f.) Loudon, *P. rigida* subsp. *serotina* (Michx. f.) R.T. Clausen], pond pine, marsh pine, pocossin pine, grows on low, wet interstream flatlands or in sandy or peat-laden swamps, marshes, pocosins, and bays along the Coastal Plain of the southeastern United States. It is usually found growing on soils with a high water table, but the species grows well on mineral soils that are not waterlogged. Pond pine growth is typically slow as a result of prolonged periods of soil water saturation and anaerobic conditions. Pond pine produces stands of fair wood quality used for pulpwood and lumber. The natural range includes soils with high acidity and accumulations of muck and peat, which do not support other pine species. Stands of pond pine support many wetland wildlife species and are considered a major sanctuary.

Fire is one of the major factors influencing the distribution of pond pine in coastal regions. It is unique among pine species in its ability to sprout readily from stumps and re-form crowns from sprouts arising from dormant buds following fires. Pond pine has serotinous cones that persist for many years and may become embedded in growing stems. Wild and managed fires release pond pine seeds and reduce woody competition, producing a subclimax type of forest. Most stands of pond pine are naturally regenerated, and the species is rarely planted in managed plantations (Crutchfield and Trew, 1961).

Pond pine occurs along the Coastal Plain in the southern United States from Virginia south to central and northwestern Florida (Fig. 1.7). The northern range extends along the Coastal Plain into southern New Jersey and Delaware. Individual trees have been identified in stands in North Carolina, but it is most often found in pure stands throughout its range. The extent of pond pine has historically been underestimated because it is often misidentified as loblolly pine, despite distinct taxonomic differences. Pond pine's most distinctive identification trait is its ability to grow and regenerate under soil conditions that effectively eliminate other pine species.

The native range of pond pine is characterized by a mild and humid climate. Annual precipitation ranges from 1120 to 1420 mm throughout its range. Average monthly temperatures in July range from 24°C in the northern portion of the range to 29°C in the South. In January, the average monthly temperatures range from 20° to 16°C. Temperature extremes of −23° and 43°C have been recorded.

Pond pine is typically found on Coastal Plain soils in the Ultisol Order and Aquult Suborder. The area is characterized by sandy and organic soils with perched water tables and poorly drained depressions. Anaerobic soil conditions contribute to the buildup of thick organic layers in the soil profile.

Spruce Pine

Pinus glabra Walter, spruce pine, cedar pine, Walter's pine, is one of the tallest and least common of the southern yellow pines. It is usually found growing singly or in small groups on low river terraces and hammocks, mixed with loblolly and shortleaf pines in old field sites (Dial et al., 1976; Manwiller, 1972). Spruce pine is rarely found growing in pure stands. The species is of little commercial value due to the brittle and close-grained characteristics of the wood. Commercial plantations are limited to small-scale plantings.

Spruce pine occurs in a wide band across the South that extends from the Coastal Plain in eastern South Carolina to northern Florida and west to southern Alabama and Louisiana (Fig. 1.8) (Little, 1971). Within this region, it is usually associated with river courses.

Spruce pine grows in a climate characterized by long, hot, humid summers and mild winters. Annual rainfall accumulation is 1270 mm, and average annual temperatures range from 16° to 23°C. The Soil Orders Entisols and Spodosols are commonly associated with spruce pine. These soils are sandy loams that are high in organic matter. The soil profile has an intermittently high water table and is generally poorly drained.

Sand Pine

Pinus clausa (Chapm.) Vasey, sand pine, scrub pine, spruce pine, is a minor southern pine whose native range is found almost entirely in Florida and forms one of the most distinctive plant communities in the state (Fig. 1.9). The species is divided into two geographic races. The Ocala race (var. *clausa*) has serotinous cones and includes most of the peninsular population. The Choctawhatchee race (var. *immuginata* Ward) has cones that open at maturity and encompasses most of the northwestern population (Little, 1971). Sand pine varies in size from a small, bushy, unmerchantable scrub to sawlog-size trees 26 m in height on the best sites. It often forms dense pure stands of approximately 25,000 trees per hectare (Burns, 1973). Sand pine is a pioneer species on dry, infertile, acid, sandy soils. Natural regeneration of Ocala sand pine is dependent on fire. In the absence of fire, sand pine is suppressed by more shade-tolerant scrub hardwoods.

The Ocala sand pine grows in a narrow band along the east coast of Florida from St. Augustine south to Fort Lauderdale. Scattered stands are found on the Gulf Coast from Tampa to Naples. The Choctawhatchee sand pine is found on the northern Gulf Coast in western Florida and Baldwin County, Alabama. Some scattered stands can be found on Florida's offshore islands.

The native range of sand pine is characterized by hot, wet summers and mild dry winters. Average monthly precipitation varies from 230 mm in July to 50 mm during most of the winter. Approximately 55% of the average annual rainfall of 1350 mm occurs from June through September. Average monthly temperatures range from 11° to 27°C, with temperature extremes of −17° and 42°C. Choctawhatchee sand pine grows in slightly wetter conditions with an average annual rainfall of 1520 mm.

Sand pine grows on well-drained to excessively drained, infertile, acid, sandy soils in the order Entisols. Soil parent materials are composed of marine sand deposited in terraces. The Central Highlands have a rolling hill relief and contain subsurface deposits of soluble limestone.

Pitch Pine

Pinus rigida Miller, pitch pine, is widely distributed in the eastern and northeastern United States, predominantly at low altitude but occurring at up to 1500 m elevation in the southeastern part of its range. The greatest abundance of the population is found in Pennsylvania and in New Jersey, where it is the dominant tree species in the Pine Barrens ecosystem. Dwarf populations of the species have been identified in the New Jersey Pine Barrens. Pitch pine has been commercially exploited since the 1800s because of the comparatively large resin content of the wood (513 kg/m^3). Early colonists recognized the wood's resistance to decay and utilized pitch pine in their ship-building industries for timber and distilled pitch for naval stores. This was followed by use of the wood to manufacture charcoal for iron ore smelting and later to manufacture turpentine. Pitch pine seeds are an important midwinter food source for birds, and seedlings provide browse for deer and rabbits.

Pitch pine grows over a wide geographic range (Little, 1971). In the southern United States, it is found throughout the Appalachian Mountains from Virginia to northern Georgia and western South Carolina (Fig. 1.10). Pitch pine grows in a humid climate. Average annual precipitation ranges from 890 to 1420 mm and is evenly distributed throughout the year. Temperatures range from winter lows of 2°C to a summer average of 21°C. Extremes can vary up to 38°C. In the Great Smoky and Appalachian Mountains, pitch pine is found at elevations up to 1370 m. It is usually found on low-fertility, gravelly textured soils in the Soil Order Entisols. Pitch pine is frequently found on warm, south- or west-facing slopes.

Virginia Pine

Pinus virginiana Miller, Virginia pine, scrub pine, poverty pine, Jersey pine, is an important pioneer tree species that often invades old field sites and is used in reforestation projects to reclaim agriculturally exhausted soils on the Piedmont (Kellison and Zobel, 1973). It is a principal source of pulpwood and some lumber in the Southeast. Virginia pine is intolerant of shade and is a transitional species that is replaced by the climax hardwood forest. Compared to other southern pines, it is aggressive on poorer sites and can tolerate soil acidity from pH 4.6 to 7.9. These characteristics have made it a valuable tree species for reclamation of strip-mining sites and spoil banks (McMinn and Crane, 1984; Williamson, 1978).

The native range of Virginia pine in the northeastern United States occupies a broad band from New York west to Ohio and southern Indiana. In the Southeast, it occurs throughout Virginia south to northern South Carolina and Georgia, central and northern Alabama, and northeastern Mississippi (Fig. 1.11). In the Southeast, it is found primarily in the Piedmont and at elevations below 1000 m in the Appalachians (Little, 1971). Successful establishment of seedlings within the range is dependent on exposure of soil mineral horizons related to site disturbance by fire, logging, and agricultural abandonment.

Virginia pine grows in a humid climate. The annual precipitation average is 890 to 1400 mm, which is evenly distributed throughout the year. Average monthly temperatures range from 21° to 24°C during July to −4° to 4°C in January. Virginia pine is found growing on soils classified in the Orders Spodosols and Inceptisols. Parent material of these soils varies from sandy marine deposits to crystalline rocks, shales, and limestones. These soils have often had moderate to severe erosion resulting in the loss of the soil's A horizon. Virginia pine requires moderately well-drained to well-drained soils and is intolerant of wet sites. It is often found in pure stands.

Slash Pine

Pinus elliotti Engelm. [*P. caribaea* sensu Small, non Morelet; *P. paustris* sensu Small, non Miller; *P. heterophylla* (Ell.) Sudw., 1893, non K. Koch, 1949], slash pine, swamp pine, Cuban pine, is the second most important pine species in the Southeast and is one of the most frequently planted timber species in North America. Two geographic varieties have been described: var. *densa* Little & Dorman and var. *elliotti*. South Florida slash pine (var. *densa*) has a grass-like seedling stage and a lower average number of needles per fascicle. It occurs in central and south Florida in dryish flatwoods and on thin soil over limestone (Squillace, 1966). Variety *elliottii* occurs along the Coastal Plain from central South Carolina to central Florida and west to southern Mississippi and eastern Louisiana (Fig. 1.12).

Slash pine's range has been extended by artificial regeneration programs that began in the 1930s and through forest management fire suppression practices. Originally it was confined to pond margins and other poorly drained habitats but can now be found on sandhills, scrub-oak ridges, and sandy soils of the middle Coastal Plain, where it frequently replaces longleaf pine in mixed stands. Slash pine plantations have been established in Tennessee, Louisiana, and eastern Texas.

Slash pine has supplied the world with a large portion of the resin and turpentine since colonial times. Naval stores were the principal product from most of the early slash pine forests. Slash pine's rapid volume growth at early ages has made it an important timber species today. Slash pine is intensively managed under short rotations.

Slash pine grows best in a region characterized by warm, humid weather, wet summers, and a drier spring and fall. Annual rainfall is 1270 mm, with 70% of this occurring during the 200-day growing season. Average annual temperature is 17°C, with occasional extremes of 41° and −18°C. The major environmental control on the natural range of slash pine is rainfall. The average number of days per month with rainfall equal to or greater than 127 mm in the summer, spring, and winter and greater than 25 mm in the fall is the controlling factor. Soils in the range are primarily

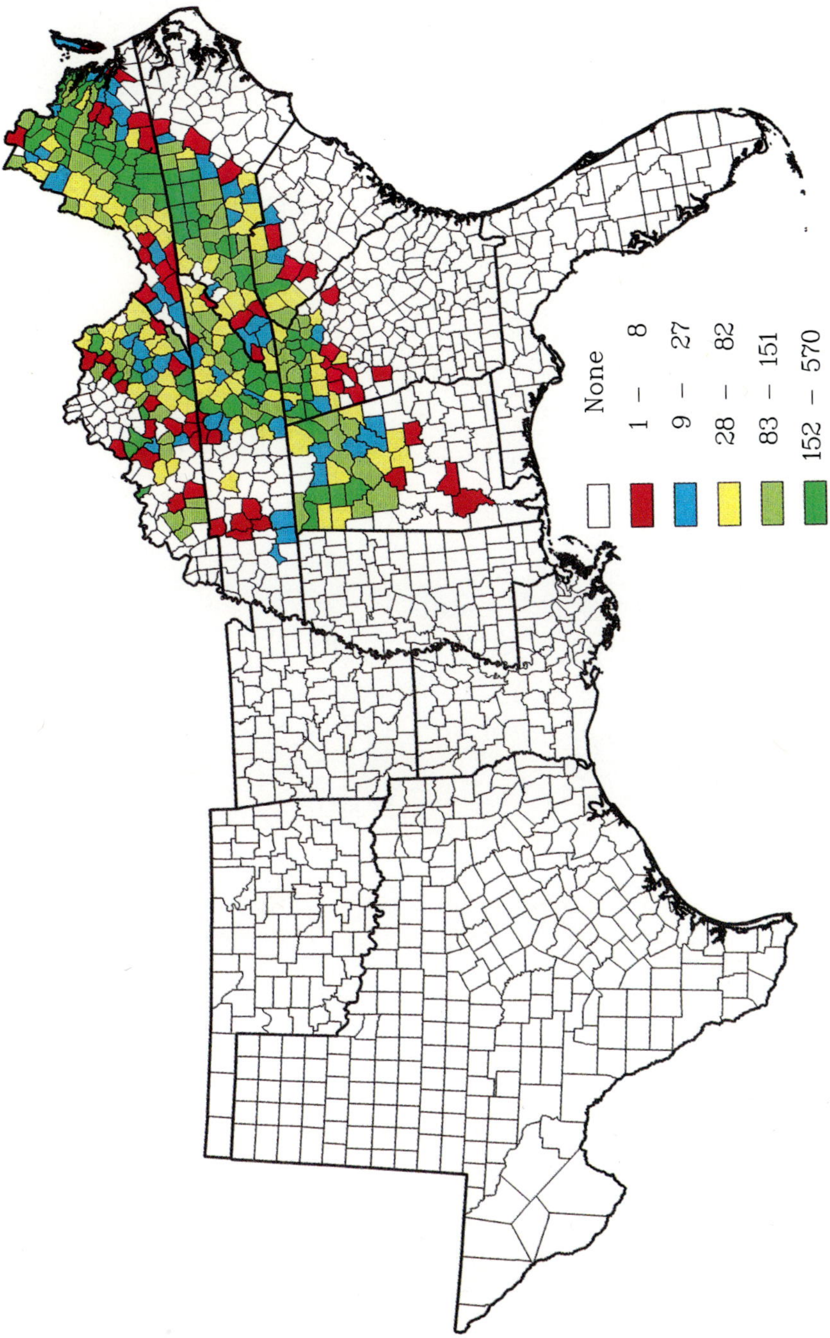

Figure 1.11. Virginia pine trees per hectare of forest land by county, 1994. Note that data for counties in central and western Texas and Oklahoma are not included in the figure. Forest inventory surveys are not conducted in these areas due to the small amount of forest land. (*Source*: USDA Forest Service, Southern Research Station, Forest Inventory and Analysis.)

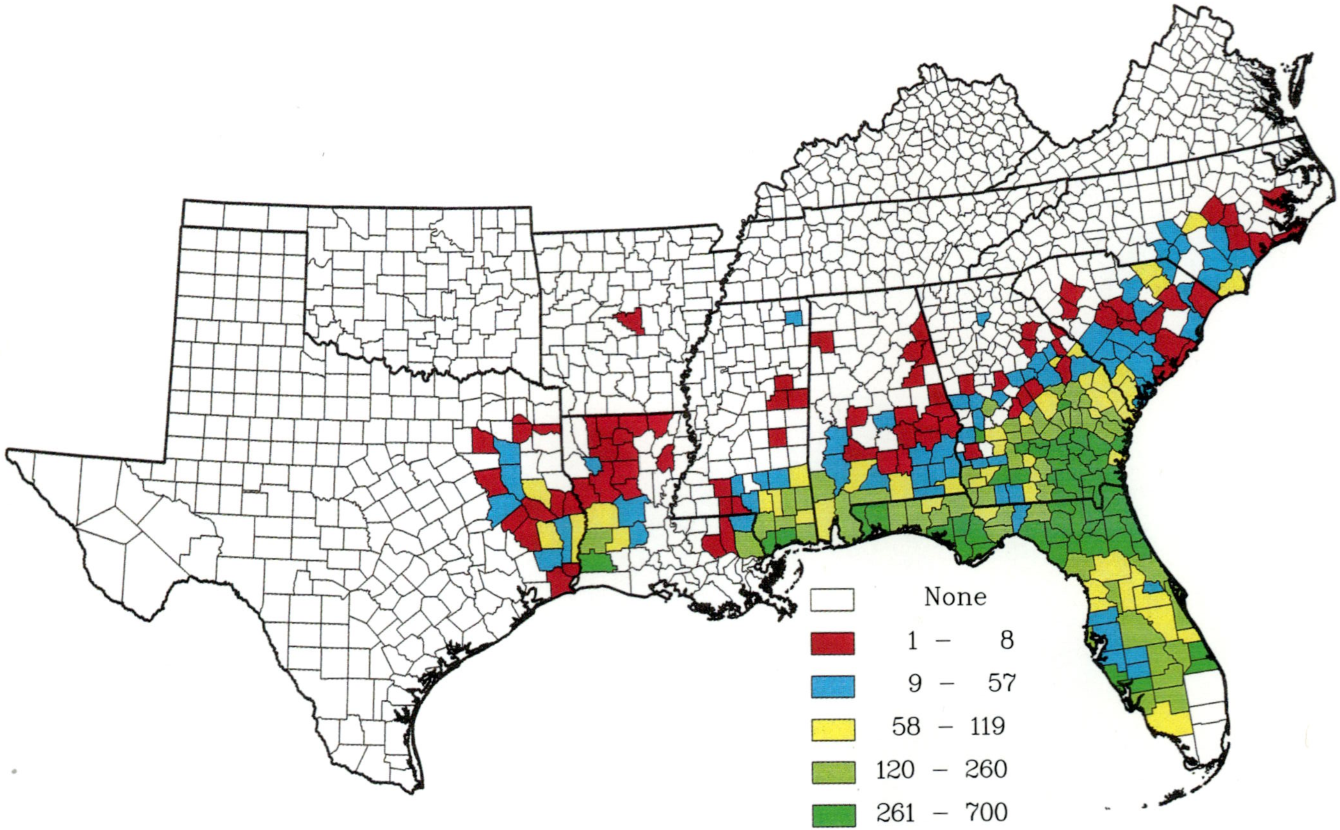

Figure 1.12. Slash pine trees per hectare of forest land by county, 1994. Note that data for counties in central and western Texas and Oklahoma are not included in the figure. Forest inventory surveys are not conducted in these areas due to the small amount of forest land. (*Source*: USDA Forest Service, Southern Research Station, Forest Inventory and Analysis.)

in the Soil Orders Spodosols, Ultisols, and Entisols. The soils are usually sandy in texture and often underlain with poorly drained hardpans or spodic horizons, 45 to 60 cm below the surface. Height growth of slash pine plantations can be modeled from three soil factors: depth to a fine-textured horizon, depth to a mottled horizon, and silt plus clay content of the finest textured horizon in the soil profile (Fowells, 1965).

Shortleaf Pine

Pinus echinata Miller, shortleaf pine, yellow pine, oldfield pine, short-straw pine, Arkansas soft pine, is the most widely distributed of the southern yellow pines. The wide distribution of shortleaf pine is partially due to its tolerance of a wide variety of soils and site conditions. Because the trees are abundant through most of their range, they are commercially important. Shortleaf pine is shade intolerant and commonly found growing in even-aged stands (Guldin, 1986). It can maintain dominance over hardwood vegetation after it overtops competing vegetation. Prescribed burning is recognized as a necessary management tool in most shortleaf pine stands. Competition from vegetation with young, slow-growing shortleaf pines exerts a profound influence on early survival and growth (Eyre, 1980; Mattoon, 1915).

In 1965, shortleaf pine, was found distributed in 22 states. In the southern United States, it occurs from Virginia south to northern Florida and west to eastern Texas, Oklahoma, Arkansas, and southern Mississippi (Fig. 1.13). The northern range extends into New Jersey and extreme southeastern New York (Little, 1971). It grows at elevations up to 1100 m in the Appalachian Mountains and often invades old field sites in the Piedmont. Smaller populations are found on the Coastal Plain.

Shortleaf pine has the largest temperature and moisture range of the southern pines. Average annual precipitation varies from 1020 mm to 1520 mm, and average annual temperatures vary from 9°C in New York to 21°C in southeastern Texas. It is found growing primarily on the Soil Order Ultisols and the Great Groups Paleudults and Hapludults. Shortleaf pine achieves its best growth on deep, well-drained soils with sandy and silt loam textures.

Natural and Human Disturbances

Southern Pine Insects and Diseases

In any effort to characterize the southern pine forest resource, one must consider the role of insects and diseases as biotic stress factors affecting forest health and productivity. Annual insect and disease mortality losses in the United States are estimated at 84.6 million cubic meters or the equivalent of approximately one fourth of the nation's timber harvest.

The reduction in growth caused by insects and diseases is estimated at another one fourth of the annual harvest. Thus, insects and diseases are reducing volume annually at the rate of one half of the nation's harvest (Stewart, 1985). In the southern United States, the annual softwood mortality is estimated at 30.2 million cubic meters, or about 18% of the softwood timber harvested in the region (Powell et al., 1993). Historical trends would indicate that one fourth of this mortality can be attributed to insects and diseases.

Southern pines are host to a diverse array of insect defoliators, bark beetles, borers, and meristem, flower, cone, and seed feeders. Pathogens range from minor root and butt decay fungi to stem rusts, which result in substantial tree mortality throughout the region. Many of the common insects and diseases of southern pine forests together with their hosts and areas of occurrence have been listed in Table 1.6. One pest and one pathogen can be separated from this list and identified as the most destructive biotic agents that contribute to tree mortality in the southern region: southern pine beetle, *Dendroctonus frontalis* Zimmermann, and fusiform rust, caused by *Cronartium quercum* (Berk.).

Southern pine beetle is a native pest that was first documented by Moravian immigrants from Austria who settled in the Piedmont of North Carolina. Foresters for their settlements recorded the "loss of many pines" (Fries et al., 1943). Further descriptive reports of damage to longleaf and yellow pine in the Coastal Plain of the Carolinas and Georgia and in eastern Tennessee were written in the early 1800s by F. Andrew Michaux. His descriptions of the beetle and tree damage leave no doubt about the cause of the southern pine mortality (Price et al., 1992).

Southern pine beetle is the most destructive insect pest of pines in the southern region. Adult beetles bore through the outer bark into the living bark, where they construct S-shaped galleries that can intersect each other and girdle the tree. Eggs deposited in individual niches on each side of the galleries hatch within four to nine days and mine a short distance before boring into the outer bark and pupating. One life cycle can be completed within 30 days under ideal conditions. There are three to seven generations per year depending on climate. The blue-stain fungus, *Ceratocstis minor* Hedg., which is introduced into the tree by the beetle, is considered to be the principal tree-killing agent. Fungal hyphae infect xylem tissue and block water-conducting tracheids. The first indication of tree mortality is discoloration of foliage, which changes from yellow to brown. Trees may be killed singly or in groups ranging up to several hundred hectares (USDA Forest Service, 1989).

In 1992, southern pine beetle outbreaks were detected on over 5.8 million hectares throughout the 13 southern states (Hofacker et al., 1993). The insect killed approximately 10,618 cubic meters of pine timber from 1973 through 1977 in the southern United States (USDA Forest Service, 1989). Numerous stand characteristics have been associated with

Table 1.6. Insects and Diseases of Southern Pines[a]

Insect		Host	Location
Defoliators			
Blackheaded pine sawfly	*Neodiprion excitans* Rohwer	Southern pines	Florida
Introduced pine sawfly	*Diprion similis* (Hartig)	Eastern white pine	North Carolina
Loblolly pine sawfly	*Neodiprion taedae linearis* Ross	Southern pines	Regionwide
Pine colaspis	*Colaspis pini* Barber	Slash pine	Regionwide
Pine webworm	*Tetralopha robustella* Zeller	Shortleaf pine	Regionwide
Redheaded pine sawfly	*Neodiprion lecontei* (Fitch)	Southern pines	Florida, Georgia, South Carolina, Tennessee
Texas leafcutting ant	*Atta texana* (Buckley)	Southern pines	Louisiana, Texas
Virginia pine sawfly	*Neodiprion pratti pratti* (Dyar)	Southern pines	North carolina, South Carolina, Virginia
Bark Beetles & Borers			
Ambrosia beetle	*Xyleborus* spp.	Southern pines	South Carolina
Black turpentine beetle	*Dendroctonus terebrans* (Olivier)	Southern pines	Regionwide
Ips engraver beetle	*Ips avulsus* (Eichhoff), *grandicollis* (Eichhoff), and *calligraphus* (Germar)	Southern pines	Regionwide
Southern pine beetle	*Dendroctonus frontalis* Zimmermann	Southern pines	Regionwide
Meristem Feeders			
Nantucket pine tip moth	*Rhyacionia frustrana* (Comstock)	Southern pines	Regionwide
Pales weevil	*Hylobius pales* (Herbst)	Southern pines	Regionwide
Scale inects	*Homoptera* spp.	Southern pines	Regionwide
Pine spittlebug	*Aphrophora parallela*	Southern pines	North Carolina, South Carolina, Tennessee, Virginia
Pitch-eating weevil	*Pachylobius picivorus* (Germar)	Southern pines	Gulf Coast from Florida to Texas
White pine weevils	*Pissodes strobi* (Peck)	Eastern white pine	North Carolina Tennessee, Virginia

Table 1.6. *Continued*

Insect		Host	Location
Flower, Cone, & Seed Feeders			
Coneworms	*Dioryctria amatella* *Dioryctria clarioualis* *Dioryctria disclusa* *Dioryctria merkeli*	Southern pines	Regionwide
Seedbugs	*Leptoglossus corculus, Tetyra bipunctata*	Southern pines	Regionwide
Slash pine flower thrips	*Gnophothrips fuscus*	Southern pines	Regionwide
White pine cone beetle	*Conopthorus coniperda*	Eastern white pine	North Carolina, Tennessee

Disease		Host	Location
Foliage			
Needle cast	*Hypoderma* spp. *Lophodermium* spp. *Ploiodema* spp. *Rhizosphaera* spp.	Southern pines	Regionwide
Brown spot needle blight	*Mycosphaerella dearnessii*	Longleaf pine	Throughout host range
Pine needle rust	*Coleosporium spp.*	*Southern pines*	*Regionwide*
Stem, Branch, & Cone			
Fusiform rust	*Cronartium quercuum f.* sp. *fusiforme*	Loblolly, slash pine	Regionwide, except Kentucky, Puerto Rico, U.S. Virgin Islands
Comandra blister rust	*Cronartium comandrae*	Loblolly, pond Virginia, shortleaf pine	Arkansas, Tennessee
Eastern gall rust	*Cronartium quercuum* f. sp. *quercuum*	Virginia, sand, shortleaf pine	Florida, Georgia, North Carolina South Carolina, Virginia
Southern cone rust	*Cronartium strobilinum*	Slash, longleaf pine	Georgia, along the Gulf Coast from Florida to Texas
Pitch canker	*Fusarium subglutinans* var. *sublutinans*	Southern pines, especially loblolly pine, slash pine, Virginia pine	Regionwide

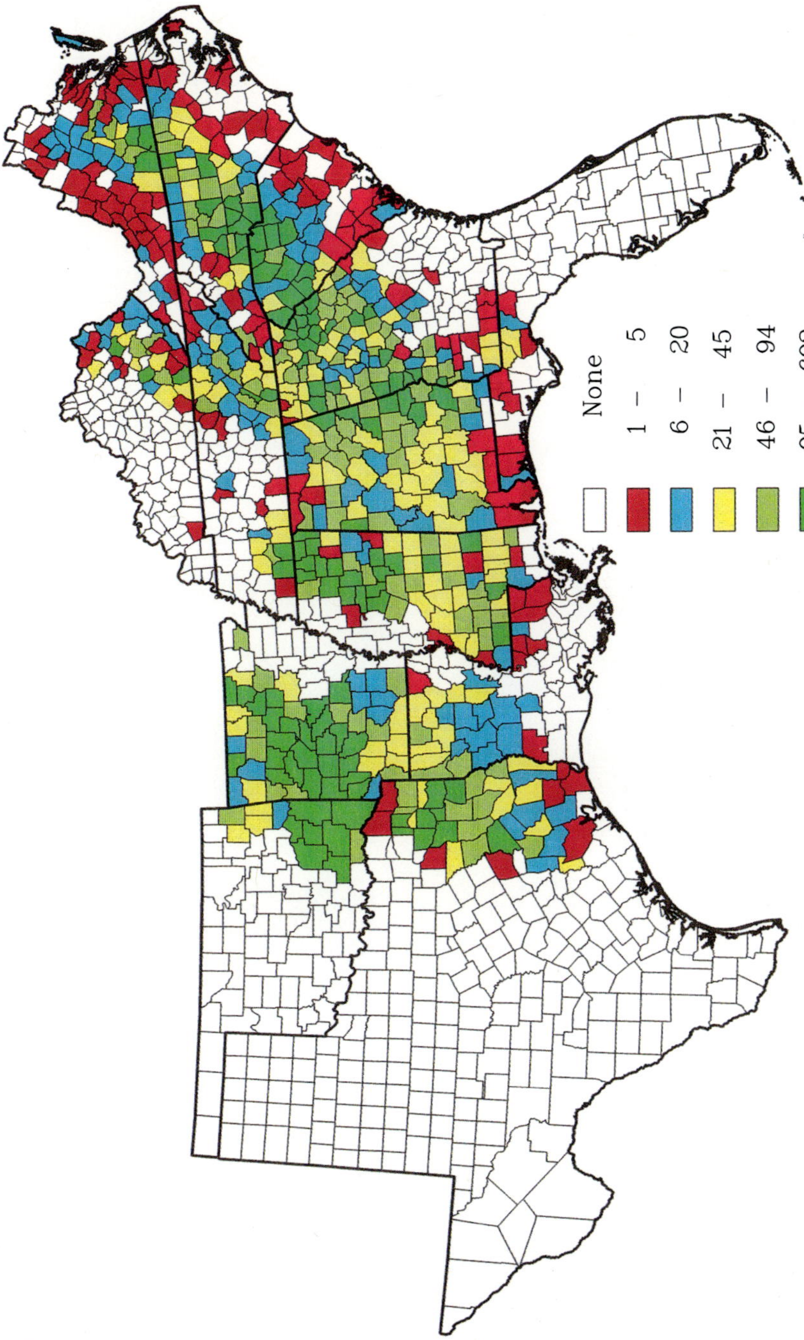

Figure 1.13. Shortleaf pine trees per hectare of forest land by county, 1994. Note that data for counties in central and western Texas and Oklahoma are not included in the figure. Forest inventory surveys are not conducted in these areas due to the small amount of forest land. (*Source:* USDA Forest Service, Southern Research Station, Forest Inventory and Analysis.)

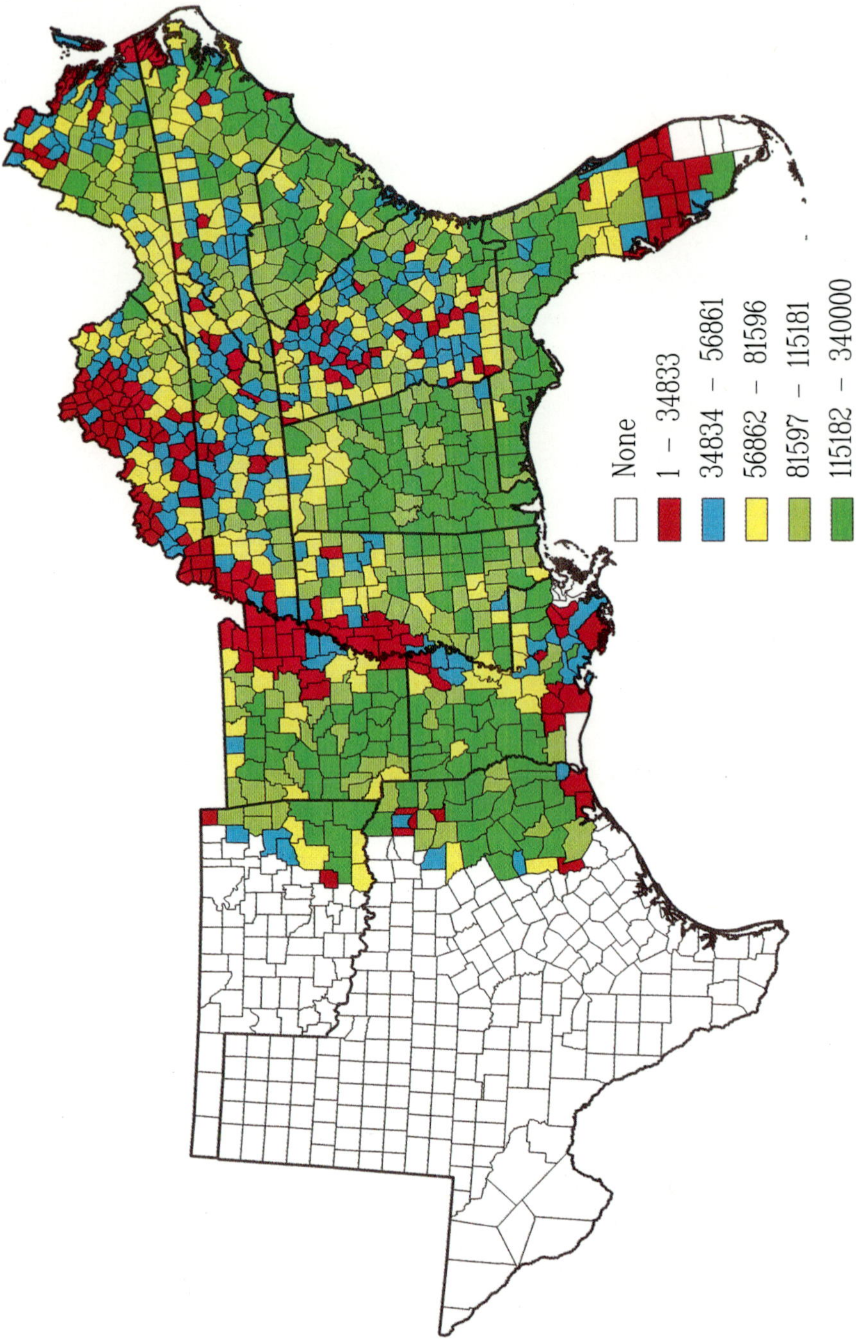

Figure 1.14. Hectares of forest land by county, 1994. Note that data for counties in central and western Texas and Oklahoma are not included in the figure. Forest inventory surveys are not conducted in these areas due to the small amount of forest land. (*Source*: USDA Forest Service, Southern Research Station, Forest Inventory and Analysis.)

Table 1.6. *Continued*

Disease		Host	Location
Root & Butt Rots & Decay			
Red heart	*phellinus pini*	Southern pines	Regionwide
Annosus root & butt rot	*Heterobasidion annosum*	Eastern white pine, southern pines	Regionwide
Red root & butt rot	*Inonotus circinatus*	Slash, sand, and shortleaf pine	South
Littleleaf disease	*Phytophthora cinnamomi* and *Pythium* spp.	Loblolly, shortleaf pine	Alabama, Georgia, Kentucky, North Carolina, South Carolina, Tennessee, Virginia
White pine root disease	*Verticicladiella procera*	Eastern white pine	Southern Appalachians

[a]*Sources*: USDA Forest Service, 1989, *Insects & Diseases of Trees in the South*; USDA Forest Service, 1993, *Forest Insect & Disease Conditions in the United States*.

outbreaks of the southern pine beetle. Natural stands attacked are characterized by high stand densities, a large proportion of shortleaf and/or loblolly pine sawtimber, and declining radial growth. Studies have not shown a single set of climatic conditions that are consistently associated with outbreaks, but Lorio and Hodges (1983) found that oleoresin exudation pressure of large loblolly pine was reduced during periods of drought-induced moisture stress, resulting in a decrease of pitch-out of boring adult beetles and increased colonization success.

Fusiform rust of southern pines is one of the most damaging forest tree diseases in the United States. In 1992, it was estimated that 6.8 million hectares in the Piedmont and Coastal Plain of 11 southern states were affected by fusiform rust. About 30% of all loblolly and slash pine hectares were infected, and approximately 10% of the trees had potentially lethal cankers. Rust causes an estimated annual loss in loblolly and slash pines in excess of $47 million in the United States (Hofacker et al., 1993). Disease incidence is estimated to be increasing at a rate of 2% per year (Schmidt et al., 1972). These increases are occurring in both natural and planted stands, despite significant improvements for resistance by genetic selection (LaFarge and Kraus, 1980; Powers, 1982; Wells and Wakeley, 1966).

Fusiform rust is a typical long-cycle rust fungus with two alternate hosts, pines and species of the black oak group (*Quercus* sp.). The fungus produces orange aeciospores on the surface of fusiform-shaped pine galls in the spring. During the late spring and summer, telia form on the oak leaves and produce spores that infect the pine. Spread of the fungus is dependent on periods of warm, moist weather in April and May. The infection of the pine is perennial and is potentially lethal if infection spreads or occurs on the main stem. Most infections take place in young planting and regeneration stands ranging from three to five years in age. Infections that occur later in the life cycle weaken the stem and result in reduced vigor and breakage. Tree mortality usually occurs prior to age 10. The disease occurs randomly in a stand, killing a proportion of infected trees, but losses in individual nurseries can exceed 80% (USDA Forest Service, 1989).

The current outlook for controlling fusiform rust by planting resistant genotypes of pines is encouraging. Forest industry projections anticipate reducing the incidence of fusiform rust by 40% in selected genotypes of loblolly pine and 60% in selected genotypes of slash pine. When combined with silvicultural control measures to eradicate or suppress oaks and apply fungicides, fusiform rust–induced mortality may be reduced to acceptable levels in managed stands and nurseries.

Fire

Of all the factors exposing ground to invasion by pines, fire has played one of the principal roles in the regeneration and succession of pine forests. For some forest ecotypes, fire has become a permanent factor in their ecology. As early as 1894, W.W. Ashe, a forester in North Carolina, complained that loblolly pine was replacing longleaf pine (Ashe, 1895). It would be much later before recognition of the dependence of longleaf pine upon fire at stages in its life cycle.

Evidence suggests that American Indian tribes of the eastern United States and those that later migrated from the Continental West set periodic fires to burn the forest. These practices likely followed those of the Shoshoni, Paiute, and other hunting and seed-gathering tribes that remained in the western United States. This practice caused historians to report that in the 1600s European sailors smelled smoke while seeking docking sites on Virginia's coast. One wrote that "the land was smelt before it was seen" (Cowdry, 1983 p. 16). This practice continued with pioneer settlers as southern pine forests were converted into productive pasture for cattle. During and following the Civil War, intentionally set fires became part of the South's culture. It is estimated that over 2 million hectares of southern forest with a value of $7 million were burned in 1880 alone (Clark, 1984).

This trend began to change with the passage of the Weeks Act of 1911 and the Clarke–McNary Act of 1924. In addition to the purchase of burnt-over and clear-cut lands, the Weeks and Clarke–McNary Acts provided funds beginning in 1928 for the Dixie Crusaders, who traveled to rural communities teaching fire prevention and forest conservation principles to over 3 million rural southerners. As a result of increased funding for forest fire protection and public education, the land area protected from wildfires increased from 28 million hectares in 1930 to more than 94 million hectares in the 1980s. The reduction in area burned by wildfires dropped from 9.5 million hectares in 1925 to as low as 112,910 hectares in 1983 (USDA Forest Service, 1988). Today, fire protection and the use of controlled or prescribed burning are the most widespread and effective timber and ecosystem management practices in the southern pine forests.

References

Adams DM, Haynes RW (1991) Softwood timber supply and the future of the southern forest economy. Southern J Applied Forestry 15(1):31–37.

Andresen JW (1964) The taxonomic status of *Pinus chiapensis*. Phytologia 10:417–421.

Andresen JW (1966) A multivariate analysis of the *Pinus chiapensis-monticola-Strobus* phylad. Phodora 68:1–24.

Ashe WW (1895) *Forest Fires: Their Destructive Work, Causes, and Prevention*. North Carolina Geological Survey Bulletin 7, Raleigh, NC, p 37.

Beltz RC, Cost ND, Kingsley NP, Peters JR (1992) *Timber Volume Distribution Maps for the Eastern United States*. USDA Forest Service, General Technical Report WO-60, 59 pp.

Buol SW, Hole FD, McCracken RJ (1989) *Soil Genesis and Classification*. Iowa State University Press, Ames, IA, 446 pp.

Burns RM (1973) Sand pine: Distinguishing characteristics and distribution. In: *Proceedings, Sand Pine Symposium*. USDA Forest Service, General Technical Report SE-2. Southeastern Forest Experiment Station, Asheville, NC, pp 13–17.

Burns RM, Honkala BH (1990) *Silvics of North America: 1. Conifers*. US Dept of Agr, Agriculture Handbook 654. Washington, DC, pp 497–512.

Clark TD (1984) *The Greening of the South*. The University Press of Kentucky, Lexington, KY, p 12.

Cowdry AE (1983) *This Land, This South*. University Press of Kentucky, Lexington, KY, p 16.

Critchfield WB, Little EL, Jr (1966) *Geographic Distribution of the Pines of the World*. US Dept of Agr Forest Serv Misc Publ 999:1–97.

Crutchfield DM, Trew IF (1961) Investigation of natural regeneration of pond pine. J Forestry 59(4):264–266.

Dial SC, Batson WT, Stalter R (1976) Some ecological and morphological observations of *Pinus glabra* Walter. Castanea 41:361–377.

Eyre FH (ed) (1980) *Forest Cover Types of the United States and Canada*. Society of American Foresters, Washington, DC, 148 pp.

Fenneman NM (1983) *Physiography of the Eastern U.S.* McGraw-Hill, New York, 714 pp.

Fowells HA (1965) *Silvics of Forest Trees of the United States*. US Dept of Agr, Agriculture Handbook 271. Washington, DC, 762 pp.

Fries AL, Hamilton LG, Rights DL, Smith MJ (1943) *Records of the Moravians in North Carolina*. North Carolina Historical Comm, Raleigh, NC, p 2593.

Guldin JM (1986) Ecology of shortleaf pine. In: Murphy PA (ed) *Proceedings, Symposium on the Shortleaf Pine Ecosystem*, March 31–April 2, 1986, Little

Rock, AR. Arkansas Cooperative Extension Service, Monticello, AR, pp 25–40.
Hack JT, Goodlett, JC (1960) *Geomorphology and Forest Ecology of a Mountain Region in the Central Applachians*. US Dept of the Interior Geological Survey, Professional Paper 347. Reston, VA, 66 pp.
Haynes RW, Adams DM, Mills JR (1995) *The RPA timber assessment update*. General Technical Report RM-259. Fort Collins, CO. US Dept of Agr Forest Serv, Rocky Mountain Forest and Range Experiment Station, pp 1–66.
Hodgkins EJ (ed) (1965) *Southeastern Forest Habitat Regions Based on Physiograghy*. Auburn University. Forestry Departmental Series 2. Agricultural Experiment Station, Auburn, AL, 10 pp.
Hofacker TH, Fowler RF, Webster K, Riffe M (1993) *Forest Insect and Disease Conditions in the United States 1992*. USDA Forest Service Forest Pest Management, Washington, DC, 147 pp.
Kellison RC, Zobel BJ (1973) *Genetics of Virginia Pine*. US Dept of Agr Forest Serv Research Paper WO-21, 10 pp.
LaFarge T, Kraus JK (1980) A progeny test of (shortleaf × loblolly) × loblolly hybrids to produce rapid-growing hybrids resistant to fusiform rust. Silvae Genetica 29:197–200.
Little EL (1971) *Atlas of United States Trees*. Vol. 1. Conifers and important hardwoods. US Dept of Agr Misc Publ 1146.
Little EL, Critchfield WB (1969) *Subdivisions of the Genus Pinus (Pines)*. US Dept of Agr Forest Serv Misc Publ 1144:1–51.
Lorio PL Jr, Hodges JD (1983) Microsite effect on oleoresin exudation pressure of large loblolly pine. Ecology 49:1207–1210.
Manwiller FG (1972) *Wood and Bark Properties of Spruce Pine*. USDA Forest Service, Research Paper SO-78. Southern Forest Experiment Station, New Orleans, LA, p 25.
Mattoon WR (1915) *Life History of the Shortleaf Pine*. US Dept of Agr Bull 244:1–46.
McMinn JW, Crane WH (1984) Five-year performane of selected woody species on an upper coastal plain spoil bank. Southern Applied Forestry 8:207–209.
Mills JR, Kincaid JC (1992) The aggregate timberland assessment system—ATLAS: a comprehensive timber projection model. General Technical Report PNW-GTR-281. Portland, OR., US Dept of Agr Forest Service, Pacific Northwest Research Station, pp 1–160.
Mirov NT (1967) *The Genus Pinus*. Ronald Press Company, New York, 602 pp.
McMinn JW, Crane WH (1984) Five-year performance of selected woody species on an upper coastal plain spoil bank. Southern J of Applied Forestry 8:207–209.
Powell DS, Faulkner JL, Darr DR, Zhu Z, MacCleery DW (1993) *Forest Resources of the United States, 1992*. USDA Forest Service, General Technical Report RM-234, Fort Collins, CO, 132 pp.
Powers H (1982) Research may soon produce planting stock to fusiform rust. Forest Farmer 4291:6–7.
Price RA (1989) The genera of *Pinaceae* in the southeastern United States. J Arnold Arboretum 70:247–305.
Price TS, Doggett C, Pye JM, Holmes TP (1992) *A History of Southern Pine Beetle Outbreaks in the Southeastern United States*. Georgia Forestry Commission, Macon, GA, 70 pp.
Schmidt RA, Goddard RE, Hollis CA, Smith WH (1972) *Incidence and Distribution of Fusiform Rust in Slash Pine Plantations in Florida and Georgia*. Florida Agric Exp Stn Bull 763, 21 pp.

Snyder EB, Dinus RJ, Derr HJ (1977) *Genetics of Longleaf Pine*. USDA Forest Service, Research Paper WO-33. Washington, DC, 24 pp.

Squillace AE (1966) Geographic variation in slash pine. Forest Sci Monograph 10:1–56.

Stewart JL (1985) Current use and potential for implementing forest management practices in intensive forest management in the USA. The Forest Chronicle 61(3):240–242.

Thornbury WB (1965) *Regional Geomorphology of the United States*. Wiley, New York, 609 pp.

USDA Forest Service (1988) *The South's Fourth Forest: Alternatives for the Future*. Forest Resource Report No 24, Washington, DC, p 512.

USDA Forest Service (1989) *Insects and Diseases of Trees in the South*. US Dep of Agr Forest Service, Southern Region, Protection Report R8-PR 16, Atlanta, GA, 98 pp.

USDA Forest Service (1993) Forest insect and disease conditions in the United States 1992. US Dept of Agr Forest Service, Forest Pest Management, Washington, DC, pp 1–147.

Walker LC (1991) *The Southern Forest: A Chronicle*. University of Texas Press, Austin, TX, 322 pp.

Wells OO, Wakeley PC (1966) Geographicac variation in survival, growth, and fusiform rust infection of planted loblolly pine. Forest Sci Monograph 11, p 40.

Williamson RK (1978) Followup on trees. Soil Conservation 43(10):9.

Zobel DB (1969) Factors affecting the distribution of *Pinus pugens*, an Applachian endemic. Ecological Monographs 39:304–333.

Zobel DB (1970) Morphological characteristics of *Pinus pugens*. J Elisha Mitchell Scientific Soc 86:214–221.

2. Soils, Geomorphology, and Land Use of the Southeastern United States

Philip Schoeneberger

The response of southern pine forests to acidic deposition is determined by acidic deposition itself, plant genetics, and ecosystem dynamics. A major dimension of ecosystem dynamics is the nature, distribution, and ultimately the ability of the below-ground matrix (soils) to ameliorate acidic deposition. Other portions of this volume (e.g., Richter and Markewitz, Chapter 8) deal with the specifics of soil response to acidic deposition. This chapter is presented as a regional overview of relevant climatic patterns, the most extensive soil groups, and general land-use patterns, grouped by major geomorphic regions, that encompass the southern pine forests of North America.

The expression of soil patterns, geomorphology, and land-use patterns varies with the scale of observation. Most quantitative soil information is collected at a level of fine detail (e.g., pedon level), the geographical equivalent of point data. This scale has historically provided sufficient detail to describe the highly variable and diverse population of soils in natural settings. It is also the principal source of data used for soil-related modeling efforts. However, at some point it becomes necessary to move beyond point data and individual soil bodies to a broader, area basis. This extrapolation is needed to comprehend the overall mosaic of the soilscape and to obtain an assessment of the potential regional impact of atmospheric pollution. Consequently, it becomes appropriate to consider generalizations and summaries of point data

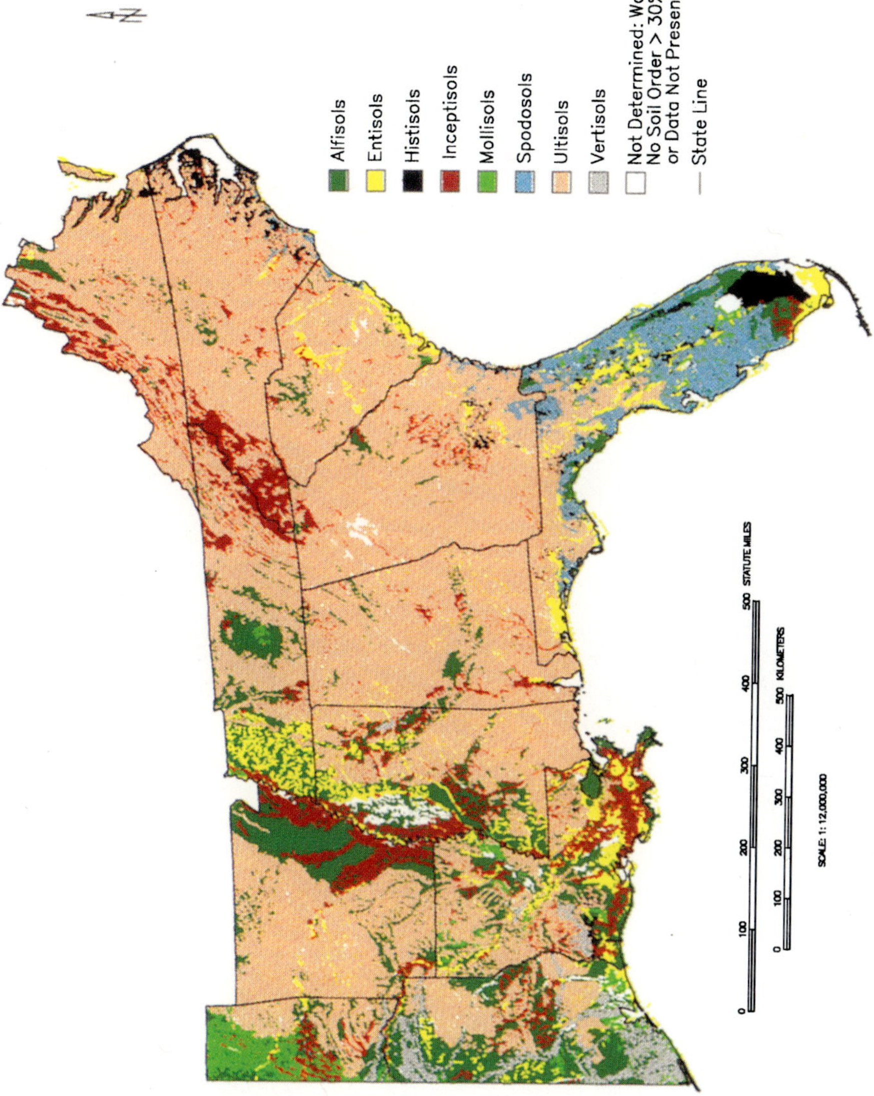

Figure 2.1. Dominant Soil-Order distributions of the southeastern United States. (*Source*: State Soil Geographic Database, 1993. Albers Equal Area Projection. Map prepared using MAPGEN 4.1. Lincoln, NE, October 1994.)

as can be found in soil taxonomy and various National Cooperative Soil Survey products.

Soil Development and Soil Taxonomy

Various studies have noted that categorical statements about expected responses of Soil Taxonomic groups to atmospheric deposition are not very reliable (Binkley et al., 1989). This should be self-evident in that Soil Taxonomy differentia (Soil Survey Staff, 1975, 1994e) are not designed for, nor based on, criteria determinant in atmospheric deposition issues. Nonetheless, some general responses may be inferred for some taxonomic groups; comments are made within the discussion of geomorphic regions later in this chapter. More importantly, soil taxonomic groups, or derivative soil map units, provide the only regional soils database. One of the most comprehensive available soils databases for the southeastern United States is the State Soil Geographic Database (STATSGO; Soil Survey Staff, 1994a–d, 1994f–m), which is the primary basis for soil groups and land-use data used in this chapter. A generalized, regional soil map of dominant Soil Orders is presented in Fig. 2.1 and Table 2.1.

General properties for the dominant Soil Orders and Suborders of the southern pine forest areas are summarized in Fig. 2.2. More detailed information for these and lower taxonomic levels, such as the Great Groups mentioned throughout the discussion of geomorphic regions, is available elsewhere (Buol et al. 1989; Soil Survey Staff, 1994e). Additional discussion of general soil properties and development is presented in Chapter 8.

Various factors influence soil development and, subsequently, the ability of the pedosphere to respond to acidic deposition (specifics of responses of southern pine forest soils to acidic deposition are covered in detail in Chapter 8). Traditionally, the main soil-forming factors are climate, parent material, living organisms, topography, and time.

The soil taxonomic groupings of the southern pine forests reflect the climate and geology that largely control soil development in this region. The temperate, continental climate of the southeastern United States results in a predominantly warm, moist soil environment. Notable exceptions are the cooler conditions at high elevations in the Appalachian Mountains and the drier conditions along the western boundary of the region. Significant, localized exceptions are common and contribute to soil and forest diversity but are not within the scope of this chapter.

The predominantly warm, moist climate combines with the general geologic stability and moderate geologic age to present a soilscape that is characterized by considerable pedogenic alteration. This is evidenced in the predominant Soil Order (Ultisols), which, by definition, is composed of well-developed soils. It is also seen within Soil Orders in which the

Table 2.1. Distribution of Dominant Soil Orders in the Southeastern United States

#) Geomorphic Region Dominant Soil Order	Hectares	% Cover
1) Coastal Plain	74,557,100	53.24
Alfisols	13,374,700	17.94
Entisols	6,540,300	8.77
Histisols	2,718,000	3.65
Inceptisols	3,634,400	4.87
Mollisols	1,036,000	1.39
Spodosols	4,974,600	6.67
Ultisols	39,227,300	52.61
Vertisols	1,304,800	1.75
Water	1,103,900	1.48
Not Determined	643,100	0.86
2) Mississippi River Valley	14,563,500	10.40
Alfisols	7,425,600	50.99
Entisols	2,329,400	15.99
Histisols	11,300	0.08
Inceptisols	2,545,700	17.48
Mollisols	91,100	0.63
Ultisols	1,013,400	6.96
Vertisols	400	0.00
Water	657,700	4.52
Not Determined	488,900	3.36
3) Piedmont	16,580,200	11.84
Alfisols	1,301,800	7.85
Entisols	27,100	0.16
Inceptisols	492,200	2.97
Ultisols	14,512,900	87.53
Water	246,200	1.48
4) Blue Ridge	4,513,200	3.22
Alfisols	62,500	1.38
Entisols	35,500	0.79
Inceptisols	2,648,100	58.67
Ultisols	1,752,400	38.83
Water	14,700	0.33
5) Ridge and Valley	7,260,700	5.19
Alfisols	572,300	7.88
Entisols	4,200	0.06
Inceptisols	1,711,700	23.57
Mollisols	5,800	0.08
Ultisols	4,787,800	65.94
Water	155,200	2.14
Not Determined	23,700	0.33
6) Appalachian Plateau	3,775,600	2.70
Alfisols	152,400	4.04
Entisols	2,400	0.06
Inceptisols	229,100	6.07
Ultisols	3,370,500	89.27
Water	21,200	0.56

Table 2.1. *Continued*

#) Geomorphic Region Dominant Soil Order	Hectares	% Cover
7) Interior Low Plateau	4,293,200	3.07
Alfisols	1,799,400	41.91
Inceptisols	177,600	4.14
Mollisols	24,800	0.58
Ultisols	2,271,300	52.90
Water	20,100	0.47
8) Ouachita Plateau	4,972,100	3.55
Alfisols	546,700	11.00
Entisols	217,400	4.37
Inceptisols	555,900	11.18
Mollisols	3,200	0.06
Ultisols	3,648,900	73.39
9) Ozark Plateau	4,127,800	2.95
Alfisols	421,800	10.22
Mollisols	96,100	2.33
Ultisols	3,578,300	86.69
Water	31,600	0.77
10) Central Lowlands	1,675,400	1.20
Alfisols	415,400	24.79
Mollisols	1,227,200	73.25
Water	32,800	1.96
11) Water*	1,978,600	1.41
Water	1,295,500	65.48
Alfisols	498,700	25.20
Entisols	15,100	0.76
Histisols	2,000	0.10
Inceptisols	159,100	8.04
Mollisols	200	0.01
Ultisols	7,600	0.38
Vertisols	200	0.01
Not Determined	200	0.01
12) Miscellaneous Areas*	1,715,200	1.23
Water	19,600	1.14
Alfisols	720,500	42.01
Entisols	13,300	0.78
Inceptisols	38,700	2.26
Mollisols	188,700	11.00
Ultisols	16,600	0.97
Vertisols	696,800	40.62
Not Determined	21,000	1.22
Total	140,012,600	100.00

* "Water" in the context of this table is a map unit composed dominantly of water but which also contains small "non-water" (soil) areas. These "non-water" areas include minor soil delineations not shown at the given map scale and inclusions (estimates of undelineated soil components) (Soil Survey Staff, 1994 a–d, f–m).

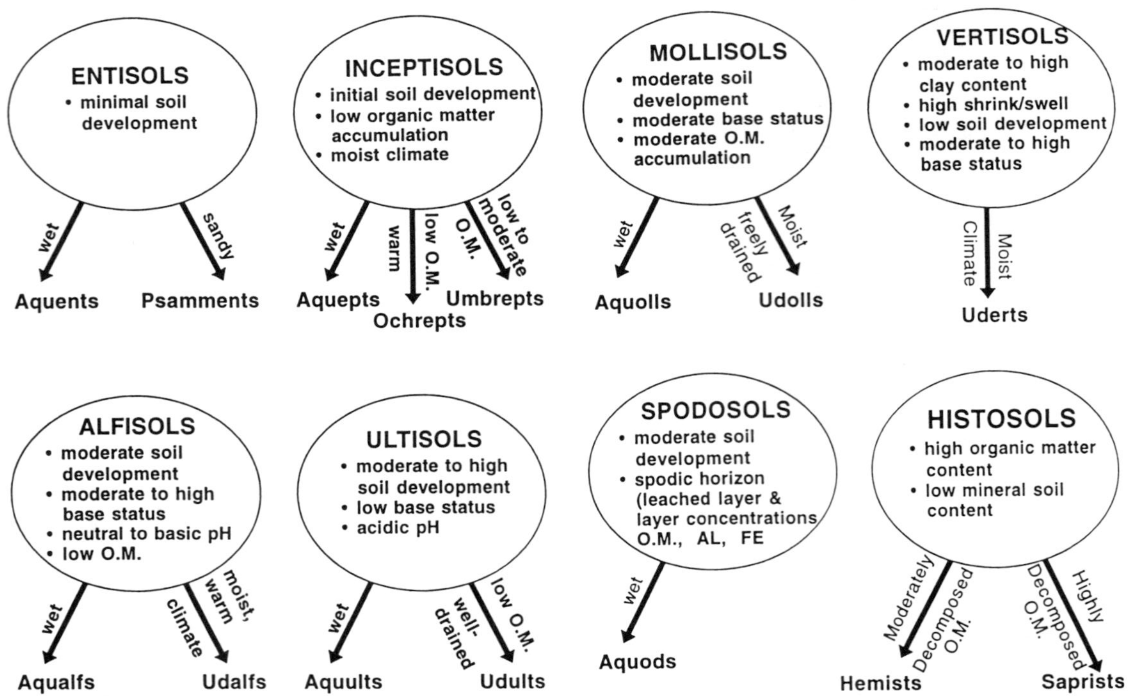

Figure 2.2. Generalized soil properties of the dominant soil taxonomic Orders and Suborders of the southeastern United States.

lower taxa are predominantly the most pedogenically altered soils allowed within the soil order (e.g., Paleudults). The most extensive exceptions to this regional trend are soils derived from relatively inert geologic material (e.g., quartz sands) and soils developing on dynamic, young portions of the landscape. These latter situations result in soils with much less development (e.g., Entisols [Psamments] and Inceptisols [Dystrochrepts]).

The material in which soils form also has a profound effect on the resultant soils. The influence of parent material is indirectly demonstrated in the discussion of geomorphic regions (later in this chapter), which separates areas of major differences in geologic materials. For example, high-base soils such as Alfisols are common only in areas containing eolian or bedrock sources of high-base geologic materials.

Soil hydrology also has an important influence on soil development and behavior that can dominate other variables. In the humid and warm southeastern United States, the most widespread expression of soil hydrology controlling soil development is "wet" soils—those soils that are saturated for some considerable part of the year. Seven of the eight widespread soil orders in the southeastern United States include substantial areas of wet members (Aquic Suborders and Histosols; Fig. 2.2).

Climate

The southeastern United States has a predominantly continental climate that grades to a maritime climate along the coastal margin. Exceptions are found in the high-elevation peaks and ridges of the Appalachian Mountains. Annual precipitation ranges from about 1000 to over 2032 mm (Fig. 2.3; Climatic Data Access Focility [CDAF], 1993; National Climate Data Center [NCDC], 1992). Several dominant trends are apparent. Precipitation is lowest in the continental interior margins and generally increases toward the coastal margins and at higher elevations. Orographic effects result in local and regional reductions in precipitation (rain shadows), as seen to the east of the Appalachian Mountains in the Piedmont of Virginia, North Carolina, South Carolina, and Georgia (Fig. 2.3). Precipitation decreases toward the western margin of the region, and this contributes to the gradual vegetation shift from woodlands to grasslands. Mean annual air temperature (Fig. 2.4; CDAF, 1993; NCDC, 1992) and potential evapotranspiration (Fig. 2.5; CDAF, 1993; NCDC, 1992; Van Wambeke et al., 1991) generally increase toward the south or with a decrease in elevation.

Geomorphic Regions and Dominant Soils

The geomorphic regions or groupings described here are well established, with only minimal argument about their boundaries and general composi-

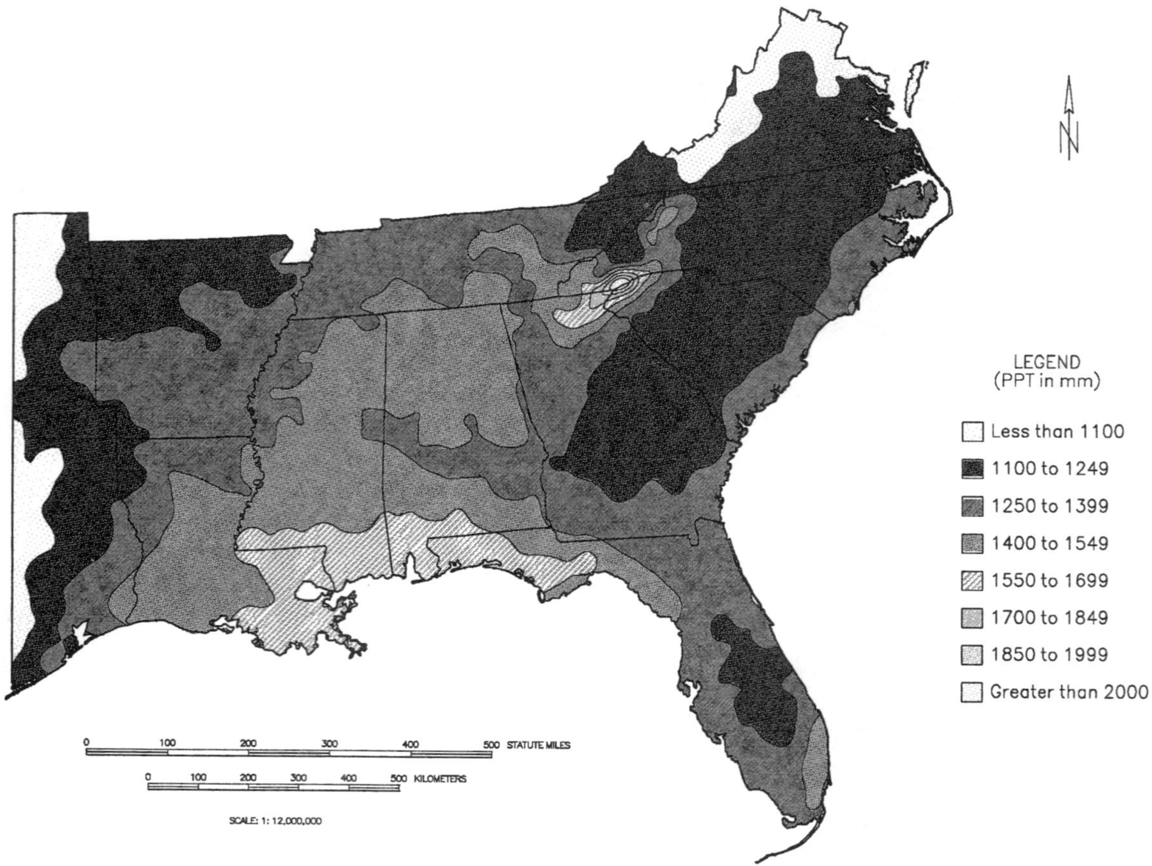

Figure 2.3. Mean annual precipitation (1961 to 1990 normal) for the southeastern Untited States. (*Source*: NCDC, 1992; CDAF, 1993. Albers Equal Area Projection. Map prepared using MAPGEN 4.13, MNTC. Lincoln, NE, December 1994.)

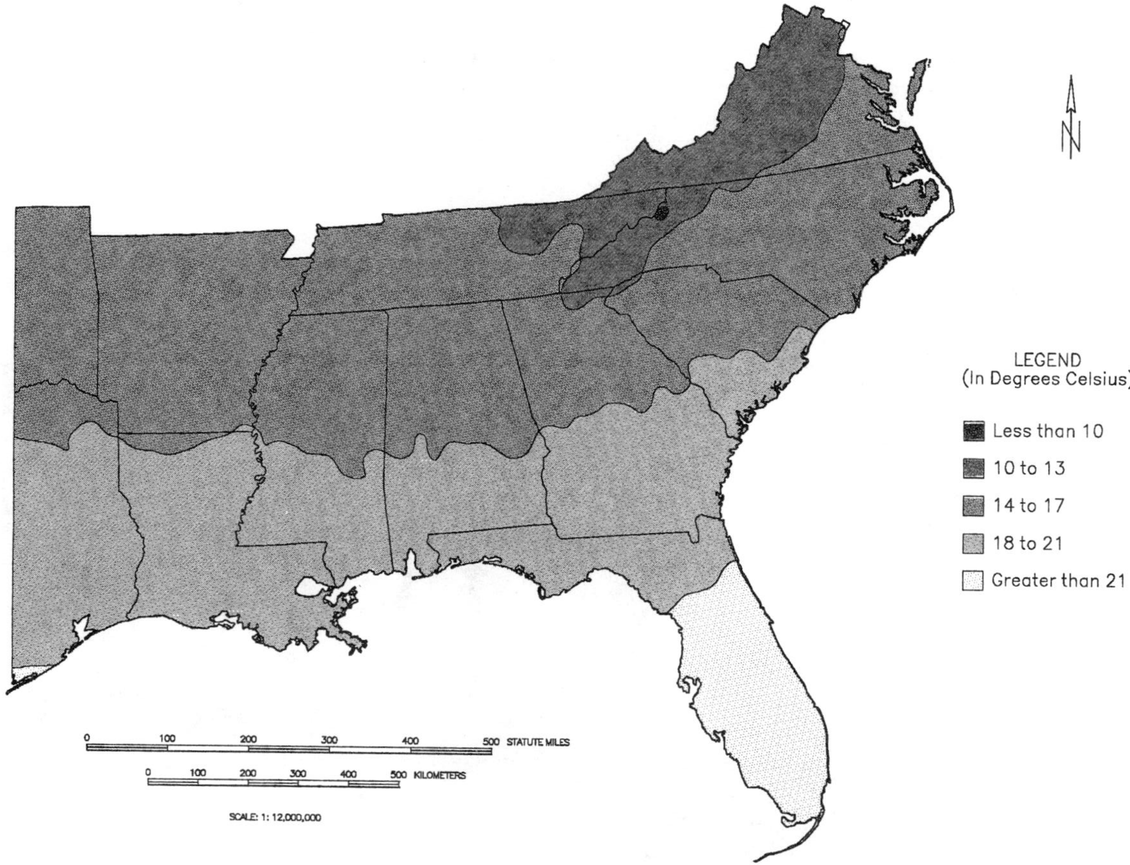

Figure 2.4. Mean annual air temperature (1961 to 1990 normal) for the southeastern United States. (*Source*: NCDC, 1992; CDAF, 1993. Albers Equal Area Projection. Map prepared using MAPGEN 4.13, MNTC. Lincoln, NE, December 1994.)

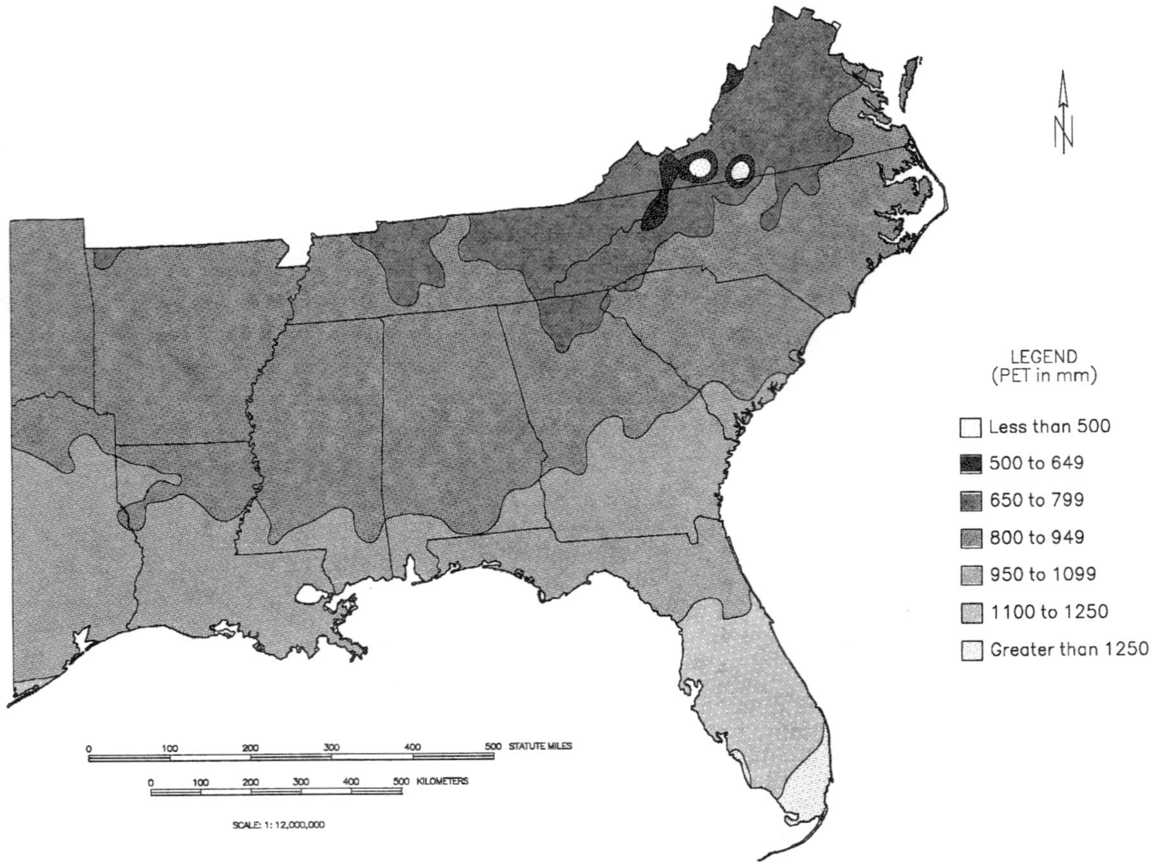

Figure 2.5. Newhall model generated potential evapotranspiration for the southeastern United States. (*Source*: NCDC, 1992; CDAF, 1993. Albers Equal Area Projection. Van Wambeke et al. 1992. Map prepared using MAPGEN 4.13, MNTC. Lincoln, NE, December 1994.)

tion (Fenneman, 1938; Soil Conservation Service, 1981; Soil Survey Staff, 1994k; Thornbury, 1965). Classical geomorphic boundaries, with their inherent emphasis on geologic materials, are often, but not always, coincident with meaningful pedosphere boundaries. Some adjustments have been made, as noted in the text, to group soils that have greater similarities to soils in adjacent areas. This displaces some traditional geomorphic boundaries.

The southeastern United States (Alabama, Arkansas, Florida, Georgia, Louisiana, Mississippi, North Carolina, Oklahoma, South Carolina, Texas, and Virginia) can be divided into 10 geomorphic regions (Fig. 2.6, p. 75). Each region contains a range of climatic, geologic, and soil characteristics. The areal distribution of geomorphic regions and their respective component subregions (Land Resource Units) is presented in Table 2.2. Regional land cover (land use) can be divided into general groups: woodland (forested), cropland, grasslands, wetlands, barren, irrigated agriculture, and water (Fig. 2.7, p. 75 and Table 2.3; Loveland et al., 1991). Comments on land cover within the geomorphic regions focuses on woodlands.

Coastal Plain

The Coastal Plain is the largest geomorphic region in the southeastern United States and is composed of broad, lowland plains from Texas to Virginia (Fig. 2.6). The region is generally characterized by broad flats and nearly level uplands of low elevations with low relief. Woodlands cover 53.7% of this geomorphic region.

Several subdivisions of the Coastal Plain are important. The Lower Coastal Plain consists of virtually level lowlands, or "flatwoods," interspersed with swamps, estuaries, and lagoons. These lowlands typically have a poorly integrated drainage network with incipient dissection, minimal local relief, and elevations of less than 30 m above mean sea level. Slopes are long and very low angle across broad interfluves. Soil drainage is poor to very poor at the center of interfluves, with progressively better drainage toward the drainageways (the "red-edge" effect). Soil materials are composed largely of sand, silt, and clay derived from river and marine sediments (Murray, 1961). The dominant soils are Ultisols: Hapludults, Ochraqults, Paleudults (Buol, 1983; Buol et al., 1989; Soil Conservation Service, 1981; Soil Survey Staff, 1975, 1994b, 1994d, 1994f, 1994g, 1994h, 1994j, 1994k, 1994m; USDA–Forest Service, 1969). Beach and dune sediments are locally extensive in shoreline and near-shore areas along the Atlantic and Gulf coasts. The dominant soils are Entisols: Quartzipsamments, Psamments, and Sulfaquents (Daniels et al., 1984). The Lower Coastal Plain is about 90% woodland (USDA–Forest Service, 1969).

Table 2.2. Geomorphic Regions—Southeastern United States

#) Geomorphic Region #: Component Land Resource Unit Name[a]	Hectares	% Cover
1) Coastal Plain	74,557,100	53.24
133A: Southern Coastal Plain	26,885,900	36.06
133B: Western Coastal Plain	13,354,100	17.91
135: Alabama, Mississippi, and Arkansas Blackland Prairies	1,529,000	2.05
137: Carolina and Georgia Sand Hills	2,806,600	3.76
138: North-Central Florida Ridge	583,800	0.78
149A: Northern Coastal Plain	26,100	0.04
150A: Gulf Coast Prairies	2,609,600	3.50
151: Gulf Coast Marsh	1,675,700	2.25
152A: Eastern Gulf Coast Flatwoods	2,436,600	3.27
152B: Western Gulf Coast Flatwoods	1,140,800	1.53
153A: Atantic Coast Flatwoods	8,891,200	11.93
153B: Tidewater Area	2,256,200	3.03
154: South-Central Florida Ridge	1,714,200	2.30
155: Southern Florida Flatwoods	5,333,000	7.15
156A: Florida Everglades & Assoc. Areas	1,818,000	2.44
156B: Southern Florida Lowlands	421,100	0.56
165B: Subhumid Intermediate Slopes	47,500	0.06
133: Not designated in database	4,800	0.01
153: Not designated in database	251,500	0.34
87A: Texas Claypan Area, Southern Part	662,000	0.89
87B: Texas Claypan Area, Northen Part	109,400	0.15
2) Mississippi River Valley	14,563,500	10.40
131: Southern Mississippi Valley Alluvium	8,614,700	59.15
134: Southern Mississippi Valley Silty Uplands	5,948,800	40.85
3) Piedmont	16,580,200	11.84
136: Southern Piedmont	15,997,400	96.48
148: Northern Piedmont	582,800	3.52
4) Blue Ridge	4,513,200	3.22
130: Blue Ridge	4,513,200	100.00
5) Ridge and Valley	7,260,700	5.19
128: Southern Apalachian Ridges and Valleys	7,256,700	99.94
147: Northern Appalachian Rigdes and Valleys	4,000	0.06
6) Appalachian Plateau	3,775,600	2.70
125: Cumberland Plateau and Mountains	1,545,700	40.94
129: Sand Mountain	2,229,900	59.06

Table 2.2. *Continued*

#) Geomorphic Region #: Component Land Resource Unit Name[a]	Hectares	% Cover
7) Interior Low Plateau	4,293,200	3.07
122: Highland Rim and Pennyroyal	3,013,500	70.19
123: Nashville Basin	1,279,700	29.81
8) Ouachita Plateau	4,972,100	3.55
118: Arkansas Valley and Ridges	2,219,200	44.63
119: Ouachita Mountains	2,752,900	55.37
9) Ozark Plateau	4,127,800	2.95
116A: Ozark Highland	2,443,800	59.20
116B: Ozark Border	500	0.01
117: Boston Mountains	1,683,500	40.78
10) Central Lowlands	1,675,400	1.20
112: Cherokee Prairies	1,674,300	99.93
78C: Central Rolling Red Plains, Eastern Part (proposed)	1,100	0.07
11) Water	1,978,600	1.41
206: Water: Atlantic, Gulf of Mexico, Lakes, etc., within the map area.	1,978,600	100.00
12) Miscellaneous	1,715,200	1.23
76: Bluestem Hills	48,000	2.80
80A: Central Rolling Red Prairies	8,500	0.50
84A: Cross Timbers	231,600	13.50
84C: East Cross Timbers	1,100	0.06
85: Grand Prairie	132,800	7.74
150B: Gulf Coast Saline Prairies	106,400	6.20
153C: Mid-Atlantic Coastal Plain	100	0.01
86A: Texas Blackland Prairie, Northern Part (proposed)	604,300	35.23
86B: Texas Blackland Prairie, Southern Part (proposed)	539,500	31.45
245: Not designated in database	17,000	0.99
246: Not designated in database	25,900	1.51
Total	140,012,600	100.00

[a]The number preceding the Component Land Resource Unit Name (e.g., 133A:) identifies the associated Land Resource Unit code (Soil Survey Staff, 1994 a–d, f–m), thereby identifying the constituent geographic areas contained within each Geomorphic Region map unit.

Table 2.3. Distribution of General Landuse Groups—Southeastern United States[a]

#) Geomorphic Region # Landuse	Hectares	% Cover
1) Coastal Plain	74,557,100	53.24
1 Cropland	29,958,500	40.18
2 Woodland	40,034,100	53.70
3 Grassland	1,720,400	2.31
4 Wetlands	1,437,900	1.93
5 Water	1,380,400	1.85
6 Barren	12,500	0.02
7 Irrigated Agriculture	13,300	0.02
2) Mississippi River Valley	14,563,500	10.40
1 Cropland	8,769,600	60.22
2 Woodland	5,285,400	36.29
3 Grassland	310,400	2.13
4 Wetlands	39,900	0.27
5 Water	139,200	0.96
6 Barren	500	0.00
7 Irrigated Agriculture	18,500	0.13
3) Piedmont	16,580,200	11.84
1 Cropland	8,750,400	52.77
2 Woodland	7,597,700	45.82
3 Grassland	125,300	0.76
5 Water	105,700	0.64
7 Irrigated Agriculture	1,100	0.01
4) Blue Ridge	4,513,200	3.22
1 Cropland	267,700	5.93
2 Woodland	4,233,700	93.81
3 Grassland	4,500	0.10
5 Water	7,300	0.16
5) Ridge and Valley	7,260,700	5.19
1 Cropland	2,061,300	28.39
2 Woodland	5,081,500	69.99
3 Grassland	53,200	0.73
5 Water	63,600	0.88
7 Irrigated Agriculture	1,100	0.02
6) Appalachian Plateau	3,775,600	2.70
1 Cropland	1,052,500	27.88
2 Woodland	2,699,100	71.49
3 Grassland	4,600	0.12
5 Water	19,400	0.51
7) Interior Low Plateau	4,293,200	3.07
1 Cropland	1,374,900	32.03
2 Woodland	2,840,200	66.16
3 Grassland	25,400	0.59
5 Water	52,600	1.23
7 Irrigated Agriculture	100	0.00

Table 2.3. *Continued*

#) Geomorphic Region # Landuse	Hectares	% Cover
8) Ouachita Uplands	4,972,100	3.55
1 Cropland	2,246,900	45.19
2 Woodland	2,597,200	52.24
3 Grassland	36,000	0.72
5 Water	90,500	1.82
7 Irrigated Agriculture	1,500	0.03
9) Ozark Plateau	4,127,800	2.95
1 Cropland	751,300	18.20
2 Woodland	3,287,400	79.64
3 Grassland	7,400	0.18
5 Water	81,200	1.97
7 Irrigated Agriculture	500	0.01
10) Central Lowlands	1,675,400	1.20
1 Cropland	1,313,900	78.42
2 Woodland	138,700	8.28
3 Grassland	199,000	11.88
5 Water	22,400	1.34
7 Irrigated Agriculture	1,400	0.08
11) Water	1,978,600	1.41
1 Cropland	742,400	37.52
2 Woodland	594,000	30.02
3 Grassland	106,800	5.40
4 Wetlands	23,100	1.17
5 Water	504,100	25.47
6 Barren	1,300	0.07
7 Irrigated Agriculture	6,900	0.35
12) Miscellaneous Areas[b]	1,715,200	1.23
1 Cropland	1,021,800	59.58
2 Woodland	514,600	30.00
3 Grassland	145,300	8.47
5 Water	33,200	1.94
7 Irrigated Agriculture	300	0.02
Total	140,012,600	100.00

[a] Loveland et al., 1991.

[b] In the context of this table, "Water" is a map unit composed of various land covers. These land covers include open, standing water (water), other wet areas that also contain some kind of vegetative cover or undelineated inclusions of "non-water" cover (e.g., woodland).

The Upper (or Interior) Coastal Plain is predominantly broad uplands and low-lying plateaus or cuestas. The Upper Coastal Plain is geologically older than the Lower Coastal Plain and exhibits greater dissection, relief, elevations (30 to 200 m), and soil development. The dominant soils are Ultisols: Paleudults and Paleaquults.

The Coastal Plain region is divided by the Mississippi River Valley. Both pieces can be further subdivided. The Eastern Coastal Plain extends from Mississippi to Virginia. It is bounded by the Atlantic Ocean, the Gulf of Mexico, and the Mississippi River Valley. The Tertiary and Cretaceous strata that form the narrow, definitive escarpments of the Upper Coastal Plain in the east broaden westward as the Tertiary sediments narrow (Murray, 1961). The interior margin is clearly defined topographically from Virginia to Georgia by a sequence of escarpments or sandy hills of higher elevation that gradually diminish to the west. The sand hills are locally extensive and dominated by Entisols (Quartzipsamments) interspersed with Ultisols (Paleudults and Fragiudults). Woodlands cover 21.2% of the sand hills.

The Coastal Plain of west Florida, Alabama, and Mississippi is sometimes referred to as the Eastern Gulf Plain (Buol, 1983) and also exhibits a westward broadening in a belted complex of low-lying cuestas with gentle slopes facing seaward and short, steep slopes facing the interior. This complex is dominated by finer textured Ultisols: Paleudults, Hapludults, Fragiudults, and Paleaquults. The geologic diversity within the belted complex is evident in Alfisols (Paleudalfs) derived from limestone. In northeastern Mississippi, Vertisols (Chromuderts) derived from shales are locally extensive. This portion of the Upper Coastal Plain is about 50 to 75% woodland.

The Florida peninsula is a group of distinct subregions. The center of the peninsula is dominated by sandy sediments overlying a broad, gently rolling limestone upland, the South-Central Florida Ridge. Elevations generally range from 20 to 80 m with minimal relief. Dominant soils are Entisols (Quartzipsamments, Psamments, and Haplaquents), Spodosols (Haplaquods), and Alfisols (Hapludalfs) where limestone occurs near the surface. Woodlands cover 26.1% of this subregion.

The eastern one third of the peninsula is a continuation of the low-elevation, low-relief flatwoods of the Atlantic Coastal Plain, the Southern Florida Flatwoods. The deep, sandy marine sediments are dominated by Spodosols (Haplaquods) interspersed with Entisols (Psamments). Organic soils (Medisaprists and Medihemists) are locally extensive, especially at very low elevations south of Lake Okeechobee. Woodlands cover 29.4% of these composite lowlands.

The other portion of the Gulf Coastal Plain lies west of the Mississippi River Valley and is similar in geomorphology to the eastern Gulf Coastal Plain. The Lower Coastal Plain is a low-elevation, low-relief plain. Near the Mississippi River Valley, it is covered by fine-textured alluvium and

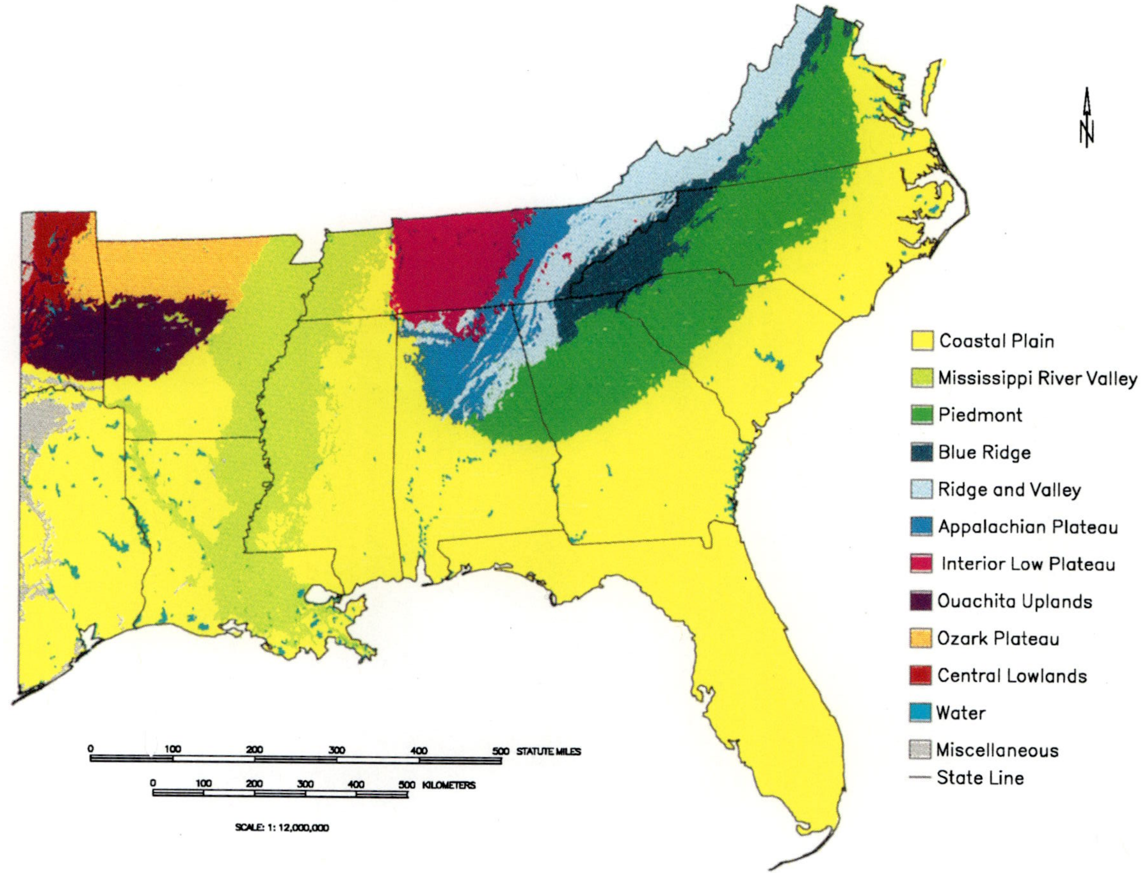

Figure 2.6. Geomorphic groupings for the southeastern United States. (*Source*: State Soil Geographic Database, 1993. Albers Equal Area Projection. Map prepared using MAPGEN 4.1. Lincoln, NE, October 1994.)

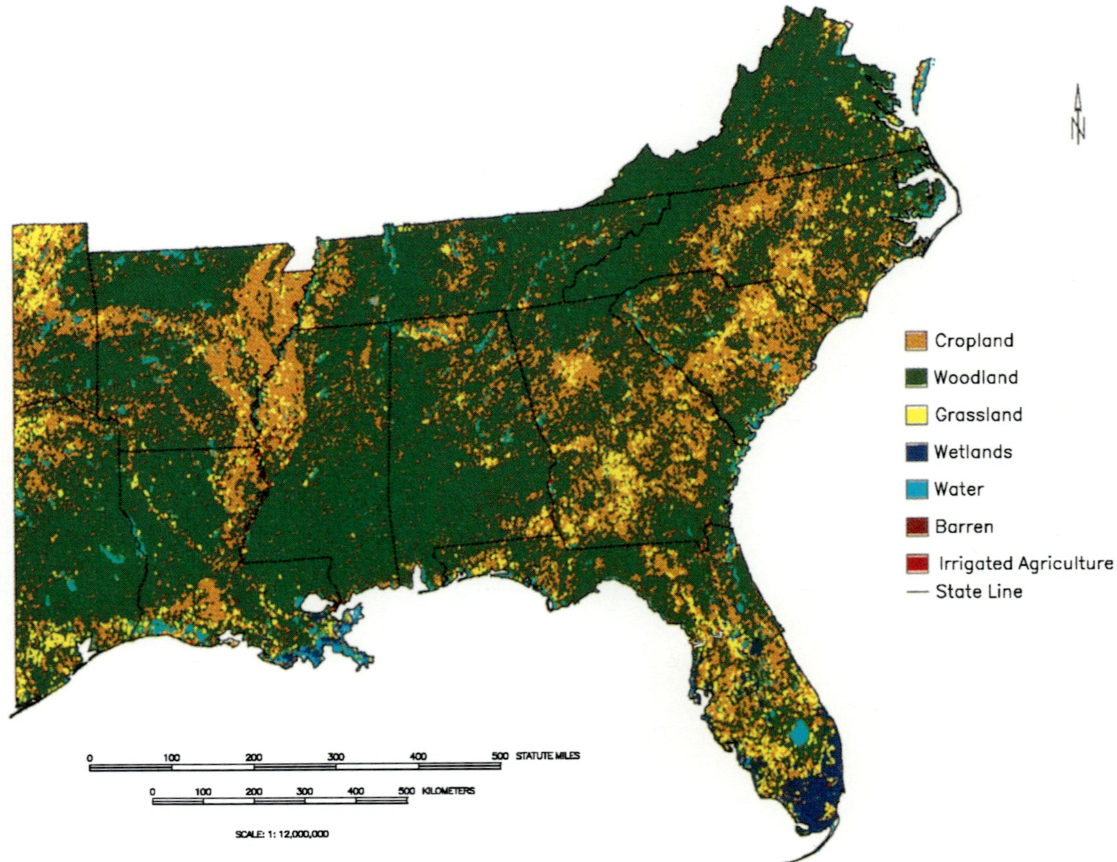

Figure 2.7. General land-use groups for the southeastern United States. (*Source*: Photo grammatic Engineering and Remote Sensing, Vol. 55, 11:1453–1463. Loveland TR, Merchant JW, Ohlen DO, Brown JF 1991. Albert Equal Area Projection. Map prepared using MAPGEN 4.1. Lincoln, NE, October 1994.)

marine sediments, which coarsen to sandier Pleistocene marine sediments to the west. The Upper Coastal Plain is a belted complex of low-lying cuestas of Tertiary and Cretaceous sedimentary strata. Permian and Pennsylvanian aged sedimentary rocks dominate its western edge.

Despite similarities to the eastern Gulf Plain in geomorphology and elevation, soils of the western Gulf Plain occur in two distinct zones. Areas of Louisiana and eastern Texas are most similar to the eastern Gulf Coastal Plain. The dominant soils are Ultisols (Paleudults and Hapludults) and Alfisols (Paleudalfs). Woodlands cover 69.8% of this subregion. Further west, soils deviate from the regional Coastal Plain pattern due to a drier climate and changes in geology. The more rolling Upper Coastal Plain of the Texas claypan subregions is dominated by Alfisols (Ochraqualfs, Paleudalfs, Hapludalfs, and Albaqualfs). The Lower Coastal Plain is dominated by Mollisols (Haplaquolls) and Vertisols (Pelluderts). Woodlands cover 70.0% of this dominant southern subregion.

Mississippi River Valley

The Mississippi River Valley region splits the southeastern United States. It is a low-elevation (0 to 170 m), low-relief (nearly level) river valley composed of a complex of river terraces, meander belts, levees, and backwater swamps. The soil materials are a variable mixture of alluvium from the Mississippi River and its tributaries and underlying marine sediments. A mantle of loess covers much of the valley; the loess thickens with increased elevation associated with older, more stable landforms such as high terraces and hills.

The active floodplain is dominated by Entisols (Fluvaquents in the active floodplains; Hydraquents in backwater, low-flow areas). Low alluvial terraces are dominated by Inceptisols (Haplaquepts). Mid-elevation terraces are dominantly Alfisols (Ochraqualfs) with intermingled Mollisols (Haplaquolls). The older, highest elevation terraces in northeastern Louisiana and eastern Arkansas are thickly covered by loess and are dominantly Alfisols (Fragiudalfs, Glossaqualfs, and Albaqualfs). Woodlands cover 27.1% of this subregion.

Included within this geomorphic region are the silt-covered uplands located to the east of the Mississippi River Valley. The physiographic western edge of the Coastal Plain adjacent to the Mississippi River system is mantled on the surface by up to 20 m of windblown silt that gradually thins to the east. The dominant soils are Alfisols (Hapludalfs and Fragiudalfs) interspersed with Ultisols (Paleudults). These thick loess soils have far greater similarities to other thick loess soils within the Mississippi River valley (e.g., Crowley's Ridge) than to the Coastal Plain proper. Consequently, for the purposes of this chapter, this physiographic subset is included within the Mississippi River Valley region. Woodlands cover 49.7% of this loess-covered subregion.

Due to their high-base status, the Alfisols that dominate this region should be resistant to acidification. Woodlands cover 36.3% of this geomorphic region.

Piedmont

The Piedmont region is the second-largest physiographic unit in the southeastern United States. This rolling, erosional plain is composed of uplands derived from diverse metamorphic and igneous rocks, with minor relief (10 to 30 m) and moderate elevations (100 to 350 m). The Piedmont flanks the eastern side of the Appalachian Highlands from Georgia through Virginia, sloping gently down to the Coastal Plain on the east and south. The stable, erosional nature of this region has produced soils dominated by Ultisols (Hapludults, Paleudults, and Rhodudults) on acidic rocks, with Alfisols (Hapludalfs and Paleudalfs) locally dominant on basic igneous rocks and Inceptisols (Dystrochrepts) dominant on steep slopes overlying mixed lithology. There are several locally extensive areas derived from metavolcanic and marine sediments of Triassic age. These "Triassic basins," referring to historical geologic setting rather than to present-day topography, are dominated by Alfisols (Hapludalfs). Woodlands cover 45.8% of the Piedmont geomorphic region.

Blue Ridge

The Blue Ridge mountain range is composed of predominantly igneous and metamorphic rocks with considerable local relief (50 to 800 m). Elevations generally range from 300 to 1200 m, with some peaks exceeding 2000 m, Moderately steep to steep slopes are extensive and dominated by Inceptisols (Dystrochrepts). Less steep areas with more stable slopes commonly found in valleys or on broad ridgetops are dominated by Inceptisols (Haplumbrepts) and Ultisols (Hapludults). Woodlands cover 93.8% of the Blue Ridge geomorphic region.

Ridge and Valley

This region is composed of a series of narrow geologic belts of moderate elevation (200 to 1000 m) and significant local relief that parallel the northeast–southwest regional trend of the Appalachian Mountains. The region slopes gently to the south and west. The alternating anticlines and synclines form valleys bounded by steep-sided ridges of varying widths. The rolling-to-hilly limestone-and-shale valleys are bounded by steep mountain ridges of more resistant sedimentary rocks. Soils are dominantly Ultisols (Paleudults and Hapludults) with some Alfisols (Paleudalfs) in areas underlain by limestone. The less stable sideslopes and steep ridges are dominated by Inceptisols (Dystrochrepts). Woodlands cover 70.0% of the Ridge and Valley geomorphic region.

Appalachian Plateaus

The Appalachian Uplands are composed of deeply dissected plateaus with gently rolling to rolling topography at moderate elevations. The region is bounded by steep escarpments (300 to 670 m) that drop away to the west and south, except at the mountainous eastern edge, where relief exceeds 1200 m. The plateaus are composed of nearly level sedimentary rocks that are typically capped by resistant sandstones and flanked by predominantly eroded shale strata on the sideslopes. The relatively stable, nearly level-to-rolling uplands and valley floors are dominated by Ultisols (Hapludults and Paleudults). The steep escarpments and valley walls are predominantly Inceptisols (Dystrochrepts). Woodlands cover 71.5% of the region.

Interior Low Plateau

This region is a rolling, low-lying interior plateau composed of a broad, extensively dissected rim surrounding a less dissected central core. It lies to the west and is elevationally lower than the Appalachian plateaus. Elevations are moderate (100 to 300 m), with the greatest local relief associated with steep slopes of the dissected rim and the more subdued relief in the rolling central plateau. The geology is predominantly limestone interbedded with diverse sedimentary rocks. The dominant soils are Alfisols (Hapludalfs and Paleudalfs) in the plateau interior and Ultisols (Paleudults, Fraqiudults, and Hapludults) interspersed with Alfisols (Hapludalfs) associated with the dissected rim. Woodlands cover 66.2% of this geomorphic region.

Ouachita Uplands

The Ouachita Uplands is a compact physiographic region that lies west of the Mississippi River and entails two subregions: the Ouachita Mountains and the Arkansas Valley and Ridges. The Ouachita Mountains subregion dominates this geomorphic region and is similar in form to the western Appalachian Uplands (Fenneman, 1938). The alternating ridge and valley structures are structurally related to and parallel the Ozark Mountains to the north. Elevations are moderate (100 to 600 m), with some peaks exceeding 700 m. There is considerable local relief (100 to 300 m) on the steep-sloped ridges and valley walls. This subregion gradually falls in elevation to the south, where it merges with the Coastal Plain. The soils are dominated by Ultisols (Hapludults) on steep sandstone or cherty-limestone ridges and rolling hills that dip southward. Soils are dominantly Inceptisols (Dystrochrepts) on the steeper, north-facing valley slopes and escarpments. The rolling valleys are composed of eroded and weathered slate, shales, and sandstones, in which the dominant soils are Ultisols (Hapludults).

The Arkansas Valley and Ridges subregion is composed of the main river valley and associated ridges, which are elevationally similar to the Ouachita Mountains to the south. Its broad valleys and ridges are dominated by soils similar to the Ouachita Mountains, except for the presence of Alfisols (Paleudalfs) on the more stable positions underlain by shallow limestone. The Arkansas River Valley itself is a gently sloping-to-rolling dissected plain dominated by Alfisols (Hapludalfs) intermingled with Ultisols (Fraqiudults) and Inceptisols (Haplaquepts in low-lying areas, Dystrochrepts on steep slopes). Overall, woodlands cover 52.2% of this geomorphic region.

Ozark Plateau

The high, deeply dissected Ozark plateau is composed of narrow, rolling ridgetops falling away to steep sideslopes of considerable relief. The northern area has moderate elevations (170 to 500 m) with significant relief (100 to 200 m). The region is composed of dissected sedimentary rocks wherein sandstone ridgetops and rolling uplands are dominanted by Ultisols (Paleudults and Fragiudults). Alfisols (Paleudalfs) dominate the valley floors overlying limestone, dolomite, and shales. Inceptisols (Dystrochrepts) dominate steep sideslopes and valley walls. The southern portion is more deeply dissected, with greater elevations (170 to 800 m) and relief (up to 300 m). The sandstone ridges and valley bottoms are primarily Ultisols (Hapludults and Paleudults), with Inceptisols (Dystrochrepts) on steep slopes. Woodlands cover 79.6% of this region.

Central Lowlands

The Central Lowlands geomorphic region consists of broad flats, broad low cuestas, and rolling hills. It has generally low elevations that have nominal local relief. It lies between the treeless plains to west, the Ozark/Ouachita Uplands to the east, and the Coastal Plain to the south.

This geomorphic region is primarily composed of the Cherokee Prairies. It has low elevations (130 to 400 m) and low relief (less than 50 m) of dissected sedimentary rocks. The Pennsylvanian-aged sandstones, shales, and clays are dominated by Mollisols (Paleudolls and Argiudolls) and interspersed with Alfisols (Albaqualfs and Hapludalfs) in alluvial areas. Woodlands cover only 8.3% of this geomorphic region.

Water

This is a composite map unit, not a separate geomorphic region, that includes all detected bodies of water within each of the geomorphic regions. These water bodies include large lakes, reservoirs, estuaries, lagoons, and coastal water-body margins. These areas contain nominal vegetative cover (e.g., $<15\%$ woodland cover).

Miscellaneous

This is a composite map unit, not a discrete geomorphic region, that occurs within most of the geomorphic regions. It is a diverse group of areas (Land Resource Units) whose commonality and principal feature is minimal woodland cover (<301%; e.g., marsh, prairie areas). These areas are of minor extent and are not relevant to the main focus of this chapter and are not discussed further.

Conclusions

The southern pine forests potentially span the entire southeastern United States, except for some extremely wet coastal margins and cool, high elevation areas in the Appalachian Mountains. Throughout this moist, temperate climate, pine forests occupy a wide range in soils. Similarly, the potential response of these soils and their resident flora to atmospheric deposition spans a wide range. Clearly, due to inherent composition and pedogenic development, some soils are likely to exhibit fairly rapid changes related to atmospheric deposition, whereas other soils will reflect virtually no impact. Site-specific conditions commonly supersede regional trends. Site-specific data that provide the basis for quantifying the impact of atmospheric deposition must consider this reality to arrive at accurate regional assessments.

References

Binkley D, Driscoll CT, Allen HL, Schoeneberger P, McAvoy D (1989) *Acidic deposition and Forest Soils*. Springer-Verlag, New York, p 149.

Buol SW (ed) (1983) *Soils of the Southern States and Puerto Rico*. Southern Cooperative Series Bulletin No 174, North Carolina State University, Raleigh, p 105.

Buol SW, Hole FD, McCracken RJ (1989) *Soil Genesis and Classification*, 3rd ed. Iowa State University Press, Ames, p 466.

Climatic Data Access Facility (1993) *Climatic Data Access Network* (*CDAN*). USDA—Soil Conservation Service, West Technical Center, Portland, OR.

Daniels RB, Kleiss HJ, Buol SW, Byrd HJ, Phillips JA (1984) *Soil Systems of North Carolina*. North Carolina Agricultural Research Service Bulletin No 467, p 77.

Fenneman NM (1938) *Physiography of the Eastern U.S.* McGraw-Hill, New York, p 714.

Loveland TR, Merchant JW, Ohlen DO, Brown JF (1991) Development of a land-cover characteristics database for the coterminous US. Photogramm Eng & Remote Sensing 55:1453–1463.

Murray GE (1961) *Geology of the Atlantic and Gulf Coast Province of North America*. Harper's Geoscience Series, New York, p 692.

National Climate Data Center (1992) *Monthly Station Normals of Temperature, Precipitation, and Heating and Cooling Days, 1961–1990*. Climatology of the United States, No 81, US Departmert of Commerce, NOAA, Asheville, NC.

Soil Conservation Service (1981) *Land Resource Regions and Major Land Resource Areas of the United States*. USDA Agricultural Handbook No 296, Washington, DC, p 56.

Soil Survey Staff (1975) *Soil Taxonomy*. USDA Agricultural Handbook No 436, Washingto, DC. p 745.

Soil Survey Staff (1994a) *Alabama State Soil Geographic Database (STATSGO) Draft*. USDA, NRCS, Auburn, AL.

Soil Survey Staff (1994b) *Arkansas State Soil Geographic Database (STATSGO) Draft*. USDA, NRCS, Little Rock, AR.

Soil Survey Staff (1994c) *Florida State Soil Geographic Database (STATSGO) Draft*. USDA, NRCS, Gainesville, FL.

Soil Survey Staff (1994d) *Georgia State Soil Geographic Database (STATSGO) Draft*. USDA, NRCS, Athens, GA.

Soil Survey Staff (1994e) *Keys to Soil Taxonomy*, 6th ed. USDA Soil Conservation Service, Washington, DC, p 306.

Soil Survey Staff (1994f) *Louisiana State Soil Geographic Database (STATSGO) Draft*. USDA, NRCS, Alexandria, LA.

Soil Survey Staff (1994g) *Mississippi State Soil Geographic Database (STATSGO) Draft*. USDA, NRCS, Jackson, MS.

Soil Survey Staff (1994h) *North Carolina State Soil Geographic Database (STATSGO) Draft*. USDA, NRCS, Raleigh, NC.

Soil Survey Staff (1994i) *Oklahoma State Soil Geographic Database (STATSGO) Draft*. USDA, NRCS, Stillwater, OK.

Soil Survey Staff (1994j) *South Carolina State Soil Geographic Database (STATSGO) Draft*. USDA, NRCS, Columbia, SC.

Soil Survey Staff (1994k) *State Soil Geographic Database (STATSGO) Data Use Guide*. Miscellaneous publication 1492, USDA Soil Conservation Service, Washington, DC.

Soil Survey Staff (1994l) *Texas State Soil Geographic Database (STATSGO) Draft*. USDA, NRCS, Temple, TX.

Soil Survey Staff (1994m) *Virginia State Soil Geographic Database (STATSGO) Draft*. USDA, NRCS, Richmond, VA.

Thornbury WB (1965) *Regional Geomorphology of the United States*. Wiley, New York, p 609.

USDA–Forest Service (1969) *A Forest Atlas of the South*. Southeastern Forest Experiment Station, Asheville, NC, p 27.

Van Wambeke A, Hastings P, Tolomeor M (1991) *Newhall Simulation Model—A Basic Program for the IBM PC (DOS 2.0 or Greater)*. Department of Agronomy, Cornell University, Ithaca, NY, p 37.

3. Air Quality and Atmospheric Deposition in Southern U.S. Forests

Eric R. Allen and Henry L. Gholz

Characterizations of chemical exposures (air quality and atmospheric deposition) as well as physical conditions (meteorology and climatology) in rural environments are essential to evaluate the hypothesis that exposure of forests to acidic deposition and airborne pollutants may influence forest health in the southern United States. Air quality and atmospheric deposition in rural areas are a result of local and regional emissions of pollutants originating from anthropogenic and biogenic sources and their dispersion, transport, and transformation. Elements that require quantification are the atmospheric concentrations of gaseous chemical species in the vicinity of the forested landscapes (air quality) to evaluate exposure indexes, and the fluxes of chemicals to the forest canopy and soil are necessary to estimate wet and dry acid deposition inputs. Long-term, widespread measurements of these variables are generally not available to describe adequately a region as large as the southeastern United States. Mesoscale meteorological–chemical models are used to predict the distributions of atmospheric chemical variables over large, nonmonitored subregions. Regional models, such as the Regional Acid Deposition Model (RADM) and Sulfate Transport Eulerian Model-II (STEM-II), are useful in predicting air quality and acid deposition patterns because of the sparse temporal and spatial data available from field observations. Monitoring data, however, are valuable in testing and calibrating these regional models.

In this chapter, regional emissions data are presented along with historic measurements of air quality and atmospheric deposition to evaluate distributions and possible trends in these exposure variables across the Southeastern United States. Additional recent monitoring data are presented in support of the synthesis of historic measurements and to provide further insight into the nature of chemical exposure of forests in the region.

Pollutant Emissions in the Southeast

Local and regional levels of pollutants in the southeastern United States mainly depend on energy demands resulting from weather and climate patterns and transportation activity. Unlike the northeastern United States, however, these demands are frequently greater in the summer than in the winter months. Electric utilities can account for as much as two thirds of sulfur dioxide emissions and one third of nitrogen oxide emissions annually. Most of the emissions data, however, are not obtained from measurements but are derived from the use of emission factors, and with source activity levels for a given time frame to provide an estimate of emissions from specific sources. This procedure can lead to uncertainties in emissions inventories when these emissions estimates are aggregated over a region; also it is assumed that all major sources have been taken into account.

The following discussion is based on the National Acid Precipitation Assessment Program (NAPAP) 1985 Emissions Inventory, which is considered by many to be the most complete and accurate large-scale inventory currently available. Annual emissions estimates for large point sources are reliable, but annual estimates for area sources are less certain.

Relationships between emissions and regional air quality or precipitation chemistry are not direct or simple. Area-specific emissions can influence the levels of transported pollutants by overwhelming concentrations or changing atmospheric chemistry (Goklany and Hoffnagle, 1984).

Sulfur Oxides

Sulfur dioxide (SO_2) is a primary pollutant emitted mainly (~95% of total) by anthropogenic sources (Placet et al., 1990). Stationary sources account for approximately 91% of the total emissions, with 71% of the total emitted from tall stacks (>120 m). The average residence time of sulfur dioxide in a midlatitude atmosphere is 25 hours (Rodhe, 1978). The most significant chemical reaction of sulfur dioxide in the atmosphere involves oxidation to sulfate (SO_4^{2-}) by gaseous phase oxidation with hydroxyl (OH) radicals (Calvert and Stockwell, 1983) and aqueous phase oxidation by hydrogen peroxide (H_2O_2) and ozone (National Academy of Sciences [NAS], 1983). The main removal paths for SO_2 are dry deposition to surfaces (vegetation and human-made) and wet and dry deposition as SO_4^{2-}.

Ambient sulfur dioxide concentrations result mainly from stationary source coal and oil combustion and from nonferrous smelters (U.S. EPA, 1989). Electric utilities produce two thirds of national SO_2 emissions, of which 96% originates from coal-burning plants (U.S. EPA, 1989). Many electric power plants are located in rural areas and have substantial impact on rural airsheds. Natural sources of sulfur compounds include biological activity in the oceans, soils, vegetation, and geothermal activity. Natural sulfur emissions include the reduced forms: hydrogen sulfide, dimethyl sulfide, and dimethyl disulfide (Aneja, 1990). Nationally, anthropogenic emissions are estimated at 21 $TgSO_2$/year (1 Tg = 10^{12} g = 1 million metric tons), whereas natural sources of sulfur are estimated at 4% of the anthropogenic total (Westberg and Lamb, 1984).

In the southeastern states, anthropogenic sources of sulfur dioxide are scattered throughout the region. Total sulfur dioxide emissions in the Southeast in 1985 were estimated at 4.8 $TgSO_2$/year, almost a quarter of the national estimate and about one third of the total for the eastern United States (Placet and Streets, 1987). Primary statewide contributions are by Texas (1.1 $TgSO_2$/year), Florida (1.0 $TgSO_2$/year), Tennessee (0.98 $TgSO_2$/year), and Louisiana (0.84 $TgSO_2$/year). The state of Arkansas produced the lowest emission (0.09 $TgSO_2$/year). Most of the midwestern states produce more than 1 $TgSO_2$/year.

From 1960 to 1980, southeastern regional sulfur dioxide emissions more than doubled from 2.0 $TgSO_2$/year to 4.4 $TgSO_2$/year, a substantial increase when compared to that for the entire eastern states from (10 to 13 $TgSO_2$/year) during the same period (Husar, 1986). During the five-year period from 1980 to 1985, SO_2 emissions in the Southeast increased by only 10% to 4.8 $TgSO_2$/year (Placet and Streets, 1987). From 1985 to 1992, SO_2 emissions in the Southeast had not changed substantially (U.S. EPA, 1993a, 1993b, 1994) (Fig. 3.1).

Nitrogen Oxides

Nitrogen oxides (NO_x), which consist of nitric oxide (NO) and nitrogen dioxide (NO_2), are mainly (86% of total) emitted from anthropogenic combustion processes (Placet et al., 1990). Nitric oxide is the principal (>90% of total) nitrogen oxide emitted as a primary pollutant. Nitrogen dioxide is a secondary pollutant resulting mainly from atmospheric oxidation of NO (Sellars et al., 1985). Area sources account for approximately 54% and utility sources for about 32% of the total anthropogenic NO_x emissions. Transportation is identified as the principal (~43%) anthropogenic source of NO_x (U.S. EPA, 1992). NO*x* emissions are great in and around metroplitan areas and much less in rural and remote areas, except near large point sources.

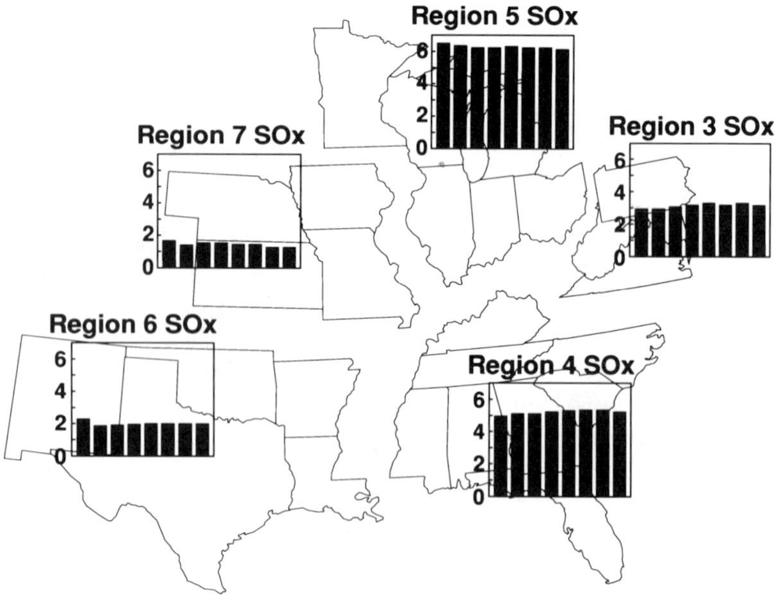

Figure 3.1. Emissions of sulfur oxides (SO_x) by EPA region, 1985 to 1992, in millions of metric tons per year (from U.S. EPA, 1994).

Chemical reactions involving NO_x in the atmosphere are numerous and complex (Finlayson–Pitts and Pitts, 1986). Reactions can involve other primary pollutants, such as hydrocarbons and carbon monoxide (CO), as well as reactive secondary pollutants, such as the hydroperoxy (HO_2) and hydroxyl (OH) free radicals and ozone (O_3). Ultimately, NO_x can be converted to nitric acid (HNO_3), peroxyacyl nitrates (PAN) and particulate nitrates (p-NO_3) (Logan, 1983). The average residence time for NO_x in the atmosphere depends on the season, about 24 hours in summer (Calvert and Stockwell, 1983) and several days in winter (Logan, 1983). NO_2 is converted to HNO_3 by reaction with OH in daytime and via formation of the nitrate radical (NO_3) and/or N_2O_5 at night (Leaitch et al., 1988). The existence of a nonphotolytic mechanism for HNO_3 formation may explain why nitrate ion (NO_3^-) in precipitation does not decrease from summer to winter in the same way as sulfate ion (SO_4^{2-}). The contribution of nitrate compounds to areosol and precipitation acidity. therefore, may be more important in winter (NAPAP, 1990). In addition to oxidation to HNO_3, NO_2 is removed by dry deposition to a variety of natural surfaces. HNO_3 removal mechanisms include both wet and dry deposition (NAPAP, 1990).

Nationwide anthropogenic emissions of nitrogen oxides are estimated at 19 TgNO_x/year for 1985 (Placet and Streets, 1987). Natural sources include lightning, biogenic processes in soils, stratospheric intrusion, and photolysis in the oceans. Natural emission inventories for nitrogen oxides are incomplete but could account from 3 to 25% of total NO_x emissions in North America.

An estimated 6.5 TgNO_x/year were emitted in the southern United States in 1985 (4.2 TgNO_x/year excluding Texas), corresponding to about one third of the nationwide estimate (Placet and Streets, 1987) and about half of that estimated for the eastern United States. The highest statewide emissions (estimated at 2.3 TgNO_x/year) were for Texas, followed by Louisiana (0.84 TgNO_x/year) and Florida (0.59 TgNO_x/year). The lowest statewide emission was for Arkansas (0.20 TgNO_x/year). Midwestern states had NO_x emissions in the range from 0.70 to 1.0 TgNO_x/year.

From 1960 to 1980, southern regional NO_x emissions increased from 2.5 to 6.5 TgNO_x/year (Husar, 1986). Since 1980, the emission rates of NO_x have leveled off and may have decreased slightly from 1985 to 1992 (U.S. EPA, 1993a, 1993b, 1994) (Fig. 3.2).

Volatile Organic Compounds and Ozone

Volatile organic compounds (VOCs), carbon monoxide (CO), and oxides of nitrogen (NO_x) in the lower troposphere are important in the formation of ozone and photochemical smog. Although sulfur and nitrogen

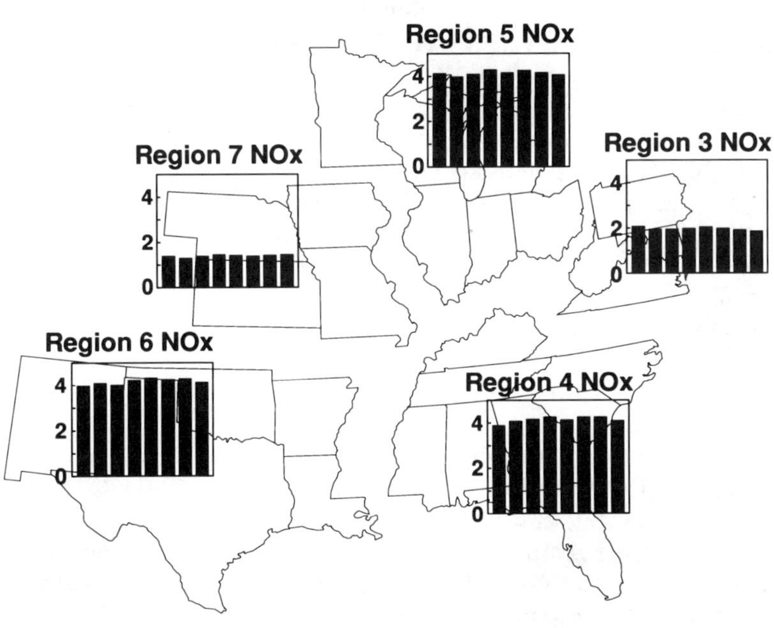

Figure 3.2. Emissions of nitrogen oxides (NO_x) by EPA region, 1985 to 1992, in millions of metric tons per year (from U.S. EPA, 1994).

oxides are primary precursors of acid precipitation, volatile organic compounds are important constituents in many atmospheric chemical reactions (Gschwandtner et al., 1986).

The largest anthropogenic source of VOCs in the United States is transportation activity, particularly that due to the combustion of gasoline in automobiles and trucks. Use of organic solvents and paints in industrial, commercial, and residential activities; storage and transfer of gasoline, crude oil, and other petroleum products; and wood burning are other major human-made sources of VOCs. Many nonmethane hydrocarbons are emitted from natural sources, principally vegetation (Altschuller, 1983; Duce et al., 1983; Tingey et al., 1991). In the United States, approximately 90% of the total estimated natural VOC emissions are attributed to trees. Deciduous species provide approximately 30%, and coniferous species contribute about 60% of the total estimated natural VOC emissions. Crops provide only 3% of the total estimated natural VOC emissions. Isoprene constitutes about 50% of the hydrocarbon emissions from deciduous forests. Conifers emit terpenoid compounds, including α-pinene (25% of terpenes). The rates, distribution, and amounts of natural VOC emissions depend on species and environmental conditions. On a nationwide basis, total natural biogenic emissions of VOCs are greatest in the Southeast due to extensive forested acreage and prevailing higher temperatures. In 1985, total natural emissions of nonmethane hydrocarbons in the United States were estimated at 30.7 Tg/year (Lamb et al., 1987). Emissions of anthropogenic volatile organic compounds were 22 Tg/year for 1985 (Placet and Streets, 1987). More recent estimates suggest that VOCs are emitted at approximately equal mass rates from natural and anthropogenic sources (Placet et al., 1990). Natural sources, however, contribute about 58% of the total VOC reactivity (O_3 production potential) budget.

Anthropogenic VOC emissions estimates for the United States have indicated a small downward trend from 21.7 Tg/year in 1985 to about 20 Tg/year in 1992 (U.S. EPA, 1994). In the South, estimates suggest that human-made VOC emissions increased to about 5.8 Tg/year by 1986 and then decreased to about 4.6 Tg/year in 1992 (Fig. 3.3). Anthropogenic VOC emissions in the southern United States for 1992, therefore, represent about 23% of the nationwide estimate (U.S. EPA, 1993a, 1993b, 1994).

Carbon monoxide emissions are mainly from transportation and industrial fossil fuel combustion sources (Wang et al., 1985). Mechanisms whereby VOCs, NO_x and CO react are complex and depend on several factors, such as the relative quantities of each pollutant present, intensity of solar radiation, and atmospheric mixing conditions. Because ozone is produced by atmospheric chemical reactions, peak O_3 concentrations can occur at considerable distances downwind from intense sources of pollutant emissions (Lefohn et al., 1987b; Reagan, 1984; Seinfeld, 1989).

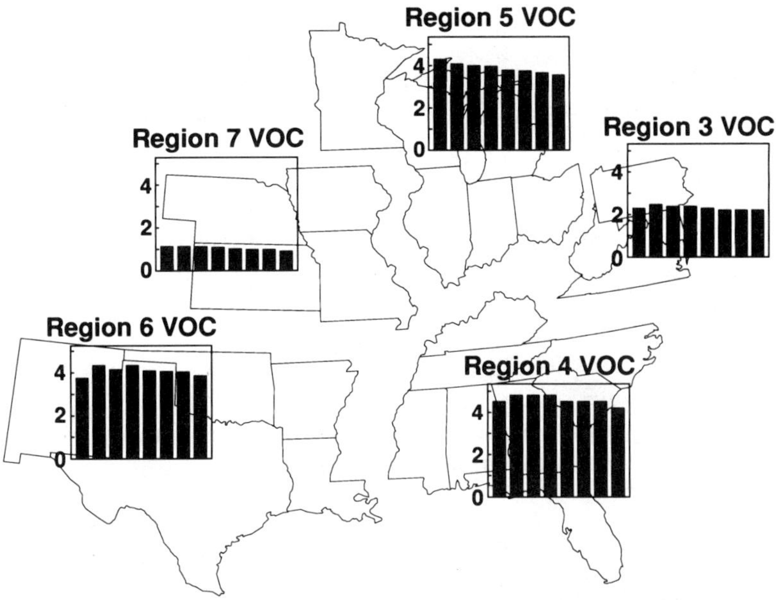

Figure 3.3. Emissions of volatile organic compounds (VOCs) by EPA region, 1985 to 1992, in millions of metric tons per year (from U.S. EPA, 1994).

Locations of high O_3 concentration areas depend on both chemical and meteorological factors. For those conditions that prevail in industrialized areas, the average residence time of ozone in the planetary boundary layer is estimated to be about 100 hours (Kleinman and Benkovitz, 1989). Removal mechanisms for O_3 include photochemical dissociation, destruction reactions with NO (Fishman et al., 1979), and dry deposition to a variety of surfaces.

Forest fires are of particular interest in assessing air quality exposure of commercial forests. Prescribed burning of logging slash and burning for vegetation control, as well as naturally occurring forest fires, produce particles, carbon monoxide, nitrogen oxides, and gaseous hydrocarbons. The ratio of hydrocarbon to nitrogen oxides is often optimal (10 to 20:1) for photochemical oxidant formation, and the contribution of forest fires to ozone production in wilderness and rural areas may be important (Bohm, 1992). Ozone has been found to accumulate near the location of a burn, and substantial increases in O_3 concentrations have been detected downwind of burn areas and near the top of burn plumes. In studies of conifer slash burns, concentrations of O_3 as high as 50% greater than ambient were found, with maximum increases observed near the top of the plumes (Bohm, 1992).

Deposition to Southeastern Forests

The nature and extent of chemical deposition to southern forests needs to be assessed to identify those pollutants that can cause stress or damage to trees. Knowledge of the direct or indirect fluxes of chemicals to the forest canopy and soils is necessary to estimate wet and dry acid deposition inputs.

Atmospheric deposition processes occur by three pathways: (1) precipitation or wet deposition, in which chemicals are dissolved in rain or snow; (2) dry deposition, involving removal of gases and particles (aerosols) at viable or nonviable terrestrial surfaces; and (3) cloud water deposition, involving chemicals dissolved in airborne water droplets. The latter process is important in montane forests, wherein cloud or fog droplets are intercepted by vegetation, including forest canopies (Mohnen, 1992). Although the latter process is an important factor in tree stress for high-elevation (>1000 m) forest ecosystems, cloud water deposition is of little importance at the lower elevation (>500 m) forest sites of the southern United States (Johnson and Lindberg, 1992), assuming a mean cloud base near 800 m (Mohnen, 1992).

Wet and dry acid deposition studies in the southern United States have been conducted at various times and locations by many groups, including (1) the U.S. Department of Agriculture and Forest Service/U.S. Environmental Protection Agency (USFS/USEPA) Atmospheric Exposure Cooperative (AEC); (2) the Florida Power Coordinating Group (FPCG); (3) the National Acid Deposition Program/National Trends Network (NADP/NTN); (4) the Electric Power Research Institute Integrated Forest Study (EPRI/IFS); (5) the Mountain Cloud Chemistry Program (MCCP); and (6) the U.S. EPA National Dry Deposition Network (NDDN) and Clean Air Status and Trends Network (CASTNET). Although these networks provide a substantial database on atmospheric deposition measurements, the spatial and temporal coverage is sparse for the Southern region as a whole. In addition, meteorological data for forest environments are usually obtained at a single location and may not represent environmental conditions for the area. Numerical modeling is used to supplement measurements and determine spatial distributions of deposition components; however, the uncertainties in these models increases when simulating deposition in complex terrain or over large, unmonitored rural areas.

The Southern Commercial Forests Research Cooperative (SCFRC) region encompasses about 5 million hectares in the southeastern United States. The size of the region coupled with the paucity of spatial and temporal measurement data makes a definitive description of the distribution and trends in deposition variables difficult. Despite these limitations, we attempt to provide estimates of wet and dry deposition inputs based on a comparison of historic and recent measurements with the realization that, in some cases, uncertainties may be large.

Wet and dry acid deposition is described in terms of the terrestrial inputs of the elemental nutrients, sulfur (sulfate ions in rain and aerosols and as gaseous SO_2), and nitrogen (nitrate and ammonium ions in rain and aerosols and as gaseous nitric acid and nitrogen oxides) in addition to hydrogen ions in precipitation.

Wet Deposition Chemistry

Several wet deposition monitoring networks were established in the United States in the early 1980s in response to the identification of acid rain as a potential nationwide environmental problem. The goal of these networks was to establish the spatial extent and identify any temporal trends for this phenomenon. The regional nature of some of the networks, however, limits their usefulness in providing data relevant to the southern United States. Only deposition data from those networks that have been operated continuously for more than five years and have documented quality assurance will be used to determine historical wet deposition distribution patterns and trends in the South. This synthesis will be supplemented with data from more recent shorter term studies that have valid quality assurance protocols for purposes of comparison.

The region of interest to SCFRC extends from North Carolina, south to Florida, and west to eastern Texas. To characterize historical wet deposition patterns in the region, data from NADP, NAPAP, and other relevant sources were used. Data from wet deposition monitoring sites, operated as part of NADP from 1979 to 1990, have been analyzed to describe spatial and temporal concentrations and deposition levels for wet deposition across the Southeast (NADP, 1991a). The locations of the sites and criteria used for these analyses have been described (Smith and Shadwick, 1992). Site estimates of annual volume-weighted precipitation concentrations (mg/l) and wet deposition fluxes (kg/ha-yr) have been subjected to simple "kriging" to display the data as geographically distributed isopleths (Barchet, 1987). Kriging of limited spatial data, however, can be a source of substantial uncertainty, particularly if urban data are kriged across large, uncharacterized rural areas.

Distribution patterns based on NADP maps of kriged annual values for 1986 to 1990 (NADP, 1987, 1988, 1989, 1990, 1991b) and interpolated values for the period 1980 to 1984 (Barchet, 1987) have been developed for pH/hydrogen ions, sulfate, nitrate, and ammonium ions. In general, precipitation-weighted mean concentrations and deposition levels of hydrogen, sulfate, and nitrate ions are higher in the North Carolina Piedmont region and steadily decline across the SCFRC region in a southwesterly direction toward the Texas Gulf Coastal Plain (Smith and

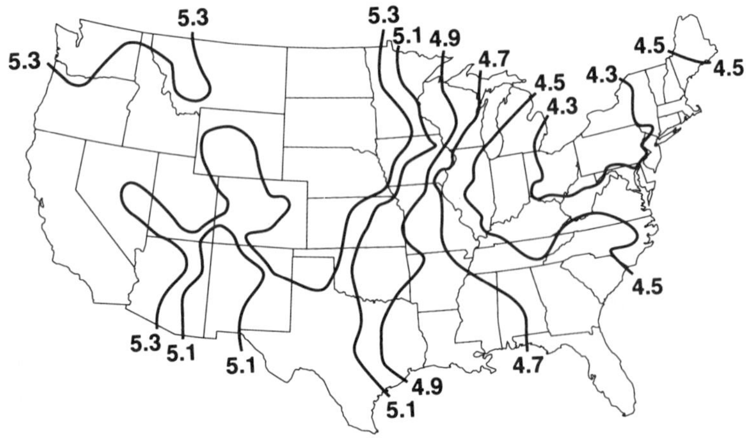

(a)

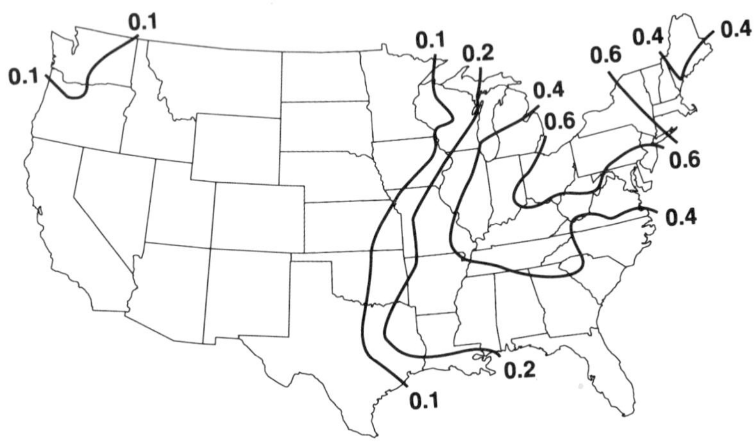

(b)

Figure 3.4. Maps of (a) precipitation-weighted mean pH and (b) H^+ deposition (kg/ha-yr) distributions for 1990 (from NADP, 1991).

Shadwick, 1992). The regional northeast-to-southwest decline is greater for hydrogen and sulfate ions and less for nitrate ions. Ammonium ion concentrations and deposition levels do not indicate any definite pattern throughout the Southeast. With the exception of hydrogen ion concentrations, the spatial decline in ion concentrations and wet deposition fluxes from northeast to southwest across the SCFRC region vary from 20 to 40%. An increase of 0.5 pH units from the Virginia Piedmont region to the Texas Gulf Coastal Plain region corresponds to almost a 70% decrease in hydrogen ion concentration. An example of the regional distribution patterns showing spatial variations in pH and hydrogen ion deposition for 1990, developed by NADP, is presented in Fig. 3.4, which shows the southern regional gradient described previously.

A more detailed assessment (Smith and Shadwick, 1992) of historic spatial and temporal trends in wet deposition component concentrations and fluxes involved dividing the southeastern United States into seven physiographic subregions of interest to SCFRC, based on climatic influences and geographic criteria (Fig. 3.5). The subregions are Piedmont (PMT), Inner Atlantic Coastal Plain (IAP), Outer Atlantic Coastal Plain (OAP), Eastern Gulf Coastal Plain (EGP), Western Gulf Coastal Plain (WGP), Eastern Gulf Coastal Flats (EGF), and Western Gulf Coastal Flats (WGF). Monthly and annual summary data for monitoring sites in the NADP network have been reviewed to examine historic wet deposition levels within these seven subregions for the period from 1979 to 1989. No monitoring sites existed in the Western Gulf Flats, and only sparse temporal and spatial coverages were found for the Eastern Gulf Flats and Outer Atlantic Plain subregions. The number of sites, site-years and site-quarters of operation for the period from 1979 to 1989 for each subregion were compiled (Table 3.1). In this data set, duplicate (colocated samplers) sampling data were retained as a measure of sampling variability but may over emphasize the influence of some sites on averaged values. Wet deposition data were excluded from analyses if (1) data failed to meet completeness criteria for the summary period, (2) site location problems existed, or (3) unacceptable samplers were used.

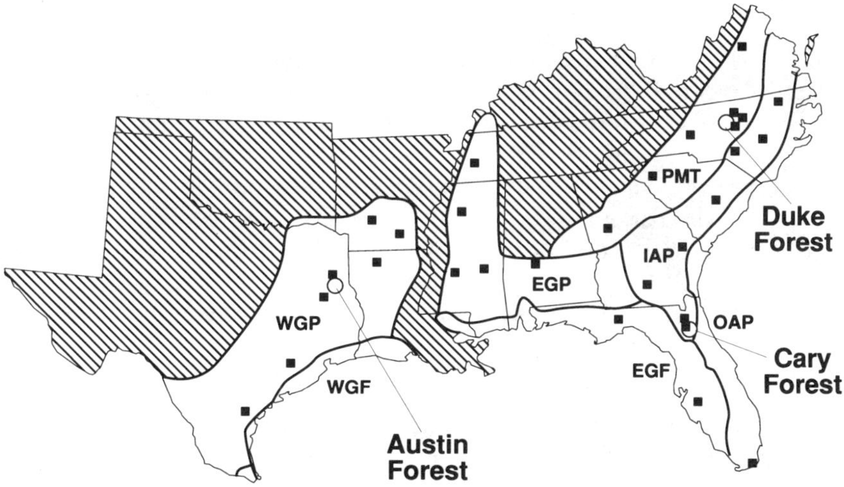

Figure 3.5. Map of Southern Commercial Forests Research Cooperative physiographic subregions, showing locations of NADP (■) and SCFRC (○) monitoring sites in the southeastern United States. The subregions are PMT = Piedmont; IAP = Inner Atlantic Coastal Plain; OAP = Outer Atlantic Coast Plain; EGP = Eastern Gulf Coastal Plain; EGF = Eastern Gulf Coastal Flats; WGP = Western Gulf Coastal Plain; and WGF = Western Gulf Coastal Flats (from Smith and Shadwick, 1992).

Table 3.1. Extent of Historical Wet Deposition Data Sources in Southern Commercial Forests Research Cooperative Subregions for 1979 to 1989

Subregion	Sites	Site Years	Site Quarters
Outer Atlantic Coastal Plain	3	11	42
Inner Atlantic Coastal Plain	9 (1)	45 (0)	163 (2)
Piedmont	8 (1)	34 (1)	143 (4)
Eastern Gulf Coastal Plain	6	26	89
Eastern Gulf Coastal Flats	3 (1)	6 (0)	28 (1)
Western Gulf Coastal Plain	7	24	101
Western Gulf Coastal Flats	0	0	0

Adapted from Smith and Shadwick, 1992. Numbers in parentheses refer to colocated samplers.

Distributions and 11-year means of annual precipitation-weighted mean major ion concentrations (Table 3.2), total precipitation, and wet deposition amounts (Table 3.3) by subregion for hydrogen (H^+), sulfate (SO_4^{2-}), nitrate (NO_3^-) and ammonium (NH_4^+) ions were determined for the period from 1979 to 1989 (Smith and Shadwick, 1992). Similar patterns were observed by analysis of monthly summary data, although there was greater variability in the latter. The tabulated data show that acid precipitation (pH < 5) is characteristic of all subregions of the SCFRC region. The 25th, 50th, 75th, and 90th percentiles for pH were 4.53, 4.66, 4.75, and 4.86, respectively (Smith and Shadwick, 1992).

The 11-year subregional data analysis suggests that the Piedmont subregion has continuously experienced distinctly higher annual averaged wet deposition levels of H^+, SO_4^{2-} and NO_3^- ions than other SCFRC subregions for the period (1979 to 1989). Statistical hypothesis testing (Dunn, 1964) of annual precipitation-weighted mean concentrations, wet deposition amounts, and precipitation totals for all subregions, excluding the Eastern Gulf Coastal Flats, Outer Atlantic Coastal Plains, and Western Gulf Coastal Flats, was used to determine differences among the remaining four subregions (Smith and Shadwick, 1992). Significant differences (at the 5% level) were found in the following cases:

1. The Piedmont experienced significantly higher levels than each of the other subregions for hydrogen, sulfate, and nitrate ions.
2. The Inner Atlantic Coastal Plain and the Eastern Gulf Plain region exhibited greater hydrogen ion levels than the Western Gulf Plain.
3. Levels of ammonium ion in the other three subregions exceeded that for the Inner Atlantic Coastal Plain.
4. For concentration only, the Piedmont received higher levels of ammonium ions in precipitation than the Western Gulf Plain.

Statements 1, 2, and 3 apply to both concentration and deposition data.

Table 3.2. Ranges and Average Values of Annual Precipitation-Weighted Mean Concentrations (mg/l) and Precipitation (cm) by Southern Commercial Forests Research Cooperative Subregions for 1979 to 1989

	pH			SO_4 (mg/l)			NO_3 (mg/l)			NH_4 (mg/l)			Total Precip (cm)		
Subregion	Min	Mean	Max	Min	Mean	Max	Min	Mean	Max	Min	Mean	Max	Min	Mean	Max
PMT	4.17	4.45	4.82	1.24	2.00	3.02	0.65	1.10	2.14	0.13	0.23	0.38	68.0	107	150
IAP	4.42	4.61	4.98	0.94	1.50	2.25	0.47	0.80	1.29	0.06	0.15	0.28	79.5	115	159
EGP	4.49	4.63	4.83	1.10	1.34	1.68	0.59	0.75	1.09	0.12	0.21	0.40	92.1	123	189
WGP	4.53	4.77	5.18	0.92	1.29	1.55	0.57	0.76	0.87	0.10	0.19	0.27	57.3	118	169
OAP	4.62	4.82	5.02	0.74	1.01	1.26	0.55	0.59	0.62	0.04	0.05	0.07	97.2	128	165
EGF	4.79	4.80	4.81	0.89	1.00	1.17	0.51	0.67	0.79	0.04	0.07	0.11	89.5	126	148

Adapted from Smith and Shadwick, 1992. Abbreviations: PMT = Piedmont, IAP = Inner Atlantic Coastal Plain, EGP = Eastern Gulf Coastal Plain, WGP = Western Gulf Coastal Plain, OAP = Outer Atlantic Coastal Plain, EGF = Eastern Gulf Coastal Flats, Tot Prcp = Total Precipitation, mg/l = milligrams per liter, cm = centimeter.

Table 3.3. Ranges and Average Values of Annual Ionic Deposition (kg/ha) and Precipitation (cm) by Southern Commercial Forests Research Cooperative Subregions for 1979 to 1989

	H (kg/ha)			SO_4 (kg/ha)			NO_3 (kg/ha)			NH_4 (kg/ha)			Total Precip (cm)		
Subregion	Min	Mean	Max	Min	Mean	Max	Min	Mean	Max	Min	Mean	Max	Min	Mean	Max
PMT	0.18	0.41	0.78	10.9	21.6	30.4	6.9	11.8	17.6	1.40	2.60	4.52	68.0	107	150
IAP	0.14	0.28	0.39	10.9	16.9	22.0	6.2	9.0	12.2	0.80	1.64	2.99	79.5	115	159
EGP	0.18	0.29	0.44	10.5	15.9	24.3	5.9	8.9	13.6	1.12	2.50	4.61	92.1	123	189
WGP	0.04	0.20	0.31	9.1	14.7	19.3	5.4	8.6	10.7	1.30	2.22	2.98	57.3	118	169
OAP	0.10	0.16	0.27	12.0	12.1	12.3	6.0	7.3	8.9	0.43	0.70	1.19	97.2	128	165
EGF	0.14	0.20	0.24	10.5	12.1	13.2	7.0	7.9	9.3	0.64	0.81	0.97	89.5	126	148

Adapted from Smith and Shadwick, 1992. Abbreviations same as in Table 3.2.

A shorter term perspective of acid deposition in the Southeast is provided by describing the acidity distribution in increments of a half pH unit (Table 3.4) from weekly collections. The data, which were supplied by NADP (1991a and b), mimic the subregional distributions from longer term volume-weighted pH levels (Table 3.2).

Distributions by subregion of precipitation-weighted mean pH, sulfate, and nitrate ion concentration values for 1979 through 1989 were estimated (Smith and Shadwick, 1992). These distributions are presented in the form of box plots, where the top, middle, and bottom lines of the box are the 75th, 50th, and 25th percentiles, respectively, and the dot (•) represents the arithmetic mean value (Figs. 3.6, 3.7, and 3.8). Whiskers, tail distributions around a box plot, extend as far as the data range, up to a distance not greater than 1.5 times the distance between the 75th, and 25th percentiles (the interquartile range). The synthesized and tabulated historic data (Tables 3.2 and 3.3) clearly show the statistically based conclusion: that the Piedmont subregion generally experiences lower pH levels in rainfall and higher H^+ deposition levels.

Mean and percentile distributions by subregion of precipitation-weighted pH values (Fig. 3.6), sulfate (Fig. 3.7), and nitrate (Fig. 3.8) ions also show that the Piedmont subregion of the SCFRC generally experienced lower pH levels, higher SO_4^{2-} and NO_3^- concentrations, and higher H^+, SO_4^{2-}, and NO_3^- deposition levels than the other subregions. Conversely, the Western Gulf Coastal Plain subregion is subject to higher pH levels and lower H^+, SO_4^{2-}, and NO_3^- concentrations and deposition values. Corresponding values for the Eastern Gulf Coastal Plain and Inner Atlantic Coastal Plain, as well as region-wide values, lie between those for the Piedmont and Western Gulf Coastal Plain.

Lower pH values and higher hydrogen ion deposition levels appear to occur during the summer months in each subregion, except the Western Gulf Coastal Plain. In the latter case, wet deposition component levels are reduced in summer, when precipitation amounts also are lower. Sulfate and nitrate ion concentration distributions showed patterns typical

Table 3.4. Counts by Southern Commercial Forests Research Cooperative Subregion of Weekly Sample pH Values, 1979 to 1989

Subregion	3.0–3.5	3.5–4.0	4.0–4.5	4.5–5.0	5.0–5.5	5.5–6.0	6.0–6.5	6.5–7.0	7.0–7.5	7.5–8.0	8.0–8.5
PMT	5	100	568	424	108	37	34	22	7	3	0
IAP	2	62	438	567	229	60	42	27	9	0	0
EGP	1	5	121	271	88	22	15	11	4	0	0
WGP	0	10	150	382	175	76	44	22	9	2	0
OAP	0	3	57	115	117	56	17	10	2	0	0
EGF	0	2	33	71	47	13	9	5	3	0	1

Categories include the upper end point in each case. Abbreviations as in Table 3.2.

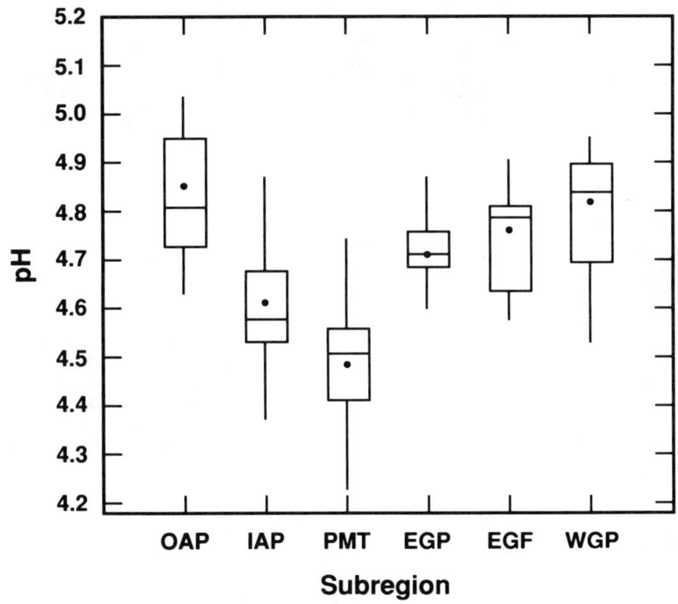

Figure 3.6. Box plot distribution by subregion of annual precipitation-weighted mean pH values for 1979 to 1989. Lines at top, middle, and bottom of box represent 75th, 50th, and 25th percentiles, respectively, and ● is the arithmetic mean (from Smith and Shadwick, 1992).

of those found for hydrogen ions. Subregions were not distinctly separated, however, in corresponding seasonal plots for ammonium ions. Precipitation amounts for the Western Gulf Coastal Plain were lower than the average for the entire SCFRC region in summer, but higher levels occurred in the fall. The same relative distribution patterns are observed when monthly precipitation-weighted mean concentration and monthly deposition values are examined over time for longer operating sites. Within subregions, north–south high–low gradients in ion concentrations and depositions are apparent (Smith and Shadwick, 1992).

There are insufficient data to establish historical temporal patterns in component concentrations or deposition rates for annual precipitation in the SCFRC region. Assuming that acid deposition mimics the pattern in regional emission trends for acid precursors (SO_2, NO_x) and allowing for the considerable uncertainties in emissions estimates, it seems reasonable to surmise that deposition levels over the SCFRC region have increased, at least for the period from 1950 to 1970. From 1970 to 1980, the trend in deposition levels is not clear. From 1980, actual wet deposition measurements suggest that, despite year-to-year variability, long-term levels have remained consistent. Moreover, it appears that wet deposition levels for individual subregions of the SCFRC have maintained

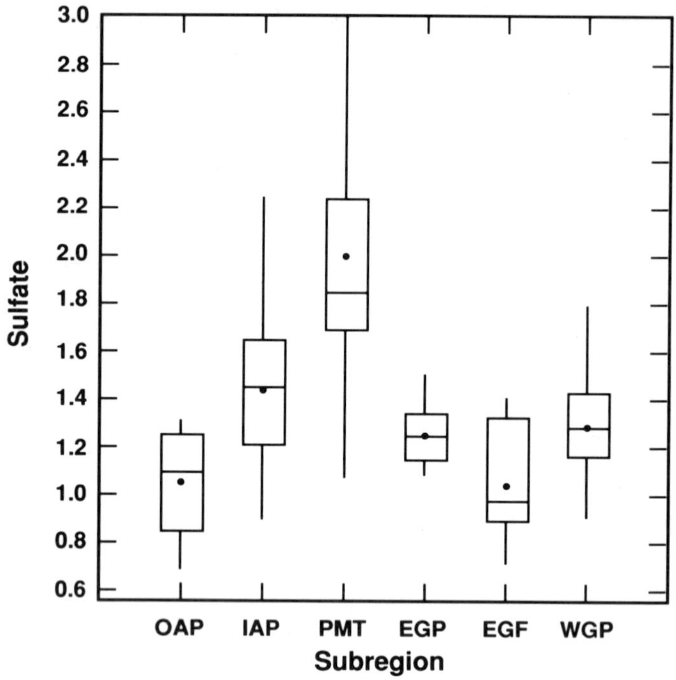

Figure 3.7. Box plot distribution by subregion of annual precipitation-weighted mean sulfate concentrations (mg/l) for 1979 to 1989. Lines at top, middle and bottom of box represent 75th, 50th, and 25th percentiles, respectively, and • is the arithmetic mean (from Smith and Shadwick, 1992).

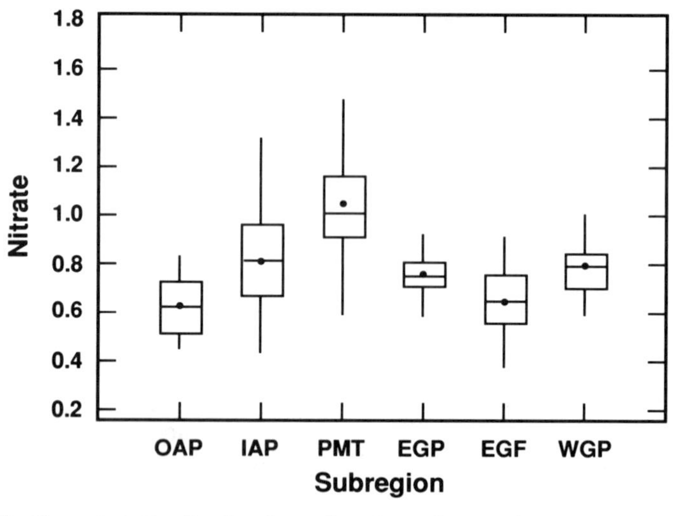

Figure 3.8. Box plot distribution by subregion of annual precipitation-weighted mean nitrate concentrations (mg/l) for 1979 to 1989. Lines at top, middle, and bottom of box represent 75th, 50th, and 25th percentiles, respectively, and • is the arithmetic mean (from Smith and Shadwick, 1992).

the same relative magnitudes—for example, deposition levels for the Piedmont have probably been higher than those in other SCFRC subregions since 1970 (Smith and Shadwick, 1992).

To illustrate the differences in acid deposition between terminal subregions in the SCFRC region, monthly precipitation-weighted mean pH values for the Charlottesville, Virginia site (Fig. 3.9) in the Piedmont are compared with similar data for the Attwater, Texas site (Fig. 3.10) in the Western Gulf Coastal Plain subregion. At Charlottesville, almost all values are below the median pH for the entire multiyear SCFRC region database, and most values are less than the 25th percentile level. For the Texas site, the situation is reversed, with almost all values above the regional median and most above the 75th percentile of all values observed for the SCFRC region. While differences in sampling frequency at the two sites may have influenced the observations, it is unlikely that these differences could account for the observed pH differences, given an interquartile range of 0.32 pH unit (Sisterson et al., 1985; Smith, 1987).

Recently, spatial wet deposition patterns and relative error estimates for sulfate (Figs. 3.11a and b) and nitrate (Figs. 3.12a and b) have been reported for 1992 (U.S. EPA, 1994). Quality assured data from two major monitoring networks were merged for spatial analysis to establish

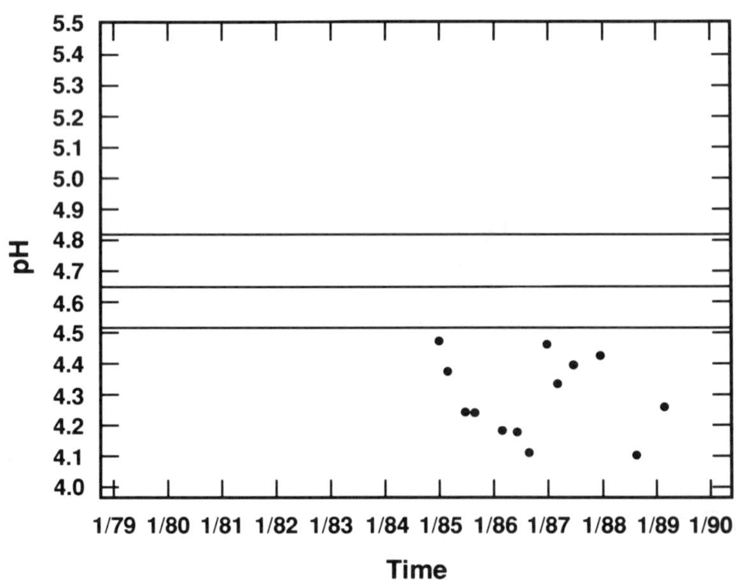

Figure 3.9. Monthly mean precipitation-weighted pH values for wet deposition at Charlottesville, VA, Piedmont (MAP3S) site. Horizontal lines are 25th, 50th, and 75th percentiles of SCFRC regional values for same time frame (from Smith and Shadwick, 1992).

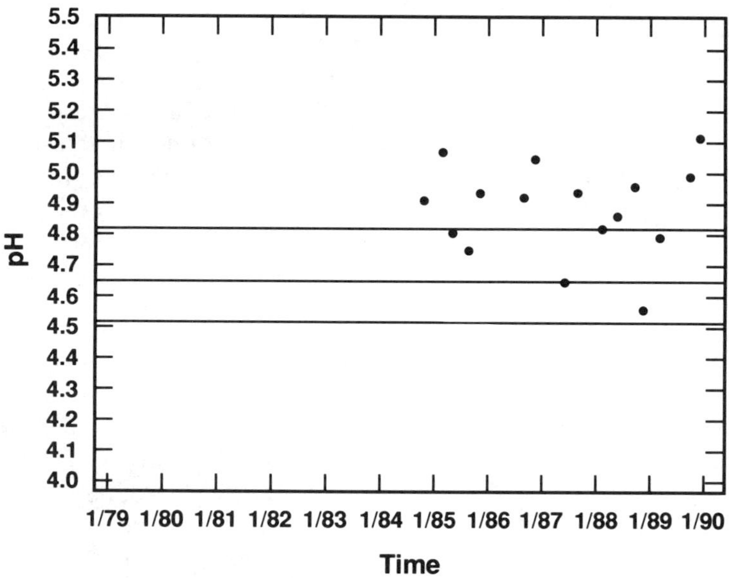

Figure 3.10. Monthly mean precipitation-weighted pH values for wet deposition at Attwater, TX, Western Gulf Plain (NADP) site. Horizontal lines are 25th, 50th, and 75th percentiles of SCFRC regional values for same time frame (from Smith and Shadwick, 1992).

the geographical extent of wet deposition components. The largest network is NADP/NTN, with approximately 200 primarily rural monitoring sites in current operation. Data from the Clean Air Status and Trends Network were utilized along with NADP/NTN data for 1992 to provide for spatial interpolation across the continental United States.

With only about 200 monitoring sites covering the conterminous United States, it is necessary to be able to predict wet deposition in nonmonitored areas. A spatial interpolation technique developed by Haas (1990a, 1990b) was used to produce estimates of deposition over broad areas of the United States for 1992. Only those sites that met NADP/NTN annual data completeness criteria (NADP, 1993) were included in the spatial analysis. Interpolation uncertainty is expressed as a percentage of the interpolated value. The maps should be used with caution, however, because the displayed values may vary from true values due to random variations in data, limited number of sites, completeness of data records at each site, site representativeness, and interpolation error. There is general agreement between the mean historic deposition data reported for sulfate and nitrate (Table 3.3) and the spatial distributions determined for 1992 (Figs. 3.11 and 3.12). The regional northeast to southwest gradient in sulfate and nitrate deposition is less evident in 1992 because of the relatively coarse increments in deposition units used in the maps.

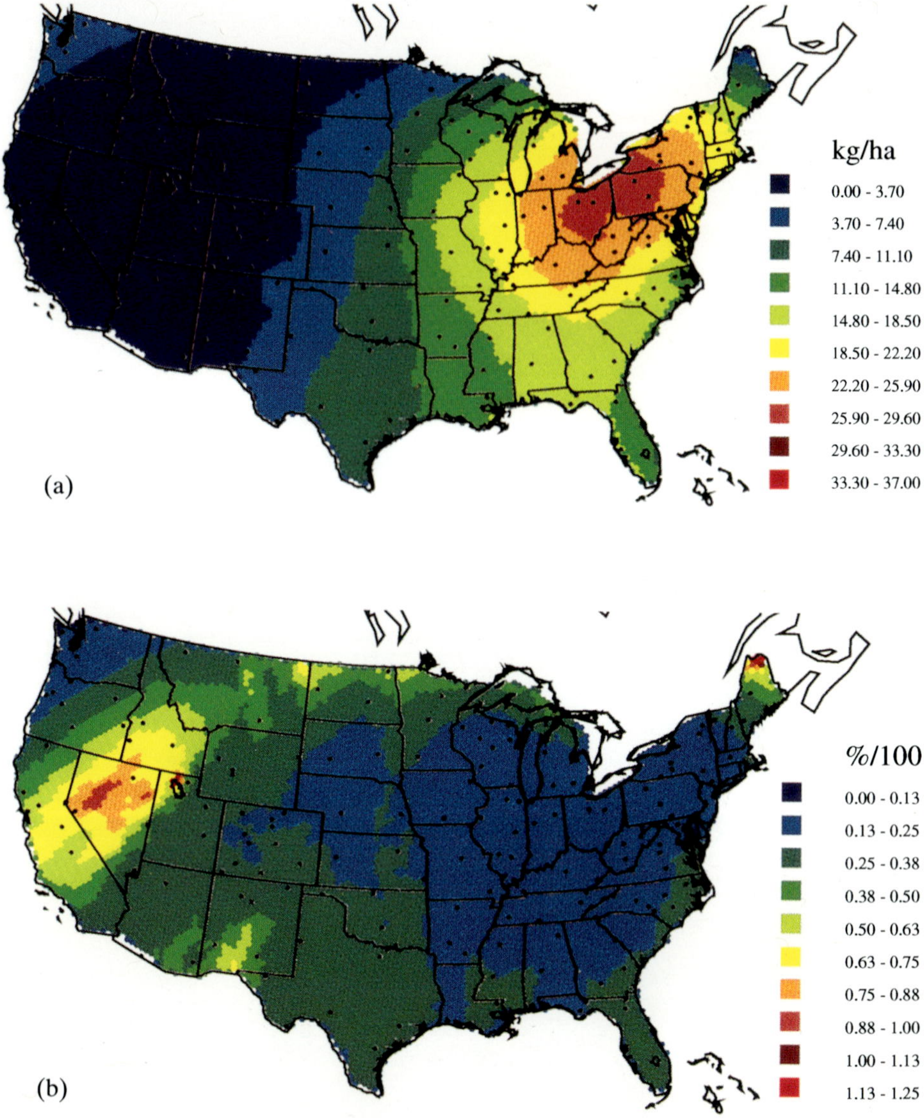

Figure 3.11. (a) Annual spatial distribution of wet sulfate deposition (kg/ha-yr), 1992. (b) Relative error for annual spatial distribution of wet sulfate deposition, 1992 (from U.S. EPA, 1994).

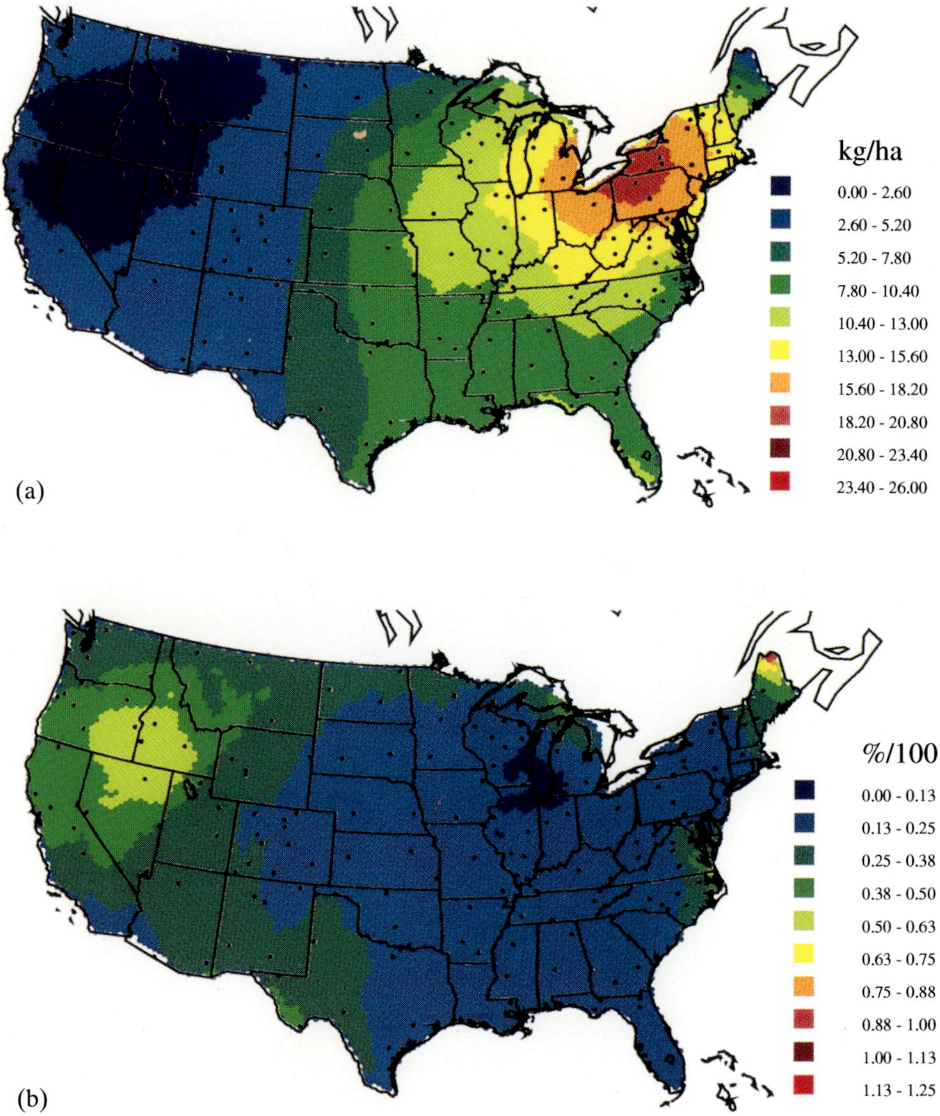

Figure 3.12. (a) Annual spatial distribution of wet nitrate deposition (kg/ha-yr), 1992. (b) Relative error for annual spatial distribution of wet nitrate deposition, 1992 (from U.S. EPA, 1994).

Wet Deposition at SCFRC Intensive Sites

Precipitation chemistry data were obtained at three SCFRC intensive research sites from October 1987 to June 1991. These sites were located near Durham, North Carolina (Duke Forest), Gainesville, Florida (Cary Forest), and Nacogdoches, Texas (Austin Forest) (Table 3.5). A fourth SCFRC site (Florida IFS), which was operated under EPRI/IFS protocols was located near (3 km) the Cary Forest, Florida site. Air quality, meteorology, and acid deposition monitoring support for pollution effects studies carried out by SCFRC investigators at these intensive research sites was provided by the University of Florida Atmospheric Exposure Monitoring Program. A primary objective of the program was to provide SCFRC investigators with detailed, accurate, and precise continuous measurements of local meteorological, air quality, and wet and dry acid deposition variables for the duration of their investigations. These quality assured data may also provide valuable insights on conclusions resulting from examination of historic regional data, as well as site-specific details and additional environmental factors not available previously. The North Carolina, Florida, and Texas SCFRC sites were situated in the Piedmont, Inner Atlantic Coastal Plain, and Western Gulf Coastal Plain subregions, respectively, of the SCFRC. Well-defined and accepted protocols for pollutant and atmospheric monitoring (U.S. EPA, 1976–1983, 1986a) and deposition sampling and analysis procedures used by NADP/NTN, EPA/NDDN, and EPRI/IFS (U.S. EPA, 1986b) were followed. Monitoring, sampling, and analytic protocols are detailed in an EPA-approved quality assurance project plan (Allen, 1987). The quality assured data were used to determine pollutant exposures and deposition inputs to the sites (Allen and Sutton, 1991a, b; Allen et al., 1992; Sutton and Allen, 1990).

Wet deposition sampling and analysis was established at the North Carolina site in October 1987 and at the Florida and Texas sites in mid-1988. Wet deposition collection and analysis was terminated at these sites in mid-1990 except at the Florida site, where monitoring was terminated in June, 1991. Quality assurance objectives for reported wet deposition sampling and analysis data—accuracy (100 ± 10% recovery), precision (<5% relative standard deviation), and completeness (>90%)—were met or exceeded (Allen and Sutton, 1991a,b). Most of the precipitation samples were collected and analyzed on an event basis.

Seasonal (quarterly) and annual precipitation-weighted concentration (mg/l) values for major ions (H^+, NH_4^+, Na^+, SO_4^{2-}, NO_3^- and Cl^-) were estimated for the North Carolina, Florida and Texas SCFRC sites (Table 3.6). Corresponding site specific values for seasonal ($kg\,ha^{-}yr^{-1}$) and annual ($kg\,ha^{-1}yr^{-1}$) wet deposition fluxes were derived from concentration and precipitation values (Table 3.7). Meteorological seasons used here are winter—first quarter (January, February, March); spring—

Table 3.5. Locations and Descriptions of Southern Commercial Forests Research Cooperative Intensive Research Air Quality and Acid Deposition Monitoring Sites Operated from 1987 to 1989

Site Name	Location	Elevation (masl)	Nearest Town	Physiography	Climate	Trees	Mean Annual Rainfall (cm)
Duke Forest, NC	36°02′N 78°59′W	152	Durham, NC, 8 km ESE	Transition zone Atlantic Coastal Plain—Piedmont	Continental most of year, subtropical in summer	Loblolly pine (*P. taeda* L.)	106
Cary Forest, FL	29°45′N 82°13′W	48	Gainesville, FL, 10 km SW	Inner Atlantic Coastal Plain	Subtropical marine elements—spring, summer Continental—fall, winter	Slash pine (*P. elliotti* var. *elliotti* Engelm)	135
Florida, IFS	29°44′N 82°11′W	50	Gainesville, FL, 13 km SW	Inner Atlantic Coastal Plain	Subtropical marine elements—spring, summer Continental—fall, winter	Slash pine (*P. elliotti* var. *elliotti* Engelm)	135
Austin Forest, TX	31°31′N 94°46′W	70	Nacogdoches, TX, 11 km NE	Western Gulf Coastal Plain	Subtropical, continental in winter	Short leaf pine (*P. echinata*) Loblolly pine (*P. taeda* L.)	117

masl = meters above sea level; cm = centimeters.

Table 3.6. Quarterly and Annual Summaries of Volume-Weighted Averages of Major Wet Deposition Component Concentrations (mg/l) at Southern Commercial Forests Research Cooperative Monitoring sites from 1987 to 1989

Site	(Qtr/Yr)[b]	Precip (cm)	Specie Concentration (mg/l)[a]					
			Na^+	NH_4^+	H^+	Cl^-	NO_3^-	SO_4^{2-}
Duke Forest, NC	4/87	23.1	0.51	0.14	0.014	0.71	0.71	1.31
	1/88	14.5	0.61	0.37	0.046	0.68	1.50	2.56
	2/88	27.9	0.38	0.63	0.091	0.36	2.67	4.74
	3/88	32.0	0.50	0.26	0.039	0.42	0.89	2.02
	4/88	21.6	0.21	0.10	0.025	0.29	0.79	1.09
Annual	1988	96.0	0.42	0.34	0.051	0.42	1.44	2.64
Duke Forest, NC	1/89	36.8	0.35	0.28	0.034	0.52	1.17	1.98
	2/89	41.4	0.38	0.39	0.052	0.39	1.38	2.55
	3/89	44.2	0.48	0.24	0.047	0.30	1.22	2.21
	4/89	33.3	0.37	0.15	0.029	0.29	0.86	1.63
Annual	1989	155.7	0.40	0.27	0.041	0.38	1.17	2.12
Cary Forest, FL	3/88	73.7	0.52	0.11	0.021	0.35	0.62	1.16
	4/88	14.7	0.37	0.07	0.020	0.67	0.48	1.16
	1/89	10.2	0.87	0.23	0.025	1.08	0.86	1.80
	2/89	37.1	0.42	0.30	0.029	0.59	1.13	1.79
	3/89	46.7	0.46	0.16	0.038	0.33	1.02	1.58
	4/89	15.7	0.44	0.07	0.013	0.64	0.39	0.90
Annual	1989	109.7	0.49	0.20	0.030	0.53	0.93	1.55
Austin Forest, TX	3/88	21.3	0.54	0.16	0.023	0.41	1.03	1.39
	4/88	32.3	0.48	0.25	0.025	0.68	0.63	1.12
	1/89	44.7	0.43	0.20	0.028	0.63	0.78	1.58
	2/89	52.3	0.55	0.28	0.024	0.66	0.89	1.38
	3/89	17.0	0.41	0.16	0.043	0.79	0.78	1.73
	4/89	13.7	0.41	0.24	0.025	0.51	0.69	1.47
Annual	1989	127.7	0.48	0.23	0.028	0.65	0.82	1.51

[a] mg/l = milligrams per liter.
[b] Qtr/Yr = Quarter/Year. Quarter = meteorological season: (1) January, February, March, (2) April, May, June, (3) July, August, September, (4) October, November, December.

second quarter (April, May, June); summer—third quarter (July, August, September); and fall—fourth quarter (October, November, December). Although ion concentrations and deposition estimates are frequently reported in units of weight per unit volume (mg/l) and weight per unit area per unit time (kg $ha^{-1} yr^{-1}$); respectively, these units do not reflect charge neutrality relationships between ions or allow for material balance estimates. Corresponding estimates of wet deposition component variables in concentration units of microgram equivalents per liter (μeq/l) and deposition units of gram equivalents per hectare per year (eq $ha^{-1} yr^{-1}$) were made for all the SCFRC sites to establish local quarterly and annual ionic relationships in wet deposition (Figs. 3.13, 3.14, and 3.15).

Analysis of data for individual precipitation events indicated that higher ion concentrations were associated with smaller quantities of

Table 3.7. Quarterly (kg ha^{-1} qr^{-1})[a] and Annual Summaries of Wet Deposition Fluxes (kg ha^{-1} yr)[b] of Major Ions at Southern Commercial Forests Research Cooperative Monitoring Sites from 1987 to 1989

Site	(Qtr/Yr)[c]	Wet Deposition (kg ha qr)[a]					
		Na^+	NH_4^+	H^+	Cl^-	NO_3^-	SO_4^{2-}
Duke Forest, NC	4/87	1.11	0.30	0.033	1.57	1.60	2.96
	1/88	0.89	0.53	0.067	1.00	2.18	3.74
	2/88	1.07	1.81	0.265	1.03	7.86	13.86
	3/88	1.70	0.87	0.123	1.41	2.96	6.73
	4/88	0.50	0.20	0.051	0.67	1.62	2.33
Annual (kg/ha-yr)	1988	4.16	3.41	0.506	4.11	14.6	26.7
Duke Forest, NC	1/89	1.29	1.04	0.125	1.94	4.31	7.33
	2/89	1.56	1.61	0.212	1.60	5.77	10.61
	3/89	2.18	1.08	0.208	1.34	5.39	9.74
	4/89	1.21	0.51	0.084	0.99	2.88	5.43
Annual (kg/ha-yr)	1989	6.24	4.24	0.629	5.87	18.4	33.1
Cary Forest, FL	3/88	3.87	0.78	0.155	2.59	4.58	8.59
	4/88	0.54	0.11	0.103	0.99	0.71	1.69
	1/89	0.90	0.23	0.025	1.09	0.87	1.83
	2/89	1.57	1.12	0.108	2.20	4.21	6.67
	3/89	2.16	0.77	0.176	1.55	4.79	7.40
	4/89	0.69	0.11	0.021	1.01	0.62	1.42
Annual (kg/ha-yr)	1989	5.32	2.23	0.330	5.85	10.5	17.3
Austin Forest, TX	3/88	1.16	0.35	0.049	0.89	2.24	3.00
	4/88	1.56	0.82	0.087	2.27	2.05	3.67
	1/89	1.93	0.90	0.126	2.83	3.50	7.04
	2/89	2.89	1.47	0.126	3.47	4.68	7.25
	3/89	0.70	0.27	0.072	1.35	1.35	2.97
	4/89	0.56	0.33	0.034	0.70	0.95	1.99
Annual (kg/ha-yr)	1989	6.08	2.97	0.358	8.35	10.5	19.3

[a] kg ha^{-1} qr^{-1} = kilograms per hectare per quarter.
[b] kg ha^{-1}yr^{-1} = kilograms per hectare per year.
[c] Qtr/Yr = Quarter/Year. Quarter = meteorological season: (1) January, February, March, (2) April, May, June, (3) July, August, September, (4) October, November, December.

precipitation, particularly after a long dry period. This observation confirmed reports (Sisterson et al., 1990) that most of the ion deposition in precipitation events occurs in the early stages of the event from "rain-out" of acidic cloud water. Volume weighting between precipitation samples, therefore, is necessary to account for dilution of cloud water components in larger rain events. Very low ambient concentrations of sulfur and nitrogen oxides and particulate matter observed during dry periods in these rural forest air sheds suggest that "wash-out" below

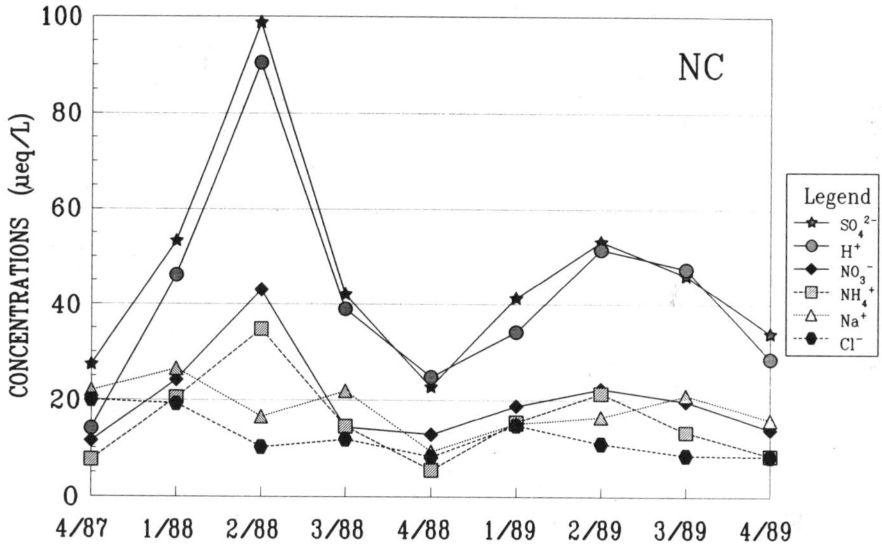

(a)

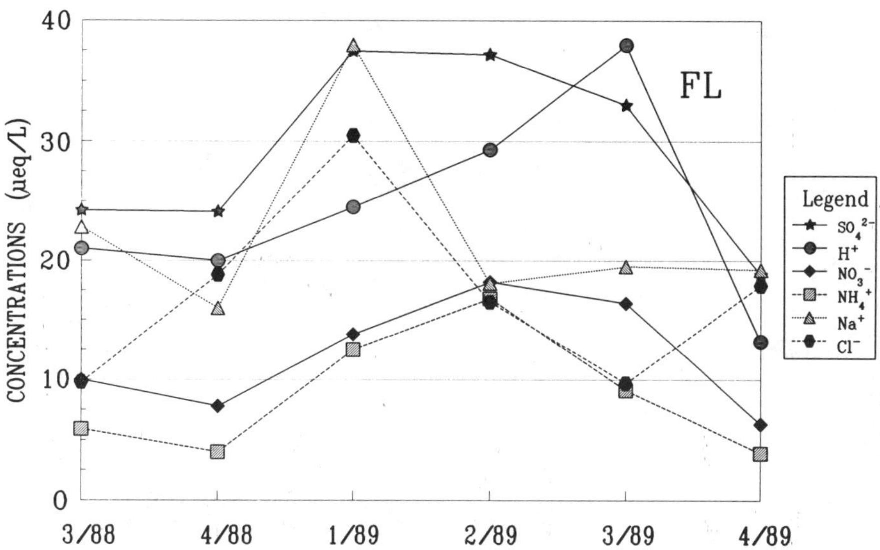

(b)

Figure 3.13. (a) Quarterly (seasonal) precipitation-weighted major ion concentrations (microgram-equivalents per liter) at Duke Forest (NC), (b) Cary Forest (FL), and (c) Austin Forest (TX) SCFRC monitoring sites for 1987 to 1989.

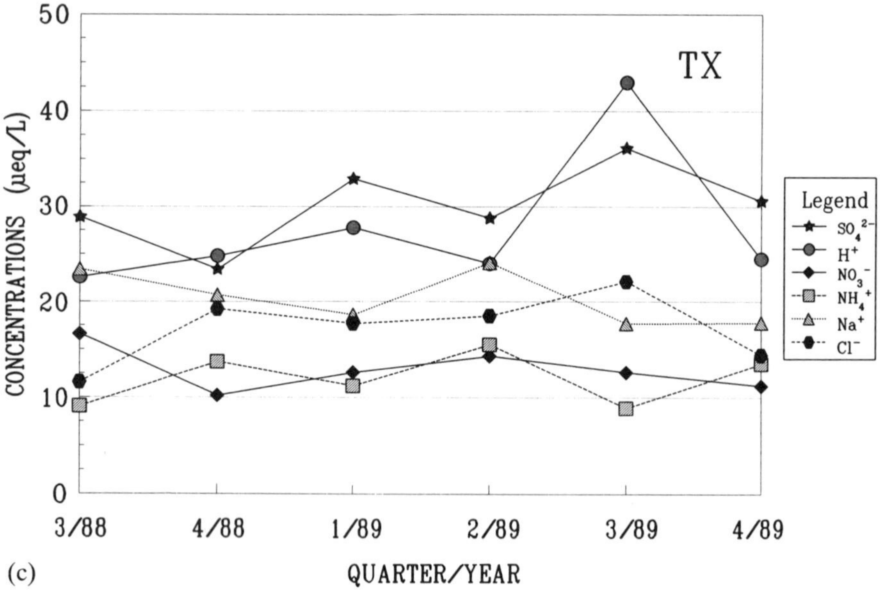

(c)

Figure 3.13. *Continued*

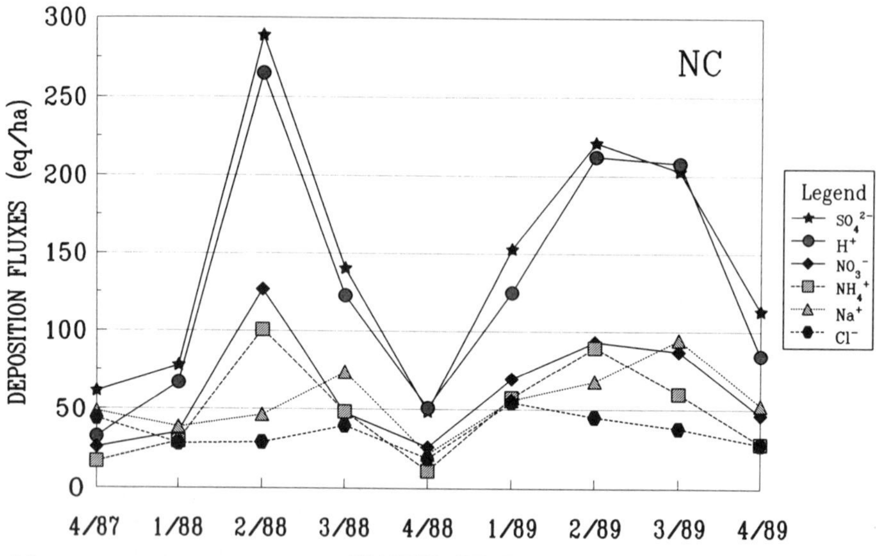

(a)

Figure 3.14. (a) Quarterly (seasonal) major ion fluxes (gram-equivalents per hectare per quarter) at Duke Forest (NC), (b) Cary Forest (FL), and (c) Austin Forest (TX) SCFRC monitoring sites for 1987 to 1989.

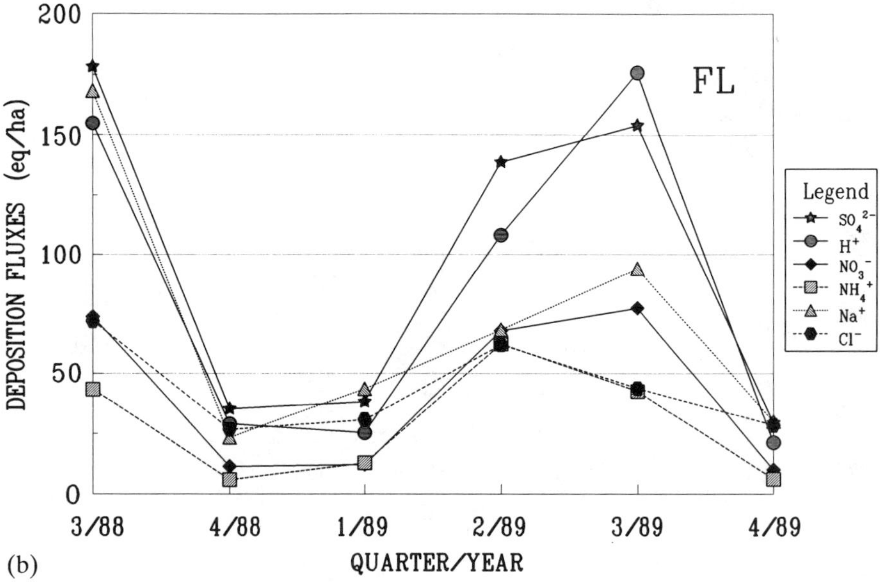

(b)

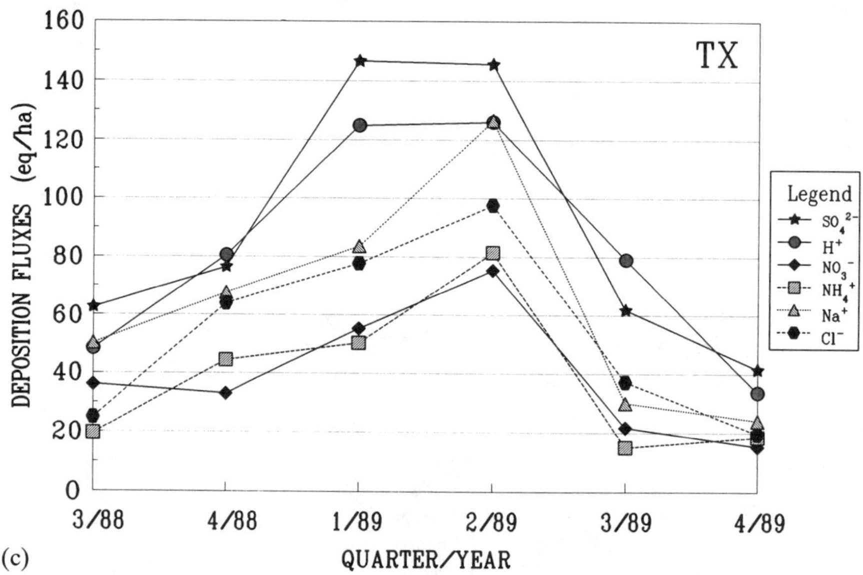

(c)

Figure 3.14. *Continued*

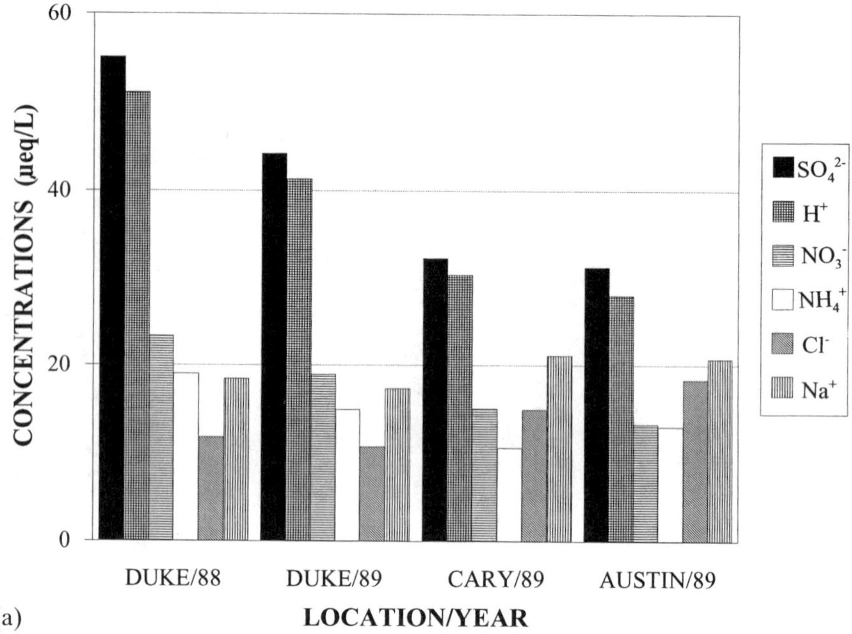

(a)

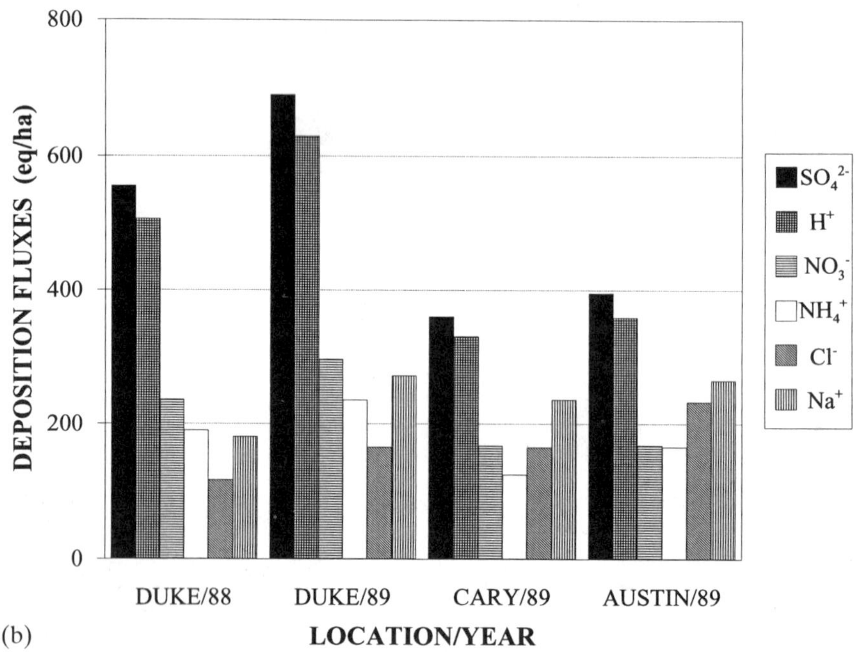

(b)

Figure 3.15. Annual distributions of ion (a) concentrations (microgram-equivalents per liter) and (b) fluxes (gram-equivalents per hectare per year) for precipitation at Duke Forest (NC), Cary Forest (FL), and Austin Forest (TX) SCFRC monitoring sites for 1987 to 1989.

precipitating clouds at these sites provided a negligible contribution to the observed concentrations of major anions (SO_4^{2-}, NO_3^-) in wet deposition.

A comparison of ion concentration and wet deposition data obtained at the three SCFRC sites from 1987 to 1989 (Tables 3.6 and 3.7) with historic data from 1979 to 1989 compiled for the corresponding subregions (Tables 3.2 and 3.3) showed that recent site-specific data are generally consistent with the range of annual data reported for a specific subregion for all major ions and for the SCFRC sites: Duke Forest (Piedmont), Cary Forest (Inner Atlantic Coastal Plain), and Austin Forest (Western Gulf Coastal Plain). Sulfate and nitrate deposition at SCFRC sites in 1988 and 1989, however, tended to be higher than the corresponding subregion ranges reported by the EPA for 1992 (Figs. 3.11 and 3.12).

High concentrations of major ionic components in precipitation at the North Carolina site in 1988 relative to 1989 were probably a reflection of the extensive drought experienced by the subregion in the former year and the much larger rainfall amount received in the latter year. In terms of wet deposition, however, the greater amount of precipitation (62%) in 1989 more than offset the lower volume-weighted ion concentration values found for the latter period.

Some important features are discernible from examination of the wet deposition data:

1. Precipitation amounts in 1988 and 1989 at the North Carolina site were approximately 10% less and 50% greater than normal, respectively. The annual precipitation-weighted average concentrations of ions, however, were typically only 10 to 15% lower for 1989 than 1988 values. Thus, the overall fluxes of individual ions deposited at the North Carolina site were about 25% greater in 1989 than in 1988. It is likely that the 1988 values represent a more typical annual flux of ions deposited at this site by wet deposition than the excessive values reported for 1989, despite the drought conditions existing in 1988. Precipitation amounts for 1989 at the Florida and Texas sites were 20% less and 10% greater than normal, respectively. As a result, the wet deposition flux values for major ions at the Florida and Texas sites may be representative for the corresponding southern SCFRC subregions. It is noteworthy that the annual average concentrations of H^+, NO_3^- and SO_4^{2-} at the two southern SCFRC sites differed by less than 15%. This concentration difference is further offset by a 16% difference in precipitation amounts at the Florida and Texas sites, which ultimately yields remarkably similar values (<5% difference) for the major ion wet deposition fluxes at the two sites (Fig. 3.15).

2. Annual average values for precipitation-weighted H^+ concentrations were highest at the North Carolina Piedmont site (0.046 mg/l, 2-year average) but were similar at the Florida and Texas sites (0.033 mg/l, 2-site average). Moreover, the annual average H^+ concentration was 25% higher in 1988 (0.051 mg/l) than that for 1989 (0.041 mg/L) at the North

Carolina site. This variation was probably a result of the higher (25%) annual NO_3^- and SO_4^{2-} concentrations observed in 1988 when compared to 1989 values. The highest seasonal concentration values for H^+, NO_3^-, and SO_4^{2-} at the North Carolina site occurred in the second quarters of 1988 and 1989, with particularly high values occurring in the second quarter of the former year (H^+ = 0.091, NO_3^- = 2.67, and SO_4^{2-} = 4.74 mg/l). The highest seasonal concentrations for H^+ at the southern coastal plain sites in Florida and Texas, however, occurred during the third quarter, but this observation was not true for both NO_3^- and SO_4^{2-}. The data suggested that the absolute seasonal wet deposited quantities of H^+, NO_3^-, and, SO_4^{2-} found at all of the sites were a strong function of the quantity of precipitation received at the site.

3. At all SCFRC monitoring sites, it was found that quarterly and annual volume-weighted SO_4^{2-} concentrations alone accounted for most of the H^+ concentrations in wet deposition (Figs. 3.13 and 3.15a). This should not be taken as implying, however, that all sulfate found in wet deposition at these locations was originally in the form of sulfuric acid. Moreover, quarterly and annual volume-weighted concentrations of NH_4^+ in precipitation correlated well with, and were similar to, concentrations of NO_3^- in rainfall (Figs. 3.13 and 3.15a). This observation, however, does not necessarily imply direct ammonium nitrate uptake by cloud water. Imbalances between the major cations (H^+ and NH_4^+) and anions (SO_4^{2-} and NO_3^-) appeared to be offset by Na^+ and Cl^- distributions (Fig. 3.15a). The ratio of SO_4^{2-} to NO_3^- , in ion equivalents, for wet deposition throughout the SCFRC region was 2.2.

4. Annual volume-weighted ammonium ion concentrations found at the North Carolina site were approximately 40% higher than for the Florida site and only 10% higher than for the Texas site in 1989. A higher annual average NH_4^+ concentration was observed at the North Carolina site in 1988 compared to 1989, which could be explained by the lower precipitation amount received during the former period. The highest seasonal concentrations of NH_4^+ were observed at all sites in the second quarter (April, May, June), which could be a result of enhanced biological activity, nitrogenous fertilizer application to agricultural soils, and animal waste decomposition during the period.

5. Higher annual average concentrations of Na^+ and Cl^- were observed at the two southern Coastal Plain sites in Florida and Texas than at the northernmost Piedmont site in North Carolina, although annual average K^+ concentrations (not given in the tables) were similar for all sites with an annual precipitation weighted mean value of 0.051 mg/l (standard deviation 0.009 between sites). The highest seasonal precipitation-weighted concentrations of Na^+ and Cl^- were observed during the first quarter at the Florida site and the second quarter at the Texas site, corresponding to the lowest and highest seasonal precipitation amounts, respectively, at the two sites. The predominance of these ions at the two southern coastal

plain sites during these periods suggested a significant marine contribution to the ionic content of rainfall. The annual equivalence ratios for Na^+ to Cl^- varied from 1.5 (excess Na^+) at the North Carolina and Florida sites to 1.1 at the Texas site. Precipitation at the latter site showed unusually high values for Na^+ and Cl^- during the second quarter of 1989; these large values led to higher annual average concentrations of sea salt ions at the southwestern site.

6. Ca^{2+} and Mg^{2+} base cation concentrations are not presented (Tables 3.6 and 3.7) because of space limitations. Individual concentrations were found to vary significantly with season and location. The ion equivalence ratios of Ca^{2+} to Mg^{2+}, however, for all sites provided a mean value of 2.27 (standard deviation 0.16), and appeared to be seasonally consistent and independent of the observing site. These ions are presumably soil derived, and such consistency in the equivalence ratios might be expected in the region. Significantly higher concentrations of these alkaline earth ions were observed in the winter and spring (first and second quarters) at the North Carolina site in 1988 and in the spring (second quarter) at the Texas site in 1989. There appeared to be little or no variation in the concentrations or equivalencies of these species with season at the North Carolina and Florida sites in 1989. The lowest concentrations of these soil-based species were consistently observed during the fall (fourth) quarter at all sites.

7. A consistent discrepancy existed in total anion and total cation equivalent material balances for wet deposition samples collected at all SCFRC sites when the base cations, calcium (Ca^{2+}), magnesium (Mg^{2+}), and potassium (K^+), were included in material balance estimates. Typically the cation budget exceeded the anion budget by 15 to 30%. During a four-month period aliquots of precipitation samples from all SCFRC sites were treated with chloroform to minimize biological degradation of precipitation components and analyzed for the presence of weak organic acids by ion chromatography. A consistent, substantial contribution by formate and acetate ions was found in these treated samples averaging 20 and 10 μeq/l, respectively. If the latter species were included in material balance estimates, the discrepancies were reduced to less than 10%. Routine measurements of weak organic acid components in wet deposition samples may be necessary for a complete accounting of ions in wet deposition to rural forested areas.

Dry Deposition

Previous investigations (Hicks et al., 1989; Sehmel, 1980) of dry deposition suggest that the input of airborne acidic gaseous and particulate pollutants to the lithosphere and hydrosphere by this mechanism occurs at an annual rate comparable to that for wet deposition. Dry deposition parameters were monitored at SCFRC sites because of the potential relative importance of dry acid deposition in estimating total acid input to forested ecosystems. Triple filter pack (TFP) sampling and analytical protocols used were identical to those selected by the EPA for NDDN (Edgerton et al., 1991), except for sampling flow rate (2 standard liters per minute). Colocated sampler studies at West Point, New York (Edgerton et al., 1991) and the Cary Forest (Kim, 1992) have demonstrated precision (<5% relative standard deviation) of the method and comparability of data using a range of flow rates (1.5 to 5 lpm) for all species measured. Field comparisons of the TFP method with more sophisticated techniques have shown that the method provides reliable data (Benner and Eatough, 1991; Quillian and Lundgren, 1992). The TFP sampler employs three filters in series to collect pollutants selectively from ambient air flowing through the filters. A Teflon filter removes fine particles (<5 μm diameter); a nylon filter quantitatively removes nitric acid (HNO_3) and some sulfur doxide; and a potassium carbonate impregnated paper filter removes the remaining sulfur dioxide and probably reactive nitrogen compounds, such as NO_2, PAN, and organic nitrates. Chemical analyses of aqueous extracts from these filters for sulfate and nitrate ions were provided by ion chromatography. Weekly integrated field samples provided weekly averaged ambient air concentrations of fine particulate sulfate (p-SO_4) and fine particulate nitrate (p-NO_3) (Teflon filter analysis), sulfur dioxide and nitric acid (Nylon filter analysis), and sulfur dioxide and nitrogen dioxide (K_2CO_3 impregnated paper filter analysis). Weekly averaged concentrations (μg/m^3) of individual species were combined with pollutant and receptor specific weekly averaged dry deposition velocities (m/s) to estimate dry deposition fluxes (μg/m^2/s). Dry deposition estimates using an inferential modeling technique (Baldocchi et al., 1987; Hicks et al., 1985, 1987, 1991; Hicks and Matt, 1988; Matt et al., 1987; Meyers et al., 1989) for the gaseous pollutants SO_2, NH_3, and O_3 and for fine particles are considered acceptable, but values for NO_x are not reliable. The model requires inputs for individual receptor species and characteristics (leaf area index, LAI, and stomatal conductance) as well as local hourly averaged meteorological data (wind speed and direction, directional frequency variation, solar radiation, relative humidity, temperature, and surface wetness) for the sampling period. Calculated hourly averaged deposition velocities were averaged over the sampling period (one week) for combination with weekly averaged pollutant concentrations to provide estimates of weekly averaged dry deposition fluxes.

Dry deposition monitoring of chemical species has only recently been initiated by regional networks in some subregions of interest to the SCFRC. These networks include the Electric Power Research Institute Integrated Forest Study established in 1986 and terminated in 1991, the National Dry Deposition Network (NDDN) established in 1987, and the Clean Air Status and Trends Network (CASTNET) established in 1991 to replace NDDN. Some measurements of dry deposition have also been reported by the National Oceanographic and Atmospheric Administration (NOAA). In addition to measuring fine particle deposition by the TFP technique, the EPRI/IFS network has measured coarse particle (>10 μm diameter) dry deposition (fallout). The latter measurements, however, are not widely regnized as valid. Coarse particle measurements are based on the relatively high settling velocities of large particles in ambient air to treated plates. Because of the long collection times necessary to accumulate adequate material for analysis, coarse particle samples are subject to contamination. Fine particle collection by filtration (TFP sampler) is less subject to contamination by fallout due to the inverted configuration of the TFP. Dry deposition estimates include specific trapping efficiencies for individual pollutants and specific receptor types, providing an appropriate estimate for vegetation exposure to individual pollutant gases; whereas ccoarse particle deposition estimates do not account for pollutant receptor exposure specificity or interaction. Moreover, EPRI/IFS network estimates of dry deposition are derived for pollutant interception by the forest canopy rather than by the ground. At SCFRC sites, airborne concentrations using TFP sampling were determined for cleared areas in grass, and the derived dry deposition fluxes refer to grass surfaces at ground level.

Limited dry deposition information for the southern United States was available prior to 1988. Measurements made at the Duke Forest IFS site in the North Carolina Piedmont region (Lindberg et al., 1992) from September 1986 to October 1987 provided the following annual total deposition estimates and wet–dry deposition ratios:

Sulfur deposition: 24.6 kg S ha^{-1} yr^{-1}; 56% dry, 44% wet
Nitrogen deposition: 15.0 kg N ha^{-1} yr^{-1}; 62% dry, 38% wet

These values, however, may not allow for comparison with non-EPRI/IFS data for the region because the site experienced below-normal precipitation during the sampling period, which could explain the dominance of dry deposition values. More importantly, the dry deposition fractions included contributions from coarse as well as fine particles, a characteristic measurement of the EPRI/IFS network. Coarse particle chemistry measurements were not required at SCFRC sites.

Dry deposition estimates for the fall of 1985 through the summer of 1987 for the NOAA Piedmont site at Panola Mountain, Georgia (Meyers et al., 1989) for sulfur and nitrogen species were 5.90 kg S ha^{-1} and 2.14 kg N ha^{-1}, respectively. These estimates, however, did not include contributions by coarse particles. Sulfur dry deposition was higher in summer, whereas nitrogen dry deposition was higher in spring. Wet deposition data were not available to assess relative contributions of wet and dry components at Panola Mountain. Thus, the need for additional detailed studies of this important terrestrial input variable was established in 1987.

Dry deposition component flux data for SCFRC sites have been reported elsewhere, but using arbitrarily selected values for deposition velocities (Allen et al., 1992). The values presented here for SCFRC sites were estimated using a recent version of the DRYDEP model (Hicks et al., 1987). Deposition velocities (V_d) for reactive vapors (SO_2, NO_2, O_3, HNO_3) and fine particles/aerosol at SCFRC sites were estimated using the hourly inferential model based on local measurements of micrometeorological variables and surface wetness, and default values for LAI and surface roughness for a grass surface, supplied in the Hicks model. Coarse particle chemical deposition measurements were not made at SCFRC sites, except for the Florida IFS site. Gravimetric estimation of suspended coarse particles using high-volume and dichotomous samplers was done at the North Carolina and Florida SCFRC sites.

Typical annual average values, with corresponding standard deviations in parentheses, for deposition velocities (V_d) to a grass surface determined using the DRYDEP model for annual conditions existing at SCFRC sites were 0.09 (0.02) cm/s for particular nitrate and sulfate; 0.53 (0.09) cm/s for ozone; 0.79 (0.06) cm/s for sulfur dioxide and nitrogen dioxide; and 1.18 (0.14) cm/s for nitric acid vapor.

Seasonally and annually averaged concentrations of dry deposition components were estimated for the three sites in the SCFRC network (Table 3.8), and corresponding dry deposition flux estimates were obtained by combining weekly averaged species concentrations with weekly averaged V_d values for an open grass surface (Table 3.9).

Some important features were identified from dry deposition component measurements. The highest levels for total dry deposition of sulfur components (p-SO_4 and SO_2) in the ambient air were found at the northernmost site (North Carolina). These observations were consistent with the higher levels of sulfate and acidity found in precipitation at this site and supported the hypothesis that the Duke Forest area should be classified as an impacted rural site. The airshed of the North Carolina Piedmont site is probably influenced by local stationary sources as well as by regional long range transport and transformation of sulfur compounds from industrialized and heavily populated regions of the midwestern United States. Moreover, the highest p-SO_4:SO_2 weight ratios observed at the North Carolina site occurred in the third quarter, summer drought of 1988 (4.4), and the lowest (0.48) occurred in the fourth quarter, wet fall season of 1989.

This variation in sulfate:sulfur dioxide ratios suggested that enhanced conversion of SO_2 to p-SO_4 occurred during the higher ambient air temperatures and solar radiation intensities available during the summer months. The higher total dry-deposited sulfur amounts found in 1989 did not, however, reflect the much higher precipitation amounts for that year.

Table 3.8. Quarterly and Annually Averaged Ambient Concentrations (μg/m³) of Dry Deposition Components at Southern Commercial Forests Research Cooperative Monitoring Sites from 1988 to 1989

Site	(Qtr/Yr)[b]	Concentration (μg/m³)[a]				
		p-SO_4	p-NO_3	HNO_3	SO_2	NO_2
Duke Forest, NC	1/88	2.40	0.46	1.74	4.14	0.16
	2/88	5.38	0.39	3.26	2.86	0.57
	3/88	6.95	0.33	2.88	1.58	0.93
	4/88	2.15	0.28	1.30	2.48	0.27
Annual	1988	4.22	0.37	2.30	2.77	0.48
Duke Forest, NC	1/89	4.02	0.19	3.35	6.79	0.35
	2/89	6.31	0.09	2.47	2.18	0.50
	3/89	5.97	0.17	1.87	1.24	0.58
	4/89	3.87	0.27	2.18	8.07	0.25
Annual	1989	5.04	0.18	2.47	4.57	0.42
Cary Forest, FL	3/88	2.70	0.59	1.29	1.86	0.46
	4/88	3.16	0.89	1.85	1.03	0.31
	1/89	3.31	1.12	2.01	1.11	0.30
	2/89	3.93	0.96	1.67	1.27	0.38
	3/89	3.88	0.48	1.35	1.00	0.34
	4/89	2.67	0.74	1.29	1.44	0.27
Annual	1989	3.45	0.83	1.58	1.21	0.32
Austin Forest, TX	3/88	4.72	0.36	2.15	1.68	0.85
	4/88	2.48	0.52	2.10	0.92	0.34
	1/89	3.22	0.69	2.27	0.63	0.27
	2/89	3.84	0.84	1.87	0.62	0.27
	3/89	4.84	0.40	1.35	0.50	0.36
	4/89	3.21	0.56	1.48	1.53	0.26
Annual	1989	3.78	0.62	1.74	0.82	0.30

[a] μg/m³ = micrograms per cubic meter.
[b] Qtr/Yr = Quarter/Year. Quarter = meteorological season: (1) Janury, February, March, (2) April, May, June, (3) July, August, September, (4) October, November, December.

Table 3.9. Estimates of Annual Dry Deposition Component Fluxes (kg/ha-yr) to Open Grass Areas at Southern Commercial Forests Research Cooperative Monitoring Sites from 1988 to 1989

Site	Year	Dry Deposition Flux (kg/ha-yr)				
		p-SO_4	p-NO_3	HNO_3	SO_2	NO_2
Duke Forest, NC	1988	1.17	0.10	8.63	6.91	1.19
	1989	1.40	0.05	9.40	11.4	1.05
Cary Forest, FL	*1988	0.41	0.10	2.91	1.80	0.48
	1989	0.95	0.24	6.25	3.02	0.85
Austin Forest, TX	*1988	0.50	0.06	3.95	1.62	0.74
	1989	1.01	0.17	7.14	2.04	0.87

Key: kg/ha-yr = kilograms per hectare per year, * = 6 month (July–December) deposition, p-SO_4 = particulate sulfate, p-NO_3 = particulate nitrate, HNO_3 = nitric acid vapor, SO_2 = sulfur dioxide, NO_2 = nitrogen dioxide.

At neither of the two southern sites was the p-SO_4:SO_2 ratio observed to fall below unity, although maximum values were consistently observed in the third quarter (summer season) of the year. It is also interesting to note that p-NO_3 concentrations were lower at the North Carolina site than at the Florida and Texas sites, whereas HNO_3 concentrations were greater by at least 40% at the northern location. It is possible that total atmospheric p-NO_3 may have been underestimated at all sites, particularly at the North Carolina site. If p-NO_3 is mainly associated with larger airborne particles (>2 μm diameter), then p-NO_3 would not have been collected efficiently using an inverted TFP at the low sampling flow rate employed. Particulate sulfate, on the other hand, is associated with submicron particles in the accumulation mode and should be representatively collected by the TFP system. It is apparent that the high HNO_3 dry deposition fluxes presented (Table 3.9) were strongly influenced by the high deposition velocities provided by the DRYDEP model (~1.2 cm/s). For example, frequently the ambient concentrations of particulate sulfate, and occasionally sulfur dioxide levels, were greater than HNO_3 levels (Table 3.8), yet p-SO_4 fluxes were always much lower than those for HNO_3. Otherwise, the values of dry deposition fluxes for each gaseous or particulate specie at the three sites mainly reflected the differences in concentrations observed at the sites because similar deposition velocities were estimated for the same specie at different sites. This lack of variation in estimated deposition velocities for the region was presumably in part due to the smoothing effect of long-term averaging of hourly deposition velocity data and the nature of the DRYDEP model used, including the use of default values for LAI and stomatal conductance for an open grass surface.

Wetfall and Dryfall to a Pine Canopy in North Central Florida

The deposition of chemicals from the atmosphere to a forest canopy can be different from that to an open, grassy area or an inert surface, depending on the site and activity of the specific chemical, physical, and physiological characteristics of the canopy (Johnson and Lindberg, 1992). Actual estimates of seasonal LAI, stomatal conductance, and environmental conditions (including surface wetness and canopy roughness) were made in a north central Florida forest from 1988 to 1991. The Florida IFS site, located 3 km due east of the Cary Forest SCFRC monitoring site, was established in 1987 to estimate wet and dry deposition to a typical forest canopy and to the soil of the region. The contribution of coarse particle (>5 μm diameter) fluxes to total dry deposition was also assessed at the Florida IFS site, although the validity of such estimates is questionable. Preliminary data from the Florida IFS site were included as part of the EPRI/IFS synthesis (Johnson and Lindberg, 1992).

Five pine plantations in the southeastern United States—*Pinus taeda* stands in Tennessee, North Carolina, and Georgia; a *P. strobus* stand in North Carolina; and the *P. elliotti* stand reported here (Florida IFS)—are described in the IFS synthesis (Johnson and Lindberg, 1992). Interestingly, the southern pine sites spanned almost the entire range of the IFS sites in terms of total strong acid inputs. For example, only two Appalachian mountain forests, both with high deposition from clouds or fog, received higher sulfate loadings than one of the *P. taeda* stands in the upper Piedmont of North Carolina (IFS Duke Forest), which experienced high sulfate deposition probably due to its proximity to electric power and industrial point sources. The Florida IFS pine site was near the lowest of all the sites (Mitchell et al., 1992) in sulfate and acid inputs. Deposition via clouds or fog was not significant at any of the low-elevation (<500 m) pine sites due to minimal cloud immersion. The fraction of total strong acid deposition in wetfall averaged about 50% for all the sites, except for the Florida IFS site, where the proportion was closer to 70% by wetfall on an equivalent basis. Likewise, the relative sensitivity of soils to added sulfate at the IFS sites, based on laboratory column studies (Mitchell et al., 1992), ranged from the greatest value for the North Carolina Piedmont site to the least for the Florida IFS site in the Inner Atlantic Coastal Plain. Thus, apparently there is a wide range of possible deposition and potential ecosystem effects even within stands of a single genus and for a relatively small geographic area.

Deposition in wetfall and dryfall to the Florida IFS site forest canopy was monitored over three complete years, from May 1988 through May 1991. This site was dominated by an unthinned, unfertilized plantation stand of *P. elliotti* var. *elliotti*, 25 years old at the beginning of the study. Details of the site and the types of research conducted can be found elsewhere (Cropper and Gholz, 1993a, 1993b; Gholz et al., 1991).

At all the IFS sites, the methodologies for estimating both wet and dry deposition were identical, and these standardized procedures have been described in detail (Lindberg et al., 1989, 1992). Wetfall and dryfall collection and analysis techniques were the same as those employed at the nearby Cary Forest SCFRC site, except that EPRI/IFS protocols were followed at the Florida IFS site. These protocols included measurements of coarse particles by deposition to treated plates, although the validity of such data is controversial. The results for coarse particle deposition are included for comparison with other EPRI/IFS sites.

Dry deposition of coarse particles ($>5\,\mu$m diam.) at the IFS site was computed through extrapolation of deposition measurements to inert petri plates at the top of the canopy using a scaling factor based on a regression between net enrichment of Ca^{2+} in throughfall over that in precipitation from a rainfall event and the number of antecedent dry hours associated with that event (Lindberg et al., 1988; Lovett and

Lindberg, 1984). Measured meteorological data taken above the canopy, seasonal leaf area index (Gholz et al., 1991), and stomatal conductance (Teskey et al., 1994) were input to the Hicks (1987) model for dry deposition velocity estimates. The latter values were used in conjunction with TFP sampling of dry deposition components at 10 m above the canopy to provide gaseous deposition fluxes. Fine-particle aerosol deposition was calculated from knowledge of cover characteristics and radionuclide tracer methods (Bondietti et al., 1984).

Wet deposition to the canopy, throughfall, and stemflow were all monitored at the Florida IFS site on an event basis using standard automated wet/dry and manual replicated collectors. Fine particles, coarse particles, and wetfall, throughfall, and stemflow were all analyzed for concentrations of SO_4^{2-}, NO_3^-, Cl^-, HPO_4^{2-}, H^+, NH_4^+, K^+, Na^+, Ca^{2+}, and Mg^{2+} ions. In all three years, 99% of the wetfall was collected and analyzed. TFP sampling covered 65% of the total dry period in the first year and 100% in the subsequent two years. Petri plate coverage for large particle collection was less, because it was not automated, and accounted for 29%, 31%, and 32% of the total dry period over the three consecutive years of study, respectively.

Estimated deposition to the Florida IFS site pine canopy over the three-year observation period followed similar annual patterns (Table 3.10) even though the number, intensity, and duration of rain events varied markedly and the fractional time that the canopy was wet varied from 28 to 37%. Precipitation increased slightly each year over the three years of study from 115 to 120 to 123 cm/yr, respectively, but all annual amounts were less than the long-term annual mean of 132 cm (NOAA, 1991). The total cation–anion balances indicated that weak acids were an important component of wetfall, because differences decreased from an anion deficit of 316 to 184 to $-51\,\text{eq}\,\text{ha}^{-1}\,\text{yr}^{-1}$ over each of the three years of study, as the weak acid analyses became available (total cation deposition varied less than 10% around an average of $1032\,\text{eq}\,\text{ha}^{-1}\,\text{yr}^{-1}$ over three years). The contribution of strong acids to total deposition at the Florida IFS site was low compared to those at all other IFS sites, as shown in Table 3.10 (Schaeffer et al., 1992). A high proportion of the total ion deposition at the Florida IFS site occurred as a result of Na^+ and Cl^-, distributed about equally in wetfall and coarse particle deposition (the latter may be indicative of uncertainty in the coarse particle deposition measurement technique). Although this high input is no doubt influenced by the proximity of the Atlantic Ocean and the Gulf of Mexico, the ratios of ions in wetfall bore little resemblance to ion ratios in sea water. This discrepancy indicated that mixing, interaction, and recombination of marine air masses with terrestrial source emissions may have occurred before the airstream reached the site, which is located centrally in the peninsula although less than 100 km from either coast.

Table 3.10. Summary of Annual Deposition Component Fluxes at Florida IFS Slash Pine Forest Site from 1988 to 1991

Deposition Component	H_2O (cm)	SO_4^{2-}	NO_3^-	Cl^-	PO_4^{3-}	H^+	NH_4^+ (eq ha^{-1} yr^{-1})	K^+	Na^+	Ca^{2+}	Mg^{2+}	Weak Acid
Vapors												
1988–89	—	26.0	184	—	—	—	—	—	—	—	—	—
1989–90	—	34.1	155	—	—	—	—	—	—	—	—	—
1990–91	—	33.6	89.6	—	—	—	—	—	—	—	—	—
Fine Particles												
1988–89	—	24.8	6	8.0	0.1	1.9	16.0	3.7	13.4	—[b]	—[b]	—
1989–90	—	28.7	5	4.3	0.0	2.8	17.8	3.7	14.8	—	—	—
1990–91[a]	—	21.8	5	4.3	0.0	2.8	17.8	3.7	14.8	—	—	—
Coarse Particles												
1988–89	—	120	98.1	196	10.2	9.9	9.5	10.0	343	160	55.3	—
1989–90	—	136	115	347	0.0	19.9	19.4	31.6	420	162	78.6	—
1990–91	—	111	108	283	0.0	25.9	129	48.5	275	158	90.9	—
Total Dry Deposition												
1988–89	—	171	289	204	10.3	43.0	25.5	13.7	356	160	55.3	—
1989–90	—	198	275	351	0.0	22.6	37.2	35.3	435	162	78.6	—
1990–91	—	167	203	287	0.0	28.7	147	52.2	272	158	90.9	—
Wet Deposition												
1988–89	115	366	164	204	9.9	280	121	12.9	251	260	136	—
1989–90	120	339	166	187	4.6	308	114	9.2	239	282	137	108[c]
1990–91	123	326	155	168	2.5	351	157	7.0	117	213	102	321
Total Deposition												
1988–89	—	537	452	408	20.2	323	146	26.6	607	420	191	—
1989–90	—	537	441	538	4.6	331	151	44.5	674	444	216	108
1990–91	—	519	357	455	2.5	380	303	59.2	389	371	193	321

cm = centimeters, eq ha^{-1} yr^{-1} = gram equivalents per hectare per year.
[a] Except for measured SO_4^{2-} and NO_3^-, assumed equal to values for 1989–90.
[b] Insufficient samples for Ca^{2+} and Mg^{2+} analysis.
[c] Estimated for 6 months in 1989–90.

The slash pine canopy had a large apparent effect on both the amount and composition of wet deposition as it moved to the ground as stemflow and throughfall (Table 3.11). Stemflow contributed little to the hydrology of the ecosystem or to ion fluxes, even though concentrations of ions in stemflow were often as much as 10 times higher (e.g., H^+) than in throughfall. Canopy effects were readily seen when total deposition was subtracted from throughfall plus stemflow (see "Net Canopy Effect," Table 3.11). High negative NCE values indicated that the canopy was a strong sink for both NH_4^+ and NO_3^-, as well as Na^+ and SO_4^{2-}.

The importance of coarse particle deposition estimates either to a pine canopy or to the soil cannot be overemphasized, but there is little agreement on a suitable technique to provide reliable data. The particle flux for some ions is highly variable from year to year and is likely to be influenced by regional patterns of prescribed biomass burning in nearby forests as well as by the indirect methods of estimation. The large apparent uptakes of both SO_4^{2-} and Na^+ ions by the pine canopy suggest that this component (coarse particle) of dry deposition may have been overestimated. However, the NCE for SO_4^{2-} for the other two IFS southern pine sites were substantially more negative than the values reported here, and all five pine IFS sites provided values in the range from -70 to over $-200\,\mathrm{eq\,ha^{-1}\,yr^{-1}}$ NCE (Mitchell et al., 1992). The Florida IFS Na^+ NCE value is indeed extraordinarily negative, but the estimated coarse particle deposition of Na^+ at this site is also much larger than at the other IFS sites. The magnitudes of both the coarse particle deposition and the NCE estimates, as well as the apparent canopy retention of SO_4^{2-} and Na^+, need further investigation. In most cases, only fine particle deposition estimates are made, such as at the Florida SCFRC site.

The relatively large and previously unmeasured contribution of dry deposition to the Florida IFS pine canopy had a particularly large effect on calculations of plant uptake and canopy leaching. For example, Gholz et al. (1985) measured large increases in base cation concentrations in throughfall under similar conditions to the current study. These increases were originally attributed to cation leaching from the pine canopy, and relatively high rates of cation uptake by the pine stand were estimated to account for this phenomenon. It is now apparent, however, that dry deposition is by far a larger source of throughfall cations than leaching. Likewise, the high degree of internal redistribution of nitrogen from senescent pine needles may be much less than that estimated previously by Gholz et al. (1985) because of the apparently much higher deposition and uptake of atmospheric N than that detected in bulk precipitation. Conclusions concerning net canopy effects on nitrogen, however, may be limited by not including organic nitrogen fluxes.

The effects of the canopy on estimates of wet and dry deposition (fine particle and gas deposition) can also be seen through a comparison of the

Table 3.11. Summary of Below-Canopy Annual Fluxes of Major Ions in Wet Deposition at Florida IFS Slash Pine Forest Site from 1988 to 1991

Below-Canopy Fluxes	H_2O (cm)	SO_4^{2+}	NO_3^-	Cl^-	PO_4^{3-}	H^+	NH_4^+	K^+ (eq ha^{-1} yr^{-1})	Na^+	Ca^{2+}	Mg^{2+}	Weak Acid
Stem Flow (SF)												
1988–89	0.6	10.1	0.7	12.0	0.1	12.3	0.1	1.0	7.8	9.3	4.6	—[b]
1989–90	0.6	9.8	0.5	10.1	0.1	16.4	0.3	0.9	9.2	9.2	—[a]	1.1
1990–91	0.5	9.8	0.6	14.5	0.2	16.7	2.8	1.4	8.3	10.0	5.5	2.4
Through Fall (TF)												
1988–89	99.7	453	184	475	11.9	290	39.0	50.4	509	420	221	—[b]
1989–90	114	433	148	414	7.9	316	44.6	51.5	493	438	230	72.6
1990–91	106	358	135	329	2.6	337	74.1	33.7	277	333	184	115
Net Canopy Effect (NCE)												
1988–89	—[c]	−73.9	−267	78.5	−8.2	−20.5	−107	24.8	−90.3	9.9	34.2	—[b]
1989–90	—[c]	−94.6	−293	−114	3.4	1.8	−106	7.9	−172	3.2	14.2	−34.2
1990–91	—[c]	−151	−222	−111	0.3	−26.7	−227	−24.1	−104	−27.6	−3.1	−204

cm = centimeters, eq ha^{-1} yr^{-1} = gram equivalents per hectare per year.
[a] Insufficient sample remaining for analysis.
[b] Not measured.
[c] Net Canopy Effect independent of precipitation amount.

wet and dry deposition component concentrations obtained at the Florida IFS site and the Cary Forest SCFRC site, located 3 km to the west, and the Bradford Forest, Florida NADP/NTN site, located about 20 km to the north. Data are presented (Table 3.12) that show a striking similarity between dry deposition data obtained at 10 m above the pine canopy (Florida IFS) and 10 m above ground (Florida SCFRC) and between wet deposition component concentration data at the Florida IFS and Florida SCFRC sites, but to a lesser extent for 1987 data at the Florida NTN site.

Total Deposition at Southern Forest Sites

A detailed analysis of the deposition components of a Florida slash pine canopy (Florida IFS site) is described in the previous section, but those estimates included coarse particle deposition. Wet and dry deposition data (gases and fine particles) from three SCFRC sites have been combined to provide estimates of the relative contributions of wet and dry deposition of sulfur and nitrogen (Table 3.13). The latter estimates, however, do not include coarse particle contributions. SCFRC deposition data represent contributions to the local terrestrial flux of the nutrients S (SO_2, SO_4^{2-}) and N (HNO_3, NO_2, NH_4^+, NO_3^-) to a cleared area in grass. These estimates suggest that wet deposition accounted for approximately 70% of the total S deposition and about 70% of the total N deposition at all three SCFRC sites in 1989. The total S and N deposition fluxes at the Piedmont (North Carolina) site, however, were substantially larger than those at the Inner Atlantic Coastal Plain (Florida) and Western Gulf Coastal Plain (Texas) sites.

Table 3.12. Comparison of Annual Average Volume-Weighted Wet and Dry Deposition Component Concentrations at Cary Forest SCFRC, Florida IFS, and Bradford Forest NAPD/NTN Sites

Wet Deposition Location/Year	H^+	NH_4^+	Na^+	Ca^{2+}	Mg^{2+} (mg/l)	Cl^-	NO_3^-	SO_4^{2-}
Cary Forest/1988–89	0.023	0.17	0.50	0.56	0.16	0.50	0.76	1.38
Florida IFS/1988–89	0.024	0.19	0.50	0.45	0.14	0.63	0.89	1.53
Bradford Forest/1987	0.019	0.10	0.24	0.08	0.03	0.42	0.65	1.08
Dry Deposition Location/Year	p-SO_4	p-NO_3	p-NH_4 ($\mu g/m^3$)	HNO_3	SO_2	NO_2		
Cary Forest/1988–89	3.28	0.89	0.93	1.71	1.32	0.36		
Florida IFS/1988–89	3.32	0.95	0.79	1.84	1.29	—		

mg/l = milligrams per liter, $\mu g/m^3$ = micrograms per cubic meter, SCFRC = Southern Commercial Forest Research Cooperative, IFS = Integrated Forest Study, NADP/NTN = National Acid Deposition Program/National Trends Network.

Table 3.13. Comparison of Sulfur (kgS/ha-yr) and Nitrogen (kgN/ha-yr) Wet and Dry Deposition Flux Estimates to an Open Grassy Surface at Southern Commercial Forests Research Cooperative Monitoring Sites from 1988 to 1989

Year	Site		Wet Deposition kg/ha-yr	(% Total)	Dry Deposition kg/ha-yr	(% Total)	Total Deposition kg/ha-yr
1988	Duke Forest	S	8.90	(70%)	3.85	(30%)	12.75
		N	5.95	(72%)	2.30	(28%)	8.25
	*Cary Forest	S	3.43	(77%)	1.04	(23%)	4.47
		N	1.89	(70%)	0.80	(30%)	2.69
	*Austin Forest	S	2.22	(69%)	0.98	(31%)	3.20
		N	1.88	(63%)	1.12	(37%)	3.00
1989	Duke Forest	S	11.0	(64%)	6.17	(36%)	17.20
		N	7.45	(75%)	2.42	(25%)	9.87
	Cary Forest	S	5.77	(76%)	1.83	(24%)	7.60
		N	4.10	(71%)	1.24	(29%)	5.80
	Austin Forest	S	6.43	(83%)	1.36	(17%)	7.79
		N	4.68	(71%)	1.88	(29%)	6.57

kg/ha-yr = kilograms per hectare per year, * = 6 month (July–December) deposition, S wet = SO_4^{2-}, S dry = p-SO_4 + SO_2, N wet = NO_3^- + NH_4^+, N dry = p-NO_3 + HNO_3 + NO_2, SO_4^{2-} = sulfate ion (wet only) p-SO_4 = particulate sulfate, SO_2 = sulfur dioxide, NO_3^- = nitrate ion (wet only), p-NO_3 = particulate nitrate, HNO_3 = nitric acid vapor, NO_2 = nitrogen dioxide.

Wet, dry, and total sulfate and nitrate deposition spatial patterns have been developed by the EPA for NDDN sites in 1991 (Figs. 3.16 and 3.17, pp. 123, 124). Only sites that met required data completeness criteria are represented. For dry deposition, valid measurements had to be available for at least 75% of 1991. For wet deposition, at least 90% of the year had to be monitored for precipitation amount, at least 75% of the year had to have valid precipitation chemistry measurements, and the rainfall sampler had to exhibit a collection efficiency of at least 75%. In the case of sulfate deposition (Fig. 3.16, p. 139) for the SCFRC region, it is seen that wet deposition accounts for about 60 to 90% of the total sulfate deposition in 1991 (15 to 25 kg SO_4 ha^{-1} yr^{-1}) and about 50 to 75% of the total nitrate (Fig. 3.17, p. 140) deposited in 1991 (8 to 15 kg NO_3 ha^{-1} yr^{-1}). These estimates are reasonably consistent with corresponding values reported for the SCFRC sites in 1989 (Table 3.13).

Air Quality

Gaseous Pollutants

The focus of most national, regional, and local long-term monitoring networks for ozone (O_3), nitrogen oxides (NO_x), and sulfur dioxide (SO_2) assessment has been the airsheds of major metropolitan and industrial areas (Lefohn and Shadwick, 1991). Primary concern is for those areas that are nonattainment for one or more of these criteria pollutants and in which public health effects are possible as a result of short or long-term exposure. Nonattainment areas are air quality control regions in which the second highest short-term (hourly) averaged pollutant concentration during the year exceeds the National Ambient Air Quality standard set by the EPA for that pollutant. Few resources have been expended in observing ambient air in rural and remote areas, where agricultural and forestry activities dominate. Monitoring of ambient air quality in the northeastern, midwestern, and southwestern United States has predominated over that in the South and Southeast due to the greater density of population centers and industrial activities in the former regions. Except for the SCFRC and NDDN monitoring sites, sparse air quality data exist for airsheds surrounding and within forested areas in the southern pines region. Moreover, in general, the local exposure of natural or cultivated stands of trees to gaseous pollutants cannot reasonably be calculated directly; nor can the behavior of these pollutants in localized rural environments be realistically derived from a knowledge of pollutant concentrations or atmospheric chemical processes occurring in remote or even nearby urban centers of population and industry. Regional models incorporating emissions data, transport, diffusion, and atmospheric chemistry, such as RAMS, RADM, and STEMII, can provide reasonable knowledge of pollutant distributions on a broader scale.

Validated ambient air quality data are archived by the EPA on the Aerometic Information and Retrieval System (AIRS). Ozone, nitrogen oxides, and sulfur dioxide concentrations are normally reported as hourly averaged values. Of those sites reporting O_3 or SO_2, only one third are classified as being rural or remote in character. Few of the sites reporting NO_x measurements are classified as the latter (Lefohn and Shadwick, 1991). Supplementary and relevant rural air quality information is available from the National Parks Service, which has established a nationwide monitoring network for hourly O_3, NO_x, and SO_2 concentrations. Beginning in the fall of 1987, criteria pollutants including O_3, NO_x, SO_2, total suspended and size-fractionated particulate matter, and lead were monitored at three sites in selected SCFRC southern pine forests. Monitoring was terminated at the North Carolina and Texas sites in June 1990 and at the Florida site in June 1991. The quality assured air quality monitoring data obtained for all SCFRC sites met accuracy (100 ± 10% recovery), precision (<5% relative standard deviation), and completeness (>90%) requirements for the criteria pollutants monitored.

Prior to 1987, there were few reports available describing southern regional rural air quality variable distributions that adequately summarized existing monitoring data in a convenient form. For air quality exposure information specifically relevant to the Southern Commercial Forest Cooperative region prior to 1988, the reader is referred to the monograph by Garner et al. (1989) and the IFS synthesis report (Johnson and Lindberg, 1992). The paucity of data available for characterizing the exposure of southern pine forest ecosystems to gaseous criteria pollutants makes any projection of exposure and resulting effects both difficult and challenging.

Sulfur Dioxide and Nitrogen Oxides

Nationwide trends in annual sulfur dioxide concentrations have been reported for the period from 1978 to 1992 by the EPA (1994). Albeit primarily for urban areas, the 1978 to 1987 data show a possible decline in ninetieth percentile and median hourly averaged values from 27 to 14 ppb and from 12 to 8 ppb SO_2, respectively, at 347 reporting sites. Statistically, the data show no increasing trend in ambient SO_2 levels during that decade. Nationwide trends in hourly averaged nitrogen dioxide levels for the 10-year period from 1978 through 1987 at 84 sites indicate that ninetieth percentile and median values show a possible declining trend from 51 ppb to 40 ppb and from 24 to 21 ppb NO_2, respectively, over that decade. These estimates of nationwide temporal trends are, however, dominated by the fact that the majority of observation stations are located in the Northeast, Midwest, and Southwest, where substantial emissions reductions in nitrogen oxides, sulfur dioxide, and volatile organic compounds (VOCs) have occurred, but they do not necessarily reflect the air quality situation in rural areas of the southeastern United States. Emissions of SO_2 and NO_x, however, have not declined in the South to the extent observed across other regions of the United States, which might influence the ambient air quality and atmospheric deposition data presented here.

Garner et al. (1989) have concluded that sulfur dioxide and nitrogen oxide concentrations in rural areas of the southeastern United States are rarely found in high enough concentrations for sufficiently extended periods of time to cause visible injury to plants and trees. Continuous monitoring of SO_2 and NO_x at the North Carolina, Florida, and Texas SCFRC sites from 1987 to 1989 appears to support this conclusion. Sulfur dioxide concentrations were less than 2 ppb more than 95% of the time, and nitrogen oxide concentrations were less than 10 ppb more than 90% of the time at all three SCFRC sites. The continuous monitoring data and the TFP observations were in agreement, inasmuch as the latter method provided estimated annual average concentrations of sulfur dioxide of 1.4, 0.5, and 0.3 ppb at the North Carolina, Florida, and Texas sites, respectively, and annual average nitrogen dioxide concentrations between 0.3 and 0.5 ppb for all three SCFRC sites.

Medians of the weekly measured SO_2 concentrations at sites operated by NDDN in 1993 have been reported by the EPA (Fig. 3.18). Sites were required to have valid measurements for 75% of the year for valid data

Figure 3.18. Median weekly sulfur dioxide concentrations ($\mu g/m^3$) at NDDN sites in 1993.

reporting. According to this distribution, median SO_2 concentrations in 1993 for the SCFRC region varied from 3.0 to 6.8 $\mu g\, SO_2/m^3$ (0.1 to 0.3 ppb SO_2) and are reasonably consistent with 1988 and 1989 data for SCFRC sites.

Significant contributions to ambient NO_x levels at two of the SCFRC sites (North Carolina and Florida) were a result of substantial local nitric oxide emissions from soil. These emissions, although occurring continuously, were observed by NO_x continuous monitoring instruments only when local ozone levels had been titrated out during strong surface inversions at night. Under these conditions, high levels of NO (>50 ppb) were periodically observed at night and in the early morning (Allen and Evans, 1992; Allen and Sutton, 1991b).

Photochemical Oxidants and Ozone

The photochemical oxidant vapors, peroxyacetyl nitrate (PAN) and peroxypropionyl nitrate (PPN), been have not been monitored routinely in rural or urban areas. Previous studies have usually been confined to urban areas with photochemical smog problems (U.S. EPA, 1986a). Atmospheric hydrogen peroxide (H_2O_2) levels have usually been monitored and studied more extensively in ambient air. In the Mountain Cloud Chemistry Project (MCCP), however, hydrogen peroxide was monitored extensively as it occurred in cloud water at Whiteface Mountain and in the ambient air on Whitetop Mountain (Mohnen, 1992). Nitric acid vapor concentrations and deposition have been estimated routinely in the EPRI/IFS study (Johnson and Lindberg, 1992) and the National Dry Deposition Network (NDDN) using filter pack samplers (Edgerton et al., 1991) similar to those used at SCFRC sites.

Currently, ozone is considered the most phytotoxic air pollutant that can cause stress to trees (Woodman and Cowling, 1987). Diffusion of ozone into leaves can reduce energy conversion and affect cellular processes. Moreover, ozone exposure can inhibit photosynthesis and affect the partitioning of photosynthates (Chappelka and Flagler, 1991; Reich and Amundson, 1985). At present, there are some questions concerning the levels or doses of ozone that can significantly influence stress in various southern pine species. Levels greater than 70 ppb ozone have received increasing attention as being damaging to certain species (Chappelka and Flagler, 1991; Reich and Amundson, 1985; Woodman and Cowling, 1987). However, there is little agreement on which exposure indexes best describe the impact of ozone exposure on tree stress (Lefohn et al., 1990a). Because characterization of ozone exposures at the three SCFRC rural forest sites will aid understanding and contribute to the development of ozone climatology models necessary to assess its potential contribution to forest decline, several exposure indexes are provided to assist in the assessment of ozone measurement data.

Primary sources of ozone data include databases developed by the EPA (SAROAD/AIRS), the Electric Power Research Institute (EPRI/SURE), the Tennessee Valley Authority (TVA), and the National Parks Service (NPS). There are several reports that have summarized and analyzed ozone monitoring data on a regional scale. One of the largest compilations and assessments of O_3 data has been carried out by Pinkerton and Lefohn (1986a, 1987) for the National Council for Air and Stream Improvement (NCASI) and the Southern Forest Atlas Project. The AIRS database is primarily urban area oriented, and the ozone data are usually analyzed for the growing season, April through October, to assess the impact on growing plants, crops, and trees. A few rural and remote sites, however, are contained in the AIRS database. Pinkerton and Lefohn (1986a) have concluded, from 1983 data for rural and remote areas in the southeastern United States located more than 50 km from urban areas with 50,000 or greater population, that the greatest differences between these areas is that urban areas experienced more elevated hourly average concentrations of 120 ppb O_3 (AAQS) or greater than rural or remote areas. Subsequently, Pinkerton and Lefohn (1986b, 1987) and Lefohn and Pinkerton (1988) reported five-year regional summaries of O_3 monitoring data. Validity of the data used for these extensive analyses was established by requiring site data capture of 50% or more hours of validated monitoring data for the 5136 hours in the April to October season. These results are of interest to the southern pines region because, with the exception of coastal Georgia and southern Florida, there is good spatial coverage of the southern United States (Fig. 3.19), although most of the monitoring sites are located within urban areas.

Historical air quality characteristics (Tables 3.14 and 3.15) for O_3 in the southern region of the United States, including seven-hour (09:00 to 15:59 hr) averages and 24-hour cumulative indexes, have been estimated (Pinkerton and Lefohn, 1987). For the five-year period 1978 to 1983, the seven- hour average for ozone in the Piedmont (Mountain Ridge) region (51 ppb O_3) is about 5 to 10 ppb higher than that for the Southern Coastal Plain region (43 ppb O_3). The Piedmont (Mountain Ridge) region also showed slightly greater values than the Southern Coastal Plain region for the cumulative number indexes n_{70} and n_{80} on a per site-year basis, where the index n_{70} (n_{80}) is the sum of the annual number of hourly occurrences equal to or greater than 70 ppb O_3 (80 ppb O_3). Although there does

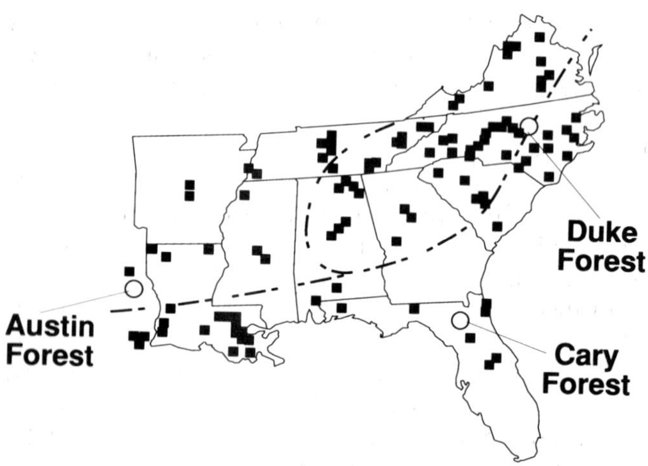

Figure 3.19. Spatial coverage of ozone monitoring stations providing data for analyses of typical ozone exposures in the southeastern U.S. (from Pinkerton and Lefohn, 1987).

Table 3.14. Summary of Ozone Monitoring Data in Southern Commercial Forests Research Cooperative Subregions for the Period 1978 to 1983

Years	Description	Piedmont/ Mountain/ Ridge–Valley	Southern Coastal Plain	Entire South
1978–1983	# Site-yrs	197	153	422
	7 hr	51	43	46
	n_{70}	328	205	267
	n_{80}	162	97	129
	n_{100}	30	22	26
	n_{120}	7	6	6

Adapted from Pinkerton and Lefohn, 1986b. Abbreviations: 7 hr = April–October mean of the 09:00–15:59 hours daily means, ppb; n_{70} = number of hourly occurrences ≥0.07 ppm; n_{80} = number of hourly occurrences ≥0.08 ppm; n_{100} = number of hourly occurrences ≥0.10 ppm; n_{120} = number of hourly occurrences ≥0.12 ppm.

appear to be a physiographic variation in ozone levels from north to south in the southern region, there is no evidence of a temporal trend in ozone levels for the five-year observation period. Statistical reviews of ambient ozone monitoring data are available for the southern pine region through 1989 (Lefohn et al., 1992; Pinkerton and Lefohn, 1986a). Other sources of pre-1985 ambient ozone data trend analyses are the NAPAP Interim Assessment Document (Barchet, 1987) and the Ozone Criteria Document (U.S. EPA, 1986a).

A regional overview of ozone levels can be expressed appropriately in a graphical form. Maps of ozone levels for states east of approximately 100°

Table 3.15. Southern Commercial Forests Research Cooperative Subregional Averages of Ozone Monitoring Data During the Full Growing Season, April to October

	Piedmont/Mountain/ Ridge–Valley		Southern Coastal Plain		Lower Mississippi Valley[a]
	1978–85	1978–83	1978–85	1978–83	1978–85
# Site-yrs[b]	270	197	225	153	67
7 hr	50	51	43	43	46
n_{70}	308	328	196	206	260
n_{80}	146	162	89	97	119
n_{100}	25	30	19	22	21
n_{120}	5	7	5	9	4

Adapted from Pinkerton and Lefohn, 1987.

[a] Data were not analyzed in previous analysis by Pinkerton and Lefohn (1986b) for 1978–1983.

[b] 7 hr = April–October mean of the 09:00–15:59 hours daily means, ppb. n_{70} = number of hourly occurrences ≥0.07 ppm; n_{80} = number of hourly occurrences ≥0.08 ppm; n_{100} = number of hourly occurrences ≥0.10 ppm; n_{120} = number of hourly occurrences ≥0.12 ppm.

west longitude are presented (Fig. 3.20a–e, pp. 141–145) for the years 1985 to 1989. In addition, a map of areas in which ozone levels were persistently relatively high or low for the 1985 to 1989 period is presented (Fig. 3.21, p. 146) as a summary of the relative levels of ozone in the yearly maps. Considerations are presented below for the construction and interpretation of these ozone level maps and the composite map. Further technical details concerning the map construction can be found in Lefohn et al. (1991).

The hourly average ozone data used in the construction of the maps are from ambient ozone monitoring. The hourly average ozone data were obtained from three EPA databases: the Aerometric Information Retrieval System (AIRS); the National Dry Deposition Network (NDDN); and the Mountain Cloud Chemistry Program (MCCP). For each of the yearly ozone level maps, the locations of the ambient ozone monitoring sites contributing hourly averaged ozone data represented on the maps are shown by black dots. The statistic chosen to represent ozone levels on the maps is the seasonal W126 ozone exposure index. The units of the seasonal W126 index are parts per million hours (ppm. h). All of the months April through October comprise the ozone season. A discussion of the use of the W126 index in gaseous air pollutant exposure reasearch is summarized in Lefohn et al. (1990a, 1992). The ozone levels are presented as smoothed contours of interpolated (by simple kriging) values on a $1/2^0 \times 1/2^0$ grid over the region of interest in 10 ppm.h intervals. The color was added to enhance cool (blues and greens) and hot (reds) regions. Visually, the yearly maps show a systematic heating up of ozone level from 1985 to 1988 and then a sudden cooling in ozone for 1989.

The composite map (Fig. 3.21) was referred to as the Student Analysis for describing ozone gradient patterns in Lefohn et al. (1991). For each year, (1) grid cell estimates for the ozone level map were standardized to remove the overall ozone level for a particular year according to (cell estimate–mean)/(standard deviation); (2) the standardized grid cell estimates were averaged over the five years by grid cells; and (3) the averaged standardized grid cell estimates were ranked with the lowest 10% (decile) assigned to category 0, the next highest 10% of the grid cells assigned to category 1, etc. Visually, the color scale runs from cold (blue) to hot (dark red). The dark red area represents a region that was relatively persistently hot in terms of the W126 ozone exposure index for 1985 to 1989. The other colored regions on the map receive similar interpretations.

Assessment of ozone exposures in southern rural forested areas is made difficult by the lack of extensive spatial and temporal rural monitoring in the past. Moreover, the undue influence of data obtained from urban monitoring sites, which may not be located near forested areas, on existing data assessments must be considered, particularly when kriging of limited monitoring data is used to obtain regional distributions. It is known that large annual variations in ozone concentrations occur at sites influenced by urban activities (Lefohn et al., 1989; U.S. EPA, 1989). For example, urban monitoring sites experienced higher levels of ozone in 1983 and 1988 than in intermediate years (Lefohn and Pinkerton, 1988). Thus, extensive temporal and spatial variations in ozone concentrations are observed at monitoring sites within a region. In addition, local meteorological factors, which can be different at urban and rural locations, play a significant role in determining observed ozone concentrations. Monitoring of ozone at higher elevations (>1500 m) has shown that different ozone exposure profiles exist than are observed at lower elevations (<500 m) and near sea level (Aneja and Li, 1990; Aneja et al., 1990, 1991a, 1991b; Lefohn et al., 1989). For example, at higher elevations, diurnal variations in O_3 levels are much less apparent or even nonexistent, and daily average ozone concentrations are much greater than at lower elevations, where daytime high and nighttime low levels of O_3 (large diurnal variations) are commonly observed. In some cases, diurnal variations in ozone do not occur at lower elevations. In these instances, the daily average ozone levels are low when strong winds and cloudy skies persist and the site is remote from anthropogenic sources of O_3 precursors.Because of the potential for vegetation and crop damage when exposed to higher concentrations of ozone, caution is necessary when attempts are made to average existing ozone values across geographic regions and over extended time periods (years). For example, averaging may disguise the identity of those sites with high ozone exposures as well as overestimate rural exposure because of the predominance of data from urban monitoring sites.

In the case of ozone exposure effects on vegetation, considerable attention has been given to the derivation of an optimum exposure statistic that is biologically relevant to the observed effects. For example, a variably weighted cumulative index (W126) has been developed, that, although biased to higher concentration exposures, does not ignore lower, less biologically significant exposures. Although various simple approaches for weighting air pollution concentrations have been proposed, Lefohn and Runeckles (1987) have suggested a more complex exposure index (W126) that applies a sigmoidal weighting function to individual hourly averaged ozone concentrations, in which the products of weighting function and hourly O_3 concentration are summed over time. In this case, lower, less biologically important concentrations, between 30 and 100 ppb O_3, are increasingly weighted for inclusion in the aggregate exposure indexes. Levels of O_3 less than 30 ppb are given a small but finite weighting in this index, but maximum weight is given to concentrations exceeding 100 ppb O_3. Currently, it is believed that a more realistic ozone index for assessing plant damage should involve a combination of factors: cumulative exposure and number of hourly average concentrations above a threshold level (Musselman et al., 1994).

Many ozone exposure indexes have been used to compare and characterize hourly average ozone concentrations over daily, monthly, seasonal, and annual periods. These indexes include the following:

1. The sum of all hourly mean ozone concentrations in units of ppm.h for a specific period. This index, which assumes no response threshold and gives equal weight to all hourly concentrations, is referred to as total dose (SUM0).
2. The sum of all hourly averaged ozone concentrations in units of ppm.h, where each hourly concentration is weighted by a sigmoidal weighting function (W126) as described previously (Lefohn et al., 1988).
3. The sum of all hourly averaged ozone concentrations equal to or greater than 70 ppb (SUM07) in units of ppm · h, which assumes a threshold level of 70 ppb O_3.
4. The total number of occurrences of hourly averaged ozone concentrations equal to or greater than 60 ppb (n_{60}) or 80 ppb (n_{80}), which assumes threshold levels of 60 or 80 ppb ozone, respectively.
5. The seasonal (quarterly) means of the average of the daily seven-hour (09:00 to 15:59 local standard time, LST) ozone concentrations in ppb, which assumes that only exposure during optimum daylight hours and during the growing season is important.
6. The seasonal (quarterly) means of the average of the daily 12-hour (07:00 to 18:59 LST) ozone concentrations in ppb, which assumes that only exposure in daylight during the growing season is important.
7. The one-hour daily maximum O_3 concentration, in units of ppb, which assumes that exposure to low concentrations of O_3 is less important.

Aggregate ozone exposure values at rural forest sites are of particular interest to forest effects researchers. These values have been reported as annual and growing season (April through October) 24-hour (total), 12-hour (daytime: 07:00 to 18:59 hours, LST), and seven-hour (optimum daylight: 09:00 to 15:59 hours, LST) aggregate values. The latter two parameters are considered to be more important because the mode of attack of ozone on vegetation and trees is believed to be through stomatal conductance into the leaves or needles. Stomatal conductance is activated by photosynthetically active solar radiation as well as vapor pressure deficit and soil moisture content variables. Plant scientists prefer to use a seven-hour daylight mean, whereas atmospheric scientists prefer 24-hour and 12-hour mean statistics of ozone data (Taylor et al., 1992). There is a strong correlation between the one-hour maximum and longer term mean ozone statistics in urban areas, which justifies the use of ozone maximums as valid surrogates in addressing the effects of ozone on human health (Taylor et al., 1992). The disparity in ozone patterns between urban and rural landscapes (Altschuller, 1988; Guderian, 1985), however, suggests that the one-hour maximum is not a valid surrogate for characterizing ozone concentrations in more remote, rural landscapes. Taylor et al. (1992) have confirmed, through statistical analyses of ozone data from low-elevation IFS sites, that only weak correlations exist between the one-hour maximum and other indexes of ozone exposure. Strong correlations were found at IFS sites between cumulative exposure indexes (SUM06 and SUM08) and longer term averages (24-hour, 12-hour, and 7-hour). Seasonal ozone exposures at SCFRC sites have been characterized in terms of several of the aforementioned indexes for purposes of comparison.

Continuous measurements of ozone were made at the North Carolina site from October 1987 to June 1990; the Florida site from July 1988 to June 1991 and the Texas site from July 1988 to June 1990. Daily averages and daily maximum ozone concentrations for the North Carolina SCFRC site in 1988 and 1989, and for Florida and Texas SCFRC sites in 1989, are presented graphically (Fig. 3.22). The data exhibited the classical seasonal cycle for ozone observations at low elevation sites (<500 m). The highest daily maximum and average concentrations occurred in the second quarter (April to June, spring season), and minimum concentrations were observed in the fall and winter months (November to January). Similar seasonal variations have been reported for widespread regions of the eastern and southern United States (Altschuller, 1988; Logan, 1989; Evans et al., 1983; Lefohn and Pinkerton, 1988; Lefohn et al., 1990a; Meagher et al., 1987) The SCFRC sites, however, are located in the southeastern United States at lower latitudes. Not all of the sites described in previous studies are located in the southeastern United States, most are situated in midwestern, mideastern, and central eastern locations. Presumably, as a result of their more southern location, the highest daily ozone levels at SCFRC sites occurred earlier in the year and the lowest daily ozone levels occurred later in the year when compared to central and northern U.S. sites described in these earlier reports. The lowest ozone levels observed at SCFRC sites were of shorter seasonal duration than those measured at higher latitude sites. Detailed examination of the daily averages and maxima at the SCFRC sites revealed monthly and weekly variations, which may have been due to cycles in meteorological variables (monthly) and to anthropogenic (weekly) cycles in urban activity influencing the transport and generation, respectively, of ozone and its precursors near these rural locations.

Ozone measurements at the two southernmost SCFRC sites (Florida and Texas) are similar (differences < 10%) and may represent typical exposures in southern rural forests. They were not substantially influenced by local urban, industrial, or commercial sources. This lack of local influences may not be the case for the North Carolina site, where air quality and deposition measurements appeared to be impacted by local (suburban) and regional (industrial) sources and showed more of an impacted rural nature rather than a remote rural character. It is likely that diurnal ozone concentration profiles in some remote rural areas could be different from those observed in urban centers of population, although it has been suggested that higher ozone levels can occur at

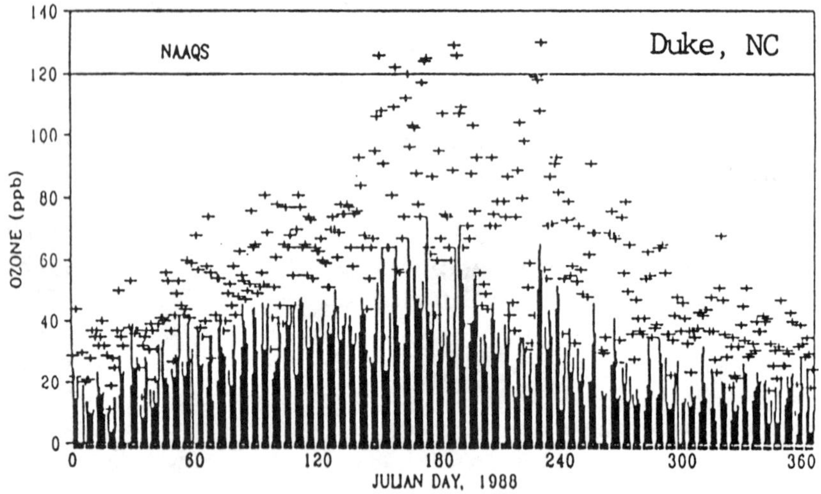

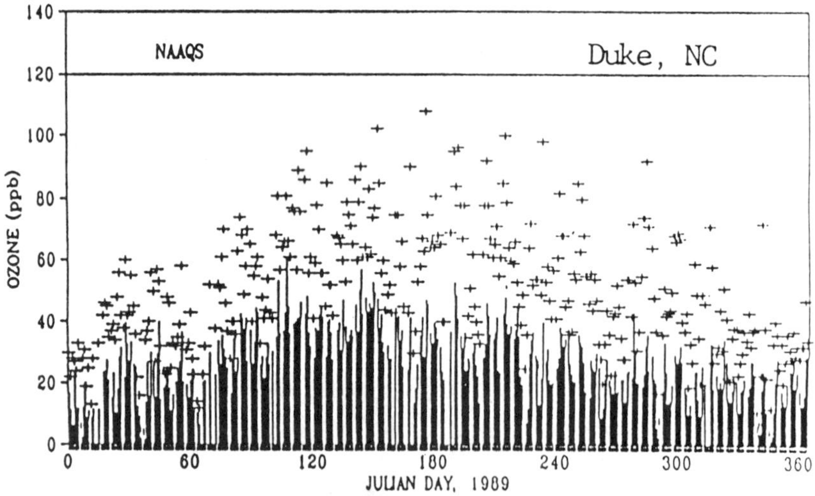

Figure 3.22. Daily averages (bars) and maximums (crosses) for ozone concentrations (ppb) at North Carolina (1988, 1989), Florida (1989), and Texas (1989), SCFRC monitoring sites.

downwind rural sites than in urban areas (Lefohn et al., 1990a). For comparative purposes, seasonally averaged diurnal patterns for ozone levels at each of the three southern sites are presented (Fig. 3.23). Large diurnal swings were observed for all three SCFRC sites and for all seasons from a nighttime low to a daytime high. However, unlike urban

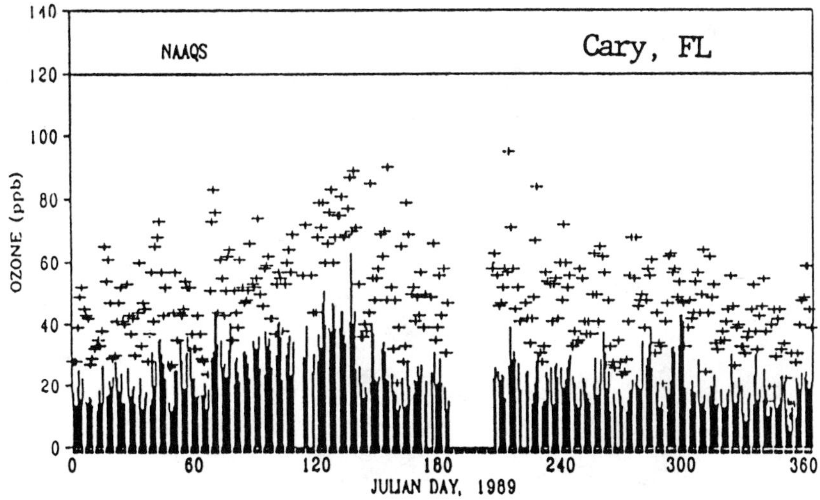

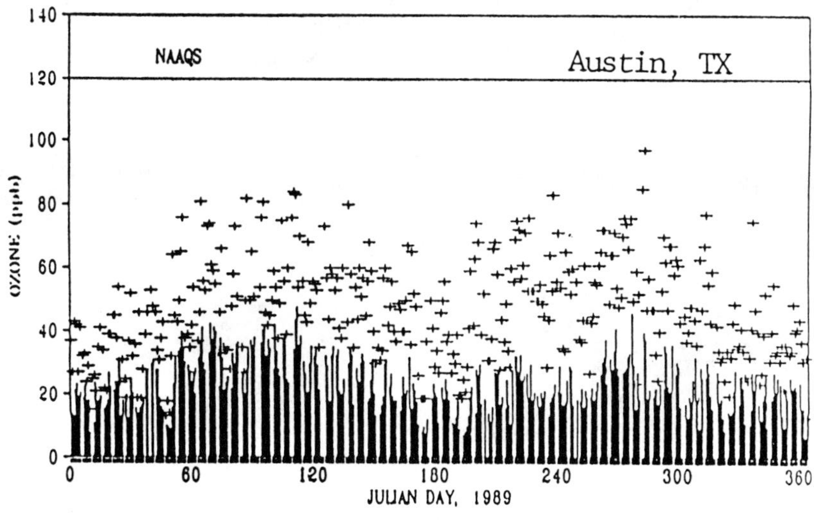

Figure 3.22. *Continued*

diurnal O_3 concentration profiles, where O_3 gradually builds up in late morning (10:00 to 12:00 h) and peaks in the early afternoon (14:00 to 15:00 h), the leading edge of the rural O_3 accumulation profile at all three SCFRC sites appeared to rise more rapidly after sunrise and reached a maximum value at about 12:00 to 13:00 hours (LST). That is, the maximum O_3 level was achieved more rapidly and earlier in the day than that found in typical urban atmospheres. In addition, a major fraction (~75%) of

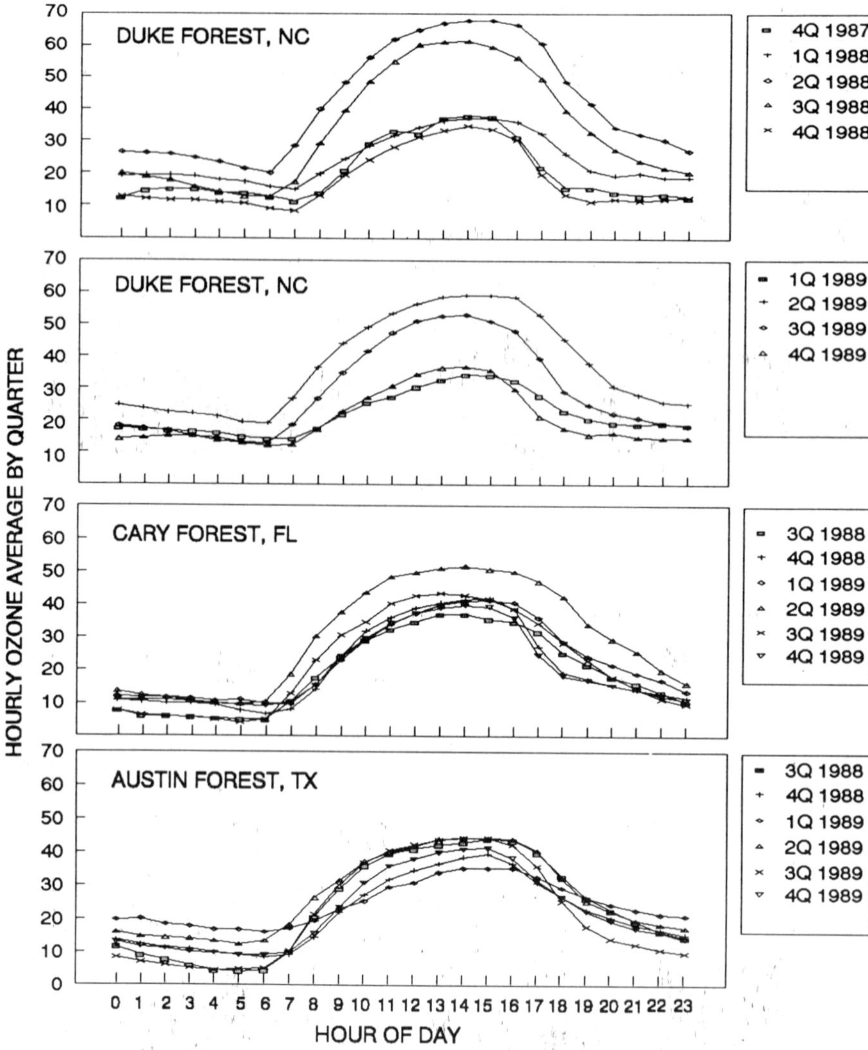

Figure 3.23. Quarterly (seasonally) averaged hourly ozone concentrations (ppb) at Duke Forest, NC, Cary Forest, FL, and Austin Forest, TX, SCFRC monitoring sites from 1987 to 1989.

the maximum ozone generated daily was routinely achieved by 10:00 hours (LST), when only about 20 to 25% of the total daily solar radiation had been received at the ground.

Previous observations of vertical ozone profiles above and within the Florida IFS site pine canopy (Cavender and Allen, 1991) suggest that the early morning ozone increase could be a result of convective turbulence after sunrise causing downward vertical mixing (fumigation) and advection

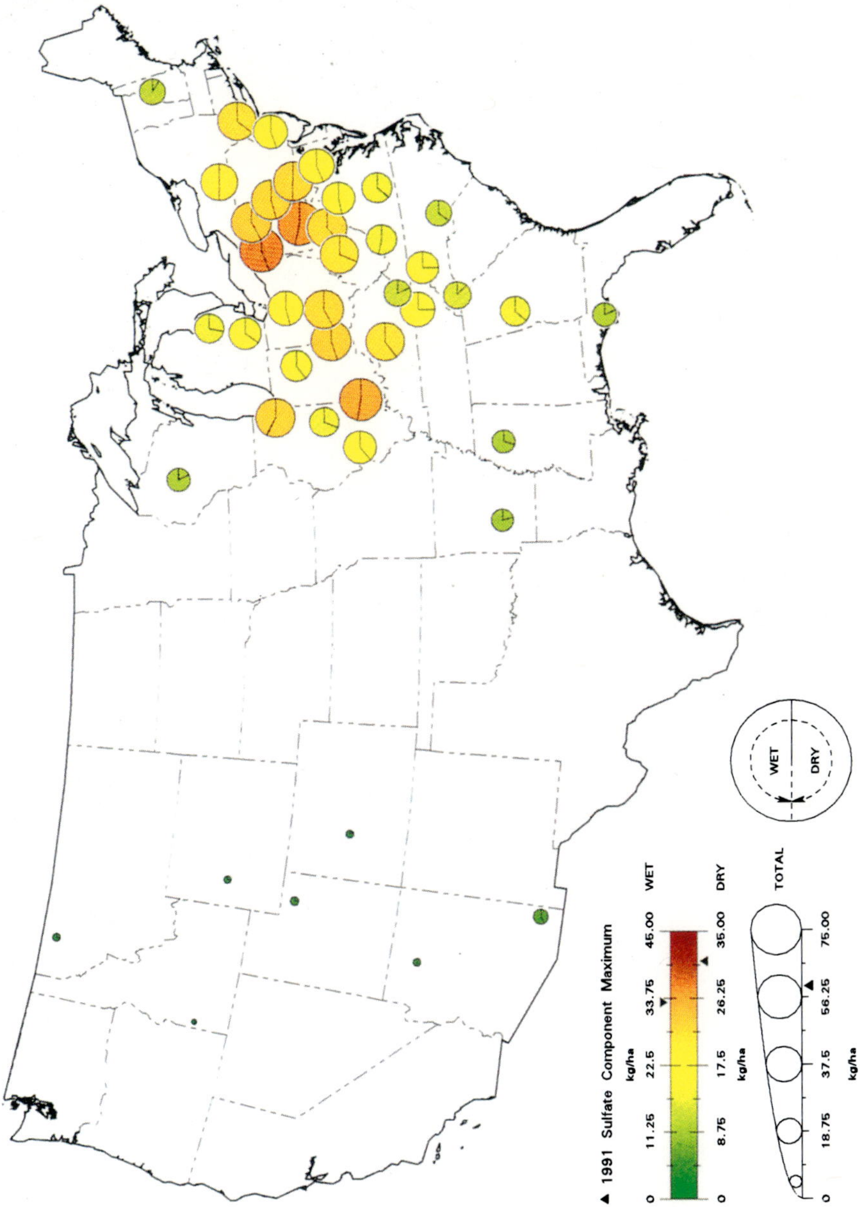

Figure 3.16. Wet, dry, and total sulfate deposition (kg/ha-yr) at NDDN sites in 1991.

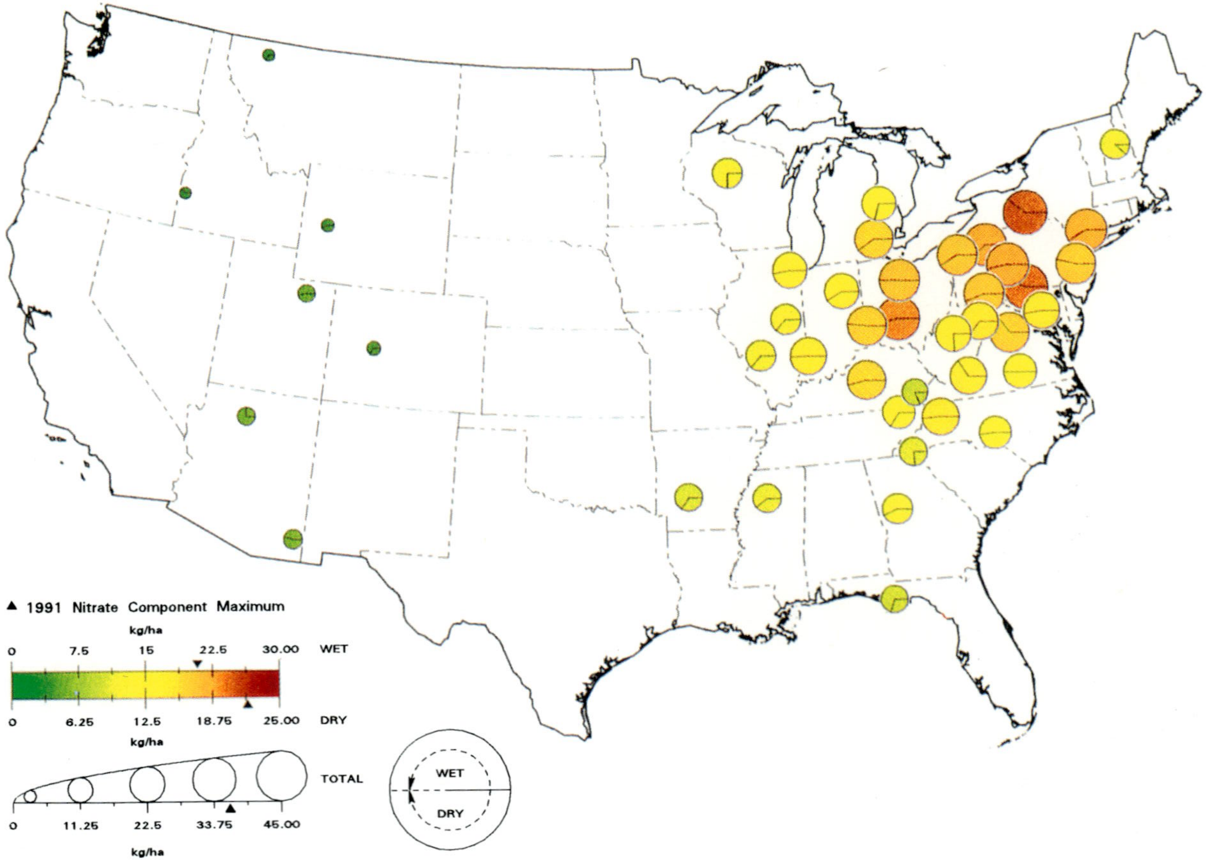

Figure 3.17. Wet, dry, and total nitrate deposition (kg/ha-yr) at NDDN sites in 1991.

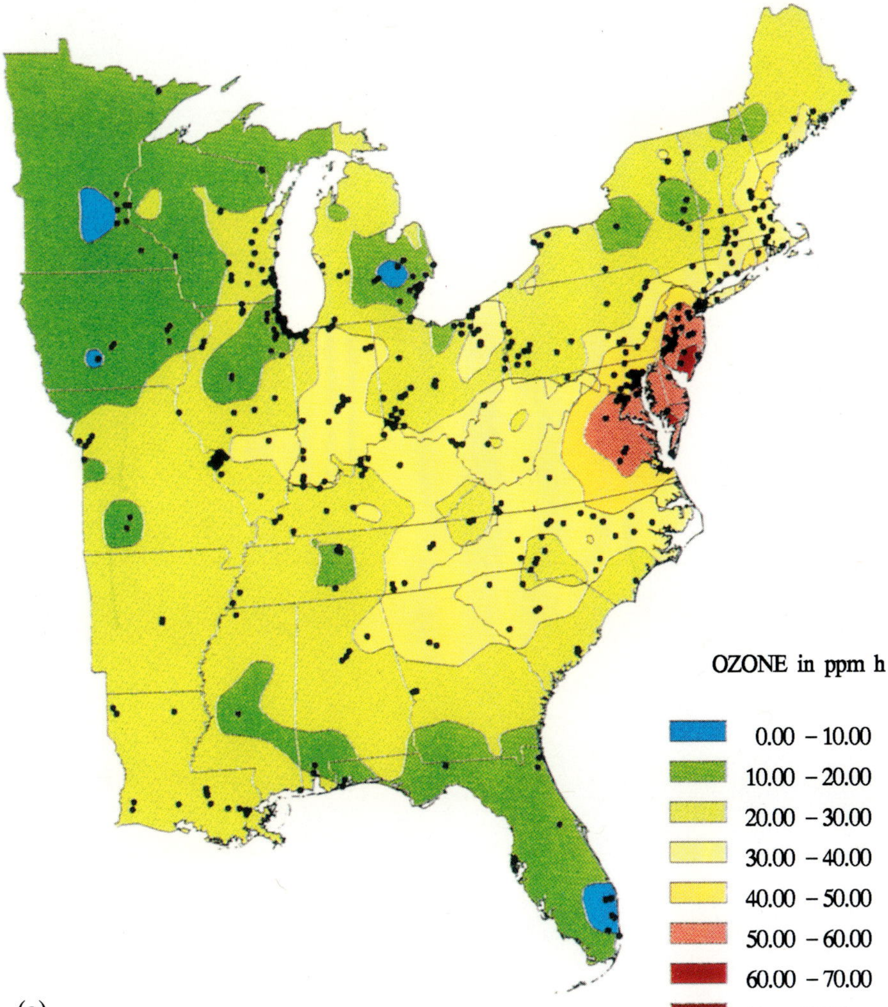

(a)

Figure 3.20. Maps of seasonal W126 ozone exposure index (ppm.hr) for (a) 1985, (b) 1986, (c) 1987, (d) 1988, and (e) 1989. (Produced by Environmental Monitoring and Assessment Program—Forests, U.S. EPA, Research Triangle Park, NC.)

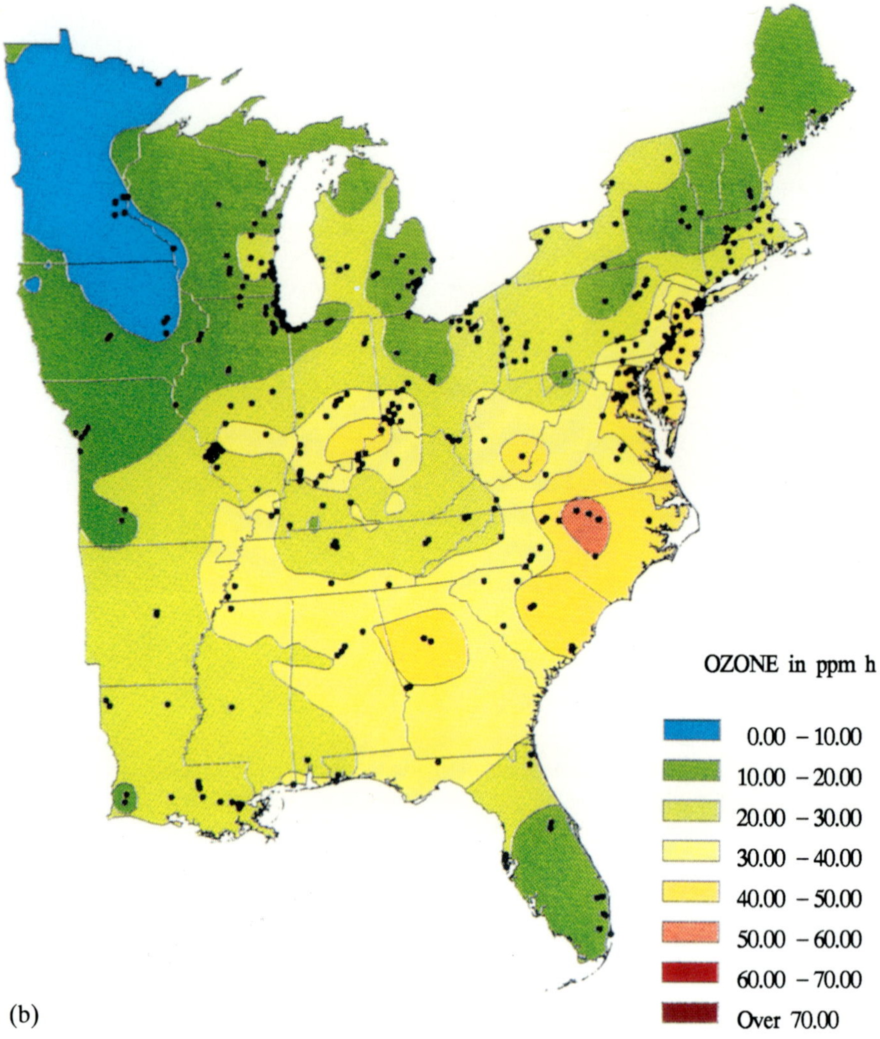
OZONE in ppm h
0.00 – 10.00
10.00 – 20.00
20.00 – 30.00
30.00 – 40.00
40.00 – 50.00
50.00 – 60.00
60.00 – 70.00
Over 70.00

(b)

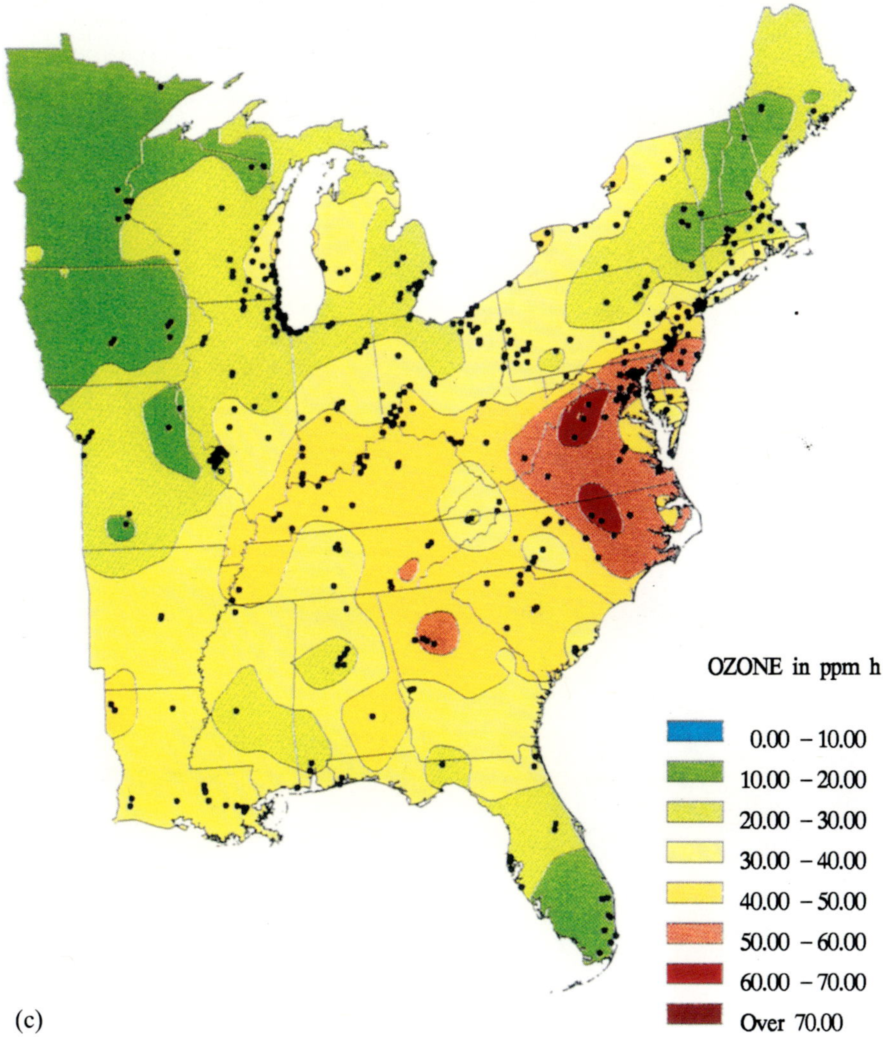
OZONE in ppm h
0.00 – 10.00
10.00 – 20.00
20.00 – 30.00
30.00 – 40.00
40.00 – 50.00
50.00 – 60.00
60.00 – 70.00
Over 70.00

(c)

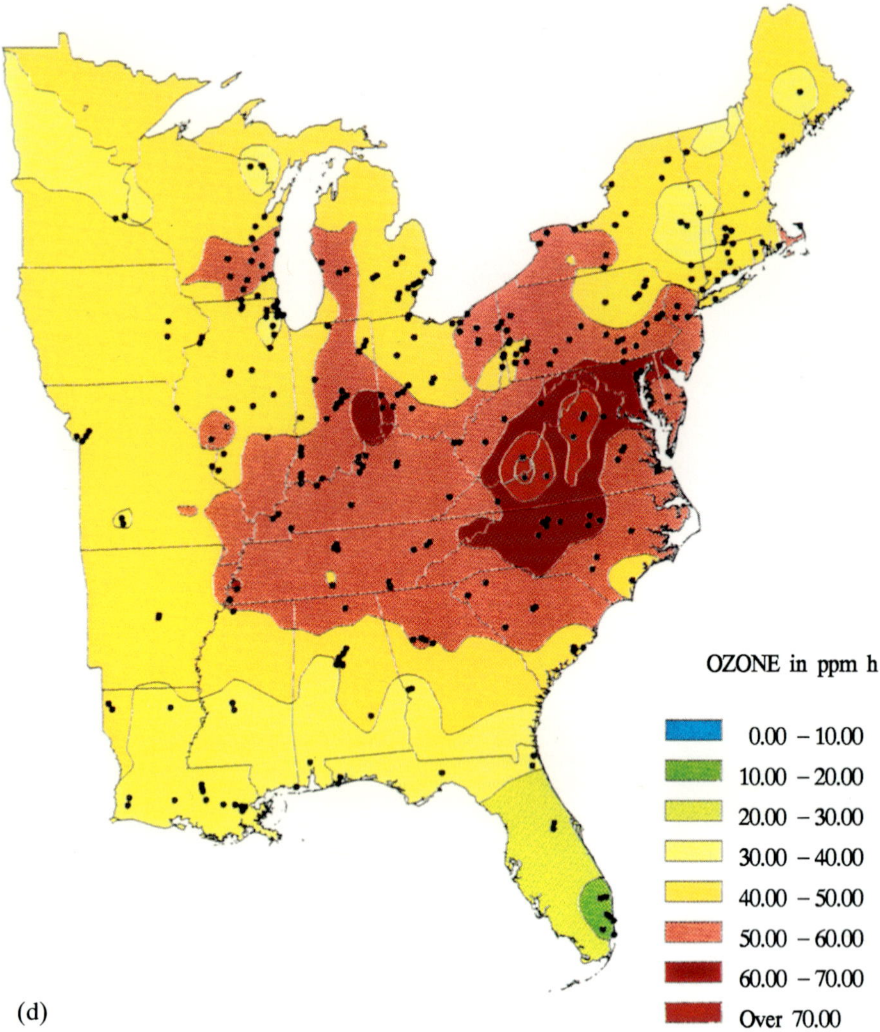
OZONE in ppm h
0.00 – 10.00
10.00 – 20.00
20.00 – 30.00
30.00 – 40.00
40.00 – 50.00
50.00 – 60.00
60.00 – 70.00
Over 70.00

(d)

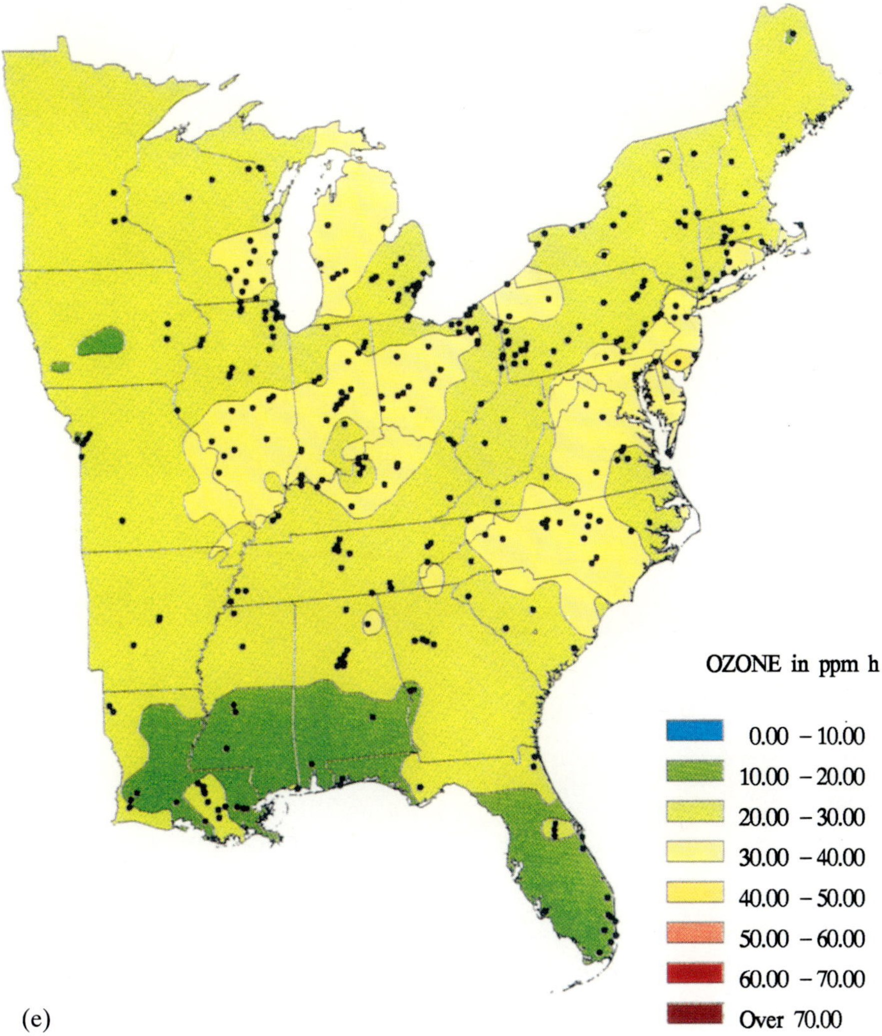
OZONE in ppm h
0.00 – 10.00
10.00 – 20.00
20.00 – 30.00
30.00 – 40.00
40.00 – 50.00
50.00 – 60.00
60.00 – 70.00
Over 70.00

(e)

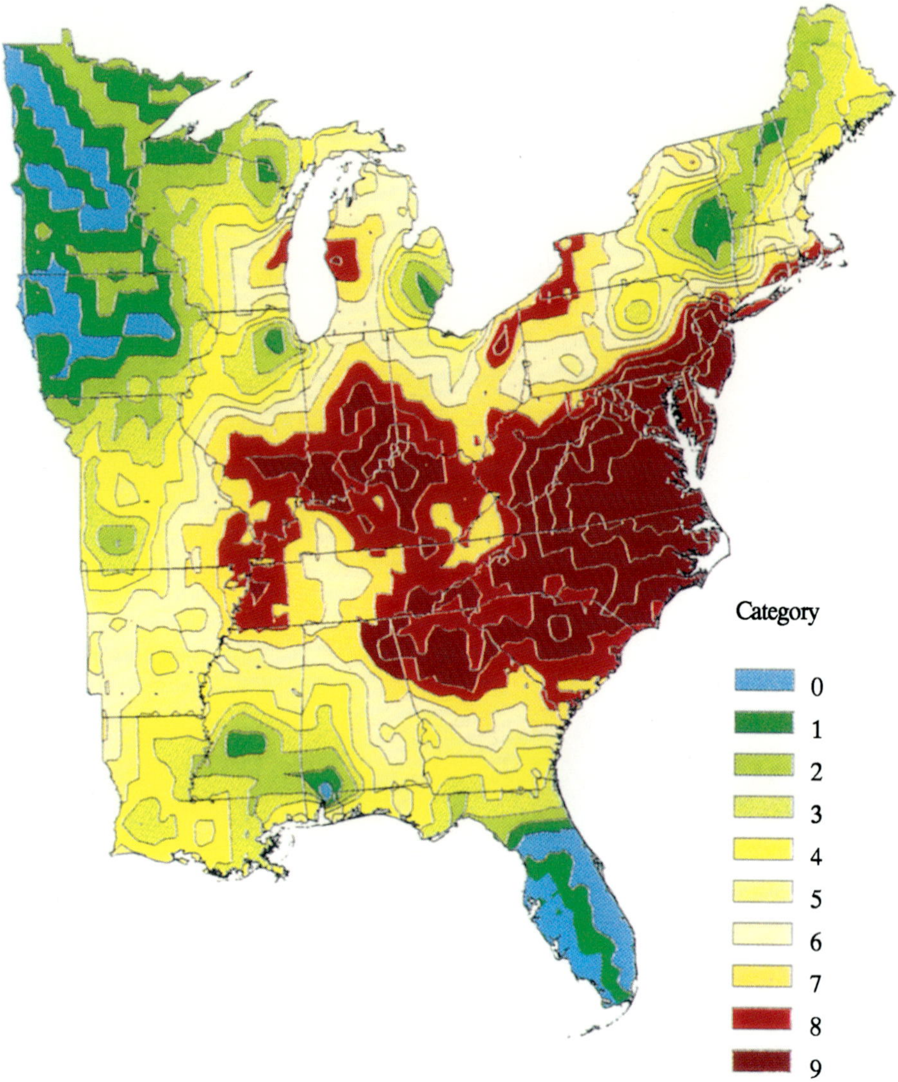

Figure 3.21. Composite map of regions that have persistent relatively low ozone exposure (Category 0) through persistent relatively high ozone exposure (Category 9) for 1988–89. (Produced by Environmental Monitoring and Assessment Program—Forests, U.S. EPA, Research Triangle Park, NC.)

of ozone isolated above the inversion layer, following break up of an overnight surface inversion layer. After 09:00 hours (LST), Cavender and Allen (1991) found that O_3 was well mixed and independent of elevation to a height of at least 27 meters; but separation of O_3 concentration profiles at different elevations above ground occurred after 16:00 hours due to the onset of a weak surface inversion layer (Fig. 3.24). This interpretation of ozone accumulation near the surface of a rural site after sunrise suggests that less than 50% of the ozone produced daily at the Florida rural site is a result of local photochemistry. This early morning accumulation of nonphotochemical O_3 near the surface is possible for other rural sites but requires experimental confirmation. The largest diurnal variations in rural ambient ozone levels are associated with clear sunny days and clear nights with strong inversions, which are indicated by little or no wind (<0.5 m/s). These conditions produce strong convective mixing in the daytime and induce the formation of a strong stagnant surface boundary layer at the ground at night. Under these conditions, ozone is rapidly mixed by convection, advected into, and photochemically generated at the site during the daytime, but it is lost at night by diffusion to the ground and by chemical reaction within the stable surface inversion layer.

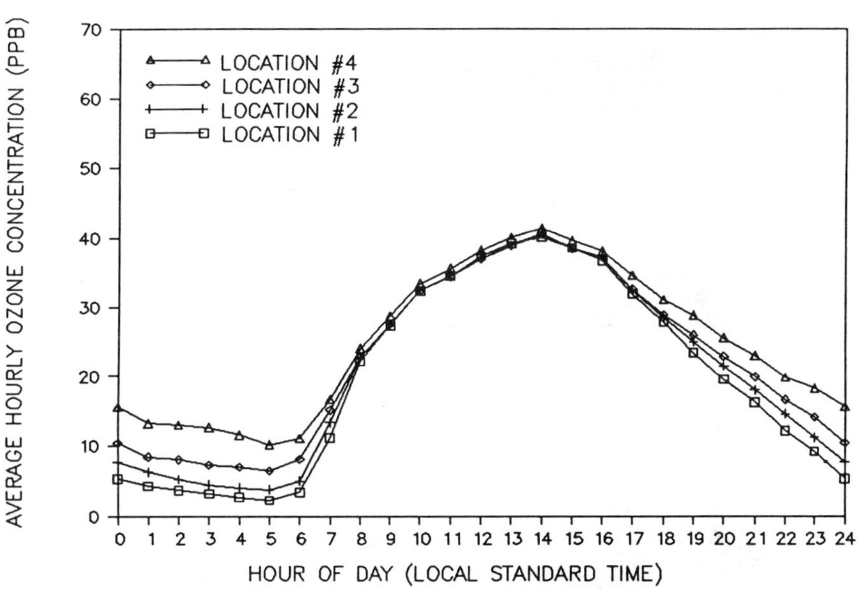

Figure 3.24. Simultaneously measured vertical profiles of monthly averaged ozone concentrations in a Florida IFS slash pine forest at (1) 3 m above ground, (2) 10 m above ground, bottom of canopy, (3) 17 m above ground, top of canopy, and (4) 27 m above ground, 10 m above canopy, for July, 1988.

On occasion, there was little or no perceptible diurnal variation in ozone concentrations at the SCFRC sites. These conditions occurred when strong winds (>5 m/s) either persisted or appeared at night, preventing formation or disrupting the inversion layer by mechanical turbulence, respectively. Such conditions enabled ozone to persist or become well mixed vertically at night as well as promoted advection into the area. Similarly, persistent diurnal O_3 concentrations were observed during thunderstorms and precipitation events.

Seasonal variations in hourly ozone averages for the three SCFRC sites are presented (Fig. 3.23) and indicate that the highest daily maximum

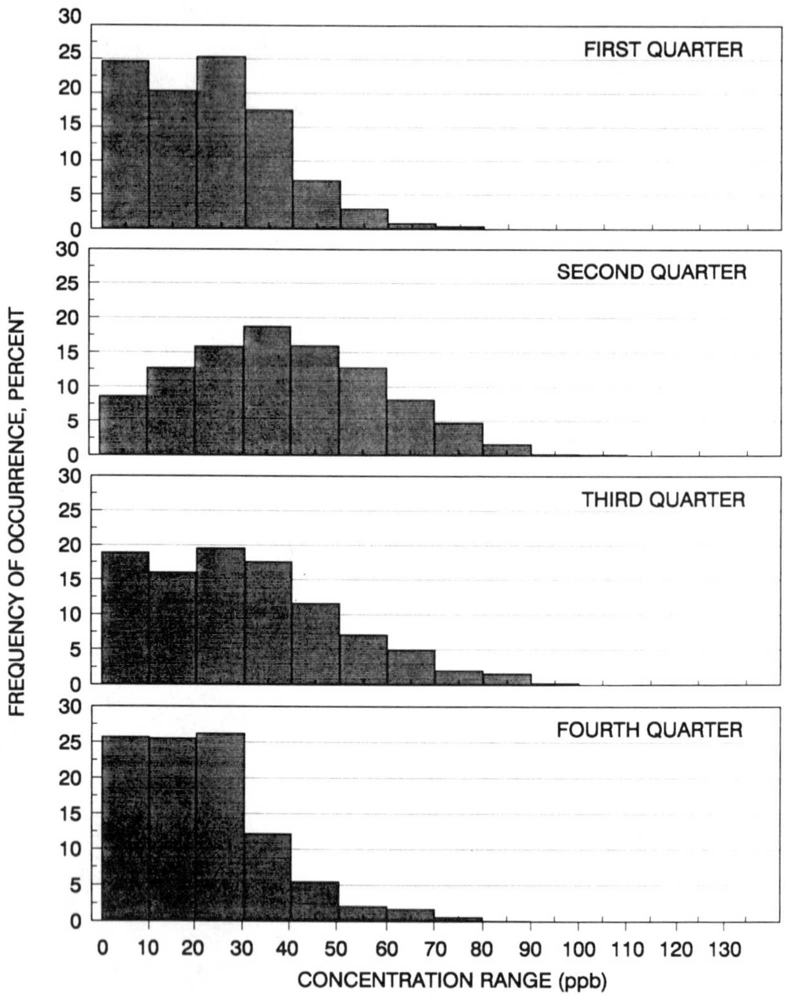

Figure 3.25. Quarterly (seasonal) distributions of ozone concentration range occurrences at Duke Forest, NC in 1989.

values (55 ppb O_3) occur during the second quarter and the lowest daily maximum values during the first and fourth quarters at the more northerly SCFRC site (North Carolina). Seasonally averaged O_3 levels at the Florida and Texas SCFRC sites were similar; even though these sites were approximately at the same latitude (30°N), they were 1300 km apart. Moreover, the largest seasonal differences in averaged hourly O_3 values were observed at the North Carolina site. Both Florida and Texas SCFRC sites showed only small differences in seasonal diurnal averages between winter and spring seasons and much lower seasonal average ozone values than those observed at the northern site in the second and third quarters, despite higher average seasonal temperatures and similar solar radiation intensities.

Typical seasonal variations in the frequency of occurrence of ozone concentration distribution for the North Carolina site in 1989 are presented (Fig. 3.25). The data show that, at least for 65% of the time, O_3 concentrations were less than 30 ppb in the first quarter, 50 ppb in the second quarter, 40 ppb in the third quarter, and 20 ppb in the fourth quarter. The unusually high frequency of occurrence of O_3 concentrations in the range 0 to 10 ppb at the North Carolina site in the summer and fall seasons may be a result of strong local emissions of NO from soil, which can rapidly titrate out ambient ozone, particularly that existing in the nighttime stable inversion layer (Allen and Evans, 1992).

Statistical analyses of quarterly and annual averaged and aggregated ozone data over 24-, 12-, and 7-hour periods for the three sites are summarized (Table 3.16). If these results are compared with previously reported (Lefohn et al., 1990a) rural forest ozone data for the region, it appears that long-term O_3 exposure levels may have declined slightly during an eight-year period ending in 1989. For example, data obtained for the low ozone years 1982 and 1989, show that the 7-hr/12-hour average values of 54/46, 45/40, and 55/48 for North Carolina, Florida, and Texas SCFRC locations in 1982, were substantially higher than the corresponding values of 41/36, 39/35, and 37/33 ppb obtained at the same locations, respectively, in 1989. The data presented (Table 3.16) also suggest that actual specific and integrated ozone exposure levels observed in southern forested landscapes are much lower than corresponding exposures at elevated locations reported in MCCP and EPRI/IFS studies.

Aggregate values in excess of a specific ozone threshold (60/80 ppb O_3) were estimated (Table 3.17) for those researchers who prefer to use statistical analyses of exposure indexes that have a definite threshold. The data indicate that ozone exposure may be cause for concern at the North Carolina site, but the low aggregate values and number of occurrences of O_3 at these levels (n_{60}, n_{80}) for the two southern sites suggest that lower O_3 exposures may be distinctive of rural areas in the southern part of the SCFRC region.

It is important to note that high levels of nitric oxide (NO) were periodically observed at the North Carolina and Florida SCFRC sites at

Table 3.16. Summary of Ozone Exposure Statistics at Southern Commercial Forests Research Cooperative Monitoring Sites, 1987 to 1989

Site	Quarter/Year	24-Hour		12-Hour		7-Hour	
		Average (ppb)	Aggregate (ppm-hr)	Average (ppb)	Aggregate (ppm-hr)	Average (ppb)	Aggregate (ppm-hr)
Duke Forest, NC	4/87	21	45.6	27	29.7	33	21.5
	1/88	24	52.4	30	32.8	33	20.8
	2/88	42	92.5	57	61.9	62	39.5
	3/88	34	75.2	49	53.9	55	35.7
	4/88	18	38.9	24	26.1	29	18.7
	Annual 1988	30	259	40	175	45	115
	1/89	22	48.1	26	28.9	29	18.7
	2/89	37	81.6	50	54.2	54	34.6
	3/89	29	64.9	41	45.4	47	30.3
	4/89	20	44.9	27	29.5	32	20.6
	Annual 1989	27	240	36	158	41	104

Cary Forest, FL	3/88	20	43.5	30	32.8	33	21.3
	4/88	21	45.6	30	33.1	36	23.4
	1/89	23	49.7	31	33.8	35	22.2
	2/89	30	65.6	43	47.4	48	30.4
	3/89	22	49.3	35	38.3	39	25.4
	4/89	20	45.0	29	32.4	35	22.3
	Annual 1989	24	210	35	152	39	100
Austin Forest, TX	3/88	24	52.3	35	38.7	30	25.4
	4/88	21	47.1	29	32.0	33	21.0
	1/89	24	52.7	29	31.7	31	19.6
	2/89	27	58.2	37	40.4	40	25.7
	3/89	22	47.8	35	38.2	40	26.0
	4/89	23	50.1	31	34.3	36	23.2
	Annual 1989	24	209	33	145	37	95

12-Hour values from 07:00 to 18:59 hours (LST); 7-hour values from 09:00 to 15:59 hours (LST); Aggegates: sum of hourly averaged concentrations in exposed period; Quarter: meteorological season—(1) January, February, March, (2) April, May. June, (3) July, August, September, (4) October, November, December.

Table 3.17. Summary of Ozone Exposure Indexes at Southern Commercial Forests Research Cooperative Monitoring Sites, 1987 to 1989

Site	Quarter/Year	SUM 06 (ppm-hr)	n_{60} No. Events	SUM 08 (ppm-hr)	n_{80} No. Events
Duke Forest, NC	4/87	3.2	46	0.3	3
	1/88	2.2	34	0.0	0
	2/88	36.9	486	13.3	136
	3/88	27.8	361	11.0	116
	4/88	1.3	21	0.0	0
	Annual 1988	68.2	902	24.3	252
	1/89	2.0	30	0.0	0
	2/89	22.6	321	3.9	45
	3/89	14.1	98	3.7	43
	4/89	3.8	56	0.3	4
	Annual 1989	42.5	505	7.9	92
Cary Forest, FL	3/88	1.8	28	0.1	1
	4/88	0.9	15	0.0	0
	1/89	3.4	51	0.2	3
	2/89	16.1	233	1.6	19
	3/89	2.9	43	0.4	4
	4/89	1.2	19	0.0	0
	Annual 1989	23.6	346	2.2	26
Austin Forest, TX	3/88	8.4	125	0.3	4
	4/88	2.3	35	0.0	0
	1/89	2.1	31	0.0	0
	2/89	5.5	80	0.7	8
	3/89	6.7	101	0.1	8
	4/89	5.4	77	0.8	9
	Annual 1989	19.7	289	1.6	25

SUM 06 = aggregate dose for ozone concentration equal to or greater than 60 ppb; SUM 08 = aggregate dose for ozone concentration equal to or greater than 80 ppb; No. events = Number of hourly concentration values (n_{60}/n_{80}) equal to or exceeding threshold values (60/80 ppb) in time period; Quarter = meteorological season—(1) January, February, March, (2) April, May, June, (3) July, August, September, (4) October, November, December.

night, when O_3 concentrations were depleted below 2 ppb. By a process of elimination it was found that significant emissions of NO occurred at the local ground surfaces. Confirmation of this conclusion was obtained by soil gas flux measurements made at the Florida SCFRC site, where a soil gas emission rate of 4 kg No $ha^{-1}\,yr^{-1}$ was determined during a three-month summer study in 1989 (Kardono, 1990). From the preceding discussion it is evident that when O_3 is monitored at 4 m above ground level at a rural site, the daytime levels after 09:00 hours will reflect O_3 concentrations several tens to hundreds of meters above the ground; whereas at night after 17:00 hours, the O_3 concentrations observed are a strong function of the height of the monitoring inlet above the ground when surface inversions occur. Moreover, emission of NO from soils in the vicinity of the monitoring inlet can cause substantial and rapid reduction of local O_3 levels within the surface inversion layer at night. It is well documented that for most trees (all of the southeastern species of importance), stomata are open in daylight and closed at night. Because the stomatal conductance of needles of southern pines is much larger during daytime hours the potential exposure to high nighttime O_3 levels or depletion of ozone concentrations at the canopy level of mature pine trees in rural areas may not be significant. The persistent existence of high O_3 concentrations at elevations above the nighttime surface inversion layer, which may extend to several hundred meters above the surface, however, does appear to provide a substantial early morning source of ozone near ground level during early morning fumigation.

Ozone Episodes

Exceedances of the ozone NAAQS (120 ppb O_3) were observed at the North Carolina SCFRC site on seven days during the second and third quarters of 1988. These episodic events coincided with even greater exceedances observed at Mount Mitchell, North Carolina (Aneja and Li, 1990). The latter was an MCCP site situated approximately 260 km west of the Duke Forest SCFRC site, but at much higher elevation (1950 meters above mean sea level). The daily ozone maximums at the Duke Forest SCFRC site exceeded 100 ppb on 17 days during the episodic period. These episodic events are attributed to air stagnation associated with the occurrence of stationary high-pressure systems in the region (Aneja and Li, 1990). During these events, ambient temperatures were generally much higher than normal. July and August 1988 produced temperatures near the highest on record, and relative humidities and wind speeds were lower than normal. Wind direction at the North Carolina SCFRC site was generally from the northwest during these events. These particular observations of high ozone levels associated with specific meteorological conditions suggested that excessive ozone contributions may have occurred at the North Carolina site by horizontal and vertical transport and trapping of ozone and its precursors (Abelson, 1988; Kelly et al., 1984; Logan, 1989; Vukovitch et al., 1977; Wolff and Liaz, 1980; Wolff et al., 1977; 1982), from the industrialized midwestern region of the United States, as well as possible contributions from stratospheric ozone intrusion. There was no evidence of unusual ozone occurrences at the two southern sites in the SCFRC network during the second and third quarters of 1988 (Tables 3.16 and 3.17).

Ozone episodes in 1988 that were observed in the eastern United States, have been reported extensively elsewhere (Aneja and Li, 1990;

Johnson and Lindberg, 1992; Mohnen, 1992) and have caused atmospheric scientists to label 1988 as a high ozone and severe to extreme drought year. Such observations, particularly during the second and third quarters of 1988, suggested that ozone, and possibly wet and dry deposition values for the North Carolina site, may not have been representative of normal years. A comparison of annual average and aggregate ozone exposure levels for 1988 and 1989 (Table 3.16) show that exposure indexes were significantly affected by the few undesirable excursions in ozone concentrations experienced at the North Carolina site in 1988. The wet and dry deposition component concentration data for North Carolina in 1988 and 1989 (Tables 3.6 and 3.8) suggest that these variables were not substantially affected by the unusually severe drought and high ozone levels observed in 1988. The North Carolina SCFRC site experienced cooler weather and much higher precipitation levels in 1989, when daily maximum ozone levels did not exceed 100 ppb.

Co-Occurrence of Multiple Gaseous Pollutants (O_3, SO_2, NO_2)

The potential for stress on trees due to multiple pollutant exposure can be determined by inspecting the sequential or simultaneous co-occurrence of gaseous pollutants. This type of investigation has been conducted for ambient monitoring data collected during the months of May to September by Lefohn et al. (1987a). The principal conclusions of their study for individual gaseous hourly average concentrations at 30 ppb or greater are as follows:

1. Simultaneous co-occurrences of any pair of gases and sequential co-occurrences of sulfur dioxide and ozone, or sulfur dioxide and nitrogen dioxide, are infrequent.
2. Sequential co-occurrences of nitrogen dioxide and ozone or complex sequential co-occurrences of nitrogen dioxide and ozone, or sulfur dioxide and ozone, occur fairly frequently.

Similar results were obtained when individual gaseous concentrations were 50 ppb or greater (Lefohn and Tingey, 1984). In the current study of three selected forested landscapes in the South, the possibility of gaseous pollutants co-occurring simultaneously or sequentially at such levels was extremely low. Although O_3 levels consistently exceeded 30 ppb during daytime hours, NO_x and SO_2 concentrations exceeded 10 ppb less than 10% and 1% of the time, respectively.

Co-Occurrence of Ozone and Acid Precipitation

Historically, prior to 1987, only a few sites have been identified in which wet deposition and ozone have been comonitored in the SCFRC region. These sites are Finley Farm (near Raleigh) in the North Carolina Piedmont and Longview, Texas in the Western Gulf Coastal Plain. Measurements were carried out for different monitoring programs at these sites, and the databases have not been combined. It is worth noting, however, that both ozone and wetfall concentrations of hydrogen, sulfate, and nitrate ions tended to be higher at these sites in the summer season. More recently, comonitored low-elevation ozone concentrations and wet/dry deposition data have been made available from the EPRI/IFS study (Johnson and Lindberg, 1992) and from the EPA/NDDN network (Edgerton et al., 1991).

Observations at the three SCFRC sites provided information on the simultaneous occurrence of ozone and acid deposition. In general, higher levels of acid deposition (wet and dry) and ozone concentrations were observed in the second (spring) and third (summer) quarters at the three sites.

Sulfur Dioxide and Nitrogen Oxides

It is important to realize that, although annual mean values of hourly averaged O_3 exposures were about 30 ppb at the North Carolina site and about 25 ppb at the Florida, and Texas sites, exposures to nitrogen oxides (NO_x) and sulfur dioxide (SO_2), were neither excessive nor extensive. At all sites, nitric oxide (NO), nitrogen dioxide (NO_2), and nitrogen oxides (NO_x) hourly averaged concentrations exceeding 10 ppb occurred less than 10% of the time. For SO_2, more than 95% of the time continuous monitoring data indicated concentrations equal to or less than 2 ppb, the monitoring instrument detection limit. The latter observations were supported by TFP measurements, which provided annual average SO_2 concentrations of 1.4, 0.5, or 0.3 ppb SO_2 at the North Carolina, Florida and Texas SCFRC sites, respectively. Moreover, TFP sampling indicated that annual average NO_2 concentrations were in the range from 0.3 to 0.5 ppb at all sites. It was concluded that continuous monitoring instruments with detection limits at least one order of magnitude less than those used (2 ppb) are necessary to obtain measurable data on airborne SO_2 and NO_x concentrations at rural sites in the southern region.

Particulate Matter and Lead

Total Suspended Particles. Gravimetric measurements of total suspended particulate (TSP) matter using the high-volume sampling method (U.S. EPA, 1976–1983) were conducted routinely for a 24-hour sampling period every sixth day at two SCFRC sites. Sampling commenced in November 1987 and ended in December 1989 at the North Carolina site and was started in October 1988 and ended in June 1991 at the Florida site. These measurements were made to obtain estimates of gravimetric values for total suspended particles (TSP $< 50\,\mu m$ diameter) at these rural low-elevation forest sites in the southeastern United States. Quarterly (seasonal) and annual averaged values for TSP at the two sites are presented (Table 3.18). It can be seen that there was little evidence of a seasonal cycle or trend in TSP concentrations at either site. There was some indication, however, that suspended particle concentrations were higher in spring and lower in winter at the North Carolina site, which may be indicative of the stronger influence of local urban/industrial sources and the impacted rural nature of this Piedmont site.

Lead. Half ($n = 74$) of the total suspended particle samples collected on filters by the high-volume sampling method were analyzed by furnace atomic absorption spectroscopy for lead content by the EPA standard reference method (U.S. EPA, 1976–1983). Quarterly and annual averaged data are presented (Table 3.18) for the Duke Forest, North Carolina and Cary Forest, Florida sites. It is seen that the seasonal and annual averaged concentrations observed at both sites were consistent throughout the year at levels substantially below the National Ambient Air Quality standard for lead ($1.5\,\mu g\,m^{-3}$ at 25°C, 1 atm). The data

Table 3.18. Quarterly and Annual Averaged Total Suspended Particles (TSP), Size-Fractionated Particles (SFP), and Lead Concentrations at Two Southern Commercial Forests Research Cooperative Monitoring Sites, 1987 to 1989

Site	Qtr/Yr	TSP µg/m³	SFP (µg/m³) Coarse	SFP (µg/m³) Fine	Lead µg/m³
Duke, NC	4/87	27.4	—	—	0.025
	1/88	28.0	14.1	21.7	0.028
	2/88	46.6	17.8	19.6	0.024
	3/88	45.1	11.1	19.1	0.023
	4/88	25.0	10.0	14.6	0.020
Annual	1988	36.2	13.3	18.8	0.024
Duke, NC	1/89	26.3	4.3	9.2	0.020
	2/89	22.0	10.7	15.8	0.019
	3/89	32.0	6.7	13.1	0.026
	4/89	19.1	6.8	10.6	0.022
Annual	1989	24.4	7.1	12.2	0.022
Cary, Fl	4/88	22.3	11.9	14.7	0.021
	1/89	25.6	8.2	9.9	0.022
	2/89	23.9	9.5	12.3	0.018
	3/89	25.2	7.7	10.6	0.022
	4/89	16.9	6.9	10.3	0.023
Annual	1989	22.9	8.1	10.8	0.021

Qtr/Yr = Quarter/Year; Quarter = meteorological season—(1) January, February, March, (2) April, May, June, (3) July, August, September, (4) October, November, December; TSP = Total suspended particulate matter; SFP = Size fractionated particles; Coarse = Particles having aerodynamic diameters in range 2.5 to 10 µm; Fine = Particles having aerodynamic diameters less than 2.5 µm; µg/m³ Micrograms per cubic meter.

obtained support the contention that airborne lead levels have decreased substantially in recent years as a result of the planned phasing out of leaded gasoline fuels used in automobiles.

Size-Fractionated Particles. Gravimetric estimates of coarse (2.5 to 10 μm diameter) and fine (<2.5 μm diameter) fractions of suspended particles in southern forest environments were made at the North Carolina and Florida sites using dichotomous samplers with PM10 inlets. Samples were collected for 24 hours every sixth day in conjunction with high-volume sampling for TSP. Although large variations were observed from one sampling period to another, quarterly averaged values were more consistent. The data presented (Table 3.18) show that, in general, coarse and fine particles concentrations at the North Carolina site were similar to or slightly higher than corresponding measurements made at the Florida site. Moreover, at both sites, suspended particles with diameters equal to or less than 10 μm appeared to account for at least 70% of total suspended particle concentrations. Differences between size-fractionated particle (SFP) concentrations obtained at the two sites (148 samples) were less than for TSP concentrations (148 samples), suggesting that there was a greater influence by larger particles (>10 μm diameter) on TSP measurements at the North Carolina site.

Particle Chemistry. Over a six-month period (December 1988 to June 1989), fine particles collected on the Teflon (inlet) filter of the TFP were analyzed for major water-soluble cations and anions by wet chemistry techniques described previously. It was assumed that the TFP method collects particles with diameters equal to or less than 2 μm diameter with an efficiency greater than 95%, for the sampling conditions and the sampler configuration employed. Preliminary data obtained for three SCFRC sites and for the Florida IFS site, where the TFP sampler was located 10 m above the slash pine canopy, are presented (Table 3.19). It is important to note the spatial consistency of these data at individual sites in the physiographic areas of the SCFRC as well as for elevation above ground at the Florida sites. Moreover, it was apparent that particles collected (26 samples at each site) at the North Carolina site contained higher proportions of sulfate and ammonium ion but lower proportions of nitrate, chloride, potassium, and sodium ions when compared to those fine particle samples collected by the TFP at the two southernmost SCFRC sites in the region. The higher proportions of sulfate and ammonium ions at the northernmost forest site were again indicative of the influence of local and regional industrial and local biogenic sources of air emissions as well as the occurrence of sulfate aerosols in the accumulation mode (<2 μm diameter). Deficiencies in nitrate, chloride, and sodium ions in particles at the North Carolina site may be a result of the inability of the TFP to collect coarse particles, in which nitrate aerosols are normally observed, as well as the absence of a strong marine influence.

Table 3.19. Summary of 6-Month Averaged (24 samples/site) TFP Particle Analysis for Major Cations and Anions in Aqueous Extracts at Southern Commercial Forests Research Cooperative and Florida IFS Monitoring Sites

	Concentration ($\mu g/m^3$)			
Ion	NC	TX	FL	FL IFS
NO_3	0.17	0.72	1.02	0.97
SO_4	5.12	3.45	3.67	3.23
Cl	0.12	0.21	0.26	0.35
HPO_4	0.03	0.02	0.04	0.01
H	—	—	—	0.05
NH_4	1.54	1.06	0.93	0.79
K	0.32	0.4	0.46	0.4
Na	0.38	0.7	0.74	0.86
Total	7.67	6.55	7.12	6.66

$\mu g/m^3$ = micrograms per cubic meter, NC = North Carolina, TX = Texas, Fl = Florida, FL IFS = Florida Integrated Forest Study.

Anion/cation equivalence ratios in these TFP particulate samples varied from 0.91 to 1.05 for all samples analyzed, which suggests almost complete material balances for the major constituent ions studied. Sulfate accounted for more than 70% of the anion content, and ammonium accounted for at least 50% of the cation content on an equivalent basis at all four locations. These analyses, along with the TSP and SFP data, suggest that water-soluble major anions and cations may account for about 40 to 60% of the fine particle (<2 μm diameter) concentrations found in the airsheds of these southern rural forest sites. The remaining fractions of fine particulate material probably consisted of insoluble and inert mineral or organic constituents

Conclusions

Precipitation Chemistry

Acidic precipitation is characteristic of each subregion in the SCFRC. Annual precipitation-weighted concentrations and wet deposition levels for hydrogen, sulfate, and nitrate ions were significantly higher in the Piedmont than in the Inner Atlantic, Eastern Gulf, or Western Gulf Coastal Plains. Annual precipitation-weighted concentrations and wet deposition levels for hydrogen ion were significantly greater in the Inner Atlantic Coastal Plain and the Eastern Gulf Plain subregions than in the Western Gulf Plain. Precipitation-weighted concentrations of hydrogen, sulfate, and nitrate ions were higher in summer in each subregion. Wet deposition amounts of hydrogen, sulfate, and nitrate ions were higher in summer in each subregion except for the Western Gulf Plain, where levels were lower as a result of lower precipitation amounts. Precipitation-weighted mean concentrations and deposition amounts for ammonium ions were higher in spring for all SCFRC subregions. Throughout the year, hydrogen, sulfate, and nitrate ion precipitation-weighted concentrations and depositions were persistently greatest in the Piedmont subregion and lowest in the Western Gulf Plain subregion.

The SCFRC sites studied in North Carolina, Florida, and Texas appeared to be representative of their physiographic regions with respect to wet deposition. Historical wet deposition data indicated that major ion concentrations and deposition levels have been stable throughout the SCFRC region from the late 1970s through the middle 1980s. Based on historical emission estimates, it is surmised that deposition levels over the region probably increased from about 1950 to 1970. Moreover, the subregions within the SCFRC have probably maintained their relative order with respect to deposition levels based on historical emission estimates. Insufficient data are available for the Western Gulf Flats, the Eastern Gulf Flats, and the Outer Atlantic Coastal Plain to define wet deposition quantitatively in these subregions. Significant quantities of organic acid anions have been found in precipitation samples, indicating the importance of weak acids in southern forest environments.

Recent quality assured observations of wet deposition components in selected low-elevation southern pine forests have shown that precipitation-weighted pH values were lower at a site in the northern Piedmont region (4.29 to 4.39) than at sites in the southern Inner Atlantic and Western Gulf Coastal Plain regions (4.46 to 4.52). Similarly, concentrations of the major ions (NH_4^+, H^+, SO_4^{2-}, NO_3^-) in wet deposition were significantly higher (50%) at the northernmost forest site in the SCFRC region. These results support the reported SCFRC regional acidic species gradient, of wet deposition major ion components decreasing from northeast to southwest, which was established from an analysis of historic data. There was a striking similarity in major ion concentrations found in precipitation at the two southern SCFRC sites in Florida and Texas, which also agreed with historic projections. At all SCFRC sites, it appeared that SO_4^{2-} concentrations alone accounted for most of the H^+ concentrations in wet deposition. However, this statement should not be taken to imply that all sulfate taken up in wet deposition was originally in the form of sulfuric acid. The ratio of SO_4^{2-} to NO_3^- ion equivalent in wet deposition was approximately 2.2. Concentrations of NH_4^+ in precipitation correlated well with, and were similar to, equivalent concentrations of NO_3^-. This observation does not necessarily imply direct ammonium nitrate uptake by precipitation. Annual wet deposition fluxes for H^+ and the major anions (NO_3^- and SO_4^{2-}) at Duke Forest, North Carolina were approximately 90 and 75% greater, respectively, than corresponding values estimated for the Cary Forest, Florida and Austin Forest, Texas sites in

1989. However, the annual volume-weighted concentrations and wet deposition fluxes of the major species at the latter sites were unusually similar considering their regional disposition. Thus, the annual wet deposition fluxes may depend primarily on the quantity of precipitation at each location.

Dry Deposition Chemistry

Triple filter pack sampling was used successfully at three SCFRC sites to provide reliable quality assured estimates of weekly averaged particulate sulfate (p-SO_4), particulate nitrate (p-NO_3), nitric acid vapor (HNO_3), sulfur dioxide (SO_2), and nitrogen dioxide (NO_2) concentrations in ambient air. Dry deposition component data obtained at the Duke Forest, North Carolina SCFRC site showed that concentrations were typically at least 40% greater than those observed at the Florida and Texas SCFRC sites. These dry deposition measurements, along with much higher concentrations of SO_4^{2-} found in wet deposition, suggested that the air shed at the northernmost site was influenced more by local and regional industrial and population sources than those of the two southernmost sites. Ratios of p-SO_4 to SO_2 and estimates of total airborne sulfur compounds at the three SCFRC sites indicated more efficient conversion of SO_2 to p-SO_4 during the summer months but reflected increased energy usage at the northernmost site in the winter, in contrast to summer energy demand excesses for the two southernmost sites. The former may be attributed to winter heating requirements, whereas the latter were probably a result of seasonal cooling demands.

Estimates of the relative contributions of wet and dry (gases and fine particles, $<2\,\mu m$ diameter) deposition to the overall terrestrial flux of nutrients S (SO_2, SO_4^{2-}) and N (HNO_3, NO_2, NH_4^+, NO_3^-) to open grassy surfaces at SCFRC sites have been provided (Table 3.13). The data suggest that wet deposition accounts for approximately 70% of the total S deposition and about 70% of the total N deposition at all sites in 1989.

More information is required in terms of the relative contribution of coarse particle deposition to overall dry deposition estimates. Measurements of vegetation physiological conditions (LAI, stomatal conductance) and local meteorological conditions (wind speed, direction, frequency, surface wetness, relative humidity, etc.) are important local inputs to a dry deposition model and minimize the need to use model default values for these variables.

Gaseous Pollutants in the Southeast

Ambient monitoring for gaseous pollutants has traditionally been conducted predominantly in urban areas or near point sources. Regional summaries displayed as maps or statistical tables have been produced with good coverage for the period from 1978 to 1990. Sulfur dioxide and nitrogen oxide concentrations are low (<2 ppb) for a large percentage of the hourly measurements. Extensive compilations of ozone statistics are available in the literature, and the current work is generally in agreement with those published exposure indexes for rural areas in the South. However, the years of coverage and types of statistics (averages, fractiles, percentiles) that have been compiled may not be sufficient or suitable to explain biological stress on trees or vegetation. Long-term trends in historic annual levels of ozone concentrations are not readily apparent from monitoring data. Monitoring data for 1989 support this conclusion, considering that 1988 was an exceptional year for ozone at the North Carolina site. Mechanisms explaining unique phenomena related to ozone accumulation and depletion and nitric oxide production at these rural forest locations have been presented. There is a lack of data concerning levels of the important photochemical oxidants (PAN, PPN, and hydrogen peroxide) and biogenic hydrocarbons at southeastern rural forest sites, which could yield new insights on atmospheric chemical processes occurring at or near these locations.

Air Quality

Typical 24-hour maximum quarterly and annual aggregate ozone exposure indexes (SUM0) are slightly higher at Duke Forest (82/240 ppm · h) than typical values obtained at the two southernmost sites (62/210 ppm · h). Long-term ozone exposure indexes appear to be consistent at the two southern sites in Florida and Texas. The northern boundary of the SCFRC region is better represented by the 1989 long-term O_3 exposure indexes for the Duke Forest, North Carolina site, in view of several O_3 episode occurrences in 1988. The annual seven-hour aggregate exposure indexes for 1989 at the North Carolina, Florida, and Texas sites were similar at 104, 100, and 95 ppm.h, respectively. However, annual SUM06 exposure indexes at the same sites for 1989 show greater differences (41.5, 23.6, and 19.7 ppm · h). Exposure indexes for the Duke Forest, North Carolina site in 1988 were not considered to be temporally representative of the region because of unusually severe regional drought conditions and high-pressure stagnation episodes in July and August, which clearly biased the annual ozone exposure indexes to much higher values (SUM06 = 68.1 ppm · h). These observations generally indicate that lower mean ozone concentrations exist at low-elevation (<500 m) sites in the southern United States than at high-elevation (>1000 m) sites in the Northeast, where daily averaged concentrations are higher and large diurnal variations in ozone concentrations are uncommon (Johnson and Lindberg 1992; Lefohn et al., 1990c; Mohnen, 1992).

Frequent occurrences of low nighttime O_3 levels at two of the sites (North Carolina and Florida) were attributed to O_3 titration in the inversion layer by local NO emissions from soil, which enhanced typical O_3 removal at night by diffusion to the ground. Exposure to nitrogen oxides or sulfur dioxide or even the coexposure of ozone with either or both of the former pollutant species was not of concern because of the low peak and long-term average concentrations of NO_x and SO_2 observed. Annual average nitrogen dioxide and sulfur dioxide concentrations at all sites were in the ranges from 0.3 to 0.5 and from 0.3 to 1.4 ppb, respectively. Peak NO levels exceeding 50 ppb were occasionally observed at night at two of the sites (North Carolina, and Florida), but these unusual excursions were attributed to local soil emissions and should not pose a threat to vegetation under these nocturnal conditions. It is important to note, however, that soil NO emissions must have occurred during daytime hours at these sites but were not detected by monitoring instruments due to the presence of substantial levels of ozone during daylight hours. Such emissions could, however, contribute to local ambient NO_x levels and indirectly to local O_3 levels in rural environments.

Annual average total suspended and size-fractionated particulate concentrations at both northern (North Carolina) and southern (Florida) sites in the SCFRC region were low, in the ranges from 14 to 45 and 18 to 32 $\mu g\,m^{-3}$, respectively. Analyses of particles collected by filter packs showed that approximately one third of the fine particle mass consisted of water-soluble materials, with sulfate and ammonium being the dominant anion and cation, respectively. Daily (24-hour) averaged lead concentrations at the two sites were low, in the range from 0.007 to 0.039 $\mu g\,m^{-3}$. These levels of airborne lead observed in southern rural forests were substantially less than the National Ambient Air Quality Standard for lead.

The values reported here for acid deposition and ozone exposure levels in North Carolina, Florida, and Texas forests fall within the ranges pH $>$ 4 and SUM0 $<$ 300 ppm h, respectively. If these values are compared to the treatments used by SCFRC researchers in open-top exposure chamber studies of pine seedling growth, the wet deposition acidity would appear to have little or no effect on the growth and health of southern pine seedling species. However, it has been suggested that even ambient levels of ozone (~50 ppb) can cause stress to pine species (Garner et al., 1989; Woodman and Cowling, 1987) and aggregate exposures greater than 100 ppm.h may cause stress or visible damage to pine species. Thus, local ozone exposures at SCRFC sites may be cause for concern.

Two important points must be stressed as a result of dry deposition studies. There are substantial differences in estimates of the dry deposition of particles and gases depending on the nature of the surface studied and types of particles collected. In the former case, differences are mainly attributed to the nature of the receptor being observed (e.g., open grassy area or mature pine canopy), even though the same type of deposition model was used. Little information is available on the contributions that coarse particles make to overall dry deposition estimates in terms of water-soluble cations and anions. Preliminary estimates suggested that these contributions could be much greater than for fine particles. For example, at the Florida IFS site, coarse particle deposition was estimated to contribute about five times as much sulfate and about 20 times as much nitrate as that estimated for fine particles and gases, although the technique used to collect coarse particles is questionable. More work is needed to determine the contributions of coarse airborne particles to dry deposition chemistry.

References

Abelson PH (1988) Rural and urban ozone. Science 241:1569.

Allen ER, (1987) *Quality Assurance Project Plan for Atmospheric Exposure Monitoring in the Southern Commercial Forests.* Environmental Engineering Sciences Department, University of Florida, Gainesville, FL.

Allen ER, Evans GF (1992) Ozone exposure characteristics at rural southern commercial forest sites. In: Berglund RL (ed) *Tropospheric Ozone and the Environment, II: Effects, Modeling and Control.* TR-20. Air and Waste Management Association, Pittsburgh, PA, pp 565–579.

Allen ER, Sutton RL (1991a) *Atmospheric Exposure Monitoring in Southern Commercial Forests.* Cooperative agreement CR-813613, Final Report to US Environmental Protection Agency. Environmental Engineering Sciences Dept, University of Florida, Gainesville, FL.

Allen ER, Sutton RL (1991b) Acid deposition to north Florida forest ecosystems. In: Watkins CE (ed) *Proceedings of the Florida Acid Deposition Conference*, Florida Department of Environmental Regulation, Tallahassee, FL, pp 100–124.

Allen ER, Sutton RL, Evans GF (1992) Air quality characterization and acid deposition monitoring at three southern commercial forest sites. In: Flagler RB (ed) *The Response of Southern Commercial Forests to Air Pollution.* TR21. Air and Waste Management Association, Pittsburgh, PA, pp 95–110.

Altschuller AP (1983) Review: Natural volatile organic substances and their effect on air quality in the United States. Atmosph Environ 17:2131–2163.

Altschuller AP (1988) Meteorology—Atmospheric chemistry and long range transport. In: Heck WW, Taylor OC, Tingey DG (eds) *Assessment of Crop Loss from Air Pollutants.* Elsevier Applied Science, New York, pp 65–89.

Aneja VP (1990) Natural sulfur emissions into the atmosphere. J Air and Waste Manage Assoc 40:469–476.

Aneja VP, Li Z (1990) Characterization of ozone at high elevations in the eastern United States. #90-95.9. *Proceedings of the 83rd Annual Meeting*, Air and Waste Management Association, Pittsburgh, PA.

Aneja VP, Claiborn CS, Li Z, Murphy A (1990) Exceedances of the National Ambient Air Quality Standard for ozone occurring at "pristine" area sites. J Air Poll Control Assoc 40:217–220.

Aneja VP, Businger S, Li Z, Claiborn CS, Murphy A (1991a) Ozone climatology at high elevations in the southern Appalachians. J Geophysical Research 96:1007–1021.

Aneja VP, Claiborn CS, Li Z, Murphy A (1991b) Trends, seasonal variations, and analysis of high elevation surface ozone, hydrogen peroxide, and nitric acid. #91-66.4. *Proceedings of the 84th Annual Meeting*, Air and Waste Manage Assoc Pittsburg, PA.

Baldocchi DD, Hicks BB, Camara P (1987) A canopy stomatal resistance model for gaseous deposition to vegetation surfaces. Atmosph Environ. 21:91–101.

Barchet WR (1987) Acidic deposition and its gaseous precursors. In: *NAPAP Interim Assessment: The Causes and Effects of Acidic Deposition. Vol. III Atmospheric Processes and Deposition*. National Acid Precipitation Assessment Program, Washington, DC, pp 5/43–5/51.

Benner CL, Eatough DJ (1991) Comparison of annular denuder and filter pack collection of HNO_3(g), HNO_2(g), SO_2(g) and particulate-phase nitrate, nitrite and sulfate in the southwest desert. Atmosph Environ 25A:1537–1545.

Bohm M (1992) Air quality and deposition. In: Olson RK, Binkley D, Bohm M (eds) *Air Pollution in Western Forests*, Ecological Studies 97. Springer-Verlag, New York, pp 63–152.

Bondietti EA, Hoffman FO, Larsen IL (1984) Air-to-vegetation transfer rates of natural submicron aerosols. J Environ Radioactiv 1:5–27.

Calvert JG, Stockwell WR (1983) Acid generation in the troposphere by gas-phase chemistry. Environ Sci Technol, 17:428A–443A.

Cavender KA, Allen ER (1991) Vertical ozone profile measurements in a rural forest environment. In: Berglund RL, Lawson DR, McKee DJ (eds) *Tropospheric Ozone and the Environment*. TR-19. Air and Waste Management Association, Pittsburgh, PA, pp 673–691.

Chappelka AH, Flagler RB (1991) Future directions in ozone forestry research. In: Berglund RL, Lawson DR, McKee DJ (eds) *Tropospheric Ozone and the Environment*. TR-19. Air and Waste Management Association, Pittsburgh, PA, pp 321–328.

Cropper WP Jr, Gholz HL (1993a) Constructing a seasonal carbon balance for a forest ecosystem. Climatol Res 3:7–12.

Cropper WP Jr, Gholz HL (1993b) Simulation of the carbon dynamics of a Florida slash pine plantation. Ecolog Model 66:231–249.

Duce RA, Mohnen VA, Zimmerman PR, Grosjean D, Cautreels W, Chatfield R, Jaenicke R, Ogren JA, Pellizzari ED, Wallace GT (1983) Organic material in the global troposphere. Rev Geophys Space Phys 21:921–952.

Dunn OJ (1964) Multiple comparisons using rank sums. Technometrics 6:241–252.

Edgerton ES, Lavery TS, Prentice HS (1991) *National Dry Deposition Network Third Annual Progress Report*. EPA-600/3-91-018. Atmospheric Research and Exposure Assessment Laboratory. US Environmental Protection Agency, Research Triangle Park, NC.

Evans GF, Finkelstein P, Martin B, Possiel N, Graves M (1983) Ozone measurements from a network of remote sites. J Air Pollut Control Assoc 33:291–296.

Finlayson–Pitts BJ, Pitts JN Jr (1986) *Atmospheric Chemistry: Fundamentals and Experimental Techniques*. John Wiley and Sons, New York.

Fishman J, Solomon S, Crutzen PJ (1979) Observational and theoretical evidence in support of a significant in-situ photochemical source of tropospheric ozone. Tellus 31:432–446.

Garner JHB, Pagano T, Cowling EB (1989) *An Evaluation of the Role of Ozone, Acid Deposition and other Airborne Pollutants in the Forests of Eastern North America*. General Technical Report SE-59. USDA Forest Service, Southeastern Forest Experiment Station, Asheville, NC, pp. 11, 130–136.

Gholz HL, Fisher RF, Pritchett WL (1985) Nutrient dynamics in slash pine plantation ecosystems. Ecology 66:647–659.

Gholz HL, Vogel SA, Cropper WP Jr, Mckelvey K, Ewel KC, Teskey RO, Curran PJ (1991) Dynamics of canopy structure and light interception in *Pinus elliotti* stands, north Florida. Ecol Monographs 61:33–51.

Goklany IM, Hoffnagle GF (1984) Trends in emissions of PM, SO_x and NO_x, and VOC:NO_x ratios and their implications for trends in pH near industrialized areas. J Air Pollut Control Assoc 34:844–846.

Gschwandtner G, Gschwandtner K, Eldridge K (1986) Historic emissions of sulfur and nitrogen oxides in the United States from 1900 to 1980. J Air Pollut Control Assoc 36:139–149.

Guderian R (1985) *Air Pollution by Photochemical Oxidants*. Ecological Studies 52, Springer-Verlag, New York.

Haas TC (1990a) Kriging and automated variogram modeling within a moving window. Atmosph Environ 24A:1759–1769.

Haas TC (1990b) Log-normal and moving wndow methods of estimating acid deposition. J Amer Statistic Assoc 85:950–963.

Hicks BB, Baldochi DD, Hosker RP Jr, Hutchinson BA, Matt DR, McMillen RT, Satterfield LC (1985) *On the Use of Monitored Air Concentrations to Infer Dry Deposition*. NOAA Technical Report Memorandum, ERL-ARL-141. NOAA-ARL, Oak Ridge, TN.

Hicks BB, Baldochi DD, Meyers TP, Hosker RP Jr, Matt DR (1987) A preliminary multiple resistance routine for deriving dry deposition velocities from measured quantities. Water, Air and Soil Pollut J 36:311–330.

Hicks BB, Hosker RP Jr, Meyers TP, Womack JD (1991) Dry deposition inferential measurement techniques—1. Design and tests of a prototype meteorological and chemical system for determining dry deposition. Atmosph Environ 25A:2345–2359.

Hicks BB, Matt DR (1988) Combining biology, chemistry and meteorology in modeling and measuring dry deposition. J Atmosph Chem 6:117–131.

Hicks BB, Meyers TP, Fairall CW, Mohnen VA, Dolski DA (1989) Ratios of dry to wet deposition of sulfur as derived from preliminary field data. Global Biogeochem Cycles 3:155–162.

Husar RB (1986) Emissions of sulfur dioxide and nitrogen oxides and trends for eastern North America. In: *Acid Deposition: Long Term Trends*. National Academy Press, Washington, DC, pp 48–92.

Johnson DW, Lindberg SE (eds) (1992) *Atmospheric Deposition and Forest Nutrient Cycling: A Synthesis of the Integrated Forest Study*. Ecological Studies, 91. Springer-Verlag, New York.

Kardono (1990) *Investigation of soil nitric oxide emission fluxes at Austin Cary Forest, Gainesville, Florida*. Master of Engineering thesis, University of Florida, Gainesvill, FL.

Kelly NA, Wolff GT, Ferman MA (1984) Sources and sinks of ozone in rural areas. Atmosph Environ 18:1251–1266.

Kim JC (1992) *Comparison Study of Dry Acid Deposition Sampling at Austin Cary Forest, Gainesville, Florida*. Master of Engineering thesis. Environmental Engineering Sciences Dept., University of Florida, Gainesville, FL.

Kleinman LI, Benkovitz CM (1989) Source attribution for hydrocarbons and ozone. In: Wolff GT, Hanisch JL, Schere K (eds) *Scientific and Technical Issues Facing Post-1987 Ozone Control Strategies*. Transactions of Air Pollution Control Association International Specialty Conference. Air and Waste Management Association, Pittsburgh, PA, pp 306–319.

Lamb B, Guenther A, Gay D, Westberg H (1987) A national inventory of biogenic hydrocarbon emissions. Atmosph Environ 21:1695–1705.

Leaitch WR, Bottenheim JW, Strapp JW (1988) Possible contribution of N_2O_5 scavenging to HNO_3 observed in winter stratiform cloud. J Geophys Res 93:12569–12585.

Lefohn AS, Benkovitz CM, Tanner M, Shadwick DS, Smith LA (1990a) *Air Quality Measurements and Characterization for Terrestrial Effects Research.* NAPAP State of Science and Technology Report No 7. National Acid Precipitation Assessment Program, Washington, DC.

Lefohn AS, Davies CE, Jones CK, Tingey DJ, Hogsett WE (1987a) Co-occurrence patterns of gaseous air pollutant pairs at different minimum concentrations in the United States. Atmosph Environ 21:2435–2444.

Lefohn AS, Knudsen HP, Logan JA, Simpson J, Bhumralkar C (1987b) An evaluation of the kriging method to predict 7-h seasonal mean ozone concentrations for estimating crop losses. J Air Pollut Control Assoc 37:595–602.

Lefohn AS, Knudsen HP, Shadwick DS, Hermann K (1992) Surface ozone exposures in the eastern United States (1985–1989). In: Flagler RB (ed) *The Response of Southern Commercial Forests to Air Pollutants.* TR-21. Air and Waste Management Association. Pittsburgh, PA, pp 81–93.

Lefohn AS, Krupa SV, Winstanley D (1990b) Surface ozone exposures measured at clean locations around the world. Environ Pollut 63:189–224.

Lefohn AS, Lawrence JA, Kohout RJ (1988) A comparison of indices that describe the relationship between exposure to ozone and reduction in the yield of agricultural crops. Atmosph Environ 22:1229–1240.

Lefohn AS, Pinkerton JE (1988) High resolution characterization of ozone data for sites located in forested areas of the United States. J Air Pollut Control Assoc. 38:1504–1511.

Lefohn AS, Runeckles VC, Krupa SV, Shadwick DS (1989) Important considerations for establishing a secondary ozone standard to protect vegetation. J Air Pollut Control Assoc 39:1039–1045.

Lefohn AS, Shadwick DS (1991) Ozone, sulfur dioxide and nitrogen dioxide trends at rural sites located in the United States. Atmosph Environ 25A: 499–501.

Lefohn AS, Shadwick DS, Foley JK (1991) The quantification of surface level ozone exposures across the United States. In: Berglund RL, Lawson DR, McKee DL (eds) *Tropospheric Ozone and the Environment.* TR-19. Air and Waste Management Association, Pittsburgh, PA, pp 197–224.

Lefohn AS, Shadwick DS, Mohnen VA (1990c) The characterization of ozone concentrations at a select set of high-elevation sites in the eastern United States. Environ Pollut 67:147–178.

Lefohn AS, Tingey DT (1984) The cooccurrance of potentially phytotoxic concentrations of various gaseous air pollutants. Atmosph Environ 18:2521–2526.

Lindberg SE, Johnson DW, Bondietti EA (1992) Background on research sites and methods. In: Johnson DW, Lindberg SE (eds) *Atmospheric Deposition and Forest Nutrient Cycling: A Synthesis of the Integrated Forest Study.* Ecological Studies 91, Springer-Verlag, New York, pp 9–26.

Lindberg SE, Johnson DW, Lovett GM, Taylor GE, Van Miegroet H, Owens JG (1989) *Sampling and Analysis Protocols and Project Description for the Integrated Forest Study.* ORNL/TM 11214. Oak Ridge National Laboratory, Oak Ridge, TN.

Lindberg SE Lovett GM, Schaeffer DA Bredemeier M (1988) Coarse aerosol deposition velocities and surface-to-canopy scaling factors from forest canopy throughfall. J Aerosol Sci 19:1187–1190.

Logan JA (1983) Nitrogen oxides in the troposphere. Global and regional budgets. J Geophys Res 88:10785–10807.

Logan JA (1989) Ozone in rural areas of the United States. J Geophys Res 94:8511–8532.

Lovett GM Lindberg SE (1984) Dry deposition and canopy exchange in a mixed oak forest determined from analysis of throughfall. J Appl Ecol 21:1013–1028.
Matt DR, McMillan RT, Womack JD, Hicks BB (1987) A comparison of estimated and measured SO_2 deposition velocities. Water, Air and Soil Pollut J 36:331–347.
Meagher JF, Lee NT, Volente RJ, Parkhurst WJ (1987) Rural ozone in the southern United States. Atmosph Environ 21:606–615.
Meyers TP, Hicks BB, Hosker RP, Womack JD, Satterfield LC (1991) Dry deposition inferential measurement techniques—II. Seasonal and annual deposition ratios of sulfur and nitrate. Atmosph Environ 25A:2361–2376.
Mitchell MJ, Lindberg SE, Harrison RE, Fitzgerald JW, Johnson DW, Zhang Y, Autry A (1992) Sulfur chemistry, deposition and cycling in forests. In: Johnson DW, Lindberg SE (eds) *Atmospheric Deposition and Forest Nutrient Cycling; a Synthesis of the Integrated Forest Study*. Ecological Studies 91, Springer-Verlag, New York, pp 72–149.
Mohnen VA (1992) Atmospheric deposition and pollutant exposure of eastern U.S. forests. In: Eagar C, Adams MB (eds) *Ecology and Decline of Red spruce in the Eastern United States*. Ecological Studies 96. Springer-Verlag, New York, pp 64–124.
Musselman RC, McCool PM, Lefohn AS (1994) Ozone descriptors for an air quality standard to protect vegetation. J Air and Waste Manage Assoc 44: 1383–1390.
NADP (1987) National Atmospheric Deposition Program, NADP/NTN Annual Data Summary. Precipitation Chemistry in the United States 1986, Natural Resource Ecology Laboratory, Colorado State University, Fort Collins, CO, pp 25–34.
NADP (1988) National Atmospheric Deposition Program, NADP/NTN Annual Data Summary. Precipitation Chemistry in the United States 1988, Natural Resource Ecology Laboratory, Colorado State University, Fort Collins, CO, pp 25–34.
NADP (1989) National Atmospheric Deposition Program, NADP/NTN Annual Data Summary. Precipitation Chemistry in the United States 1988, Natural Resource Ecology Laboratory, Colorado State University, Fort Collins, CO, pp 25–34.
NADP (1990) National Atmospheric Deposition Program, NADP/NTN Annual Data Summary. Precipitation Chemistry in the United States 1989, Natural Resource Ecology Laboratory, Colorado State University, Fort Collins, CO, pp 25–34.
NADP (1991a) National Atmospheric Deposition Program (IR-7). National Trends Network Tape of Weekly Data: July 1978–April 1991. NADP/NTN Coordination Office, National Resource Ecology Laboratory, Colorado State University, Fort Collins, CO.
NADP (1991b) National Atmospheric Deposition Program, NADP/NTN Annual Data Summary. *Precipitation Chemistry in the United States, 1990*. Natural Resource Ecology Laboratory, Colorado State University, Fort Collins, CO, pp 1–27.
NADP (1993) National Atmospheric Deposition Program, NADP/NTN Annual Data Summary. *Precipitation Chemistry in the United States, 1992*. Natural Resource Ecology Laboratory, Coloado State University, Fort Collins, CO.
NAPAP (1990) National Acid Precipitation Assessment Program. *Existing and Historical Conditions—Causes and Effects,* State of Science and Technology Report No 16. National Acid Precipitation Assessment Program, Washington, DC.

National Academy of Sciences (1983) *Acid Deposition Atmospheric Processes in Eastern North America.* National Academy Press, Washington, DC.
NOAA (1991) *Climate Data*—Florida. National Climate Data Center. National Oceanographic and Atmospheric Administration, Asheville, NC.
Pinkerton JE, Lefohn AS (1986a) *Review of Ambient Ozone Monitoring Data for the Years 1978–1983.* Technical Bulletin No 502. National Council for Air and Stream Improvement, New York.
Pinkerton JE, Lefohn AS (1986b) Characterization of ambient ozone in commercial timberlands using available monitoring data. TAPPI J 69:58–62.
Pinkerton JE, Lefohn AS (1987) The characterization of ozone data for sites located in forested areas of the eastern United States. J Air Pollut Control Assoc 37:1005–1010.
Placet M, Battye RE, Fehsenfeld FC, Bassett GW (1990) Emissions involved in acid deposition processes. Report 1. In: *NAPAP State of Science and State of Technology.* National Acid Precipitation Assessment Program, Washington, DC.
Placet M, Streets DG (1987) Emissions of acidic deposition precursors. In: *NAPAP Interim Assessment. Volume II: Emissions and Controls.* National Acid Precipitation Assessment Program, Washington, DC.
Quillian AM, Lundgren DA (1992) Field measurements of dry deposition compounds using the transition flow reactor. J Air and Waste Manage Assoc 42:36–39.
Reagan J (1984) Air quality interpretation. In: *National Crop Loss Assessment-Network (NCLAN) 1982. Annual Report.* EP-600/3-84-049. US Environmental Protection Agency, Corvallis, OR, pp 198–219.
Reich PB, Amundson AG (1985) Ambient levels of ozone reduce net photosynthesis in tree and crop species. Science 230:566–570.
Rodhe H (1978) Budget and turn-over times of atmospheric sulfur compounds. Atmosph Environ, 12:671–680.
Schaeffer DA, Conklin P, Knoerr KP, Lindberg SE, Lovett GM, Binkley D (1992) Processing of acid deposition. In: Johnson DW, Lindberg SE (eds) *Atmospheric Deposition and Forest Nutrient Cycling.* Ecological Studies 91. Springer-Verlag, New York, pp 426–466.
Sehmel C (1980) Particle and gas dry deposition. A review. Atmosph Environ. 14:983–1011.
Seinfeld JH (1989) Urban air pollution. State of the science. Science 243:745–752.
Sellars FM, Fitzgerald TE, Lennon JM, Maiocco LJ, Monzione NM, Neal DR (1985) National acid precipitation assessment program emission inventory allocation factors. EPA-600/7-85-035. US Environmental Protection Agency, Research Triangle Park, NC.
Sisterson DJ, Bowersox VC, Olsen AR, Meyers TP, Vong RL (1990) *Deposition Monitoring Methods and Results.* Section 6.2.4.4, "Sub-event sampling." State of Science and Technology Report No 6. National Acid Precipitation Assessment Program, Washington, DC, pp 6.188–6.191.
Sisterson DL, Wurfel BF, Lesht BM (1985) Chemical differences between event and weekly precipitation samples in northeastern Illinois. Atmosph Environ 19:1453–1469.
Smith LA (1987) *Analysis of Precipitation Chemistry from National Atmospheric Deposition Program Sites in North Carolina: 1979–1985.* Atmospheric Impacts Research Program, North Carolina State University, Raleigh, NC.
Smith LA, Shadwick DS (1992) Spatial and temporal variability of wet deposition components in the Southern Commercial Forest region. In: Flagler RB (ed) *The Response of Southern Commercial Forests to Air Pollution.* TR-21. Air and Waste Management Association, Pittsburgh, PA, pp 111–122.

Sutton RL, Allen ER (1990) Atmospheric deposition parameters in southern commercial forests. #90-109.3. *Proceedings of the 83rd Annual Meeting.* Air Pollution Control Association, Pittsburgh, PA.

Taylor GE, Ross–Todd BM, Allen ER, Conklin P, Edmonds B, Joranger E, Miller E, Ragsdale L, Shepard T, Silsbee D, Swank W (1992) Patterns of trospospheric ozone in forested landscapes of the Integrated Forest Study. In: Johnson DW, Lindberg SE (eds) *Atmospheric Deposition and Forest Nutrient Cycling—A Synthesis of the Integrated Forest Study.* Ecological Studies 91. Springer-Verlag, New York, pp 50–71.

Teskey RO, Gholz HL, Cropper WP Jr (1994) Influence of climate and fertilization on net photosynthesis of mature slash pine. Tree Physiol 14:1215–1228.

Tingey DT, Turner DP, Weber JW (1991) Factors controlling the emissions of monoterpenes and other volatile organic compounds. In: Sharkey TD, Holland EA, Mooney HA (eds) *Trace Gas Emissions by Plants.* Physiological Ecology Series, Academic Press, New York, pp 93–118.

U.S. EPA (1976–1983) *Quality Assurance Handbook for Air Pollution Measurement Systems.* Vol. I–V. EPA 600/9-76-005, EPA-600/4-77-027a, EPA 600/4-82-042a & b. Environmental Monitoring and Support Laboratory. US Environmental Protection Agency, Research Triangle Park, NC.

U.S. EPA (1986a) *Air Quality Criteria for Ozone and Other Photochemical Oxidants. Vol. II.* EPA-600/8-84/0206F. US Environmental Protection Agency, Research Triangle Park, NC.

U.S. EPA (1986b) *Development of Standard Methods for the Collection and Analysis of Precipitation.* EPA-600/4-86-024. Illinois State Water Survey Division, Champaign, IL.

U.S. EPA (1989) *National Air Quality and Emissions Trends Report, 1987.* EPA-450/4-89-001. Office of Air Quality Planning and Standards, Research Triangle Park, NC.

U.S. EPA (1992) *National Air Quality and Emission Trends Report, 1991.* EPA-0450-R.92-001. United States Environmental Protection Agency, Office of Air Quality Planning and Standards, Research Triangle Park, NC.

U.S. EPA (1993a) *National Air Pollution Emissions Estimates, 1990–1992.* EPA-454/R.93. Office of Air Quality Planning and Standards, Research Triangle Park, NC.

U.S. EPA (1993b) *National Air Quality and Emissions Trends Report, 1992.* EPA-454/R93-031. Office of Air Quality Planning and Standards, Research Triangle Park, NC.

U.S. EPA (1994) *A Clean Air Act Exposure and Effects Assessment 1993–1994. A Prototype Biennial Ecological Assessment.* EPA 600/X-94-020. United States Environmental Protection Agency, Office of Research and Development, Washington, DC.

Vukovich FM, Bach WD, Crissman BW, King WJ (1977) On the relationship between the high ozone in the rural surface layer and high pressure systems. *Atmosph Environ* 11:967–983.

Wang WC. Wuebbles DJ, Washington WM (1985) Potential climatic effects of perturbations other than carbon dioxide. In: McCracken MC, Luther FM (eds) *Projecting the Climatic Effects of Increasing Carbon Dioxide.* DOE/ER-0237. US Department of Energy, Washington, DC.

Westberg H, Lamb B (1984) Estimation of biogenic sulfur emissions from the continential U.S. In: Aneja VP (ed) *Environmental Impact of Natural Emissions.* Air and Waste Management Association, Pittsburgh, PA, pp 41–53.

Wolff GT, Kelly NA, Ferman MA (1982) Source regions of summertime ozone and haze episodes in the eastern U.S. Water, Air, Soil Pollu J 18:65–81.

Wolff GT, Liaz PJ, (1980) Development of an ozone layer associated with synoptic scale episodes in the eastern U.S. Environ Sci Technol 14:1257–1260.
Wolff GT, Liaz PJ, Wright GD, Meyers RE, Cederwall RT (1977) An investigation of long-range transport of ozone across the midwestern and eastern U.S. Atmosph Environ 11:797–802.
Woodman JN, Cowling EB (1987) Airborne chemicals and forest health. Environ Sci Technol. 21:120–126.

2. Abiotic and Biotic Stress Factors on Southern Pine Forests

4. Response of Loblolly Pine to Moisture and Nutrient Stress

Phillip M. Dougherty

Loblolly pine (*Pinus taeda* L.), because of its extensive distribution, is exposed to a wide range of climatic and edaphic conditions (see Chapter 1). In addition to the natural distribution of loblolly, many plantations have been established outside of its native range. For example, plantations of loblolly pine have been successfully established and maintained in Illinois, 240 km north of the natural range of loblolly pine (Gilmore et al., 1966); and in Oklahoma, 370 km northwest (Posey, 1967) and 485 km west (Woods et al., 1988) of the native range. At the planting 370 km northwest of the natural range of loblolly, annual rainfall averages only 765 mm and temperature extremes have reached as high as 46°C and as low as −27°C. Natural regeneration has occurred at this location on microsites, where soil depths are at least 30.5 cm. The occurrence of natural regeneration at this site suggests that seed production and seedling establishment are not the factors that limit the northwestern extension of loblolly pine.

Large-scale movements of seed source within the range of loblolly pine have also created unique combinations of loblolly pine genetics and environment. Such seed source movements are commonly done by forest industries to enhance growth potential, tolerance to drought, and resistance to fusiform rust (Lambeth et al., 1984). Western sources of loblolly are reported to be more drought and rust resistant, and East Coast sources one reported to have higher growth potential (Lambeth et al., 1984). The combination of an extensive natural distribution, extended range plantings, and seed source movements has created a broad genetic × environment base for observing loblolly pine responses to its environment.

These extensive combinations of genetic, climatic, and edaphic conditions have resulted in a wide range of observed productivity potential of loblolly forests. The site index (base age 25) can be as much as 29 m (Allen et al., 1990) in intensively managed plantations grown along the East Coast, where rainfall is adequate and evenly distributed throughout the growing season. The site index for a plantation located in southwest Oklahoma, 485 km west of the natural range of loblolly, was only 9.5 m (Woods et al., 1988). Annual rainfall at this location is only 584 mm. This range in site index is probably due to both genetic variation in growth potential and climatic and edaphic influences. Whether this range in growth potential of loblolly represents the maximum range in growth potential of loblolly in the southern United States is not certain. However, the observed range in site index and the associated climatic and edaphic conditions do help to bracket the range of environmental adaptability of loblolly pine.

The purpose of this chapter is to summarize what is known about the response of loblolly pine to moisture and nutrient stress. Comparative references to the response of other southern pine species to water and nutrient stresses will be made where information is available for these species. Specific responses to moisture and nutrient stress considered in this chapter include (1) carbon gain, (2) tree growth and forest productivity, (3) wood quality, and (4) mortality.

Water Stress Effects on Carbon Gain

Water stress can affect carbon gain of loblolly pine (1) directly, through its influence on stomatal closure; and (2) indirectly, due to its influence on carbon fixation processes (nonstomatal limitations) and on leaf area development and duration. Each of these effects will be considered separately.

Water Stress Effects on Stomatal Conductance and Photosynthesis

When water uptake lags behind water loss, internal water stress develops. Turgidity in the stomatal guard cells will decrease as water potential decreases and stomatal closure is initiated. The relationship between leaf conductance to water vapor and leaf water potential is illustrated in Fig. 4.1 for seedlings of loblolly pine (Arrington, 1988). No specific threshold water potential at which stomatal closure occurs exists for loblolly pine.

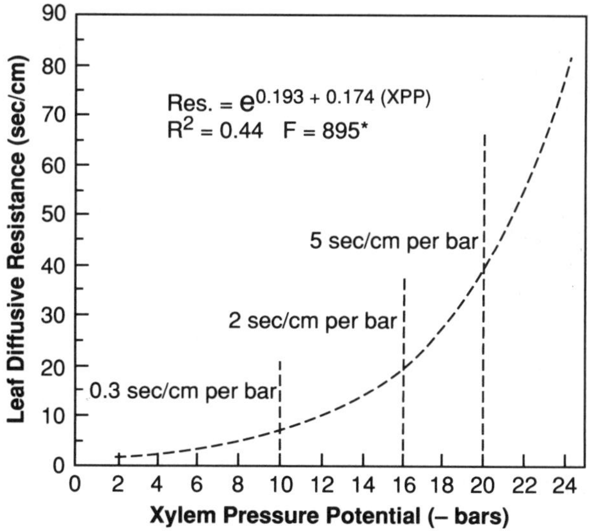

Figure 4.1. Relationship of leaf diffusive resistance to changes in xylem pressure potential developed for one- and two-year-old loblolly pine seedlings (adapted from Arrington, 1988).

The rate of stomatal closure does begin to increase rapidly at a xylem pressure potential of $-1.0\,\text{MPa}$ (Arrington, 1988; Teskey et al., 1986). Because water uptake lags behind water loss on most days, some decrease in stomatal conductance would be expected almost every day. The extent that leaf water potential decreases is dependent on several factors: (1) the degree to which available soil moisture has been depleted; (2) the vapor pressure gradient, or deficit, that exists between the foliage and the atmosphere; and (3) the resistance to water uptake by the root system. Root resistance to water uptake appears to be important during the winter period, whereas soil moisture supply and vapor pressure gradient dominate in the growing season.

Increased resistance to root water uptake and flow can occur due to (1) low soil temperature–induced increases in root resistance to water flow (Kramer, 1942), (2) damage to root systems (Carlson et al., 1988), (3) suberization of roots (Kaufmann, 1968), and (4) reduced or lack of new root growth (Brissette and Chambers, 1992; Carlson, 1986). Teskey et al. (1987) reported that maximum leaf conductance in loblolly pine was closely related to previous night minimum temperature. The low-temperature effect on leaf conductance suggests that during the winter the major resistance to water flow through loblolly pine trees is root resistance. Low-temperature effects on stomatal conductance are more prevalent at the higher latitudes and in the more continental part of the range of loblolly pine. Thus, an increase in winter temperatures, as predicted under doubled carbon dioxide conditions, could greatly enhance the winter gas exchange capacity of loblolly pine.

In the summer, depletion of available soil water and high atmospheric moisture demand (vapor pressure deficits, VPDs) are the factors that most influence internal xylem pressure potential and leaf conductance (Teskey et al., 1987). The relationship between leaf conductance to water vapor, water potential, and vapor pressure deficit is shown in Fig. 4.2. The effect of vapor pressure deficit on leaf conductance of loblolly pine appears to be an indirect effect rather than a direct effect on needle stomata (Teskey et al., 1987). The lack of a direct stomatal response to relative humidity contrasts with a species like radiata pine (*Pinus radiata* D. Don), which appears to have a direct stomatal response to relative humidity (Dean and Sands, 1983). The relationship among leaf conductance, water availability, and vapor pressure deficit needs to be better defined and incorporated into process models so the combined simultaneous effects of increased temperature and vapor pressure deficits and decreased precipitation on gas exchange capacity can be evaluated. Reliable models of leaf conductance are necessary to determine internal doses of gaseous pollutants that plants receive under current and future climatic conditions. For example, current data suggest that if temperature increases and vapor pressure deficits increase, leaf conductance will decrease even if rainfall regimes do not change. In addition, the effects of increased atmospheric carbon dioxide on the relationship of leaf conductance to water demand and supply must be considered.

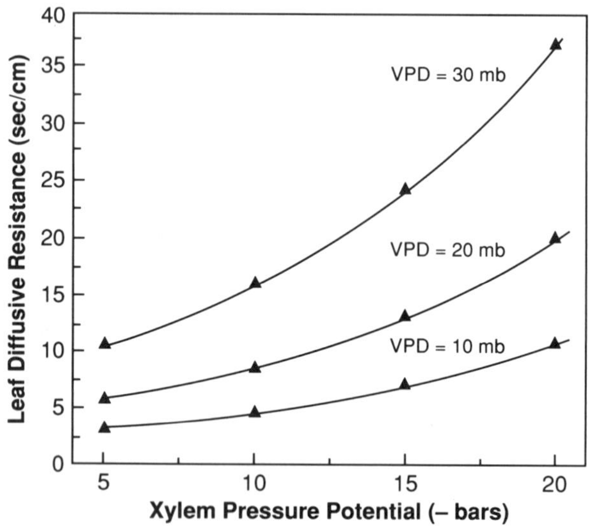

Figure 4.2. Relationship of leaf diffusive resistance with xylem pressure potential and vapor pressure deficit developed for one- and two-year-old loblolly pine seedlings (adapted from Arrington, 1988).

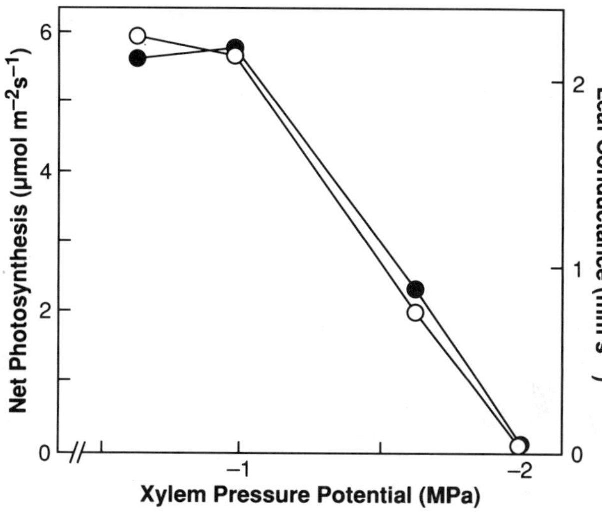

Figure 4.3. Relationship of net photosynthesis (●) and leaf conductance to water vapor (○) with xylem pressure potential established for *Pinus taeda* seedlings exposed to a drying cycle (adapted from Teskey et al., 1986).

Relationship of Net Carbon Exchange to Leaf Conductance

Net carbon exchange (NCE) and leaf conductance are linearly related and decrease in parallel as internal water stress develops (Teskey et al., 1986) (Fig. 4.3). This is to be expected since both carbon dioxide and water vapor are exchanged largely through the stomata and the driving forces for gas exchange between the leaf and the surrounding air both increase with an increase in radiation. However, some of the decrease in NCE that occurs with a decrease in leaf conductance may actually be due to internal (nonstomatal) changes in physical or physiological processes related to NCE (Teskey et al., 1986).

Drought effects on diurnal trends in leaf conductance and NCE are illustrated (Figs. 4.4a and 4.4b) for a day when soil moisture is high and vapor pressure deficit is low and for a day when soil moisture is low and vapor pressure deficit is high (Teskey et al., 1987). Under the latter conditions, only small amounts of photosynthate are produced on a daily basis, and stored carbohydrates would likely have to meet some of the tissue maintenance respiration requirements. If such conditions persist, tissue senescence and ultimately tree mortality would occur. The seasonal effect that water stress can have on NCE can be illustrated by contrasting annual trends in NCE predicted for a "wet" year versus those predicted for a "dry" year (Fig. 4.5). Not all the differences in these NCE trends are due to water stress. High temperatures associated with drought conditions also result in increased maintenance respiratory cost and

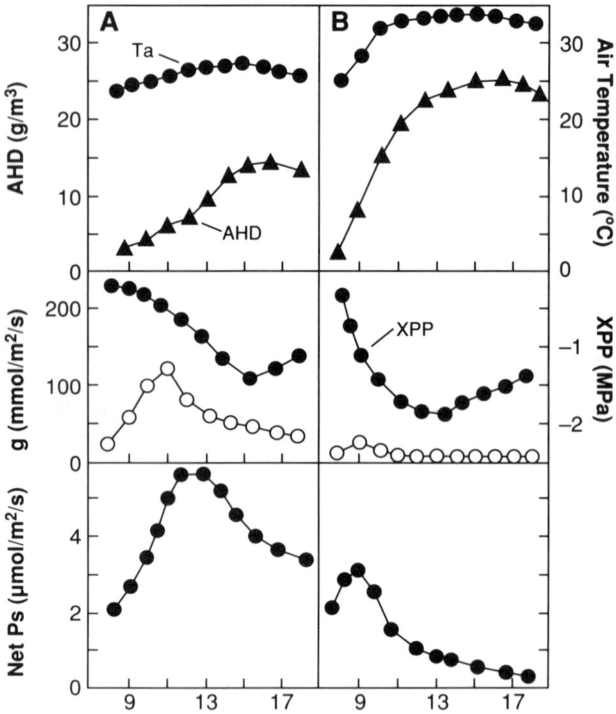

Figure 4.4. Diurnal trends in net photosynthess (Net Ps), leaf conductance, and xylem pressure potential (XPP) of loblolly pine trees growing in a plantation on days of low (A) and high (B) absolute humidity deficits (AHD). Air temperature is also shown for each of these days.

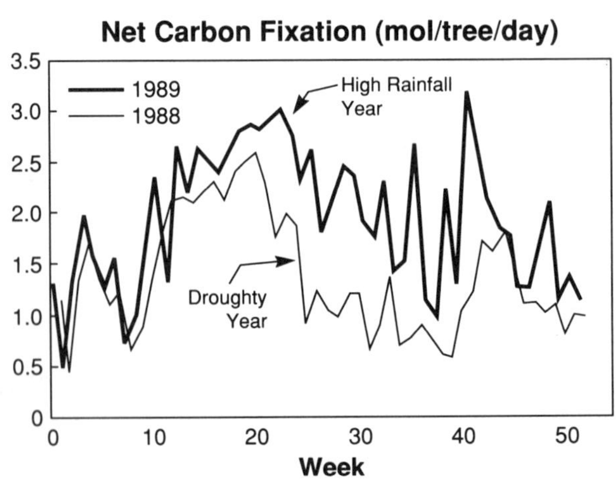

Figure 4.5. Trend in simulated weekly net carbon fixation for a loblolly pine tree using the weather records for 1988 and 1989 for Athens, Georgia (adapted from Dougherty et al., 1992).

contribute to a decline in NCE during drought periods. These trends indicate the major impact on annual NCE that drought can have. Drought years are not uncommon for much of the southern United States. For example, in a study conducted for seven years in southeast Oklahoma, four out of seven years experienced extended summer drought periods similar to the conditions that were used to predict the annual NCE trend in Fig. 4.5 (Dougherty et al., 1995).

In addition to the reduction in NCE through stomatal and nonstomatal limitations, extended periods of water stress can effect leaf biomass amount and duration. The total NCE capacity of a tree is dependent on the potential NCE rate of its foliage, the amount of foliage in the crown, and its display over time. The effects of water stress on NCE rate were discussed in the previous section. Drought effects on foliage production and retention by loblolly pine have been studied only at a few locations. In a study reported by Dougherty et al. (1995), it was observed that annual foliage biomass production was dependent on stand density and average temperature for the months of June through September. These months represent the period in which most needle expansion occurs. During this period, if average temperature increased from 24.5°C to 26.5°C, needle biomass production of a closed canopy stand was decreased 1000 kg/ha or by 27% (Dougherty et al., 1995). An increase in temperature results in both an increase in maintenance respiration and an increase in evaporative demand.

Foliage duration of loblolly pine can be reduced by as much as two months due to drought (Dougherty et al., 1995; Hennessey et al., 1992), as illustrated in Fig. 4.6. Accelerated needlefall was significantly related to an index of water balance (i.e., the average of monthly rain minus precipitation for the two months prior to when needlefall occurred). No assessment of the effect of these reductions in needle biomass production and duration on annual NCE was made by the aforementioned authors. However, such reductions in foliage biomass production and duration would have to represent a considerable loss in carbon fixation potential. Reductions in needle biomass would also result in reduced maintenance respiration cost, which may also be important for survival under extended drought periods for areas in which air temperatures are high.

Water Stress Effects on Tree Growth and Stand Productivity

Water stress is often thought to be the environmental factor most limiting productivity of forests across the southern United States. However, few long-term studies have been conducted to demonstrate the effects of water additions or removals on tree growth or productivity of loblolly pine. Most of the estimates of water stress effects on tree growth come from comparing growth rates in wet versus dry years (Bassett, 1964; Cregg et al., 1988; Moehring and Ralston, 1967). The major apparent

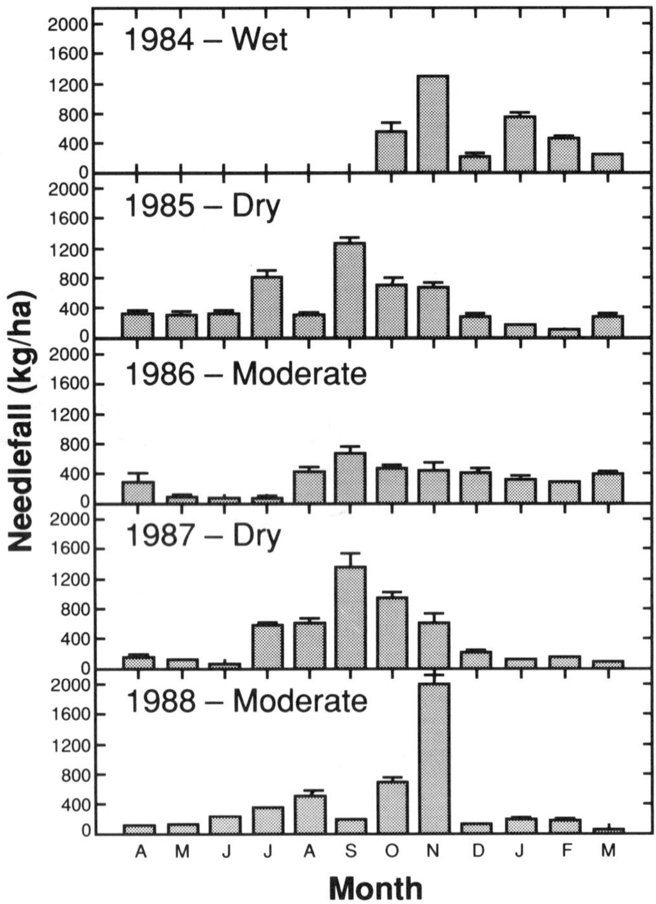

Figure 4.6. Average monthly needlefall and standard error for unthinned loblolly pine growing in southeast Oklahoma from 1984 through 1988 (adapted from Hennessey et al., 1992).

influence of a drought year on growth, as illustrated in Fig. 4.7, is probably due to a combination of drought-related influences, such as water stress effects on carbon gain, carbon allocation, nutrient availability and uptake, decreased turgor pressure effects on expansive growth, as well as elevated temperature effects on depletion of stored carbohydrates. Bassett's (1964) results indicated that basal area growth of loblolly pine trees begins to decrease when available soil moisture is depleted to near 40% of the available water in the soil profile. Moehring and Ralston (1967) reported that basal growth rate is dependent on both the status of available soil moisture and its rate of depletion (evaporative demand). Their research demonstrated that growth rates are substantially reduced

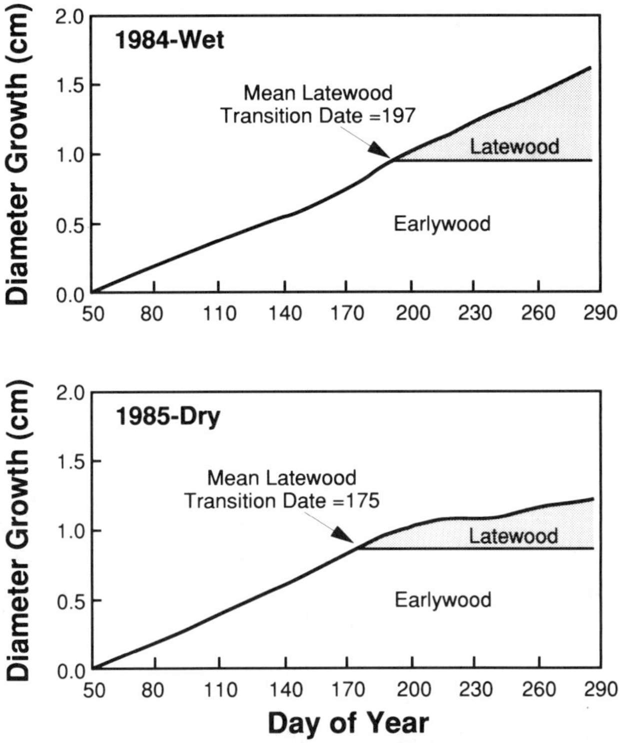

Figure 4.7. Average cumulative diameter growth observed for loblolly pine in a thinning level study located in southeast Oklahoma in 1984 (high rainfall) and in 1985 (low summer rainfall). Shaded areas represent diameter growth after latewood initiation (adapted from Cregg et al., 1988).

even when soil water is near field capacity, if evaporative demand is high. Cregg et al. (1990), using regression analysis, found that basal area growth rates of loblolly pine, in the part of the growing season following the transition from earlywood production to latewood production, are positively related to available soil moisture and negatively related to potential evapotranspiration. Cregg et al.'s results are consistent with Moehring and Ralston's (1967) results. The apparent interaction of rainfall, temperature, and evaporative demand (implied) in determining diameter growth rates of loblolly pine trees is demonstrated in Fig. 4.8, adapted from Coile (1936). He reported that only small variations in temperature for the period of February through May and early summer rainfall had large effects on growth rates of loblolly pine trees. Recognition of the role of evaporative demand on growth rates of loblolly is important. The results of these studies clearly indicate that to evaluate the potential effects of air pollution and climate change on growth of loblolly pine

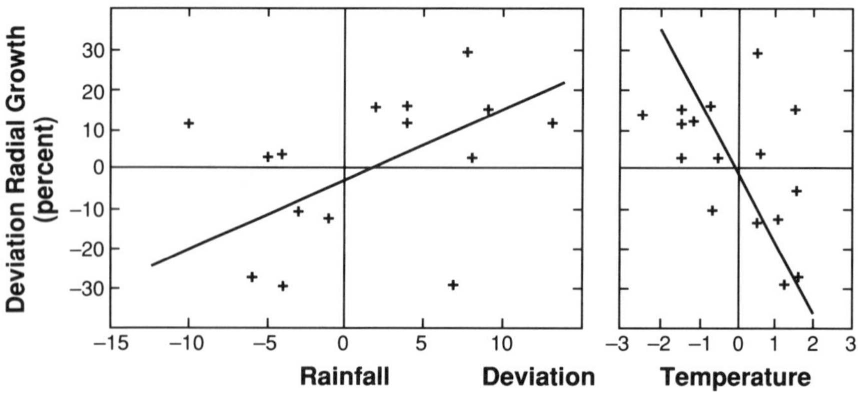

Figure 4.8. Relationship of growth of 10 loblolly pine trees located near DeRidder, Louisiana, to rainfall (left) and to temperature (right). The percentage deviation of the average annual growth of 10 trees is plotted against the deviation of observed January through May precipitation from average annual precipitation. The graph on the right is based on the same growth data plotted over the deviation of the average temperature from the average periodic temperature of the entire time series, 1918 to 1933. Both lines were fitted on the basis of the dependent variable, growth (adapted from Coile, 1936).

forests, the combined effects of changes in precipitation, absolute humidity (evaporative demand), and temperature must be considered.

Water Stress Effects on Wood Quality

Early research indicated that wood density of loblolly pine decreased from the far southeastern part of the range of loblolly to the far northwestern part of the growing range (Zobel et al., 1960). This decrease in wood density corresponds to a gradient in increasing summer water deficits. In loblolly pine, ring wood density is largely a function of the percentage of the growth ring that is made up of latewood (Megraw, 1985). Cregg et al. (1988) reported that, in a year with high soil moisture and below average potential evapotranspiration, the transition from production of earlywood to latewood was delayed by about three weeks relative to when this transition occurred in a year with lower soil moisture (Fig. 4.7). Cregg et al. (1988) predicted the date of transition from earlywood production to latewood production using soil moisture, stand density, and potential evapotranspiration. Interestingly, although the start of latewood production was delayed in a wet year, sustained growth rates in the late season offset the later transition date, resulting in a greater proportion of latewood in a wet year versus that produced in a dry year (Fig. 4.7). Blanche et al. (1992) reported that the number of cells produced before and after the transition from earlywood production to latewood production was about the same. Potential evapotranspiration in the period after the transition to latewood would be much higher than that during the earlywood production phase. Thus, water stress related to higher potential evapotranspiration does not appear to result in a reduction in cell initiation but only a reduction in cell expansion and secondary cell wall development. This is consistent with Hsaio's (1973) report, which suggests that cell expansion is much more sensitive to water stress than cell initiation.

These results imply that site water balance does affect carbon partitioning, as manifested in wood density and thus wood quality. An increase in ambient temperature, if accompanied with increased potential evapotranspiration and decreased precipitation, would be expected to reduce wood density of loblolly pine. However, the actual effects of climate change on wood density will depend on how changes in climate and associated air pollutant inputs that affect NCE and tree growth are distributed throughout the year (i.e., whether these changes favor earlywood production or latewood production).

Water Stress Effects on Seedling Mortality

Mortality of loblolly pine is most common at two stages of development: (1) at establishment and (2) after density-related mortality begins. In stage 1, water stress is often the major environmental stress factor affecting seedlings (Carter et al., 1984; Morris et al., 1993). In a study designed to investigate the site factors that determined loblolly pine establishment from seed, Dougherty (1990) reported that rapid wetting and drying during the germination period was the major factor to regulate to obtain high rates of seedling germination and establishment. Jones (1971) reported that the best month for slash pine (*Pinus elliottii* Engelm. var. *elliottii*) seed germination and establishment in the Middle and Upper Coastal Plains of Georgia was in April, when temperatures were moderate and soil moisture high. Again, this indicates that a constant high-moisture supply is critical for young seedling establishment. Derr and Mann (1971) reported that mortality for naturally seeded pines may range from less than 10% on sites with deep soils and well-distributed summer rainfall to as high as 70% in years when summer droughts of four to eight weeks occur. However, after successful establishment in year 1, they report that mortality is low until crown closure.

For planted seedlings, mortality can be henough in the first year to require replanting. The range of planted seedling mortality reported by Barnett et al. (1984) is about the same as Derr and Mann (1971) reported for seeded pines. A more typical range of first-year mortality for planted seedlings is from 10 to 45%, as reported by Matney and Hodges (1992). In the second year, planted pine mortality is usually moderate (<10%). Thereafter, mortality levels off at 1 to 2% per year. The effect of drought

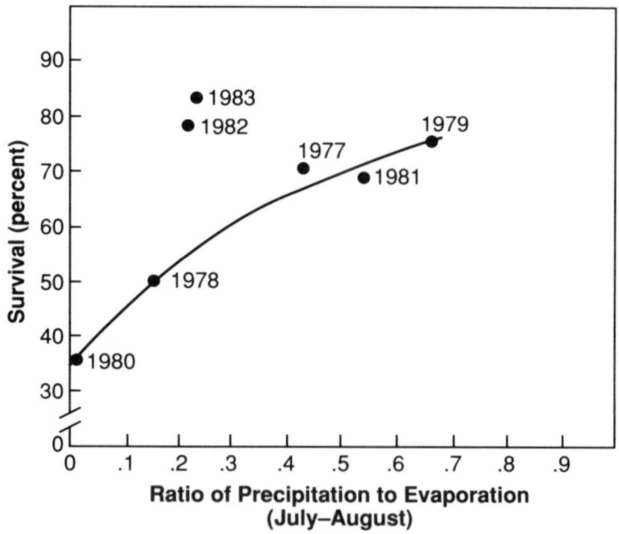

Figure 4.9. Average region survival for a large-scale industrial planting program in southeast Oklahoma plotted as a function of the ratio of precipitation and evaporation for July through August for the period 1977 through 1983. In 1982 and 1983, nursery culturing practices to produce drought hardy seedlings were applied (adapted from Barnett et al., 1984).

on seedling survival is illustrated in Fig. 4.9 for a large-scale planting program applied over a wide range of drought conditions (Barnett et al., 1984). A study conducted by Morris et al. (1993) clearly demonstrates that, in the establishment phase of an artificial regeneration program, competition-mediated water stress is the major factor controlling loblolly pine seedling success in the first year after outplanting. These results also suggest that a future hotter and drier climate may lead to increased seedling mortality during the establishment phase. Changes in nursery culturing practices to produce loblolly pine seedlings that are suited for planting in drought environments are possible. For example, the better than expected survival shown in Fig. 4.9 for the 1982 and 1983 drought-like planting years was largely the result of using nursery culturing practices to produce drought-adapted seedlings (Barnett et al., 1984).

Water Stress Effects on Established Tree Mortality

While it is well documented for seedlings that water stress is the leading stress factor causing mortality, for established trees it is harder to elucidate the extent that water stress contributes to tree mortality. No data were identified in which soil moisture was manipulated in closed stands to determine water stress effects on mortality. Mortality of established trees following an extreme drought in 1980 was reported by Yeiser and Burnett

(1982). Lambeth et al. (1984) investigated the drought conditions that resulted in mortality of established trees of different genetic sources. They developed a soil moisture deficit index using Thornthwaite and Mathers's (1955) potential evapotranspiration (PET) estimates and available soil water storage capacity calculated for each soil series/phase. Soil moisture deficit was defined as the annual sum of potential evapotranspiration after all soil moisture had been depleted, assuming that actual evapotranspiration equaled PET. Using this index, they reported that the probability of mortality increased when the soil moisture deficit exceeded 32 cm. In the Lambeth et al. (1984) study, mortality at high soil moisture deficits was also affected by variables other than water stress, such as high-temperature effects on depleting stored carbohydrates. The interacting effects of water stress and biotic agents on mortality rates related to insect and disease attacks are discussed in Chapters 6 and 7. Well-designed studies to define the causes of mortality and the capacity for trees at various stages of development to buffer against prolonged stress-induced mortality need to be conducted.

Water Management in Loblolly Pine Forests

The only widespread use of irrigation in forestry in the southern United States is in the nursery production phase. Nursery irrigation is routinely applied to seedlings to facilitate seed germination, minimize water stress, and accelerate growth of seedlings. However, the potential for applying irrigation to forest stands was explored by Manogaran (1973). He developed a model of basal area growth of loblolly pine in response to soil water availability and used it to estimate the potential response of loblolly pine to irrigation. His analysis suggests that yield could be increased by irrigation across the range of loblolly pine. The estimated growth gains from irrigation along the East Coast were low (near $0.5\,m^{-3}\,ha^{-1}\,yr^{-1}$) and were as much as $3.0\,m^3\,ha^{-1}\,yr^{-}1$ in the western part of the loblolly pine range. His analysis suggested that it was not economically feasible to apply irrigation on a wide scale. The economic feasibility of forest irrigation may change if secondary land wastewater application becomes more common.

Although irrigation may not currently be an economical treatment, other types of cultural treatments can be applied to manage the site water resource. Controlling competing vegetation at the time of planting can greatly reduce water stress that seedlings experience, reduce mortality, and enhance seedling growth (Carter et al., 1984; Morris et al., 1993; Nelson et al., 1981; Perry et al., 1994; Wittwer et al., 1986). Competing vegetation control in year 2, after outplanting, has much less effect on survival and growth than if vegetation control is practiced in year 1. The research of Sands and Nambiar (1984) with radiata pine (*Pinus radiata* D. Don) provides a good explanation of why competition control in year 1 is most effective (Fig. 4.10). In year 1, the root systems of the planted pines occupy the upper part of the soil profile just as the competing herbaceous vegetation does. Soil water is rapidly depleted from the upper part of the profile by the competing vegetation, which has very high rooting intensity (cm of roots/cm^3 soil volume) relative to the planted pine. Without the herbaceous vegetation present, pines, which have a very low rooting intensity, utilize water sparingly (Dougherty and Gresham, 1988; Morris et al., 1993; Sands and Nambiar, 1984). By year 2, roots extended into the lower soil horizons, and the water potential of the pines was not as adversely affected by competing vegetation (Morris et al., 1993; Sands and Nambiar, 1984). Competing vegetation can be reduced by using herbicides, mechanical site preparation treatments, or prescribed burning

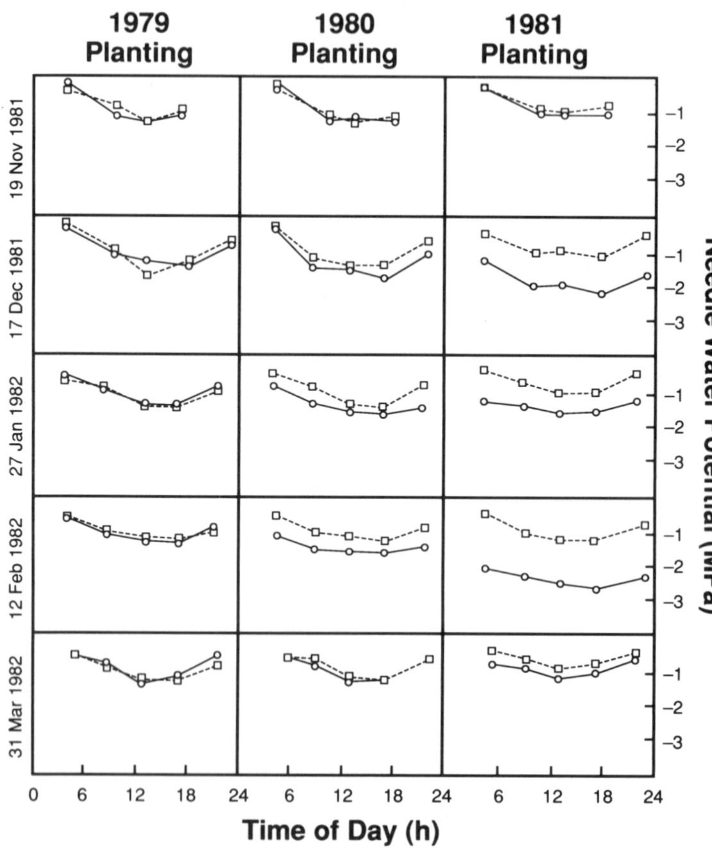

Figure 4.10. Diurnal trends in needle water potential of radiata pine that did (○) or did not (□) receive a weed competition treatment observed on five days in 1981–1982. Seedlings were planted: August 1979, August 1980, and July 1981 (adapted from Sands and Nambiar, 1984).

after a stand reaches a safe size to burn—usually when trees have attained 3 to 4 m in height (Van Lear and Waldrop, 1991).

Managing stand density by reducing initial planting density or thinning can also be used to budget soil moisture in southern pine forest (McClurkin, 1961; Zahner and Whitmore, 1960). Stogsdill et al. (1989) reported that thinning could increase throughfall by 3% of gross precipitation for each 4 m^2 reduction in basal area. Removal of 50% of the basal area of a stand that had reached stand closure resulted in 6.6 cm more throughfall in the months of May through October. At a moisture loss rate of 4 mm/day, this would represent about 16 days of moisture supply. Most of the increased soil moisture that resulted because of thinning was due to reduced interception rather than reductions in the rate of water loss. The benefits of thinning decrease as understory vegetation reinvades the forest floor (Cregg et al., 1990). Thus, it would be necessary to conduct periodic burns if the soil moisture benefits from thinning are to be maintained over time.

Nutrient Stress in Loblolly Pine Forests

Most loblolly pine stands in the southern United States will yield a biological response (Allen, 1987) if fertilized with nitrogen (N) and phosphorous (P). Based on 30 North Carolina State Forest Nutrition Cooperative research installations, which are distributed across the southern United States, 90% or more of stands fertilized with N and P produced more than 3.1 $m^3\,ha^{-1}\,yr^{-1}$ stemwood over a six-year response period (NCSFNC, 1994). The average annual response at the 30 NCSFNC installations was 5.3 $m^3\,ha^{-1}\,yr^{-1}$ over a six-year response period. Ten percent of these installations produced average annual responses of more than 8 $m^3\,ha^{-1}\,yr^{-1}$. The high frequency of stands responding to fertilization and the magnitude of response indicate the extent that nutrition is limiting across the South. It is informative to compare the NCSFNC regionwide fertilizer responses to the regionwide predicted responses to irrigation (Manogaran, 1973). The maximum response predicted for irrigation of 30- to 40-year-old loblolly stands located in the western part of the range, where water is most limiting, was only near 3 $m^3\,ha^{-1}\,yr^{-1}$. This magnitude of response would be near the low end of response obtained from fertilization of intermediate-age loblolly pine plantations.

Extreme P deficiencies are common on poorly drained acid soils in the Lower Coastal Plains (Allen, 1987). On these soils, single P fertilization at the time of planting can increase the site index by 2.5 to 4.5 m over a 25-year period (Joekla et al., 1992). Potassium and/or boron deficiencies may exist on some coastal plains' soil with deep sandy surface horizons (Allen, 1994). However, no large-scale fertilization program exists in the southern United States to apply elements other than nitrogen or phosphorous.

Nutrient demand, supply, and deficiencies are not constant over a rotation or perhaps even within a year (Fig. 4.11). At the time of establishment of a plantation, loblolly pine nutrient demands are extremely low relative to the potential supply (Morris and Campbell, 1992). For example, the total uptake of N by planted pine seedlings would only be about $5\,kg\,ha^{-1}\,yr^{-1}$. Conversely, the N-supplying capacity during the establishment phase following harvesting can approach $100\,kg\,ha^{-1}\,yr^{-1}$ (Fig. 4.11). Thus, deficiencies in the establishment phase will normally be low. Exceptions may occur when sites become severely compacted or internal drainage becomes restricted. Severe deficiencies of P may exist at the time of planting, especially on poorly drained acid soils, as has been discussed in a previous section.

After stand closure occurs and annual productivity reaches a maximum, annual N requirements can exceed $110\,kg\,ha^{-1}\,yr^{-1}$. It was estimated that 25% of the annual N requirements of a 16-year-old loblolly stand was met through internal recycling of N from senescent foliage (Wells and Jorgensen, 1975). Dalla-Tea and Joekla (1994) estimated that 13% and 26% of the annual requirements of P for loblolly pine and slash pine, respectively, were met by nutrient reabsorption from senescent foliage. Nutrient contributions from retranslocation of nutrients from other senescent tissue have not been established and were not considered by Wells and Jorgensen (1975). On most sites, the total N supply from internal retranslocation, decomposition, mineralization, and atmospheric inputs (4 to $10\,kg\,ha^{-1}\,yr^{-1}$) is not adequate to meet stand uptake needs during the rapid growth phase. Atmospheric nutrient inputs (especially

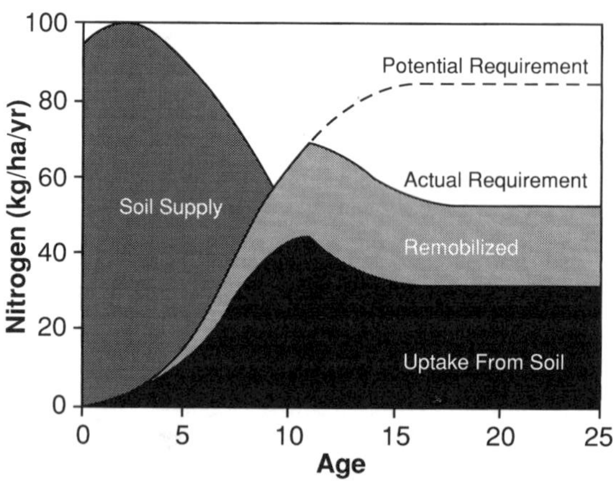

Figure 4.11. Conceptualized soil nitrogen supply, potential nitrogen requirement, actual nitrogen requirement, remobilized nitrogen, and nitrogen uptake from the soil for a site prepared loblolly pine plantation for the period from establishment to age 26 years (adapted from Allen et al., 1990).

N), even though small on an annual basis, are important in contributing to stand productivity and building the nutrient capital over a forest rotation.

Nitrogen Stress Effects on Carbon Gain

Nutrition affects carbon gain through its effects on (1) carbon fixation rates, (2) leaf area quantity and distribution within the crown, and (3) promoting high sink demand for the products of photosynthesis. The relationship between sink strength and carbon gain has not been experimentally demonstrated for loblolly pine. Nutrient effects on leaf area duration of loblolly pine are uncertain at this time. Nitrogen will be the only element considered here because it is the only element that has been investigated for its effect on carbon exchange. However, other elements often change in concert with N and may be responsible for part of the effects ascribed to N treatments.

The maximum carbon exchange rate of loblolly pine foliage has been shown to increase linearly with an increase in N (Fig. 4.12). These results are similar to those reported for *Pinus sylvestris* (Smolander et al., 1990). A linear increase in carbon gain with an increase in leaf nitrogen concentration is not surprising because foliar N concentration has been found

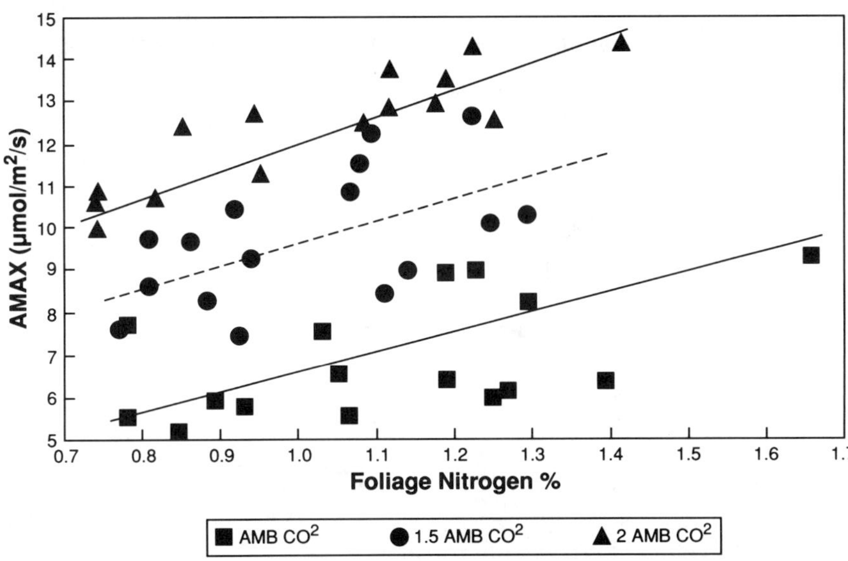

Figure 4.12. The relationship of light-saturated net photosynthesis and leaf nitrogen concentration (% oven dry weight basis) at three concentrations of atmospheric carbon dioxide concentration: ambient, ambient plus 175 VPM carbon dioxide, and ambient plus 350 VPM carbon dioxide. Trend lines were hand drawn.

to be strongly correlated with rubulose-1,5-bisphosphate carboxylase/oxygenase (Rubisco) concentration (Evans, 1989). In addition to the role of N in promoting higher levels of rubisco, N is necessary to maintain high sink activity for the carbohydrates being produced through the photosynthetic process.

The total carbon gain response from nutrient additions depends on both the effects of fertilization on (1) net photosynthesis rates and (2) the quantity of foliage produced. Vose and Allen (1988) demonstrated that the leaf area of loblolly pine could be increased by up to 60% with N and P fertilization. Gholz et al. (1991) reported that fertilization of slash pine increased Leaf Area Index (LAI) by 40%. Zhang (1993) determined that, on an annual basis, carbon fixation was increased by 70% due to fertilization; 41% was due to increased leaf area; and 29% was due to increased net photosynthesis rates per unit leaf area. Zhang's results are for young loblolly pine trees that had not reached stand closure and were maintained near field capacity with a drip irrigation system. Gillespie et al. (1994) reported that N fertilization also increases the quantity of foliage and the quantity of foliage that is retained on branches at lower depth in the crown. The combined effects of fertilization on loblolly pine carbon gain are a major reason that fertilization is becoming widely used to increase productivity of loblolly pine forests. In the future, as atmospheric carbon dioxide concentration increases and annual carbon fixation potential is increased, N deficiencies in loblolly forests are likely to be even higher. Thus, the need for and benefits from fertilization may also increase.

Nutrient Stress Effects on Growth and Productivity

Allen (1987) reported that the typical response to fertilization with N or N and P of established closed-canopy loblolly stands ranges from 2.8 to $3.0\,m^3\,ha^{-1}\,yr^{-1}$ and can reach as high as $7.0\,m^3\,ha^{-1}\,yr^{-1}$. A major underlying factor that results in increased yield due to fertilization is the increase in net carbon exchange capacity and leaf area, as discussed earlier. The empirical relationship between annual yield and peak leaf area reported by Vose and Allen (1988) is illustrated in Fig. 4.13. This relationship suggests that for each unit increase in LAI, annual stemwood production could be increased by as much as $6.5\,m^3\,ha^{-1}\,yr^{-1}$. LAI of closed-canopy loblolly stands in the southern United States lies between 1.5 and 3.0, which is far below the optimum LAI of 6.0 determined by Gholz (1986). Thus, the potential for increasing LAI and yield is great.

Fertilization may also affect yield by changing carbon allocation patterns of trees. In a recent review of carbon allocation in pines, Dewar et al. (1994) concluded that the availability of carbohydrates and labile N must be considered simultaneously to define carbon partitioning patterns in pines. Nutrient effects on carbon accumulation in loblolly seedlings have been investigated (Brissette and Tiarks, 1990; Li et al., 1991). For

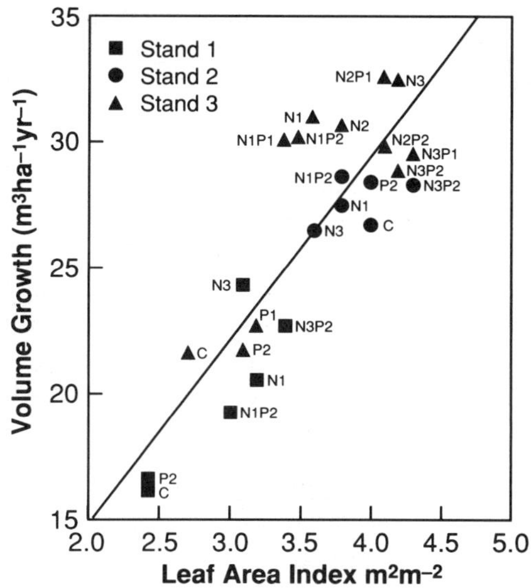

Figure 4.13. Relationship between annual stemwood volume growth and leaf area index for loblolly pine stands that received no fertilization (C), phosphorous at two levels (P1 or P2), nitrogen at three levels (N1, N2, N3), or nitrogen and phosphorous combined (N1P1, N2P1, N3P1, N1P2, N2P2, or N3P2) (adapted from Vose and Allen, 1988).

seedlings, N additions promote foliage development more than root or stem development (Brissette and Tiarks, 1990; Li et al., 1991). However, no reports about the effects of fertilizer on carbon accumulation of loblolly pine trees or stands were located. A good mechanistic understanding of how N and carbohydrate availability interact to control carbon allocation and growth of loblolly pine trees is needed before process models can be developed to predict growth and yield of loblolly pine. In addition, because of the current N limitations to growth on many sites, expected annual inputs of nutrients from air pollution must be considered in assessments of how the future growth potential of loblolly forest will change. For example, if atmospheric N inputs increase from $10\,kg\,ha^{-1}\,yr^{-1}$ to $30\,kg\,ha^{-1}\,yr^{-1}$, the yield response would be expected to be large. Moreover, estimates of how productivity will change under doubled atmospheric carbon dioxide concentration and associated climate must be considered in light of how we expect N availability to change over time.

Conclusions

A review of the literature on nutrient and water stress in southern pine forests indicates that, on most sites, additional inputs of nitrogen and phosphorous would be beneficial to the growth of loblolly pine forests and that such additions would likely be retained in the vegetation. Accelerated leaching of basic cations, which may be occurring on some soils with the addition of acidic rain, has not been demonstrated to have a detrimental effect on the productivity of loblolly pine forests. However, Binckley et al. (1989) reported that about 15% of the pine plantations assessed for potassium had foliar concentrations that were below the established critical value for potassium of 0.35%.

The extent that southern pine forests undergo water stress varies annually, with stage of development, geographic location, and with soil-topographic conditions. Historical levels of water stress have been sufficient to cause reduced tree growth, decreased wood quality, and even established tree mortality. Either a substantial improvement or degradation in the average annual water balance would affect the productivity of loblolly pine forests.

In this chapter, the responses of loblolly pine forests to water and nutrient stress were considered separately. However, it is impossible to alter either the moisture or nutrient regimes of forests without affecting the other. The close linkage of water and nutrient regimes and their interactive effects on pine forests indicates that evaluation of the effects of nutrient additions or climate alterations related to air pollution will require an integrated approach that considers nutrition and water effects simultaneously. In addition, the role of moisture and nutrient stress in altering host–pest relationships must be considered.

References

Allen HL (1987) Forest fertilizers, nutrient amendment, stand productivity and environmental impact. J For 85:37–46.

Allen HL (1994) Enhancing the productivity of southern pine stands with fertilization. Consultant (Summer 1994):12–17.

Allen HL, Dougherty PM, Campbell RG (1990) Manipulation of water and nutrients—practice and opportunity in southern U.S. pine forest. For Ecol and Mgmt 30:437–453.

Arrington SW (1988) *An Evaluation of Ripping and Herbicide Treatment Effects on Soil-Seedling Water Relations and Resulting Seedling Growth for Loblolly Pine on an Upper Coastal Plain site in northwestern Alabama.* M.S. thesis, Department of Forestry, Mississippi State University.

Barnett JP, Campbell TE, Dougherty PM (1984) Seedling establishment—artificial methods. In: Karr BL, Baker JB, Monaghan T (eds) *Proceedings of the Symposium on the Loblolly Pine Ecosystem (West Region)*, March 20–22, Jackson, MS, pp 105–125.

Bassett JR (1964) Tree growth as affected by soil moisture availability. Soil Sci Soc Amer Proc 28:436–438.

Binckley D, Driscoll CT, Allen HL, Schoeneberger P, McAvoy D (1989) *Acidic Deposition and Forest Soils—Ecological Studies 72.* Springer-Verlag, New York.

Blanche CA, Lorio PL, Sommers RA, Hodges JD, Nebeker TE (1992) Seasonal cambial growth and development of loblolly pine: Xylem formation, inner bark chemistry, resins ducts, and resin flow. For Ecol and Mgmt 49:151–165.

Brissette JC, Chambers JL (1992) Leaf water status and root system water flux of shortleaf pine (*Pinus echinata* Mill.) seedlings in relation to new root growth after transplanting. Tree Physiol 11:289–303.

Brissette JC, Tiarks AE (1990) Nitrogen fertilization affects the partitioning of dry matter growth between shoots and roots of loblolly pine nursery stock. Proceedings of the 6th biennial southern silviculture conference; 1990, October. USDA Forest Service, SE Gen Tech Rep 70:108–117.

Carlson WC (1986) Root system considerations in the quality of loblolly pine seedlings. South J Appl For 10:87–92.

Carlson WC, Harrington CA, Farnum P, Halgren SW (1988) Effects of root severing treatments on loblolly pine. Can J For Res 18:1376–1385.

Carter GA, Miller JH, Davis DE, Patterson RM (1984) Effect of vegetation competition on the moisture and nutrient status of loblolly pine. Can J For Res 14:1–9.

Coile TS (1936) The effect of rainfall and temperature on the annual radial growth of pine in the southern United States. Ecol Monograph 6:533–563.

Cregg BM, Dougherty PM, Hennessey TC (1988) Growth and wood quality of young loblolly pine trees in relation to stand density and climatic factors. Can J For Res 18:851–858.

Cregg, BM, Hennessey TC, Dougherty PM (1990) Water relations of loblolly pine trees in southeastern Oklahoma following precommercial thinning. Can J For Res 20:1508–1513.

Dalla-Tea F, Joekla EJ (1994) Needlefall returns and resorption rates of nutrients in young intensively managed slash and loblolly pine stands. For Sci 40: 650–662.

Dean CA, Sands R (1983) Stomatal response to evaporative demand and soil water status in families of radiata pine. Aust For Res 13:179–182.

Derr HJ, Mann WF (1971) *Direct-Seeding pines in the South*. USDA Agricultural. Handbook 391.

Dewar RC, Ludlow AR, Dougherty PM (1994) Environmental influences on carbon allocation. Ecological Bull 43:64–75.

Dougherty PM (1990) *A Field Investigation of the Factors which Control Germination and Establishment of Loblolly Pine Seeds*. Georgia Forestry Commission Report 7, p 5.

Dougherty PM, Gresham CA (1988) Conceptual analysis of southern pine plantation establishment and early growth. South J Appl For 12:160–166.

Dougherty PM, Hennessey TC, Zarnoch SJ, Stenberg PT, Holeman RT, Wittwer RF (1995) Effects of stand development and weather on monthly leaf biomass dynamics of a loblolly pine (*Pinus taeda* L.) stand. For Ecol and Mgmt 72:213–227.

Dougherty PM, Teskey RO, Jarvis PG (1992) Development of MAESTRO, a process based model for assessing the impact of ozone on net carbon exchange of loblolly pine trees. In: Flagler RB (ed) *The Response of Southern Commercial Forest to Air Pollution*. November 4–7, Atlanta, GA, Air and Waste Management Association, pp 303–312.

Evans JR (1989) Photosynthesis and nitrogen relationships in leaves of C-3 plants. Oecologia 78:9–19.

Gholz HL (1986) Canopy development and dynamics in relation to primary productivity. In: Fujimori T, Whitehead D (eds) *Crown and Canopy Structure in Relation to Productivity*. Forest and Forest Products Research Institute, Ibaraki, Japan, pp 224–242.

Gholz HL, Vogel SA, Cropper WP Jr, McKelvey K, Ewel KC, Teskey RO, Currans PT (1991) Dynamics of canopy structure and light interception in Pinus Elliottii stands, north Florida. Ecol Monograph 61:33–51.

Gillispie AR, Allen HL, Vose JM (1994) Amount and vertical distribution of foliage of young loblolly pine trees as affected by canopy position and silviculture treatment. Can J For Res 24:1337–1344.

Gilmore AR, Boyce SG, Ryker RA (1966) The relationship of specific gravity of loblolly pine to environmental factors in southern Illinois. For Sci 12:399–405.

Hennessey TC, Dougherty PM, Cregg BM, Wittwer RF (1992) Annual variation in needlefall of a loblolly pine stand in relation to climate and stand density. For Ecol and Mgmt 51:329–338.

Hsaio TC (1973) Plant response to water stress. Annual Rev of Plant Physiol 24:519–570.

Joekla EJ, Allen HL, McFee WW (1992) Fertilization of southern pines. In: Duryea ML, Dougherty PM (eds) *Forest Regeneration Manual*. Netherlands: Kluwer Academic, Netherlands, pp 263–277.

Jones EP (1971) *Season for Direct Seeding Slash Pine in the Middle and Upper Coastal Plains of Georgia*. USDA Forest Service Research Note SE-151.

Kaufmann MR (1968) Water relations of pine seedlings in relation to root and shoot growth. Plant Physiol 43:281–288.

Kramer PI (1942) Species differences with respect to water absorption at low soil temperature. Amer Jour Bot 29:828–832.

Lambeth CC, Dougherty PM, Gladstone WT, McCullough RB, Wells OO (1984) Large-scale planting of North Carolina loblolly pine in Arkansas and Oklahoma: A case of gain versus risk. J For 82:736–741.

Li B, Allen HL, McKeand SE (1991) Nitrogen and family effects on biomass allocation of loblolly pine seedlings. For Sci 37:271–283.

Manogaran C (1973) Economic feasibility of irrigating southern pines. Water Resources Research 2:1485–1496.

Matney TG, Hodges JD (1992) Evaluating regeneration success. In: Duryea ML, Dougherty PM (eds) *Forest Regeneration Manual*. Kluwer Academic, Netherlands, pp 321–334.

McClurkin DG (1961) Soil moisture trends following thinning in shortleaf pine. Soil Sci Soc Amer Proc 25:135–138.

Megraw RA (1985) *Wood Quality in Loblolly Pine*. TAPPI Press, Atlanta, GA, p 88.

Moehring DM, Ralston CW (1967) Diameter growth of loblolly pine related to available soil moisture and rate of moisture loss. Soil Sci Soc Amer Proc 31:560–562.

Morris LA, Campbell RG (1992) Soil and site potential. In: Duryea ML, Dougherty PM (eds) *Forest Regeneration Manual*. Kluwer Academic, Netherlands, pp 183–206.

Morris LA, Moss SA, Garbett WS (1993) Competitive interference between selected herbaceous and woody plants and *Pinus taeda* L. during two growing seasons following plantings. For Sci 39:166–187.

Nelson LR, Pedersen RC, Autry LL, Dudley S, Walstad JD (1981) Impacts of herbaceous weeds in young loblolly pine plantations. South J Appl For 5:153–158.

North Carolina State Forest Nutrition Cooperative (1994) *Six-Year Growth and Foliar Nutrient Responses of Midrotation Loblolly Pine Plantations to N and P Fertilization*. Report No 29, College of Forest Resources, North Carolina State University; Raleigh, NC, p 38.

Perry MA, Mitchell RJ, Zutter BR, Glover GR, Gjerstad DH (1994) Seasonal variation in competitive effect on water stress and pine responses. Can J For Res 24:1440–1449.

Posey C (1967) Natural regeneration of loblolly pine 230 miles northwest of its native range. J For 65:732.

Sands R, Nambiar EKS (1984) Water relations of *Pinus radiata* in competetion with weeds. Can J For Res 14:233–237.

Smolander H, Oker–Blom P, Kellomaki S (1990) The effect of nitrogen concentration on needle photosynthesis and within shoot shading in Scots pine. Silva Fennica 24:123–128.

Stogsdill WR, Wittwer RF, Hennessey TC, Dougherty PM (1989) Relationships between throughfall and stand density in a *Pinus taeda* Plantation. For Ecol and Mgmt 29:105–113.

Teskey RO, Bongarten BM, Dougherty PM, Hennessey TC (1987) Physiology and genetics of tree growth response to moisture and temperature stress: An examination of the characteristics of loblolly pine (*Pinus taeda* L.). Tree Physiol 3:41–61.

Teskey RO, Fites JA, Samuelson LJ, Bongarten BC (1986) Stomatal and nonstomatal limitations to net photosynthesis in *Pinus taeda* (L.) under different environmental conditions. Tree Physiol 2:131–142.

Thornthwaite CW, Mather JR (1955) The water balance. Climatology 8:1–86.

Van Lear DH, Waldrop TA (1991) Prescribed burning for regeneration. In: Duryea ML, Dougherty PM (eds) *Forest Regeneration Manual.* Kluwer Academic, Netherlands, pp 235–250.

Vose JM, Allen HL (1988) Leaf area, stemwood growth and nutrient relationships in loblolly pine. For Sci 34:547–563.

Wells C, Jorgensen JR (1975) Nutrient cycling in loblolly pine plantations. In: Bernier B, Winget CH (eds) *Forest Soils and Forest Land Management.* Laval University Press, Quebec, pp 137–158.

Wittwer RF, Dougherty PM, Crosby D (1986) Effects of ripping and herbicide site preparation treatments on loblolly pine seedling growth and survival. South J Appl For 10:253–257.

Woods ED, Wittwer RF, Dougherty PM, Crockett JJ, Tauer CG (1988) *Influence of Site Factors on Growth of Loblolly and Shortleaf Pine in Oklahoma.* Research Report P-900, Agriculture Experiment Station Division of Agriculture, Oklahoma State University.

Yeiser JL, Burnett F (1982) Fate of forest trees stressed by heat and drought in southeastern Arkansas. South J Appl For 6:194–195.

Zahner R, Whitmore FW (1960) Early growth of radically thinned loblolly pine. J For 58:628–634.

Zhang S (1993) *The Effects of Nitrogen Availability on Leaf Area, Photosynthesis and Foliar Nutrient Status of Loblolly Pine.* Ph.D. dissertation, College of Forest Resources, North Carolina State University.

Zobel B, Thorbjornsen E, Henson F (1960) Geographic site and individual tree variation in wood properties of loblolly pine. Silvae Genetica 9:149–158.

5. An Overview of Responses of Southern Pines to Airborne Chemical Stresses

Paul Berrang, James S. Meadows, and John D. Hodges

The devastation of plant communities around some primitive smelters made it apparent even centuries ago that anthropogenic pollutants could severely damage forests. At first, it appeared that the effects of these pollutants diminished rapidly with increasing distance from their sources. However, as knowledge of plant communities increased, so did the realization that the effects of some pollutants were more widespread than was obvious. Over time, the emphasis of research on the impact of air pollutants on plants has shifted from the localized effects around point sources to the effects of regionally distributed pollutants.

The best documented example of a regional pollutant affecting forests is the impact of oxidants in the San Bernardino Mountains. Miller (1973) and Miller et al. (1963, 1969) showed that ozone reduced the rates of photosynthesis and increased mortality in ponderosa pine (*Pinus ponderosa* Laws.), leading to the replacement of this species with more ozone-resistant trees. Other regional pollutants that have been implicated in forest problems include acid mist, which is thought to affect the cold hardiness (DeHayes et al., 1991) and soil chemistry (Johnson and Fernandez, 1992) of red spruce (*Picea rubens* Sarg.) stands in the eastern United States; and ozone and sulfur dioxide, which are thought to affect the genetic composition of white pine (*Pinus strobus* L.) (Bennett et al., 1994; Kriebel and Leben, 1981) and quaking aspen

(*Populus tremuloides* Michx.) (Berrang et al., 1991) in the northeastern United States.

Controlling emissions of pollutants entails costs as well as benefits, and as society becomes aware of ever more subtle environmental problems, it will become increasingly difficult to determine whether the benefits of reducing emissions justify the costs. Making these decisions will require continuous improvement in our understanding of air pollutant action. This will be especially true in the southeastern United States, where increases in population and industry are integrated with an increasingly valuable resource—the southern pine forests. The concentrations of ozone in the Southeast are much lower than those in California (which have been linked to the decline of ponderosa pine), and the precipitation in the Southeast is less acid than that in the Northeast (which has been linked to the decline of red spruce). However, if levels of these or other pollutants in the Southeast are sufficient to diminish the productivity of southern pine forests, there would be tremendous economic and ecological implications. This chapter will provide an overview of the effects of airborne chemical stresses on pine forests in the Southeast.

We focus on six pollutants: (1) acid precipitation, (2) sulfur dioxide, (3) ozone, (4) nitrogen oxides, (5) fluorides, and (6) heavy metals. Sources and distribution of these pollutants and their effects on pines, particularly pines from the Southeast, are discussed. Research findings on other tree species are cited when data for a particular pollutant are lacking for southern pines. More in-depth research findings are discussed in subsequent chapters in this book.

Acid Precipitation

Acid precipitation became a major environmental concern in both Europe and North America during the 1980s. In general, acid precipitation is a secondary pollutant, composed primarily of hydrogen, sulfate, and nitrate ions, and represents the wet deposition of acid-forming sulfur and nitrogen compounds. Dry deposition of these compounds is discussed later in this chapter.

The widespread nature of acid precipitation and the importance of its potential effects on vegetation spawned a tremendous research effort. Thousands of studies were performed that addressed ecological, physiological, and biochemical aspects of the potential effects of acid precipitation on plant species in agriculture, horticulture, and forestry. Many reviews have been published, including those by Binkley et al. (1989), Burgess (1984), Cape (1993), Garner et al. (1989), Hakkarinen and Allan (1986), Jacobson (1991), Mason (1990), McLaughlin (1985), and Morrison (1984).

Sources and Distribution of Acid Precipitation

In a pollution-free world, the pH of rain would be about 5.7, which is approximately the pH of pure water in equilibrium with atmospheric carbon dioxide. Although this pH value is technically acid, the term *acid precipitation* generally refers to precipitation with a pH below 5.6 that contains significant amounts of acidic ions, primarily H^+, SO_4^{2-}, and NO_3^- (Morrison, 1984). The major acidifying pollutants are SO_2, NO_x, and HCl (Mason, 1990). Emissions of SO_2 in the Southeast increased sharply between 1950 and 1980, then started to decline, and are considerably lower than in the Northeast. Coal-burning electrical facilities produce almost 70% of the SO_2 emissions in the United States. Trends for NO_x emissions in the Southeast have been similar to those for SO_2: They increased sharply between 1960 and 1980 and then started to decline. Transportation and electrical generation are the principal sources of NO_x in the United States (Placet et al., 1991). The acids formed from these precursors eventually return to the earth's surface in the form of acid precipitation, which may fall hundreds of miles from the source of the NO_x or SO_2.

In 1992, the average pH of rain samples in most of the Southeast was about 4.7 but ranged from 5.3 in west Texas to 4.5 in Tennessee and Virginia. While pH values in the Southeast were generally higher than those in the Northeast, they were generally lower than those in the West. While the vast majority of pH measurements were well above 4.0, minimum readings for the year were as low as 3.2 and 3.4 in West Virginia and Maryland, respectively (National Atmospheric Deposition Program [NADP], 1993).

General Effects of Acid Precipitation on Pine Growth

Because of its potential economic and ecological importance, the effect of acid precipitation on the growth of pines has been studied extensively. Early research attempted to establish a relationship between tree growth in existing stands and historical levels of acid precipitation. One of the earliest such observational studies was first reported by Jonsson and Sundberg in 1972. Their initial analyses of tree ring data from several thousand Scots pines (*P. sylvestris* L.) and spruces (*Picea* sp.) in Sweden suggested that radial growth of trees growing in stands judged to be susceptible to the effects of acid precipitation (based on soil characteristics) started to decrease around 1950. The authors postulated that acid deposition may have been responsible for these perceived decreases in growth (Jonsson and Sundberg, 1972). After additional analyses, they (Jonsson and Svensson, 1982) concluded that no general changes in growth rate had occurred between 1950 and 1974, but by that time their work had attracted considerable interest in the United States.

Observational studies conducted with pines native to the southeastern United States have also produced ambiguous results. One of the earliest was a study of radial growth of pitch (*P. rigida* Mill.), shortleaf (*P. echinata* Mill.), and loblolly (*P. taeda* L.) pine in New Jersey. The authors (Johnson et al., 1981) hypothesized that acid precipitation may have been responsible for decreases in growth rates during the past 25 years. They based their conclusion on a strong statistical relationship between growth rates and stream pH over 17 years. They also found stream pH to be correlated with precipitation pH over a recent three-year period at their site. Johnson et al. (1984) emphasized that it is unlikely that acid precipitation was the sole cause of this growth decline but suggested that it was probably a major contributing factor.

The USDA Forest Service surveys forests in the Southeast approximately every 10 years. Based on these surveys, Sheffield et al. (1985) reported that there had been a decrease in radial growth of pines in natural stands on upland sites over the past 30 years. They suggested nine possible causes for this apparent decrease in growth, one of which was atmospheric deposition. A second group (Bechtold et al., 1991; Ruark et al., 1991; Zahner et al., 1989) also analyzed this survey data and concluded that some southeastern forests were growing more slowly. These authors showed that part of the decrease in growth was associated with competition and drought (Zahner et al., 1989) and they tentatively eliminated stand factors as a cause (Bechtold et al., 1991; Ruark et al., 1991), but they were unable to determine the relative importance of atmospheric deposition and the other factors that Sheffield et al. (1985) had proposed, such as loss of old field sites or diseases. A third group (Cleveland et al., 1992; Hyink, 1991; Zeide, 1992) maintained that there has not been a decline in growth of southern pines.

Most of the aforementioned authors indicate that the inherent variability of natural systems makes it difficult to identify a growth decline and almost impossible to determine what caused it. Experiments that included controlled treatments were needed to determine the effects of atmospheric deposition on the growth of southern pines. Starting in the mid-1980s, the Southern Commercial Forest Research Cooperative (SCFRC) established six similar experiments at five field sites scattered across the range of southern pines in the southeastern United States to determine the effects of acid precipitation and other stresses. These experiments are described in detail in Chapter 10, but briefly they involved exposing loblolly, slash (*P. elliottii* Engl.), or shortleaf pine seedlings to simulated rain and ozone in open-top chambers that had been modified to exclude ambient rain. Ambient rain was replaced with a similar quantity of simulated rain that had a constant pH. The three most common treatments of pH 3.3, 4.3, and 5.3 were applied for several years while the seedlings grew to sapling size. A number of response variables—including height, diameter, and biomass—were measured periodically. An analysis of data from all six experiments indicated that growth rates and biomass increased as the pH of the simulated rain decreased. The authors attributed this increase in growth to nitrogen fertilization (Flagler et al., 1992).

In unrelated experiments with southern pines, most other researchers reported similar results. Edwards GS et al. (1992a), Hart et al. (1986), Seiler and Paganelli (1987), and Walker and McLaughlin (1991) found no effect of simulated precipitation with pH as low as 3.0 to 3.8 on growth of loblolly pines in experiments that lasted from 90 days to two years. Schier (1987) reported that root growth of pitch and shortleaf pine seedlings decreased when they germinated in soil that had been pretreated with simulated rain with a pH as low as 3.0 for one year, but that shoot growth increased. McLaughlin et al. (1994) reported large differences among families of loblolly pine in their response to acid precipitation in a 12-week exposure. However, growth of most families was stimulated when the pH of the rain treatment decreased from 5.2 to 4.5, and growth was reduced when the pH of the rain treatment decreased from 4.5 to 3.3. Lee et al. (1990) and Ludovici (1990) reported that simulated rain with a pH of 3.0 slightly increased root growth of laboratory-grown loblolly pine. Esher et al. (1992) reported that application of throughfall acidified to a pH of 3.6 for two years resulted in decreased numbers, lengths, and biomass of lateral roots of field-grown pines.

Although there is no clear evidence to suggest that growth of southern pines is being decreased by ambient levels of acidic precipitation, this pollutant could be affecting forests in other ways. As a result of these other changes, the future growth of forests may be affected. The potential effects of acid precipitation on these processes are reviewed next.

Direct Effects of Acid Precipitation on Southern Pines

Potential direct effects of acid precipitation on southern pines include (1) physical or chemical damage to foliage, including both visible and microscopic effects, physical and chemical changes in cuticle composition, and accelerated foliar leaching of cations; (2) alterations of reproductive processes and capacity; (3) disruption of various physiological processes; and (4) increased sensitivity to drought or low-temperature stress.

Traditional air pollution research has often included a description of the injury caused by the pollutant of interest. Starting in the mid-1980s, the effects of acid precipitation on loblolly, shortleaf, and slash pine were evaluated by numerous researchers. None of them reported symptoms visible to the naked eye that were associated with rain treatments. Visible symptoms of acid rain injury have been reported on other species, including jack (*P. banksiana* Lamb.) and red pine (*P. resinosa* Ait.) (Percy, 1983) and white (*P. strobus* L.) (Maurice and Crang, 1989; Wood and Bormann, 1977), Scots (Abrahamsen et al., 1976), and Austrian pine (*P.

nigra Arn.) (Zobel and Nighswander, 1991). On pines, the symptoms of injury from acid precipitation that are visible to the unaided eye often include necrotic lesions on the needles. In all but one of these experiments, visible injury occurred only when the pH of the simulated rain was below 3.0. In the one exception, Maurice and Crang (1989) reported foliar injury in white pine for all pH treatments below 5.6. There is no evidence to suggest that even the most acidic rain samples collected in the Southeast will cause visible symptoms of injury on southern pines.

Microscopy, however, can reveal injury that is otherwise not visible. This has not been reported in southern pines, but Back and Huttunen (1992a) used light microscopy to reveal collapsed cells in the outer mesophyll of apparently healthy Scots pine that had been treated with simulated rain with a pH of 3.0. Use of a scanning electron microscope (SEM) has indicated that acid rain treatment of Scots pine foliage can also change the shape and size of chloroplasts and increase lipid structures in the cytoplasm (Back and Huttunen, 1992b; Holopainen and Nygren, 1989).

Acid precipitation can cause physical and chemical changes in the composition of the cuticle of conifers by eroding existing cuticle or by interfering with the formation of new cuticle. These changes can occur when rain pH values are near ambient and when no damage is visible. They can result in increased needle wettability and deformed stomatal openings. While this phenomena is well documented for pH between 3.0 and 3.5 in spruces (Mengel et al., 1989; Percy and Baker, 1990; Percy et al., 1990; Rinallo et al., 1986), it has been shown only on a limited basis in pines (Turunen and Huttunen, 1991) and has apparently not been examined in southern pines.

Several experiments, including some with tree seedlings, have shown that acid precipitation accelerates leaching of foliar cations, particularly calcium, magnesium, and potassium (Fairfax and Lepp, 1975; Scherbatskoy and Klein, 1983; Wood and Bormann, 1975). However, other research suggests that because the quantity of cations leached is relatively small in relation to root uptake and because relatively low energy costs are involved, accelerated leaching does not lead to drastic alterations in tree nutrient status (Amthor, 1986; Turner and Tingey, 1989; Tveite, 1980). In one of the few studies of this on southern pines, Edwards GS et al. (1992b) reported that precipitation with a pH as low as 3.8 did not accelerate foliar leaching in loblolly pine seedlings.

Acid precipitation has been shown to affect the capacity for reproduction in some pines, but this has received little attention in southern pines. Cox (1988) found that while the pollen of most pine species was relatively resistant to the effects of media pH, pollen germination of white pine was reduced by 50% at a pH of 3.3 and pollen tube elongation was reduced at a pH of about 4.0. Seed germination of pines is generally not inhibited by decreasing media pH (MacDonald et al., 1986; Percy, 1986; Raynal et al.,

1982). However, Percy (1986) reported that a variety of morphological traits of very young pine seedlings, including hypocotyl length, cotyledon length, primary needle initiation, and needle length, are affected by five weeks of exposure to simulated rain with a pH of 4.6 or lower. The biological significance of these effects is not known.

Acid precipitation may affect several physiological processes in plants, including photosynthesis, water relations, and nutrient relations. The SCFRC and other programs sponsored numerous studies of the effects of acid precipitation on rates of photosynthesis. Researchers tested the effects of simulated rains with pH as low as 3.0 to 3.5 on loblolly, slash, and shortleaf pine for from 19 weeks to two years. Some reported that acid precipitation had no effect on rates of photosynthesis (Chappelka et al., 1990b; Flagler et al. 1994; Richardson et al., 1992a; Sasek et al., 1991; Seiler and Pagnelli, 1987), while others reported that it increased rates of photosynthesis (Byres et al., 1992b; Hanson et al., 1988; Lee et al., 1990). No researchers reported that acid precipitation decreased rates of photosynthesis in southern pines.

Work with several spruces has suggested that pretreatment with acid precipitation can make seedlings more sensitive to drought stress. Mengel et al. (1989) reported that eight weeks of exposure to precipitation with a pH of 3.0 resulted in increased transpiration and lower water potentials under drought stress and suggested that increased cuticular transpiration was responsible. Eamus et al. (1989) conducted a similar study and obtained similar results but suggested that the increased sensitivity to drought stress was due to impaired water uptake. There are indications that these effects may occur in southern pines. Schier (1987) reported that precipitation with a pH of 3.0 increased shoot growth and decreased root growth in shortleaf and pitch pine seedlings, which may predispose them to higher levels of stress during droughts. Flagler et al., (1994) reported that treatment with precipitation with a pH of 3.3 can increase rates of transpiration in shortleaf pine saplings. Two groups of investigators that tested this hypothesis with loblolly pine reported conflicting results. Lee et al. (1990) observed that transpiration was more sensitive to decreasing needle water potential when seedlings were pretreated with precipitation with a pH of 3.0 for 10 weeks, while Walker and McLaughlin (1993) reported no difference in xylem pressure potential when seedlings pretreated with precipitation with a pH of 3.6 for 26 weeks were drought stressed.

A number of researchers have shown that the development of cold tolerance is delayed in spruce exposed to acid mist similar to that found at high elevations and in the northeastern United States (Cape et al., 1991; DeHayes et al., 1991; Fowler et al., 1989). It is not known if this effect would occur when plants are exposed to the less acidic precipitation found on most pine forests in the Southeast or if this response occurs at all in pine species. In one evaluation of frost injury on loblolly pine, precipitation with a pH as low as 3.3 did not appear to affect frost injury (Chappelka et al., 1990a).

In summary, it appears unlikely that precipitation of ambient acidity has any significant direct impact on southern pine forests. However, this does not eliminate the possibility that there may be significant indirect effects on these forests.

Indirect Effects of Acid Precipitation on Southern Pines

Potential indirect effects of acid precipitation on southern pines include (1) alterations in susceptibility to attack by insects and diseases, and (2) changes in soil chemistry.

Acid precipitation may affect insects and diseases that attack southern pines or alter pine susceptibility to these pests. A few applications of precipitation with a pH of 4.0 resulted in slightly less infection of loblolly pine with fusiform rust (*Cronartium quercuum* f. sp. *fusiforme*) (Bruck et al., 1981), but when pH 3.0 precipitation was applied to the soil for a year, mortality was increased due to damping off in pitch pine (Schier, 1987). Application of throughfall acidified to a pH of 3.6 to the soil in loblolly and longleaf (*P. palustris* Mill.) pine stands increased populations of herbivorous mites (Esher et al., 1992). Application of simulated precipitation with a pH of 3.0 to Scots pine apparently increased susceptibility to aphids (*Cinara pinea* Mordv.) (Neuvonen et al., 1992). These relationships can become complex. Although these treatments did not affect the growth of European pine sawfly (*Neodiprion sertifer* Geoffroy) (Neuvonen et al., 1990b), they did increase damage from this serious pest by decreasing susceptibility of the insect to a virus that attacks it (Neuvonen et al., 1990a). One poorly understood aspect of acid precipitation is changes in populations of epiphytes and endophytes, communities of microbes that inhabit leaves. Acid precipitation decreases the amount of epiphytic fungi on Scots pine (Helander et al., 1993; Ranta, 1990). The decreased competition from these fungi may in turn increase host susceptibility to pathogenic fungi, as shown for Sclorroderris canker (*Gremmeniella abietina*) on spruce (Barklund et al., 1984), but there is no evidence that this may occur in southern pines.

Most researchers agree that any significant long-term effects of acid precipitation on tree growth and forest productivity will be mediated indirectly through changes in the edaphic environment (Binkley et al., 1989). Major effects of acid precipitation on soils may be grouped as follows: (1) the fertilizer effect of nitrogen deposition, (2) leaching of base cations, and (3) aluminum toxicity to the roots.

Nitrogen availability limits growth of most forests in the Southeast (Binkley et al., 1989). The increase in growth rates due to acid rain at the previously described, multiyear, SCFRC-funded field sites was attributed to nitrogen fertilization (Flagler et al., 1992). Higher concentrations of nitrogen were observed in stems and branches, but not foliage (Baker et al., 1994), of loblolly pine exposed to simulated rain with a pH of 3.3 at one of these sites. Total nitrogen accumulation, but not nitrogen concentration, increased in the seedlings as pH of simulated rain decreased from 5.3 to 3.3 (Wright et al., 1990–1991). One laboratory researcher reported that exposure to simulated rain with a pH of 3.0 for one year increased foliar nitrogen (Schier, 1986), while others found no differences in foliar nitrogen among rain treatments after exposure to simulated rain with a somewhat higher pH of 3.3 for a relatively short 11-week exposure (Reddy et al., 1991).

Acid precipitation results in a significant increase in the hydrogen ion concentration of the soil solution. These H^+ ions may exchange with base cations adsorbed onto soil colloids, resulting in increases in both the exchangeable acidity of the soil and the concentration of base cations in the soil solution. Leaching of the base cations in solution may then occur whenever there is a surplus of water and there are equivalent amounts of mobile anions in the leachate to satisfy charge neutrality requirements. Furthermore, not only does acid precipitation increase hydrogen-ion exchange processes, which increase the concentration of base cations in the soil solution; it also adds sulfate and nitrate anions to the soil. Nutrient transport from forest ecosystems is strongly regulated by the availability of anions in the soil solution (Johnson and Cole, 1980).

Consequently, the degree to which sulfate and nitrate anions (deposited into the soil through acid precipitation) affect cation leaching depends on the extent to which they are adsorbed into soil particles. Because nitrate is readily immobilized in the soils of most southern pine forests by plants and microorganisms, it is usually of much less importance than sulfate in cation leaching. However, soils differ in their susceptibility to sulfate anion-induced cation leaching. Because sulfate anions are readily adsorbed by amorphous oxides and hydroxides of iron and aluminum (Chao et al., 1964), soils high in iron and aluminum sesquioxides are said to have a high sulfate adsorption capacity. The more sulfate anions that a soil can adsorb, the less subject that soil is to acid precipitation-induced cation leaching. Conversely, sulfate is readily leached from soils low in sesquioxides (Cole and Johnson, 1977; Johnson and Cole, 1977). In these soils, sulfate leaching due to acid precipitation is a major concern because it is generally accompanied by accelerated leaching of calcium, magnesium, and potassium from the soil profile. Turner et al. (1986) estimated that soils thought to be sensitive enough to basic cation depletion for reductions in tree growth to occur as a result of sulfate deposition make up about half the forested soils in the Southeast.

Abrahamsen (1984) pointed out that leached cations may be replaced by increased mobilization of base cations through weathering processes. However, there are indications that soil weathering processes are adapted to the rate at which H^+ ions are naturally produced within the system and that additional H^+ supplied through acid deposition may not be neutralized. Experimental evidence, however, is insufficient to make conclusive statements regarding the potential role of soil weathering processes in arresting acid precipitation-induced soil acidification.

One potential effect of accelerated cation leaching—especially in relatively young, unweathered soils with a low buffering capacity—is increased soil acidification. Availability of aluminum can increase as soils become more acidic. Consequently, soil acidification accelerated by acid precipitation could result in toxic levels of this element. Aluminum toxicity is a potential problem in many acid soils, particularly those in which the pH is below 5, at which point the solubility of aluminum rises dramatically. Highly weathered, acid soils, which are common in the Southeast, generally have a higher degree of buffering capacity and are much less sensitive to further acidification due to acid precipitation than slightly acid to neutral soils with a low buffering capacity. However, acidic precipitation can also increase aluminum availability, even when soil acidity is not affected, because of the increased solution strength (Reuss and Walthall, 1988). Turner et al. (1986) estimated that about 10% of the soils in the Southeast were sensitive enough to aluminum toxicity for acid precipitation to result in decreases in tree growth.

High aluminum concentrations can impact plants by inhibiting cell division in the roots and inhibiting uptake of phosphorus (Abrahamsen, 1984). These effects are generally recognized in agricultural crops but have not been widely demonstrated in southern pines. However, Schier (1987) did report apparent aluminum toxicity to the roots of young shortleaf and pitch, but not loblolly pine seedlings, when the nutrient-poor soil in which they were planted was pretreated with pH 3.0-simulated precipitation for one year.

The evaluations of Turner et al. (1986) suggest that up to 60% of the soils in the Southeast may be changed by acid precipitation, but empirical proof of such changes is difficult. Binkley et al. (1989) evaluated the effects of ambient levels of acid precipitation on typical forest soils of the Southeast using a soil chemistry simulation model. They concluded that there would be significant decreases in base saturation and increases in aluminum in solution in many of these soils over a time frame ranging from several decades to a century if 1984 deposition rates were maintained. The rate of change and the endpoints of these changes varied with the rate of deposition. However, their model did not include the effects of nutrient uptake by rapidly growing pines or removal of nutrients during timber harvests, both of which would tend to accelerate the loss of exchangeable cations. The long period of time over which these changes are predicted to occur would make them difficult to detect.

Summary of Acid Precipitation

The bulk of experimental evidence available today suggests that direct effects of acid precipitation on southern pines are probably not important.

However, indirect effects, especially those that are soil mediated, may have important, long-term effects on southern pine forests. The extent of these soil-mediated changes will depend, to some extent, on sulfur dioxide and nitrogen oxide emissions, which have been declining in the Southeast since about 1980. It will be extremely difficult to determine whether changes in soils are actually occurring and what factors are responsible if they are detected.

Sulfur Dioxide

Sulfur dioxide is a widespread gaseous pollutant that damages forests in many industrialized countries. It is referred to as a primary pollutant because it is emitted directly into the air in a gaseous form. Sulfur dioxide affects plants in many ways and has been the subject of much research. Books edited by Winner et al. (1985) and Koziol and Whatley (1984) provide comprehensive coverage of sulfur dioxide and its action on plant metabolism, growth, and community structure.

Sources and Distribution of Sulfur Dioxide

Emissions of SO_2 in the Southeast more than doubled between 1900 and 1950 and quadrupled by 1970 (Gschwandtner et al., 1986). In 1970, the U.S. Congress passed the original version of the Clean Air Act, which initiated a major decrease in SO_2 emissions. The use of low-sulfur coal and new emission control technologies have allowed this decrease in emissions to continue for more than 20 years, in spite of continued increases in coal consumption (National Acid Precipitation Assessment Program [NAPAP], 1987). Almost 70% of the SO_2 emissions in the United States today are from coal-fired power plants. Before 1930, most of the SO_2 in the United States was released from stacks less than 120 feet tall. Since 1980, about half the SO_2 has been released from stacks more than 480 feet tall. Higher stacks increase the potential for long-range transport of this pollutant (Gschwandtner et al., 1986). Lefohn and Shadwick reported in 1991 that many sites in the South had experienced decreases in ambient SO_2. Apparently, the decreases in SO_2 emissions coupled with the distribution of this pollutant over a larger area are reducing the concentrations of this pollutant.

Little is known of SO_2 concentrations in forested areas in the Southeast. With the exception of Tennessee, most states have few SO_2 monitors (Lefohn and Benkovitz, 1989), and most monitors are located in urban areas or near point sources (Heck, 1984). Available data fluctuate widely, but most SO_2 levels in the Southeast are relatively low. Even monitors close to point sources only show occasional high readings (Lefohn and Benkovitz, 1989). Lefohn and Benkovitz (1989) analyzed the 1987 data from two high-elevation monitors in the Southeast: Mount Mitchell in North Carolina and Whitetop Mountain in Virginia. The maximum monthly concentrations ranged from 2 to 47 ppb SO_2, and 99% of the readings were below 2 to 15 ppb depending on the monitor and month of sampling. Allen et al. (1992) summarized the data from three forested SCFRC sites in Florida, Texas, and North Carolina. They reported that 95% of the readings were less than 2 ppb SO_2.

Effects of Sulfur Dioxide on Southern Pines

Most early tests of the effects of SO_2 on pines used exposures to extremely high concentrations for short time periods. For example, Constantinidou et al. (1976) reported that growth of red pine needles was decreased by exposure to 500 ppb SO_2 for 15 minutes, and Berry (1974) reported that Virginia, shortleaf, slash, and loblolly pine seedlings were injured by exposure to 250 ppb SO_2 for two hours. These concentrations are obviously well above those that have been reported recently for forested areas in the Southeast. In addition, others have reported that the inhibition of photosynthesis caused by acute fumigations like these was usually reversible if it did not cause visible injury (Darrall, 1989).

Fumigations with lower concentrations of SO_2 that last for longer periods of time would be more representative of ambient conditions. The lowest concentrations of SO_2 that have been reported to decrease rates of growth or photosynthesis in pines are somewhat higher than almost all the concentrations recently measured in most forested areas in the Southeast. Leininger et al. (1991) reported that diameter growth and needle weight of ponderosa pine was reduced by average concentrations of SO_2 as low as 35 ppb for 60 days. Height (Farrar et al., 1977) and diameter (Garsed and Rutter, 1984) growth of Scots pine was reduced by exposure to concentrations of SO_2 as low as 50 ppb for six months and 40 ppb for two years, respectively. Relatively few chronic exposures to SO_2 have been conducted with southern pines. Winner et al. (1987) reported that when 12 families of loblolly pine were exposed to 60 ppb SO_2 for 5 h/day for 40 days, the growth of only one family was reduced, the growth of three families was stimulated, and eight families were not affected. Not only were these levels higher than those being measured in forested areas in the Southeast, but the durations were considerably longer than the periods of high SO_2 measured there as well.

Summary of Sulfur Dioxide

Because the concentrations of SO_2 and the durations of exposure required to injure plants or reduce growth are higher and longer than those that have recently been reported in most forested areas, SO_2 does not appear to be an important problem in the southeastern United States today. However, SO_2 is an important problem in some countries (Bell, 1984;

Shu-Wen, 1984), and the low concentrations of SO_2 that do exist in this country are the precursor for a widely distributed secondary pollutant—acidic deposition.

Ozone

Ozone is a major constituent of the stratosphere, where it absorbs some frequencies of ultraviolet light, but it is also an important pollutant in the troposphere, close to the earth's surface. It is widespread throughout the industrial regions of the world and reaches high concentrations in eastern North America, the southwestern United States, and central and northwestern Europe. Ozone is currently the only regionally dispersed air pollutant that has conclusively been shown to injure forest trees (Garner et al., 1989). Many research efforts conducted over a number of years have provided conclusive evidence that ozone damages white pine stands in eastern North America (Berry, 1961; Berry and Ripperton, 1963; Costonis and Sinclair, 1969; Dochinger and Seliskar, 1970; Linzon, 1966) and Jeffrey (*P. jeffreyi* Grev. and Balf.) and ponderosa pine stands in southern California (Miller et al., 1963, 1969, 1983).

As a result of the potential for widespread ozone damage to vegetation, numerous reviews have described the consequences of tropospheric ozone. Some recent notable ones include those written by Adams et al. (1986), Chappelka and Chevone (1992), Garner and Smith (1986), Heagle (1989), Heath (1994), Krupa and Manning (1988), and Runeckles and Chevone (1992).

Sources and Distribution of Ozone

Ozone is one of several photochemical oxidants in the earth's atmosphere, but, based on a review of reported concentrations and effects on plants (Garner et al., 1989), it appears to be the most common of the phytotoxic compounds. Tropospheric ozone is not a primary pollutant and appears to originate from a variety of sources. One theory hypothesizes that ozone is introduced into the troposphere from the stratosphere by tropopause-folding events (Chameides and Lodge, 1992). Recent analyses suggest that this is one important source of ozone near the earth's surface (Vukovich et al., 1985). Photochemical generation of ozone is also an important source of this pollutant in the troposphere. The complex reactions leading to the formation of ozone were reviewed by Chameides and Lodge (1992) and basically involve the reaction of volatile organic compounds (VOCs) and nitrogen oxides (NO_x) in the presence of sunlight. The relationships between these two precursors and production of ozone are nonlinear. Concentrations of NO_x tend to control ozone production when concentrations of NO_x are relatively low, and concentrations of

VOCs tend to control ozone production when concentrations of NO_x are relatively high.

Almost 90% of the NO_x that serves as a precursor to ozone originates from anthropogenic rather than natural sources. The major source of anthropogenic NO_x is transportation, followed by utilities. Transportation is the major source of anthropogenic VOCs as well, followed by solvent evaporation and gasoline marketing and storage (Placet et al., 1991). There is some disagreement regarding the relative importance of anthropogenic and natural sources of hydrocarbons (e.g., Altshuller, 1983, versus Chameides et al., 1988). However, Chameides and Lodge (1992) argued that biogenic emissions of hydrocarbons are at least as high as anthropogenic emissions for the nation as a whole and presumably even higher in the Southeast, where extensive forests and warm conditions favor their production. In addition, some of the biogenic hydrocarbons are more reactive than ones from anthropogenic sources (Logan, 1989).

A discussion of the distribution of ozone requires a parameter for comparing ozone exposures from different locations or different periods of time. A good parameter should be correlated with the response of vegetation over a broad range of concentrations, durations of episodes, and differences in vegetative susceptibility. An appropriate parameter would be useful for comparing the work of different researchers and for comparing ambient and experimental ozone profiles. It would also be useful for developing an ozone standard to protect vegetation. Many parameters have been proposed, but none is universally applicable. Some evidence suggests that short-term, high concentrations of ozone have more of an impact on plants than long-term, low concentrations of ozone. For this reason, Lefohn (1992) proposed several indexes that use sigmoidal weighting schemes to lend more importance to peak ozone episodes. However, almost no research has compared the effects of exposure dynamics on the growth of southern pines. Using data from a loblolly pine fumigation, Lefohn et al. (1992b) evaluated the performance of a variety of indexes and concluded that the sum of all hourly ozone concentrations did not describe differences among treatments but that they were unable to identify a superior index among the other options evaluated. Additional research is needed in this area to help develop a secondary standard for ozone.

There is disagreement about the levels of ozone that would exist in the absence of anthropogenic air pollutants. Altshuller (1987) suggested concentrations in the range of 10 to 20 ppb, and Afonsi et al. (1991) estimated that average monthly ozone concentrations were 5 to 10 ppb in the late 1800s. However, Lefohn et al. (1990) argued that concentrations of 30 to 45 ppb are a more appropriate reference point for North America. It is also difficult to forecast ozone levels in the future. Thompson (1992) predicted, through the use of several models, that ozone concentrations will rise 0.5% per year for the next 50 years, while Chameides et al.

(1994) suggested that the frequency of ozone concentrations high enough to damage plants will triple over the next 30 years.

Lefohn et al. (1987) noted that the typical diurnal ozone pattern for the rural southeastern United States reaches a minimum ozone concentration of about 12 ppb in the early morning, rises to a peak of about 50 ppb in the early afternoon, and drops back again in the late afternoon and evening. The diurnal pattern is different at high elevations, where there tend to be fewer low concentrations of ozone and a flatter curve (Berry, 1964; Lefohn, 1992). There is a great deal of temporal and spatial variation in ozone concentrations. For example, Lefohn (1992) reported that in the South, 1988 was a particularly high ozone year and 1982 was a particularly low ozone year. He also reported that about a tenth of the total forested land in the South was in counties that did not attain the U.S. EPA standard for ozone between 1986 and 1988. Lefohn et al. (1992a) used a statistical technique known as kriging to show that between 1985 and 1989, areas in the Southeast with typically high ozone levels were located in north Georgia, South Carolina, North Carolina, and Virginia and that areas with typically low ozone levels were located in southern Mississippi and Florida.

Visible Symptoms of Ozone and Other Foliar Effects

Davis (1984) gave a general description of the symptoms of ozone injury visible to the unaided eye on conifers. Acute injury typically occurs as death of the needle tip, resulting in a distal necrosis of the needle. Chlorotic mottle may be observed when exposures to O_3 are less severe. The mottle is characterized by alternating areas of yellow or brown injured tissue and green healthy tissue. Older needles often fall off prematurely, giving the crown a tufted appearance as the current foliage remains on the tree. Ozone damage to individual species may vary somewhat from these general descriptions, but is generally similar for white pine (Berry and Ripperton, 1963; Costonis, 1970; Costonis and Sinclair, 1969), ponderosa, and Jeffrey pines (Parmeter and Miller, 1968; Parmeter et al., 1962).

Southern pines generally do not show visible symptoms of ozone injury at the concentrations of ozone found in the Southeast. Visible injury can develop when they are exposed to above-ambient concentrations of ozone (Kress et al., 1992; Meier et al., 1990), and certain sensitive families may show symptoms in ambient air (Chappelka et al., 1990b; Shafer et al., 1987). One exception is a variety of slash pine restricted to the southern tip of Florida (*P. elliotti* var. *densa*), on which Evans and Fitzgerald (1993) observed chlorotic mottle that closely resembled ozone injury when it was growing in the field in ambient air.

Microscopy can reveal injury that is not visible to the unaided eye. Evans and Fitzgerald (1993) reported increased numbers of dead cells in slash pine needles fumigated with ambient levels of ozone for one month. These needles did not show visible symptoms and had been exposed to only 11 ppm.h of ozone at the time they were examined. Ultrastructural changes in the biosynthesis and crystallization of wax in the cuticle of spruce needles occurred after 11 weeks of exposure to 70 ppb ozone for 7 h/day (Percy et al., 1990). Krause et al. (1987) also observed injury to epicuticular wax of loblolly pine when it was exposed to ambient and above-ambient levels of ozone. This pollutant can also cause ultrastructural changes in the interior of pine needles. Sutinen (1989) observed a decrease in the size of chloroplasts and in starch accumulation when Scots pine were exposed to 20 ppb ozone for 9 h/day for three months. McQuattie and Schier (1993) noted that while pitch pine needles exposed to 50 ppb ozone for 8 h/day for 13 weeks did not show injury visible to the unaided eye, light microscopy revealed more densely stained mesophyll cells and more intercellular debris than in control needles. They noted more injury near the substomatal chamber and less in the vascular tissue. Krause et al. (1987) observed cytolysis of subsidiary and mesophyll cells of loblolly pine seedlings exposed to ambient and above-ambient levels of ozone.

Ambient ozone can cause premature needle loss in white pine (Dochinger and Seliskar, 1970; McLaughlin et al., 1982) and ponderosa pine (Axelrod et al., 1980; Miller et al., 1963), two species known to show visible symptoms of injury when exposed to ambient levels of ozone. Similar results have been found for loblolly (Stow et al., 1992), shortleaf (Flagler et al., 1994), and slash pine (Byres et al., 1992a) even in the absence of visible injury. Intensive observations of needle formation and loss have indicated that only needle longevity and not needle production is diminished by ozone (Byres et al., 1992a; Stow et al., 1992). Stow et al. (1992) found that needle abscission in loblolly pine was initiated after exposure to between 130 and 220 ppm · h of ozone and that ambient levels of ozone (96, 113, and 138 ppm · h in 1988, 1989, and 1990, respectively) reduced leaf area duration by 22 to 25%.

Because ozone reacts with cell membranes and can increase their permeability, there has been some concern that it may increase the leaching of nutrients from foliage. However, work with several species of pines that included concentrations of ozone as high as 140 ppb has not shown any increase in foliar leaching (Edwards GS et al., 1992b; Lovett and Hubbell, 1991; Skeffington and Roberts, 1985).

General Effects of Ozone on Growth

A number of investigators have attempted to determine, without the use of fumigation chambers, whether ozone is reducing the growth rates of pines. Benoit et al. (1982) and McLaughlin et al. (1982) conducted tree-ring analyses of symptomatic and asymptomatic individuals of white pine growing in the southern Appalachians. The radial growth of individuals with visible ozone injury was considerably less than that of individuals without symptoms. Van Deusen and Barnard (1989) compared the radial growth of loblolly pine from measurement plots where nearby plants of other species showed symptoms of ozone injury with the radial growth of trees from other plots where other species did not show symptoms. They reported that growth was significantly less on plots where symptoms of injury were found in South Carolina. These studies are valuable because they involve mature trees under field conditions, but they do not eliminate factors other than ozone that may affect the response measurements, and they do not allow a quantifiable evaluation of ozone effects. That requires experimentally applied treatments.

Since 1985, numerous researchers at many laboratories have established that if ozone exposures are high enough, they will eventually reduce the growth of southern pines (Adams et al., 1988, 1990; Albaugh et al., 1991; Chappelka et al., 1990b; Dean and Johnson, 1992; Edwards GS, 1992a; Edwards NT et al., 1990, 1992; Flagler et al., 1992; Hogsett et al., 1985; Kress and Allen, 1991; Lee et al., 1990; McLaughlin et al., 1994; Meier et al., 1990; Mudano et al., 1992; Qiu et al., 1992; Schier et al., 1990; Shafer and Heagle, 1989; Shafer et al., 1987, 1993; Shelburne et al., 1993; Simmons and Kelly, 1989; Somerville et al., 1992; Tjoelker and Luxmoore, 1991). These experiments involved fumigating loblolly, shortleaf, and slash pine in growth chambers, continuous stirred tank reactors (CSTRs), and opentop chambers. They utilized a wide variety of exposure profiles and durations that lasted from a few weeks to a few years. Many of these studies involved proportional additions of ozone to ambient air (e.g., ambient versus twice ambient) instead of a constant exposure (e.g., 100 ppb for 6 h/day). Typical response variables were height, diameter, and biomass of the foliage, stems, and roots. Results of the various experiments varied, but, in general, ozone was more likely to reduce diameter growth than height growth and biomass was usually more responsive than dimensional variables. Specific details of many of these studies can be found in Chapters 10 and 11.

Exposure to near-ambient levels of ozone can elicit responses that are more difficult to interpret. For example, one would expect growth to decrease as ozone increased. However, McLaughlin et al. (1994) reported that height growth of loblolly pine seedlings at ambient levels of ozone was less than at slightly higher levels of ozone. Similar results have been reported for volume in this species (Kress and Allen, 1991) and biomass in shortleaf pine (Shelburne et al., 1993). In addition, Adams et al. (1988) found that biomass of seedlings at subambient levels of ozone was less than that of seedlings at ambient levels. It is difficult to offer a physiological explanation for the apparent nonlinear responses to ozone that are sometimes reported, but they seem to be widespread. Since all these responses involve aboveground portions of trees, apparent increases in growth at near-ambient levels of ozone may be the result of subtle changes in the allocation of carbon within plants. It is difficult to collect data on the belowground effects of air pollutants on large plants growing in the field, but laboratory studies with slash pine seedlings suggest that the growth of roots is more sensitive to ozone than the growth of other plant parts (Hogsett et al., 1985).

Only a few people have attempted to quantify the effects of ambient ozone on growth of southern pines, possibly because of the nonlinear responses that are seen around ambient levels of this pollutant. Shafer and Heagle (1989) developed separate response models for four families in each of three growing seasons. In some cases, they had to eliminate data for one treatment level because nonlinear responses prevented successful model development. They found stem biomass and root collar diameter to be more consistent indicators of the effects of ozone than height, which responded erratically, or needle biomass, which was complicated by simultaneous growth of young needles and abscission of old needles. Using stem biomass data, they reported that ambient levels of ozone (12 h/day seasonal mean concentrations of 22 and 29 ppb ozone in 1986 and 1987, respectively) reduced the growth of loblolly pine from 0 to 19% depending on which family and year were analyzed.

Flagler et al. (1992) used a procedure described by Rawlings and Spruill (1992) to analyze the combined data from the six large SCFRC experiments mentioned earlier. Although these data included three species and sites that differed in soil, water availability, and ambient ozone levels, they chose to combine species and sites and develop separate models for each rain pH treatment. At a rain pH of 4.3, their model predicted that ambient levels of ozone (12 h/day seasonal exposures ranged from 167 to 262 ppm.h) would result in a 14% loss of foliage biomass, small decreases in stem and total biomass and diameter, and a small increase in height when compared to charcoal-filtered air (12 h/day seasonal exposures ranged from 31 to 117 ppm.h). However, because charcoal-filtered air contains very low levels of ozone, some researchers (Lefohn et al., 1990) question its validity as a control for ozone-effects research.

Ozone Uptake and Its Effect on Stomatal Conductance

Ozone enters pines through stomata and passes into the air spaces between the mesophyll cells of the needle interior. Under low to moderate O_3 concentrations, leaf diffusive resistance accounts for about 90 to 100% of the total resistance of the leaf to O_3 flux (Heath, 1980). Mesophyll, or internal, resistance probably accounts for the remainder. The specific role of mesophyll resistance in O_3 uptake is unclear. However, in general, the magnitude of mesophyll resistance decreases as the concentration of O_3 decreases. Consequently, at ambient levels of O_3, leaf diffusive resistance probably accounts for nearly 100% of the total leaf resistance to O_3 flux, with little or no influence due to mesophyll resistance. Based on these relationships, Reich (1987) postulated that the leaf diffusive conductance of a given species directly determines the quantity of O_3 uptake and plant response for any given O_3 dose. He further concluded that ozone uptake may be the single most important factor determining plant response to ozone and that interspecific variation in ozone uptake can be explained by species differences in inherent leaf conductance.

Exposure to ozone can decrease stomatal conductance in loblolly and shortleaf (Elsik et al., 1992) as well as in other species of pine (Barnes, 1972a; Coyne and Bingham, 1982; Weber et al., 1993). At one time, agricultural researchers suggested that stomatal conductance was directly affected by ozone (e.g., Hill and Littlefield, 1969; Rich and Turner, 1972; Rosen et al., 1978), but more recent work (Reich et al., 1985; Temple, 1986) suggests that at near-ambient levels, ozone influences stomatal conductance through its effect on photosynthesis. This issue has yet to be resolved and is still being discussed in the literature (Flagler et al., 1994; Heath, 1994; Keller and Hasler, 1984; Reich and Lassoie, 1984; Skarby et al., 1987; Weber et al., 1993).

Effects of Ozone on Physiology

An early evaluation of the effects of ozone on rates of photosynthesis in southern pines (Barnes, 1972a) showed that exposing loblolly and slash pine to 50 ppb ozone for 18 weeks decreased rates of photosynthesis by 9 and 15%, respectively. A number of more recent, longer term studies, many of which are described in detail in Chapter 11, have corroborated and elaborated on these results. In a two-year field study of loblolly pine seedlings, Richardson et al. (1992b) evaluated the effects of exposure to ozone levels ranging from half to three times those found in ambient air and found that photosynthetic rates were inhibited after a cumulative exposure to between 100 and 150 ppm · h, which was usually achieved near the end of one growing season in ambient air. They found that increasing the cumulative exposure to between 250 and 300 ppm · h, which was only achieved at more than twice ambient levels of ozone, decreased rates of photosynthesis by about 90%. They further estimated that ambient levels of ozone in the South can reduce the instantaneous carbon fixation rate of loblolly pine by about 19% (Richardson et al., 1992a). In other multiyear field studies, Flagler et al. (1994) found that while there was no difference in rates of photosynthesis in shortleaf pine between charcoal-filtered air and ambient air, the rate decreased linearly as ozone concentrations increased from ambient to 1.7 and 2.25 times ambient. Byres et al. (1992b) found that increasing ozone concentrations from ambient levels to two or three times ambient levels reduced rates of photosynthesis in slash pine by 15 to 38%.

The effects of ozone on photosynthesis in shorter term studies are sometimes not significant. Barnes (1972a) reported that there was no effect of 50 or 150 ppb ozone on photosynthesis of three pine species in a 12-week laboratory study with minimal replication. Lee et al. (1990) reported no significant effect of 100 ppb ozone on loblolly pine photosynthesis when it was applied 4 h/day, 3 days/week for 10 weeks. Tjoelker and Luxmoore (1991) found no significant effects of ambient levels plus 60 ppb ozone on photosynthesis of loblolly pine in an 18-week field study, whereas Hanson et al. (1988) found inconsistent effects in a 12-week laboratory study in which up to 320 ppb ozone were applied 6 h/day for 4 days/week. The cumulative ozone doses applied in many of these studies were considerably less than, or only barely within, the minimum dose that Richardson et al. (1992b) suggested was required for reductions in photosynthesis.

Sasek et al. (1991) noted that ozone exposure during one growing season can result in decreased rates of photosynthesis in loblolly pine during the following season. This was true for needles exposed during the first season as well as for new needles produced during the second spring. However, there seems to be a different kind of carry-over effect among needle flushes within a growing season—later flushes of plants exposed to high levels of ozone appear to show increased photosynthetic potential when expressed on a per-unit-area basis.

Flagler et al. (1994) noted that leaf-level gas exchange measurements made on shortleaf pine did not describe adequately the impact of ozone on photosynthetic processes. Calculations of whole-seedling gas exchange that take into account needle surface area and differences among flushes show larger reductions in photosynthesis as a result of ozone exposure.

Richardson et al. (1992b) found that short exposures to relatively high concentrations of ozone were no more effective at reducing rates of photosynthesis than long exposures to relatively low concentrations of ozone. This suggests that ozone standards based on exposure parameters that heavily weight peak ozone standards may not be the most appropriate criteria for protecting southern pines.

The measures of photosynthesis discussed thus far actually describe net photosynthesis, which includes both photosynthesis and light respiration. Little information is available on the effects of ozone on respiration, and those data that are available appear inconsistent. Skarby et al. (1987) reported that 125 ppb ozone increased dark respiration of Scots pine shoots, and Edwards et al. (1992a) reported that twice-ambient levels of ozone increased total respirational losses in loblolly pine. The increased respiration may be due to repair of ozone-induced injury to the photosynthetic system. Barnes (1972a) reported that 150 ppb ozone increased dark respiration of shoots of slash pine and loblolly pine after 36 days of exposure but that there was no difference after 77 or 84 days of exposure.

Hanson et al. (1988) reported that 320 ppb ozone applied 6 h/day for 4 days/week for 13 weeks had no effect on dark respiration of loblolly pine. Edwards (1991) reported suppressed respiration of loblolly pine roots in seedlings exposed to twice-ambient levels of ozone. Obviously, additional work is needed in this area, but it is possible that reported reductions in net photosynthesis following exposure to ozone may be at least partially due to increases in respiration.

Carbohydrates fixed by photosynthesis that are not consumed by maintenance respiration, repair, or growth may be available for production of nonstructural carbohydrates. A number of researchers have reported that when southern pines are exposed to above-ambient levels of ozone, the concentration of starch in the needles is reduced (Friend et al., 1992; Horton et al., 1989; Meier et al., 1990; Paynter et al., 1991). Friend and Tomlinson (1992) found that three years of exposure to ambient ozone not only decreased the amount of carbon that went into starch formation in August but also increased the amount that went into formation of organic acids, lipids, and pigments. They inferred that this was the result of injury and repair processes. Edwards GS et al. (1992a) found that twiceambient levels of ozone also reduced carbon export from the foliage of loblolly pine to its shoots and roots. Faulkner et al. (1991) showed that thrice-ambient levels of ozone reduced the carbohydrate content in the roots. Spence et al. (1990) suggested that carbohydrates in ozone-exposed pines that were not reaching the roots were accumulating in the stems and branches.

Effects of Ozone on Metabolism

Ozone is a highly reactive gas. Most researchers agree that its mode of action is through its capacity to oxidize certain metabolites (Kozlowski and Constantinidou, 1986). It is thought that ozone cannot travel far inside a plant cell before it is decomposed or transformed. Very little is known about what compounds are produced from ozone once it enters the leaf, although the production of hydrogen peroxide, free radicals, and superoxide is often hypothesized (Heath, 1994). Ozone and its products are known to react quickly with certain components of biological membranes, changing their permeability (Mudd et al., 1984). It is usually assumed that either the double bonds of membrane lipids or thiol groups of proteins are attacked (Manderscheid et al., 1992). It is also assumed that various antioxidants, such as ascorbate, superoxide dismutase, or glutathione, can provide protection from ozone effects.

There has been relatively little work in this area on southern pines. Ozone is supposed to increase the permeability of cell membranes (Fangmeier et al., 1990), but Keller (1986) showed that exposure to 50 ppb ozone for a growing season decreased cell permeability of spruce. Fangmeier et al. (1990) demonstrated that exposure to ozone during needle elongation changed the types of fatty acids in loblolly pine needles and that exposure after needle elongation decreased the total amount of fatty acids, but most of the differences they reported were between ambient and thrice-ambient levels of this pollutant. Chameides (1989) suggested that reactions with ascorbate could absorb all the ozone entering plant cells. Barnes (1972b) found that exposure to 50 ppb ozone for as little as four weeks resulted in increased levels of ascorbic acid in shortleaf, slash, pond, and loblolly pine. Mandersheid et al. (1991) saw similar results for loblolly, but only after several months of exposure to 1.4 to 3.0 times ambient levels of ozone. Sheu and Chappelka (1993) found that long-term exposure to 1.7 times ambient levels of ozone resulted in increased levels of superoxide dismutase in an ozone-tolerant family of loblolly pine but not in an ozone-sensitive family, suggesting that this enzyme may play a role in ozone tolerance. Variable results have been found for the enzyme peroxidase in loblolly pine. Sheu and Chappelka (1993) reported that exposure to ozone had no effect on levels of peroxidase in an ozone-tolerant family but decreased it in an ozone-sensitive family. Manderscheid et al. (1991) reported increased levels of peroxidase in loblolly pine foliage exposed to twice-ambient levels of ozone. Manderscheid et al. (1992) found increases in enzymes and amino acids involved in protein breakdown in loblolly pine exposed to the same amount of ozone and proposed that these compounds play a role in reallocating nutrients from senescent needles.

Effects of Ozone on Reproductive Processes

Although ozone has been shown to have detrimental effects on reproductive processes, particularly pollen germination (Feder, 1968; Mumford et al., 1972), little of this research has been done with pines. Benoit et al. (1983) found that although 100 ppb ozone did not affect pollen production or germination in white pine, fumigation under wet conditions with 150 ppb ozone did reduce pollen germination.

Seed production may be reduced in some forest tree species by oxidant pollution. For example, Luck (1980) reported reduced cone production per tree in Jeffrey and ponderosa pine growing in areas subjected to severe oxidant pollution in southern California but failed to eliminate the possibility that other factors may have contributed to this effect.

Variation Among Pines in Sensitivity to Ozone

Plant susceptibility to ozone depends on a number of factors, including genotype, stage of development, leaf age, and environmental conditions. Interspecific variation in ozone resistance has been summarized in various lists (e.g., see Davis and Wilhour, 1976). These usually categorize species by the severity of symptoms after brief exposures to high concentrations of ozone. In southern California, interspecific differences in sensitivity to ozone led, at least in part, to the replacement of ponderosa pine with other species (Miller, 1973). Ozone levels are lower in the Southeast than in southern California, and there is no evidence that ambient ozone is responsible for similar changes in species composition there.

Only a few studies have been conducted specifically to evaluate genetic variation in ozone tolerance within southern pines. These have estimated the heritability of loblolly pine for sensitivity to visible injury (Weir, 1977). In one study of 40 open-pollinated families,there was substantial genetic variation for this trait and heritabilities were high enough to warrant a breeding program. However, a second study with 32 full-sib families found much less genetic variation for ozone tolerance.

Many fumigations of southern pines have been conducted with seedlots of known genetic origin. While not specifically designed to estimate genetic parameters, they do provide some genetic information. Taylor (1994) did an excellent job of extracting as much genetic information as possible from these loblolly pine fumigations. One of his conclusions was that ranking families according to visible foliar injury will not provide information about the effects of ozone on growth. In fact, family rankings determined with one growth-related parameter may not even be related to rankings determined with other growth-related parameters. This would make breeding specifically for ozone tolerance difficult. Taylor (1994) also found that ozone × family interactions account for a large amount of the variation within these experiments. He suggested that multiple genes control resistance to ozone injury and that some of these genes are closely linked to other traits that affect seedling performance. Thus, gains in air pollution tolerance may come at the cost of other important traits.

There is ample evidence that populations of plants in areas with high levels of SO_2 are more tolerant of this pollutant than populations of the same species in pristine areas (Bell et al., 1982; Horsman et al., 1978; Taylor and Murdy, 1975). There are only a few indications that this interspecific effect may occur in areas with high levels of ozone (Berrang et al., 1991; Dunn, 1959) and none that are specific for southern pines. Taylor (1994) suggested that ambient levels of ozone are high enough to select against particularly sensitive individuals of loblolly pine in natural populations. His estimates of selection pressure coupled with the long generation time of loblolly pine led him to guess that ambient levels of ozone have not yet altered the frequency of alleles in natural populations. However, he speculated that such changes may have already occurred in the intensively bred, managed stands of this species. An ongoing breeding program for optimum growth may be the best way to ensure that seedlings for commercial forests are tolerant of the air quality in their environment.

In addition to genetic considerations, the sensitivity of pines to ozone is affected by their stage of development, which may refer not only to the age of the plant but also to a stage in the life cycle of a plant organ, such as a leaf. Seedlings in the cotyledon stage are especially sensitive because of their limited food reserves. In general, conifers in the cotyledon or primary needle stages of development are more sensitive to ozone than those in the secondary needle stage. For example, Davis and Wood (1973) observed that cotyledons of Virginia pine were more sensitive to 250 ppb ozone than primary needles, which, in turn, were more sensitive than secondary needles. In fact, secondary needles of young seedlings apparently had the same degree of ozone tolerance as those three years old. Kress et al. (1982) also found that primary needles were more sensitive to 250 ppb ozone than secondary needles of loblolly pine but noted that the relative ranking of families for visible injury was about the same for the two needle types. Berry (1974) found that the maximum sensitivity to 250 ppb ozone for four species of southern pines occurred at two weeks old. The importance of the sensitivity of young seedlings is not known; natural selection is thought to be most intense in young stands, but ozone levels tend to be lowest close to the ground.

Almost all the data published on the effects of ozone on southern pines were obtained from trees less than three years old. Only a few studies have compared the responses of seedlings and mature southern pines. Kossuth and Biggs (1989) compared the effects of 150 ppb ozone on ramets of slash pine grafted from mature trees and seedlings and concluded that the response was similar. Teskey et al. (cited in Flagler et al., 1990) evaluated the effects of ambient and 2.5 times ambient levels of ozone on mature and seedling loblolly pine using branch chambers and open-top chambers. They concluded that while the response to above-ambient levels of ozone was similar for seedlings and mature trees, at near-ambient levels of ozone the seedlings experience a larger reduction in rates of photosynthesis than mature trees. Additional information on the correlation between juvenile and mature responses to ozone is needed before much of the existing experimental data can be extrapolated to field conditions.

Interactions Between Ozone and Other Factors

Because mycorrhizal fungi are affected by the supply of carbohydrates from the roots of their host (Hacskaylo, 1973), ozone-induced changes in total carbohydrates and changes in their distribution within pines may affect pines' suitability as host plants. Mycorrhizal infection was evaluated after a number of recent experiments in which southern pines were exposed to ozone. About half the published papers documented decreases in mycorrhizal infection as a result of ozone exposure. McQuattie and Schier (1992) and Meier et al. (1990) reported decreases in mycorrhizal infection in pitch and loblolly pine after exposure to 50 ppb ozone for about 12 weeks. Edwards and Kelly (1992) reported similar results for loblolly pine exposed to twice-ambient levels of ozone for two years. An equal number of papers report nonsignificant differences or differences that are not consistent over families, time, or other experimental treatments (Adams and O'Neill, 1991; Qiu et al., 1993; Simmons and Kelly, 1989). No authors reported that ozone significantly and consistently increased mycorrhizal infection. Long-term exposure to ozone may decrease mycorrhizal infection, but it is apparently difficult to detect these changes.

There is some evidence that ozone can predispose pines to attack by diseases or insects, either by weakening them or by changing the distribution or composition of carbohydrates or other compounds within the tree. Fenn et al. (1990) showed that both the frequency and the degree of infection of ponderosa pine seedlings with black-stain root disease (caused by *Leptographium wageneri*) increased as ozone exposure increased from subambient to thrice-ambient levels. Lackner and Alexander (1983) found that unchambered white pine showing symptoms of ozone injury were more likely to be infested with *L. procera* than those that did not show symptoms of ozone injury. Carey and Kelley (1994) found that exposure to 2.5 times ambient levels of ozone made loblolly pine seedlings more susceptible to pitch canker (caused by *Fusarium subglutinans*), a serious disease of southern pines. Studies of annosum root rot [caused by *Heterobasidion annosum* (Fr.) Bref.] on unchambered pines have produced inconsistent results (James et al., 1980; Leininger, 1989).

Bark beetles (*Dendroctonus* spp.) are important pests of pines in the West and Southeast, frequently attacking trees weakened by other factors such as drought. In southern California, these beetles were the cause of death for many ponderosa pine damaged by ozone (Stark et al., 1968). It is not unlikely that a similar relationship exists between the southern pine beetle and ozone, but the evidence needed to test this hypothesis is not available.

Numerous environmental factors also influence the magnitude of the response to ozone. Especially important are light intensity, humidity, temperature, and soil moisture. In general, these environmental factors influence ozone response through their effects on stomatal aperture, which controls O_3 uptake. Any environmental factor that tends to reduce stomatal aperture has the potential to reduce O_3 uptake. The effect of many of these factors on stomatal conductance of loblolly pine has been described by Teskey et al. (1986). With the exception of the effects of soil moisture, relatively little research has actually been done on the interactions between ozone and these factors in pines.

Water stress usually results in reductions in stomatal conductance, and experiments with agricultural crops (Heagle et al., 1987, 1988) have shown that although water stress itself tends to decrease yield, it also tends to decrease the effects of ozone on yield. Elsik et al. (1992) found simultaneous exposure of loblolly and shortleaf pine to 1.7 times ambient levels of ozone and water stress resulted in changes in leaf biomass in well-watered loblolly pine but not in water-stressed loblolly pine. Other water × ozone interactions generally were not significant, probably because only mild water stress was induced. The effects of ozone stress on susceptibility to soil moisture deficit are not as clear. Ozone tends to decrease stomatal conductance, which would be expected to delay the onset of water stress; but in a study with loblolly pine, Lee et al. (1990) found that preexposure to 100 ppb ozone for 4 h/day for 3 days/week made transpiration more sensitive to the effects of water stress. As noted previously, ozone sometimes causes a greater reduction in root growth than in shoot growth. Decreases in the relative size of the roots as a result of previous exposure to ozone might increase susceptibility to drought. However, there is no experimental evidence to support this hypothesis in southern pines, and Chevone et al. (1990) have suggested that the slow onset of drought typical of most natural systems may often allow normal plant processes to compensate partially for lack of water.

Competition is an important source of stress for forest trees, but we are largely ignorant of the interactions between ozone and competition. This is partly because it is difficult to fumigate groups of large organisms. Innovative techniques are needed to evaluate the importance of these interactions.

Decreases in winter hardiness have been attributed to exposure to ozone in several conifers (Barnes and Davison, 1988; Fincher et al., 1989). The plasma membrane is credited with protecting plants from intracellular ice formation (Chambers and Hale, 1932; Mazur, 1969), and ozone-induced disruptions of this membrane could result in loss of cold hardiness. Experiments with some species of pines have found no effect of ambient levels of ozone on cold hardiness (Senser and Payer, 1989; Wellburn and Wellburn, 1994), but two studies with loblolly pine suggest that it can occur. Chappelka et al. (1990a) reported that after a late-season frost, foliar injury was more severe on seedlings exposed to 1.7 times ambient levels of ozone than on seedlings exposed to below-ambient levels of ozone. Edwards GS et al. (1990) used freezing and measurement of electrolyte leakage to determine that seedlings exposed to twice-ambient levels of ozone were more sensitive to low temperatures in the fall and spring than seedlings exposed to ambient levels of ozone.

Summary of Ozone

The high concentrations of ozone present in the mountains of southern California are known to affect the growth, health, and relative abundance of pines there. Above-ambient levels of ozone are also known to cause numerous and diverse effects on pine seedlings and saplings in the Southeast. These include changes in growth and in a number of physiological processes, including photosynthesis and stomatal conductance. If these effects occur at ambient concentrations for mature southern pines, this pollutant could have important ecological as well as economic impacts. However, many aspects of the response of southern pines to ozone are still poorly understood. The apparent stimulation of growth at near-ambient levels of this pollutant; belowground responses to ozone; the response of mature trees; interactions with other stresses, including drought, insects, and diseases; the effects of ozone on respiration; the implications of genetic variation for tolerance to this pollutant; and parameters for relating diverse ozone profiles to plant effects have yet to be researched fully.

Nitrogen Oxides

It has long been known that nitrogen oxides (NO_x) may injure vegetation, but there has been little research aimed specifically at their effects relative to the efforts expended on sulfur dioxide and ozone. The most complete recent review of this primary pollutant is the U.S. EPA's Air Quality Criteria document (1993).

Sources and Distribution of Nitrogen Oxides

A number of nitrogen oxides occur in the atmosphere. By far, the most common are nitric oxide (NO) and nitrogen dioxide (NO_2) (EPA, 1993). Almost 90% of the nitrogen oxides emitted in the United States come from anthropogenic sources. Estimates of natural sources of these pollutants are uncertain, and there is a tendency for natural sources to increase in importance during warm weather in rural areas. The major anthropogenic sources of nitrogen oxides are transportation and power generation (Placet et al., 1991).

It is difficult to estimate the exposure of southern pine forests to NO_x because there are few monitors in rural areas (Lefohn and Shadwick, 1991). Nationally, NO_2 hourly average concentrations tend to be between 1 and 20 ppb in rural areas and between 10 and 30 ppb near electrical generating plants (EPA, 1993). Lefohn and Benkovitz (1989) summarized the 1987 data from Mount Mitchell, a high-elevation site in North Carolina, and reported that 90% of the readings were less than 3 ppb and the maximum was 7 ppb. Allen et al. (1992) summarized the data from monitors at three forested, but not necessarily rural, SCFRC research sites located in Florida, Texas, and North Carolina. They reported that 90% of the readings were less than 10 ppb. Data from point sources can be variable. The yearly maximum NO_2 readings from near an electrical generating facility near Atlanta, Georgia ranged from 166 ppb in 1986 to 111 ppb in 1990, but in most years 90% of the readings from that monitor were less than 50 ppb (EPA, 1993).

Effects of Nitrogen Oxides on Southern Pines

Very little research has been directed at the responses of pines to nitrogen oxides. The EPA (1993) reviewed numerous experiments with mostly agricultural crops and suggested that the relationship between exposure to NO_x and effects on growth or yield is often nonlinear and may consist of (1) a threshold exposure that must be exceeded for any effect to occur, (2) a range of concentrations above this threshold that result in a stimulation of growth, and (3) a decrease in growth once this range is exceeded. They also noted that, typically, daily exposures to concentrations of 100 to 500 ppb NO_x for two weeks or longer would be required to reduce growth or yield.

Researchers working with Scots pine indicated that exposure to 500 ppb NO_2 for 6 h/day for 66 days did not cause visible injury (Oleksyn et al., 1988) but that exposure to the same concentration for three days did cause temporary reductions in rates of photosynthesis (Lorenc–Plucinska, 1988). When three clones of white pine were exposed to 100 ppb NO_2 for 4 h/day for 35 days, one of the three clones showed visible injury on about 10% of its needles. Exposing the same three clones to 300 ppb NO_2 for the same length of time caused visible injury on about 20% of the needles from two clones (Yang et al., 1983). In one experiment that included southern pines, Kress and Skelly (1982) reported that exposure to 100 ppb NO_2 for 6 h/day for 28 days caused significant reductions in shoot biomass for pitch pine and nonsignificant reductions in growth for loblolly and Virginia pine. A follow-up study with loblolly pine that used the same procedure resulted in significant reductions in growth for one of two families (Kress et al., 1982).

There are only two published cases in which injury to pines from NO_x may have occurred in the field in the Southeast (Phillips et al., 1977; Skelly et al., 1972; Stone and Skelly, 1974). Reductions in radial growth of loblolly and white pine near a U.S. Army munitions factory in Virginia were recorded for World War II and the Korean conflict, when emissions of NO_x and SO_2 from this plant would have been high. Monitoring data are not available for these time periods, but limited data collected during the Vietnam conflict, when production of ammunition was also high, suggest that levels of one or both of these pollutants may have been high enough to reduce growth. However, since we cannot eliminate the possibility that the reductions in growth were caused by SO_2, the reductions in growth cannot be attributed solely to NO_x.

Summary of Nitrogen Oxides

The threshold concentrations of NO_x required to cause reductions in growth of southern pines are so high that they rarely occur under ambient conditions. This suggests that NO_x is not an important problem for pine growth in the southeastern United States. However, NO_x is still important as a precursor for two widely distributed secondary pollutants—acid precipitation and ozone.

Fluorides

Fluorides are the most toxic of the common primary air pollutants and are capable of injuring susceptible species of pines at concentrations as low as 0.5 ppb (Rakowski, and Zwiazek, 1992). Reviews of the effects of fluorides on forest tree species include those written by Amundson and Weinstein (1980), Weinstein and Alscher-Herman (1982), Halbwachs (1984), and Kozlowski and Constantinidou (1986).

Sources and Distribution of Fluorides

A major anthropogenic source of airborne fluorides is the manufacturing of aluminum, steel, bricks, and phosphate fertilizers (EPA, 1978). Fluorides are point source pollutants; the atmospheric concentration and amount accumulated by plant tissues decrease rapidly with distance from these sources. The bulk of fluorides affecting plants are in gaseous form, primarily hydrogen fluoride. There are a number of natural sources of fluorides, including fumaroles and volcanoes. Fluorides can also occur naturally in soils, but this form is insoluble and has relatively little influence on vegetation.

Little data are available on fluoride concentrations in the ambient air (Amundson and Weinstein, 1980) because they are localized around point sources. Between 1965 and 1970, 97% of the samples collected in nonurban areas of the United States had no detectable levels of fluoride, and the highest level measured was 0.2 ppb (EPA, 1978). In 1982, Treshow and Anderson noted that major improvements in control of fluoride emissions had recently eliminated the obvious effects of this pollutant in most areas. This suggests that ambient levels of fluorides have decreased considerably since the EPA data were collected, but occasional releases of toxic amounts of these pollutants still occur.

Effects of Fluorides on Southern Pines

Fluoride typically causes tip burn of conifer needles, but the severity of the symptoms depends on the fluoride concentration. Low-level, chronic exposure to fluorides can result in chlorosis of the needle tips (EPA, 1978) or in premature abscission of older needles (Malhotra and Blauel, 1980). Treshow (1970) rated the relative sensitivity of a number of species to fluoride on the basis of visible symptoms. All pines on the list were identified as sensitive, but southern pines were not included in the classification. Rakowski and Zwiazek (1992) found that concentrations of hydrogen fluoride (HF) as low as 0.5 ppb for 20 days or 2.0 ppb for five days could cause tip necrosis in white pine.

A number of studies have been published on the effects of high fluoride concentrations on the growth of trees, but only a few have involved pines. Few researchers have had access to data on ambient levels of fluorides, and many have had to relate effects on plants to proximity to sources or to concentrations within the plants being affected. Havas and Huttunen (1972) reported reduced xylem increment in Scots pine growing in the vicinity of a fertilizer plant that emitted fluorine dusts. Gilbert (1983) found a significant correlation between annual brick production and annual xylem increment in Austrian pine (*P. nigra* Arnold) and concluded that fluorides from the brickworks inhibited diameter growth. Wheeler (1972) found that foliar leaching and litterfall were enhanced when concentrations of fluorides in the needles of loblolly and slash pine from North Carolina reached 13 to 49 ppm, but there was no visible injury and no reduction in wood production. Doley (1989) evaluated the effects of hydrogen fluoride on slash and several other species of pines. Basal diameter growth increased in comparison to controls after exposure to 1.5 ppb F for 200 days but decreased after exposure to 5.5 ppb F for the same period of time.

Summary of Fluorides

Although fluoride is the most toxic of all the major air pollutants, relatively little is released and it is extremely localized. A small number of studies have shown that fluorides cause visible injury and reductions in growth of southern pines at very low concentrations. However, except for isolated incidents, there seems to be no evidence that ambient levels of fluorides in the southeastern United States are adversely affecting pines. In fact, the dearth of recent scientific papers and government documents relating to the effects of fluorides on plants suggests that this pollutant is no longer considered important.

Heavy Metals

Many of the so-called heavy metals, or trace metals, are extremely toxic to vegetation and are known to devastate plant communities around strong point sources. The primary metals of concern are lead, copper, nickel, zinc, cadmium, and aluminum. Numerous reviews have dealt specifically with the effects of heavy metals on vegetation: Foy et al. (1978), Hutchinson (1980), Kozlowski and Constantinidou (1986), Lepp (1981), McLaughlin (1985), Norton et al. (1980), and Smith (1991).

Sources and Distribution of Heavy Metals

The primary anthropogenic sources of heavy metals are emissions from large industrial facilities, such as iron and steel mills, base metal smelters, refineries, and foundries. The major source of lead is exhaust from motor vehicles using leaded gasoline. The distribution of most heavy metals tends to be localized around point sources. Lead tends to accumulate in urban areas and roads.

Although heavy metals originate from point sources, there is some evidence that heavy metals are more widely distributed than is generally recognized. Heavy metals tend to be associated with small airborne particles, which are more likely to escape emissions control and can be transported long distances (Smith, 1991). Even at low deposition rates, heavy metals can accumulate in soils because they tend to remain there for long periods of time. Hutchinson (1980) reported residence times for heavy metals in soils ranging from 300 to 3000 years. Fernandez and Czapowskyj (1986) reported finding higher levels of cadmium, copper, nickel, lead, and zinc in forest floor samples collected in rural areas of northern Maine than would be expected far from point sources. Friedland et al. (1984) reported that lead, copper, and zinc increased in rural areas of New England over a 14-year period. Lindberg et al. (1985) compared reported rates of heavy metal deposition from Georgia and Tennessee with other areas. Although deposition of cadmium, manganese, lead, and zinc on rural sites in the southeastern United States was considerably lower than on rural sites in the northeastern United States and Germany, it was 50 to 1000 times higher than deposition on a presumably pristine site in Greenland and Antarctica. Additional data are needed to determine how widespread elevated heavy metal deposition is in the Southeast.

Effects of Heavy Metals on Southern Pines

Although all heavy metals can be toxic at some concentration, direct toxicity seems to be relatively rare in the field, occurring mainly in the vicinity of point sources (Smith, 1991). The primary mechanism for uptake is by roots from the soil. Heavy metals deposited to the soil in various industrial dusts or in precipitation become more soluble and more available to plants as the soil becomes more acidic (Malhotra and Blauel, 1980). This raises concerns about decreases in soil pH regardless of the cause.

In conifer stands, chronic exposure to heavy metals may reduce growth and yield without the development of visible injury for several years, but acute exposures can cause rapid development of symptoms. General chlorosis and premature defoliation result in crown deterioration of affected trees. Severe heavy metal pollution may cause complete denudation of the area over the course of several years (Malhotra and Blauel, 1980).

Jordon et al. (1990) investigated the relationship between radial growth and heavy metals in mature loblolly pine growing in east Alabama. They found a significant negative correlation between the stand basal area increment at various points in time and concentrations of cadmium, lead, and zinc in the xylem formed at those times. Increment cores collected from shortleaf pine in east Tennessee showed suppressed radial growth and steadily increasing concentrations of Al, Cd, Cu, Fe, Mn, Mo, Ti, and Zn over the past 20 to 25 years (Baes and McLaughlin, 1984; Baes et al., 1983). Although field studies such as these are invaluable, they do not eliminate the possibility that other factors may have caused the reductions in growth that were measured.

A number of researchers have used controlled exposures to document the effects of heavy metals on the growth of southern pines. Kelly et al. (1979) observed that $CdCl_2$ reduced the height growth and root dry weight of white and loblolly pine when it was applied to the soil at 100 μg/g but not when it was applied at 15 μg/g. Schier et al. (1990) found that aluminum reduced height growth and biomass of all plant components in sand-cultured pitch pine when applied with a nutrient solution at 25 mg/l. Concentrations of aluminum as low as 5 mg/l have also been shown to reduce the growth of roots in several species of hydroponically grown southern pines (Pagnelli et al., 1987). However, Seiler and Pagnelli (1987) and Chappelka et al. (1991) found that growth of loblolly pine was relatively tolerant of lead even when applied to the soil at rates as high as 600 mg/kg.

Summary of Heavy Metals

The ecological devastation that can occur around strong point sources of heavy metals is not an important problem in the Southeast today. However, there is limited evidence that heavy metals may be more widespread than once thought, and the possibility that they may affect southern pines at ambient concentrations cannot be eliminated. Additional work on the distribution of these pollutants, their effects on forest growth, and the role they play in nutrient cycling is needed before even tentative conclusions can be justified.

Conclusions

During the last three decades, solid advances have been made in our understanding of the effects of air pollutants on pines and our ability to minimize the damage from these pollutants. However, many important questions remain to be answered.

In the mid-1960s, it was not unusual to find visible injury to vegetation from hydrogen fluoride or sulfur dioxide. Today, air pollution controls have largely eliminated hydrogen fluoride and sulfur dioxide damage to southern pines. It was discovered in the mid-1960s that the extremely high levels of ozone in southern California were causing widespread damage to ponderosa pine. It was also realized that somewhat lower levels of ozone in the eastern United States were injuring unusually sensitive individuals of white pine. It is now known that the high levels of ozone in southern California, in conjunction with other stressors, caused decreases in growth, reductions in rates of photosynthesis, and changes in species composition. Moreover, it is now considered likely that levels of ozone in the East have eliminated sensitive individuals of white pine from some natural populations, and it is considered possible that they have also eliminated sensitive genotypes of loblolly pine from highly selected breeding populations. It is known that ambient levels of ozone cause premature needle loss in pines in the Southeast and that slightly higher than ambient levels of ozone can cause reductions in rates of growth and photosynthesis. It is even possible that ambient levels of ozone in the South are reducing rates of pines, but inconsistencies in the response of pines exposed to these levels require additional research before a question of this importance can be answered with certainty.

A number of other important research questions remain regarding the effects of ozone on southern pines. Almost all research has been conducted on seedlings and saplings. Does ozone have the same impact on mature trees? How do the effects of ozone measured on individual plants reflect what happens to trees growing in forests under competitive conditions? How does ozone interact with other stressors, such as drought, insects, or disease? What kind of secondary ozone standard will minimize the effects of ambient ozone on this resource without unduly impacting economic development? What are the belowground effects of ozone? What are the effects of ozone on respiration and allocation of carbohydrates?

In the midst of this 30-year period, acid precipitation was first recognized as a widely distributed pollutant. Research in the Northeast and at high elevations, where precipitation tends to be more acidic than at low elevations in the Southeast, has implicated this pollutant in changes in cold hardiness of red spruce. At one time, acidic precipitation was blamed for many tree health problems in the Southeast, but numerous experiments have demonstrated that the direct effects of acid precipitation on southern pines are relatively unimportant. However, a number of indirect effects of acid rain are still considered potentially important. One of the most important indirect effects of acid precipitation may be changes in soil productivity. It appears possible that ambient rainfall could result in cation leaching or aluminum toxicity in some southeastern soils. However, these changes would require decades to develop and would be extremely difficult to detect. Techniques should be developed to determine if these hypothesized indirect effects of acid precipitation are actually occurring.

Heavy metals have long been recognized as a localized problem around point sources. There is limited evidence that these chemicals may be more widespread than once thought, but little is known about the effects on southern pines or associated organisms. Much more data are needed on the distribution of these pollutants and their effects on southern pine ecosystems before their impacts can be evaluated.

Advances in air pollution control made over the past 30 years have helped protect the productivity of one of the Southeast's most valuable natural resources, the southern pines. However, advances in our understanding of plant biology continue to make us aware of new questions and more subtle but potentially important threats. An ongoing research program will be the key to continuing to protect this resource during the next 30 years.

References

Abrahamsen G (1984) Effects of acidic deposition on forest soil and vegetation. Philos Trans R Soc Lond B Biol Sci 305:369–382.

Abrahamsen G, Bjor K, Horntvedt R, Tveite B (1976) Effects of acid precipitation on coniferous forests. In: Braekke FH (ed) *Impact of Acid Precipitation on Forest and Freshwater Ecosystems in Norway*. SNSF Res Rep FR 6/76, SNSF Project, Oslo-As, Norway, pp 36–63.

Adams MB, Kelly JM, Edwards NT (1988) Growth of *Pinus taeda* L. seedlings varies with family and ozone exposure level. Water Air Soil Pollut 38:137–150.

Adams MB, Kelly JM, Taylor GE Jr (1990) Growth of five families during three years of ozone exposure. New Phytol 116:689–694.

Adams MB, O'Neill EG (1991) Effects of ozone and acidic deposition on carbon allocation and mycorrhizal colonization of *Pinus taeda* L. seedlings. For Sci 37:5–16.

Adams RM, Garner JHB, Kress LW, Laurence JA, Oshima R, Pell E, Reinert R, Taylor OC, Tingey DT, Gumpertz M, Lang DS (1986) Effects of ozone and other photochemical oxidants on vegetation. In: *Air Quality Criteria for Ozone and Other Photochemical Oxidants*. Vol. III. EPA/600/8-84/020cF. US Environmental Protection Agency, Research Triangle Park, NC, pp 1–297.

Afonsi DH, Sandroni S, Viarengo S (1991) Tropospheric ozone in the nineteenth century: The Moncalieri series. J Geophys Res 96:17349–17352.

Albaugh TJ, Allen HL, Kress LW (1991) Biomass-D^2H relationships for young loblolly pine as affected by ozone. Biomass and Bioenergy 1:143–148.

Allen ER, Sutton RL, Evans GF (1992) Air quality characterization and acid deposition monitoring at three southern commercial forest sites. In: Flagler RB (ed) *The Response of Southern Commercial Forests to Air Pollution*. Air and Waste Management Assoc., Pittsburgh, PA, pp 95–110.

Altshuller AP (1983) Review: Natural volatile organic substances and their effect on air quality in the United States. Atmospheric Environ 17:2131–2165.

Altshuller AP (1987) Estimation of the natural background of ozone present at surface rural locations. J Air Pollut Control Assoc 37:1409–1417.

Amthor JS (1986) An estimate of the "cost" of nutrient leaching from forest canopies by rain. New Phytol 102:359–364.

Amundson RG, Weinstein LH (1980) Effects of airborne F on forest ecosystems. In: *Proc Symp Effects of Air Pollut on Mediterranean and Temperate For Ecosystems*, June 1980, Riverside, CA. USDA Forest Service Gen Tech Rep PSW-43, Pacific Southwest For and Range Exp Stn, Berkeley, CA, pp 63–78.

Axelrod MC, Coyne PI, Bingham GE, Kercher JR, Miller PR, Hung RC (1980) Canopy analysis of pollutant injured ponderosa pine in the San Bernardino National Forest. In: *Proc Symp Effects of Air Pollut on Mediterranean and Temperate For Ecosystems*, June 1980, Riverside, CA. USDA Forest Service, Gen Tech Rep PSW-43, Pacific Southwest For and Range Exp Stn, Berkeley, CA, p 227.

Back J, Huttunen S (1992a) Effects of long-term exposure to simulated acid rain on conifer needle ultrastructure and hardening status. For Ecol and Management 51:95–103.

Back J, Huttunen S (1992b) Structural responses of needles of conifer seedlings to acid rain treatment. New Phytol 120:77–88.

Baes CF, McLaughlin SB (1984) Trace elements in tree rings: Evidence of recent and historical pollution. Science 224:494–495.

Baes CF, McLaughlin SB, Hagan TA (1983) Multielemental analysis of tree rings: Temporal accumulation patterns and relationships with air pollution. In: Davi, DD, Miller AA, Dochinger L (eds) *Air Pollution and Productivity of the Forest*. Izaak Walton League of America, Arlington, VA, pp 273–286.

Baker TR, Allen HL, Kress LW (1994) Nutritional response of loblolly pine exposed to ozone and simulated acid rain. Can J For Res 24:453–461.

Barklund P, Axelsson G, Unestam T (1984) *Gremmeniella abietina* in Norway spruce, latent infection, sudden outbreaks, acid rain, predisposition. In: Manion PD (ed) *Sclerroderris Canker of Conifers*. For Sci 13:111–113.

Barnes JD, Davison AW (1988) The influence of ozone on the winter hardiness of Norway spruce (*Picea abies* (L.) Karst.). New Phytol 108:159–166.

Barnes RL (1972a) Effects of chronic exposure to ozone on photosynthesis and respiration of pines. Environ Pollut 3:133–138.

Barnes RL (1972b) Effects of chronic exposure to ozone on soluble sugar and ascorbic acid contents of pine seedlings. Can J Bot 50:215–219.

Bechtold WA, Ruark GA, Lloyd FT (1991) Changing stand structure and regional growth reductions in Georgia's natural pine stands. For Sci 37:703–717.

Bell JNB (1984) Air pollution problems in western Europe. In: Koziol MJ, Whatley R (eds) *Gaseous Air Pollutants and Plant Metabolism*. Butterworths, Boston, pp 3–24.

Bell JNB, Ayazloo M, Wilson GB (1982) Selection for sulphur dioxide tolerance in grass populations in polluted areas. In: Bornkman R, Lee JA, Seaward MRD (eds) *Urban Ecology*. Blackwell Scientific, Oxford, pp 171–180.

Bennett JP, Anderson RL, Mielke ML, Ebersole JJ (1994) Foliar injury air pollution surveys of eastern white pine (*Pinus strobus* L.): A review. Environ Monitor Assess 30:247–274.

Benoit LF, Skelly JM, Moore LD, Dochinger LS (1982) Radial growth reductions of *Pinus strobus* L. correlated with foliar ozone sensitivity as an indicator of ozone-induced losses in eastern forests. Can J For Res 12:673–678.

Benoit LF, Skelly JM, Moore LD, Dochinger LS (1983) The influence of ozone on *Pinus strobus* L. pollen germination. Can J For Res 13:184–187.

Berrang P, Karnosky DF, Bennett JP (1991) Natural selection for ozone tolerance in *Populus tremuloides*: An evaluation of nationwide trends. Can J For Res 21:1091–1097.

Berry CR (1961) *White Pine Emergence Tipburn, a Physiographic Disturbance*. USDA Forest Service, Stn Pap 130. Southeast For Exp Stn, Asheville, NC.

Berry CR (1964) Differences in concentrations of surface oxidant between valley and mountaintop conditions in the southern Appalachians. J Air Pollut Control Assoc 14:238–239.

Berry CR (1974) Age of pine seedlings with primary needles affects sensitivity to ozone and sulfur dioxide. Phytopathology 64:207–209.

Berry CR, Ripperton LA (1963) Ozone, a possible cause of white pine emergence tipburn. Phytopathology 53:552–557.

Binkley D, Driscoll CT, Allen HL, Schoeneberger P, McAvoy D (1989) *Acid Deposition and Forest Soils. Context and Case Studies of the Southeastern United States*. Springer-Verlag, New York.

Bruck RI, Shafer SR, Heagle AS (1981) Effects of simulated acid rain on the development of fusiform rust on loblolly pine. (Abstract) Phytopathology 71:864.

Burgess RL (ed) (1984) *Effects of Acidic Deposition on Forest Ecosystems in the Northeastern United States: An Evaluation of Current Evidence*. Publ No ESF 84-016. College of Environ Sci and For, State Univ NY, Syracuse.

Byres DP, Dean TJ, Johnson JD (1992a) Long-term effects of ozone and simulated acid rain on the foliage dynamics of slash pine (*Pinus elliottii* var. *elliottii* Engelm.). New Phytol 120:61–67.

Byres DP, Johnson JD, Dean TJ (1992b) Seasonal response of slash pine (*Pinus elliottii* var. *elliottii* Engelm.) photosynthesis to long-term exposure to ozone and acidic precipitation. New Phytol 122:91–96.

Cape JN (1993) Direct damage to vegetation caused by acid rain and polluted cloud: definition of critical levels for forest trees. Environ Pollut 82:176–180.

Cape JN, Leith ID, Fowler D, Murray MB, Sheppard LJ, Eamus D, Wilson RHF (1991) Sulphate and ammonium in mist impair the frost hardening of red spruce seedlings. New Phytol 118:119–126.

Carey WA, Kelley WD (1994) Interaction of ozone exposure and *Fusarium subglutinans* inoculation on growth and disease development of loblolly pine seedlings. Environ Pollut 84:35–43.

Chambers R, Hale HP (1932) The formation of ice in protoplasm. Proc R Soc B110:337–352.

Chameides WL (1989) The chemistry of ozone depletion in plant leaves: Role of ascorbic acid. Environ Sci Tech 23:595–600.

Chameides WL, Kasibhatla PS, Yienger J, Levy H (1994) Growth of continental-scale metro-agro-plexes, regional ozone pollution and world food production. Science 264:74–77.

Chameides WL, Lindsay RW, Richardson J, Kiang CS (1988) The role of biogenic hydrocarbons in urban photochemical smog: Atlanta as a case study. Science 241:1473–1475.

Chameides WL, Lodge JP (1992) Tropospheric ozone: Formation and fate. In: Lefohn AS (ed) *Surface Level Ozone Exposures and Their Effects on Vegetation*. Lewis Publ, Chelaea, MI, pp 5–30.

Chao TT, Harwood ME, Fang SC (1964) Iron or aluminum coatings in relation to sulfate adsorption characteristics of soils. Soil Sci Am Proc 28:632–635.

Chappelka AH, Chevone BI (1992) Tree responses to ozone. In: Lefohn AS (ed) *Surface Level Ozone Exposures and Their Effects on Vegetation*. Lewis Publ, Chelsea, MI, pp 271–324.

Chappelka AH, Kush JS, Meldahl RS, Lockaby BG (1990a) An ozone-low temperature interaction in loblolly pine (*Pinus taeda* L.) New Phytol 114:721–726.

Chappelka AH, Kush JS, Runion GB, Meier S, Kelley WD (1991) Effects of soil-applied lead on seedling growth and ectomycorrhizal colonization of loblolly pine. Environ Pollut 72:307–316.

Chappelka AH, Lockaby BG, Mitchell RJ, Meldahl RS, Kush JK, Jordan DN (1990b) *Growth and Physiological Responses of Loblolly Pine Exposed to Ozone and Simulated Acidic Rain in the Field*. Presented at the 83rd Annual Meeting and Exhibition, Air and Waste Management Assoc, Pittsburgh, PA, June 1990. Paper no 90-187.5.

Chevone BI, Seiler JR, Melkonian J, Amundson RG (1990) Ozone—water stress interactions. In: Alscher RG, Cumming JR (eds) *Stress Responses in Plants: Adaptation and Acclimation Mechanisms*. Wiley–Liss, New York, pp 311–328.
Cleveland GD, Haines LW, Jahromi ST, Bryant RL (1992) No evidence of growth decline in International Paper's southern commercial loblolly and slash pine forests. In: Flagler RB (ed) *The Response of Southern Commercial Forests to Air Pollution*. Air and Waste Management Assoc, Pittsburgh, PA, pp 37–47.
Cole DW, Johnson DW (1977) Atmospheric sulfate additions and cation leaching in a Douglas-fir ecosystem. Water Resources Research 13:313–317.
Constantinidou HA, Kozlowski TT, Jensen K (1976) Effects of sulfur dioxide on *Pinus resinosa* seedlings in the cotyledon stage. J Environ Quality 5:141–144.
Costonis AC (1970) Acute foliar injury of eastern white pine induced by sulfur dioxide and ozone. Phytopathology 60:994–999.
Costonis AC, Sinclair WA (1969) Relationships of atmospheric ozone to needle blight of eastern white pine. Phytopathology 59:1566–1574.
Cox RM (1988) The sensitivity of pollen from various coniferous and broad-leaved trees to combinations of acidity and trace metals. New Phytol 109:193–201.
Coyne PI, Bingham CE (1982) Variation in photosynthesis and stomatal conductance in an ozone-stressed ponderosa pine stand: Light response. For Sci 28:257–273.
Darrall NM (1989) The effect of air pollutants on physiological processes in plants. Plant Cell Environ 12:1–30.
Davi DD (1984) Description of leaf injury caused by gaseous air pollutants. In: Davi DD, Miller AA, Dochinger L (eds) *Air Pollution and Productivity of the Forest*, Proc Symp October 1983, Washington DC. Izaak Walton League Amer, Arlington, VA, pp 77–82.
Davi DD, Wilhour RG (1976) *Susceptibility of Woody Plants to Sulfur Dioxide and Photochemical Oxidants: A Literature Review*. Publ No EPA-600/3-76-102. US Environmental Protection Agency, Washington, DC.
Davi DD, Wood FA (1973) The influence of environmental factors on the sensitivity of Virginia pine to ozone. Phytopathology 63:371–376.
Dean TJ, Johnson JD (1992) Growth response of young slash pine trees to simulated acid rain and ozone stress. Can J For Res 22:839–848.
DeHayes DH (1992) Winter injury and developmental cold tolerance of red spruce. In: Eager C, Adams MB (eds) *Ecology and Decline of Red Spruce in the Eastern United States*. Springer-Verlag, New York, pp 295–337.
DeHayes DH, Thornton FC, Waite CE, Ingle MA (1991) Ambient cloud deposition reduces cold tolerance of red spruce seedlings. Can J For Res 21:1292–1295.
Dochinger LS, Seliskar CE (1970) Air pollution and the chlorotic dwarf disease of eastern white pine. For Sci 16:46–55.
Doley D (1989) Fluoride-induced enhancement and inhibition of shoot growth in four taxa of *Pinus* New Phytol 112:543–552.
Dunn DB (1959) Some effects of air pollution on *Lupinus* in the Los Angeles area. Ecology 40:621–625.
Eamus D, Leith I, Fowler D (1989) The influence of acid mist upon transpiration, shoot water potential and pressure—volume curves of red spruce seedlings. Ann Sci For 46(suppl):577s–580s.
Edwards GS, Kelly JM (1992) Ectomycorrhizal colonization of loblolly pine seedlings during three growing seasons in response to ozone, acidic precipitation, and soil Mg status. Environ Pollut 76:71–77.

Edwards GS, Friend AL, O'Neill EG, Tomlinson PT (1992a) Seasonal patterns of biomass accumulation and carbon allocation in *Pinus taeda* seedlings exposed to ozone, acidic deposition, and reduced soil Mg. Can J For Res 22: 640–646.

Edwards GS, Kelly JM, Mays PA (1992b) Ozone, acidic precipitation, and soil Mg impacts on soil and loblolly pine seedling nutrient status after three growing seasons. Water, Air, and Soil Pollut 63:281–294.

Edwards GS, Pier PA, Kelly JM (1990) Influence of ozone and soil magnesium status on the cold hardiness of loblolly pine (*Pinus taeda* L.) seedlings. New Phytol 115:157–164.

Edwards NT (1991) Root and soil respiration responses to ozone in *Pinus taeda* L. seedlings. New Phytol 118:315–321.

Edwards NT, Edwards GL, Kelly JM, Taylor GE Jr (1992) Three-year growth responses of *Pinus taeda* L. to simulated rain chemistry, soil magnesium status, and ozone. Water, Air, and Soil Pollut 63:105–118.

Edwards NT, Taylor GE Jr, Adams MB, Simmons GL, Kelly JM (1990) Ozone, acidic rain and soil magnesium effects on growth and foliar pigments of *Pinus taeda* L. Tree Physiol 6:95–104.

Elsik CG, Flagler RB, Boutton TW (1992) Effects of ozone and water deficit on growth and physiology of *Pinus taeda* and *Pinus echinata*. In: Flagler RB (ed) *The Response of Southern Commercial Forests to Air Pollution*. Air and Waste Management Assoc, Pittsburgh, PA, pp 225–245.

Environmental Protection Agency (EPA) (1978) *Diagnosing Vegetation Injury Caused by Air Pollution*. US Environmental Protection Agency 450/3-78-005.

Environmental Protection Agency (EPA) (1993) *Air Quality Criteria for Oxides of Nitrogen*. US EPA Environmental Criteria and Assessment Office, Research Triangle Park, NC, EPA 600/8-91/049bF.

Esher RJ, Marx DH, Ursic SJ, Baker RL, Brown RL, Coleman DC (1992) Simulated acid rain effects on fine roots, ectomycorrhizae, microorganisms, and invertebrates in pine forests of the southern United States. Water Air and Soil Pollut 61:269–278.

Evans LS, Fitzgerald GA (1993) Histological effects of ozone on slash pine (*Pinus elliottii* var. *densa*). Environ Exp Bot 33:505–513.

Fairfax JAW, Lepp NW (1975) Effect of simulated "acid rain" on cation loss from leaves. Nature (Lond) 255:861–875.

Fangmeier A, Kress LW, Lepper P, Heck WW (1990) Ozone effects on the fatty acid composition of loblolly pine needles (*Pinus taeda* L.). New Phytol 115: 639–647.

Farrar JF, Relton J, Rutter AJ (1977) Sulphur dioxide and the growth of *Pinus sylvestris*. J Appl Ecol 14:861–875.

Faulkner P, Schoeneberger MM, Kress LW (1991) Belowground changes in loblolly pine as indicators of ozone stress. In: Coleman SS, Neary DG (eds) *Proc 6th Biennial South Silvicultural Res Conf*, Memphis, TN, October 1990. USDA Forest Service, Southeastern For Exp Stn, Asheville, NC, Gen Tech Rep SE-70, pp 332–341.

Feder WA (1968) Reduction in tobacco pollen germination and tube elongation, induced by low levels of ozone. Science (Washington DC) 160: 1122.

Fenn ME, Dunn PH, Wilborn R (1990) Black stain root disease in ozone-stressed ponderosa pine. Plant Dis 74:426–430.

Fernandez IJ, Czapowskyj MM (1986) *Selected Relationships for Trace Metals in Maine Low Elevation Spruce-Fir Forest Floors*. Maine Agric Exp Stn Tech Bull 119, University of Maine, Orono.

Fincher J, Cumming JR, Alscher RG, Rubin G, Weinstein L (1989) Long-term ozone exposure affects winter hardiness of red spruce (*Picea* Sarg.) seedlings. New Phytol 113:85–96.

Flagler RB, Chappelka AH, Sasek TW, Kress LW (1990) Effects of regional air pollutants on southern pine species. In: *The Potential Impact of Acidic Deposition and Ozone on Southern Commercial Pine Forests* (Draft Copy). Southern Commercial Forest Research Cooperative of the Forest Response Program, pp 55–99.

Flagler RB, Lock JE, Elsik CG (1994) Leaf-level and whole-plant gas exchange characteristics of shortleaf pine exposed to ozone and simulated acid rain. Tree Physiol 14:361–374.

Flagler RB, Spruill SE, Chappelka AH, Dean TJ, Kress LW, Reardon JC (1992) Growth of three southern pine species as affected by acid rain and ozone: a combined analysis. In: Flagler RB (ed) *The Response of Southern Commercial Forests to Air Pollution*. Air and Waste Management Assoc, Pittsburgh, PA, pp 207–224.

Foy CD, Chaney RC, White MC (1978) The physiology of metal toxicity in plants. Annu Rev Plant Physiol 29:511–566.

Fowler D, Cape JN, Deans JD, Leith ID, Murray MB, Smith RI, Sheppard LJ, Unsworth MH (1989) Effects of acid mist on the frost hardiness of red spruce seedlings. New Phytol 113:321–335.

Friedland AJ, Johnson AJ, Siccama TG (1984) Accumulation of trace metals in the forest floor in Vermont: Spatial and temporal patterns. Water Air Soil Pollut 21:161–170.

Friend AL, Tomlinson PT (1992) Mild ozone exposure alters ^{14}C dynamics in foliage of *Pinus taeda* L. Tree Physiol 11:215–227.

Friend AL, Tomlinson PT, Dickson RE, O'Neill EG, Edwards NT, Taylor GE Jr (1992) Biochemical composition of loblolly pine reflects pollutant exposure. Tree Physiol 11:35–47.

Garner JHB, Pagano T, Cowling EB (1989) *An Evaluation of the Role of Ozone, Acid Deposition, and Other Airborne Pollutants in the Forests of Eastern North America*. Gen Tech Rep SE-59. USDA Forest Service, Southeastern Forest Expt Stn, Asheville, NC.

Garner JHB, Smith WH (1986) Effects of ozone and other photochemical oxidants on natural and agroecosystems. In: *Air Quality Criteria for Ozone and Other Photochemical Oxidants*. Vol III. EPA/600/8-84/020cF. US Environmental Protection Agency, Research Triangle Park, NC.

Garsed SG, Rutter AJ (1984) The effects of fluctuating concentrations of sulphur dioxide on the growth of *Pinus sylvestris* L. and *Picea sitchensis* (Bong.) Carr. New Phytol 97:175–195.

Gilbert OL (1983) The growth of planted trees subject to fumes from brickworks. Environ Pollut Ser A Ecol Biol 31:301–310.

Gschwandtner G, Gschwandtner K, Eldridge K, Mann C, Mobley D (1986) Historic emissions of sulfur and nitrogen oxides in the United States from 1900 to 1980. J Air Pollut Control Assoc 36:139–149.

Hacskaylo E (1973) The Torrey symposium on current aspects of fungal development: IV. Dependence of mycorrhizal fungi on hosts. Bull Torrey Bot Club 100:217–223.

Hakkarinen C, Allan MA (1986) Forest health and acidic deposition. Publ No EA-4813-SR, Res Project 5002, Electric Power Res Inst, Palo Alto, CA.

Halbwachs G (1984) Organismal responses of higher plants to atmospheric pollutants: Sulphur dioxide and fluoride. In: Treshow M (ed) *Air Pollution and Plant Life*. John Wiley and Sons, Chichester, UK, pp 175–214.

Hanson PJ, McLaughlin SB, Edwards NT (1988) Net CO_2 exchange of *Pinus taeda* shoots exposed to variable ozone levels and rain chemistries in field and laboratory settings. Physiologia Plantarum 74:635–642.

Hart R, Biggs RH, Webb PG (1986) Effects of simulated acid rain on growth and yield of Valencia orange, floradade tomato and slash pine in Florida. Environ Toxicology and Chem 5:79–85.

Havas P, Huttunen S (1972) The effects of air pollution on the radial growth of Scots pine (*Pinus sylvestris*). Biol Conserv 4:361–368.

Heagle AS (1989) Ozone and crop yield. Annu Rev Phytopath 27:397–423.

Heagle AS, Flagler RB, Patterson RP, Lesser VM, Shafer SR, Heck WW (1987) Injury and yield response of soybean to chronic doses of ozone and soil moisture deficit. Crop Sci 27:1016–1024.

Heagle AS, Miller JE, Heck WW, Patterson RP (1988) Injury and yield response of cotton to chronic doses of ozone and soil moisture deficit. J Environ Qual 17:627–635.

Heath RL (1980) Initial events in injury to plants by air pollutants. Annu Rev Plant Physiol 31:395–431.

Heath RL (1994) Alterations of plant metabolism. In: Alscher RG, Wellburn AR (eds) *Plant Responses to the Gaseous Environment*. Chapman and Hall, New York, pp 121–145.

Heck WW (1984) Defining gaseous pollution problems in North America. In: Koziol MJ, Whatley FR (eds) *Gaseous Air Pollutants and Plant Metabolism*. Butterworths, Boston, pp 35–48.

Helander ML, Ranta H, Neuvonen S (1993) Responses of phyllosphere microfungi to simulated sulphuric and nitric acid deposition. Mycol Res 97:533–537.

Hill AC, Littlefield N (1969) Ozone: Effects on apparent photosynthesis, rate of transpiration, and stomatal closure in plants. Environ Sci and Technol 3: 52–56.

Hogsett WE, Plocher M, Wildman V, Tingey DT, Bennett JP (1985) Growth response of two varieties of slash pine seedlings to chronic ozone exposures. Can J Bot 63:2369–2376.

Holopainen T, Nygren P (1989) Effects of potassium deficiency and simulated rain, alone and in combination, on the ultrastructure of Scots pine needles. Can J For Res 19:1402–1411.

Horsman DC, Roberts TM, Bradshaw AD (1978) Evolution of sulphur dioxide tolerance in perennial ryegrass. Nature 276:493–494.

Horton SJ, Schoeneberger MM, Reinert RA, Shafer SR, Allen HL (1989) Growth, carbohydrate reserves and nutrient content of loblolly pine seedlings exposed to ozone and simulated acidic rain. (Abstract) *Agronomy Abstr*, ASA, Madison, WI, p 305.

Hutchinson TC (1980) Impact of heavy metals on terrestrial and aquatic ecosystems. In: *Proc Symp Effects of Air Pollut on Mediterranean and Temperate Forest Ecosystems*, June 1980, Riverside, CA. USDA For Serv Gen Tech Rep PSW-43, Pacific Southwest For and Range Exp Stn, Berkeley, CA, pp 158–164.

Hyink DM (1991) Comment 1. For Sci 37:718–722.

Jacobson JS (1991) The effects of acidic precipitation on crops. In: Chadwick MJ, Hutton M (eds) *Acid Depositions in Europe: Environmental Effects, Control Strategies, and Policy Options*. Stockholm Environ Inst, Stockholm, Sweden, pp 81–98.

James RL, Cobb FW Jr, Wilcox WW, Rowney D (1980) Effects of photochemical oxidant injury on ponderosa and jeffrey pines on susceptibility of sapwood and freshly cut stumps to *Fomes annosus*. Phytopathology 70:704–708.

Johnson AH, Siccama TG, Turner RS, Lord DG (1984) Assessing the possibility of a link between acid precipitation and decreased growth rates of trees in the northeastern United States. In: Linthurst RA (ed) *Direct and Indirect Effects of Acidic Deposition on Vegetation*. Butterworth, Boston, pp 81–95.
Johnson AH, Siccama TG, Wang D, Turner RS, Barringer TH (1981) Recent changes in patterns of tree growth rate in the New Jersey pinelands: A possible effect of acid rain. J Environ Quality 10:427–430.
Johnson DW, Cole DW (1977) Sulfate mobility in an outwash soil in western Washington. Water Air Soil Pollut 7:489–495.
Johnson DW, Cole DW (1980) Anion mobility in soils: Relevance to nutrient transport from forest ecosystems. Environ Int 3:79–90.
Johnson DH, Fernandez IJ (1992) Soil-mediated effects of atmospheric deposition on eastern U.S. spruce-fir forests. In: Eager C, Adams MB (eds) *Ecology and Decline of Red Spruce in the Eastern United States*. Springer-Verlag, New York, pp 235–270.
Jonsson B, Sundberg R (1972) *Has the Acidification by Atmospheric Pollution Caused a Growth Reduction in Swedish Forests? A Comparison of Growth Between Regions with Different Soil Properties*. Inst Skogsproduction Res Note No 20, Inst Skogsproduction, Stockholm.
Jonsson B, Svensson LG (1982) *A Study of the Effects of Air Pollution on Forest Yield. A Follow-Up of the Report of Jonsson and Sundberg 1972 and a New Study Based on Forest Types*. Avdelningen for Skogsuppskattning och Skogsindelning, Sveriges Lantbruksuniversitet, No 9.
Jordan DN, Wright LM, Lockaby, BG (1990) Relationship between xylem trace metals and radial growth of loblolly pine in rural Alabama. J Environ Qual 19:504–508.
Keller T (1986) The electrical conductivity of Norway spruce needle diffusate as affected by certain pollutants. Tree Physiol 1:85–94.
Keller T, Hasler R (1984) The influence of a fall fumigation with ozone on the stomatal behavior of spruce and fir. Oecologia (Berlin) 64:284–286.
Kelly JM, Parker GR, McFee WW (1979) Heavy metal accumulation and growth of seedlings of five forest species as influenced by soil cadmium level. J Environ Qual 8:361–364.
Kossuth SV, Biggs RH (1989) Response of seedlings and mature trees of the same genotype to ozone. In: Noble RD, Martin JL, Jensen KF (eds) *Air Pollution Effects on Vegetation Including Forest Ecosystesm*. Proc 2nd US–USSR Symp September 1988, Corvallis, OR, Raleigh, NC, and Gatlinburg, TN. USDA Forest Service Northeastern Forest Exp Stn, Broomall, PA, p 287.
Koziol MJ, Whatley FR (eds) (1984) *Gaseous Air Pollutants and Plant Metabolism*. Butterworths, Boston.
Kozlowski TT, Constantinidou HA (1986) Responses of woody plants to environmental pollution. I. Sources and types of pollutants and plant responses. For Abstr 47:5–51.
Krause CR, Shafer SR, Heagle SR (1987) Ultrastructural effects of ozone to field-grown loblolly pine seedlings. In: *Abstracts of Presentations 1987 Annu Meeting Am Phytopath Soc*, August 1987, Cincinnati, OH. Abstract No 558.
Kress LW, Allen HL (1991) *Impact of Ozone and Acidic Precipitation on the Growth of Loblolly Pine Seedlings*. Presented 84th Annu Meeting Air Waste Management Assoc, June, Vancouver, BC. Paper No 91-142.9.
Kress LW, Allen HL, Mudano JE, Stow TK (1992) Impact of ozone on loblolly pine seedling foliage production and retention. Environ Toxicol Pollut 11: 1115–1128.

Kress LW, Skelly JM (1982) Response of several eastern forest tree species to chronic doses of ozone and nitrogen dioxide. Plant Disease 66:1149–1152.

Kress LW, Skelly JM, Hinkelmann KH (1982) Growth impact of O_3, NO_2 and/or SO_2 on *Pinus taeda*. Environ Monitoring Assess 1:229–239.

Kriebel HB, Leben C (1981) The impact of SO_2 air pollution on the gene pool of eastern white pine. In: *Proc XVII IUFRO World Congress, Div 2*, Kyoto, Japan, pp 185–189.

Krupa SV, Manning WJ (1988) Atmospheric ozone: Formation and effects on vegetation. Environ Pollut 50:101–137.

Lackner AL, Alexander SA (1983) Root disease and insect infestations on air-pollution sensitive *Pinus strobus* and studies of pathogenicity of *Verticicladiella procera*. Plant Dis 67:679–681.

Lee WS, Chevone BI, Seiler JR (1990) Growth and gas exchange of loblolly pine seedlings as influenced by drought and air pollutants. Water Air and Soil Pollut 51:105–116.

Lefohn AS (1992) The characterization of ambient ozone exposures. In: Lefohn AS (ed) *Surface Level Ozone Exposures and Their Effects on Vegetation*. Lewis Publ, Chelsea, MI, pp 31–92.

Lefohn AS, Benkovitz CM (1989) Air quality measurements and characterization for effects research. In: *State of Science and State of Technology*. National Acid Precipitation Assessment Program, Washington, DC.

Lefohn AS, Hogsett WE, Tingey DT (1987) The development of sulfur dioxide and ozone exposure profiles that mimic ambient conditions in the rural southeastern United States. Atmospheric Environ 21:659–669.

Lefohn AS, Knudsen HP, Shadwick DS, Hermann KA (1992a) Surface ozone exposures in the eastern United States (1985–1989). In: Flagler RB (ed) *The Response of Southern Commercial Forests to Air Pollution*. Air and Waste Management Assoc, Pittsburgh, PA, pp 81–93.

Lefohn AS, Krupa SV, Winstanley D (1990) Surface ozone exposures measured at clean locations around the world. Environ Pollut 63:189–224.

Lefohn AS, Shadwick DS (1991) Ozone, sulfur dioxide, and nitrogen dioxide trends at rural sites located in the United States. Atmospheric Environ 25A: 491–501.

Lefohn AS, Shadwick DS, Somerville MC, Chappelka AH, Lockaby BG, Meldahl RS (1992b) The characterization and comparison of ozone exposure indices used in assessing the response of loblolly pine to ozone. Atmospheric Environ 26A:287–298.

Leininger TD (1989) Root disease incidence in eastern white pine plantations with and without symptoms of ozone injury in the Coweeta basin of North Carolina. Plant Dis 74:552–554.

Leininger TD, Miller PR, Schilling SL, Dunn PH (1991) Seedling responses of five species of western conifers to simulated ambient sulfur dioxide exposures. For Sci 37:1538–1549.

Lepp NW (ed) (1981) *Effects of Heavy Metal Pollution on Plants*. Applied Sci Publ, London.

Lindberg SE, Lovett GM, Turner RR, Hoffman WH (1985) Atmospheric chemistry deposition and canopy interactions. In: Johnson DW, Van Hook RI (eds) *Analysis of Biogeochemical Cycling Processes in Walker Branch Watershed*. Springer-Verlag, New York, pp 96–163.

Linzon SN (1966) Damage to eastern white pine by sulfur dioxide, semimature-tissue needle blight, and ozone. J Air Pollut Control Assoc 16:140–144.

Logan JA (1989) Ozone in rural areas of the United States. J Geophys Res 94(D6):8511–8532.

Lorenc-Plucinska G (1988) Effect of nitrogen dioxide on CO_2 exchange in Scots pine seedlings. Photosynthetica 22:108–111.

Lovett GM, Hubbell JG (1991) Effects of ozone and acid mist on foliar leaching from eastern white pine and sugar maple. Can J For Res 21:794–802.

Luck RF (1980) Impact of oxidant air pollution on ponderosa and Jeffrey pine cone production. In: *Proc Symp Effects Air Pollut on Mediterranean and Temperate For Ecosystems*, June 1980, Riverside, CA. USDA For Serv Gen Tech Rep PSW-43. Pacific Southwest For and Range Expt Stn, Berkeley, CA, p 240.

Ludovici KH (1990) *Influence of Different Simulated Rain Chemistries on a Hapludult and the Root Growth of Pinus taeda*. M.S. thesis, North Carolina State University, Raleigh.

MacDonald NW, Hart JB, Nguyen PV (1986) Simulated acid rain effects on jack pine seedling establishment and nutrition. Soil Sci Soc Am J 50:219–225.

Malhotra SS, Blauel RA (1980) *Diagnosis of Air Pollutant and Natural Stress Symptoms on Forest Vegetation in Western Canada*. Can For Service Northern For Res Centre Information Rep NOR-X-228. Northern For Res Centre, Edmonton, Alberta.

Mandersheid R, Jager H-J, Kress LW (1992) Effects of ozone on foliar nitrogen metabolism of *Pinus taeda* L. and implications for carbohydrate metabolism. New Phytol 121:623–633.

Manderscheid R, Jager H-J, Schoeneberger MM (1991) Dose–response relationships of ozone effects on foliar levels of antioxidants, soluble polyamines and peroxidase activity of (*Pinus taeda* L.): Assessment of the usefulness as early ozone indicators. Angew Botanik 65:373–386.

Mason BJ (1990) Acid rain—cause and consequence. Weather 45:70–79.

Maurice CG, Crang RE (1989) Foliar injury and shoot growth in response to prolonged acidic misting of *Pinus strobus* seedlings. Arch Environ Contam Toxicol 18:277–284.

Mazur P (1969) Freezing injury in plants. Ann Rev Plant Physiol 20:419–448.

McLaughlin SB (1985) Effects of air pollution on forests: A critical review. J Air Pollut Control Assoc 35:512–534.

McLaughlin SB, McConathy RK, Duvick D, Mann LK (1982) Effects of chronic air pollution stress on photosynthesis, carbon allocation, and growth of white pine trees. For Sci 28:60–70.

McLaughlin SB, Layton PA, Adams MB, Edwards NT, Hanson PJ, O'Neill EG, Roy WK (1994) Growth responses of 53 open-pollinated loblolly pine families to ozone and acid rain. J Environ Qual 23:247–257.

McQuattie CJ, Schier GA (1992) Effect of ozone and aluminum on pitch pine (*Pinus rigida*) seedlings: anatomy of mycorrhizae. Can J For Res 22: 1901–1916.

McQuattie CJ, Schier GA (1993) Effect of ozone and aluminum on pitch pine (*Pinus rigida*) seedlings: Needle ultrastructure. Can J For Res 23:1375–1387.

Meier S, Grand LF, Schoeneberger MM, Reinert RA, Bruck RI (1990) Growth, ectomycorrhizae and nonstructural carbohydrates of loblolly pine seedlings exposed to ozone and soil water deficit. Environ Pollut 64:11–27.

Mengel K, Hogrebe MR, Esch A (1989) Effect of acidic fog on needle surface and water relations of *Picea abies*. Physiologia Plantarum 75:201–207.

Miller PL (1973) Oxidant-induced community change in a mixed conifer forest. Advan Chem Ser 122:101–117.

Miller PR, Longbotham GJ, Longbotham CR (1983) Sensitivity of selected western conifers to ozone. Plant Dis 67:1113–1115.

Miller PR, Parmeter JR, Taylor OC. Cardiff EA (1963) Ozone injury to the foliage of *Pinus ponderosa*. Phytopathology 53:1072–1076.

Miller PR, Parmeter JR, Flick BH, Martinez CW (1969) Ozone dosage response of ponderosa pine seedlings. J Air Pollut Control Assoc 19:435–438.
Morrison IK (1984) A review of literature on acid deposition effects in forest ecosystems. For Abstr 45:483–506.
Mudd JB, Banerjee SK, Dooley MM, Knight KL (1984) Pollutants and plant cells: effects on membranes. In: Koziol MJ, Whatley FR (eds) *Gaseous Air Pollutants and Plant Metabolism.* Butterworths, Boston, pp 105–116.
Mudano JE, Allen HL, Kress LW (1992) Stem and foliage elongation of young loblolly pine as affected by ozone. For Sci 38:324–335.
Mumford RA, Lipke H, Loufer DA, Feder WA (1972) Ozone-induced changes in corn pollen. Environ Sci Technol 6:427–430.
National Acid Precipitation Assessment Program (NAPAP) (1987) Interim assessment: the causes and effects of acidic deposition. In: Hedrick CN (ed) *National Acid Precipitation Assessment Program*, Vol. 1, Executive Summary. Office of the Director of Research, Washington, DC.
National Atmospheric Deposition Program (1993) *NADP/NTN Annual Data Summary. Precipitation Chemistry in the United States. 1992.* Natural Resource Ecology Lab, Colorado State University, Fort Collins.
Neuvonen S, Routio I, Haukioja E (1992) Combined effects of simulated acid rain and aphid infestation on the growth of Scots pine (*Pinus sylvestris*) seedlings. Ann Bot Fennici 29:101–106.
Neuvonen S, Saikkonen K, Haukioja E (1990a) Simulated acid rain reduces the susceptibility of the European pine sawfly (*Neodiprion sertifer*) to its nuclear polyhidrosis virus. Oecologia 83:209–212.
Neuvonen S, Saikkonen S, Suomela J (1990b) Effect of simulated acid rain on the growth performance of the European sawfly (*Neodiprion sertifer*). Scand J For Res 5:541–550.
Norton SA, Hanson DW, Campana RJ (1980) The impact of acid precipitation and heavy metals in forest ecosystems. In: *Proc Symp Effects Air Pollut on Mediterranean and Temperate For Ecosystems*, June 1980, Riverside, CA. USDA For Serv Gen Tech Rep PSW-43. Pacific Southwest For and Range Expt Stn, Berkeley, CA, pp 152–157.
Oleksyn J, Karolewski P, Rachwal L (1988) Susceptibility of European *Pinus sylvestris* L. populations to SO_2, NO_2, $SO_2 + NO_2$ and HF under laboratory and field conditions. Acta Soc Bot Poloniae 57:107–115.
Pagnelli DJ, Seiler JR, Feret PR (1987) Root regeneration as an indicator of aluminum toxicity in loblolly pine. Plant and Soil 102:115–118.
Parmeter P Jr, Bega RW, Neff T (1962) A chlorotic decline of ponderosa pine in southern California. Plant Dis Rep 46:269–273.
Parmeter P Jr, Miller PR (1968) Studies relating to the cause of decline and death of ponderosa pine in southern California. Plant Dis Rep 52:707–711.
Paynter VA, Reardon JC, Shelburne VB (1991) Carbohydrate changes in shortleaf pine (*Pinus echinata*) needles exposed to acid rain and ozone. Can J For Res 21:666–671.
Percy KE (1983) Sensitivity of eastern Canadian forest tree species to simulated acid precipitation. In: *Proc 12th Int Meeting Specialists in Air Pollut Damages in Forests*, IUFRO Section 2.09, August 23–30, 1982, Oulu, Finland. Aquilo Ser Bot 19:41–49.
Percy KE (1986) The effects of simulated acid rain on germinative capacity, growth and morphology of forest tree seedlings. New Phytol 104:473–484.
Percy KE, Baker EA (1990) Effects of simulated acid rain on epicuticular wax production, morphology, chemical composition and on cuticular membrane thickness in two clones of Sitka spruce (*Picea sitchensis* (Bong.) Carr.). New Phytol 116:79–87.

Percy KE, Krause CR, Jensen KF (1990) Effects of ozone and acidic fog on red spruce needle epicuticular wax ultrastructure. Can J For Res 20:117–120.

Phillips SO, Skelly JM, Burkhart HE (1977) Growth fluctuation of loblolly pine due to periodic air pollution levels: interaction of rainfall and age. Phytopathology 67:716–720.

Placet M, Battye RE, Fehsenfeld FC, Bassett GW (1991) Emissions involved in acidic deposition processes. In: Irving PM (ed) *Acid Deposition: State of Science and Technology*. National Acid Precipitation Assessment Program, Washington, DC, pp 111–148.

Qui Z, Chappelka AH, Somers GL, Lockaby BG, Meldahl RS (1992) Effects of ozone and simulated acidic precipitation on above- and below-ground growth of loblolly pine (*Pinus taeda*). Can J For Res 22:582–587.

Qiu Z, Chappelka AH, Somers GL, Lockaby BG, Meldahl RS (1993) Effects of ozone and simulated acidic precipitation on ectomycorrhizal formation on loblolly pine seedlings. Environ Exp Bot 33:423–431.

Rakowski KJ, Zwiazek JJ (1992) Early effects of hydrogen fluoride on water relations, photosynthesis and membrane integrity in eastern white pine (*Pinus strobus*) seedlings. Environ Exp Bot 32:377–382.

Ranta HM (1990) Effect of simulated acid rain on quantity of epiphytic microfungi on Scots pine (*Pinus sylvestris* L.) needles. Environ Pollut 67:349–359.

Rawlings JO, Spruill SE (1992) Response model synthesis of the southern commercial forest research data. In: Flagler RB (ed) *The Response of Southern Commercial Forests to Air Pollution*. Air and Waste Management Assoc, Pittsburgh, PA, pp 187–205.

Raynal DJ, Roman JR, Eichenlaub WM (1982) Response of tree seedlings to acid precipitation—I. Effect of substrate acidity on seed germination. Environ Exp Bot 22:377–383.

Reddy GB, Reinert RA, Eason E (1991) Effect of acid rain and ozone on soil and secondary needle nutrients of loblolly pine. In: Wright RJ, Baligar VC, Murrmann RP (eds) *Plant–Soil Interactions at Low pH*. Kluwer Academic, Boston, pp 139–145.

Reich PB (1987) Quantifying plant response to ozone: A unifying theory. Tree Physiol 3:63–91.

Reich PB, Lassoie JP (1984) Effects of low level O_3 exposure on leaf diffusive conductance and water use efficiency in hybrid poplar. Plant Cell Environ 7:661–668.

Reich PB, Schoettle AW, Amundson RG (1985) Effect of low levels of O_3, leaf age, and/or water stress on leaf diffusive conductance and water use efficiency in soybean. Physiol Plant 63:58–64.

Rich S, Turner NC (1972) Importance of moisture on stomatal behavior of plants subjected to ozone. J Air Pollut Control 22:718–721.

Richardson CJ, Sasek TW, Fendick EA (1992a) Implications of physiological responses to chronic air pollution for forest decline in the southeastern United States. Environ Toxicology Chem 11:1105–1114.

Richardson CJ, Sasek TW, Fendick EA, Kress LW (1992b) Ozone exposure–response relationships for photosynthesis in genetic strains of loblolly pine seedlings. For Ecol Management 51:163–178.

Rinallo C, Raddi P, Gellini R, di Lonnardo V (1986) Effects of simulated acid deposition on the surface structure of Norway spruce and silver fir needles. Eur J For Pathol 16:440–446.

Rosen PM, Musselman RC, Kender WJ (1978) Relationship of stomatal resistance to sulfur dioxide and ozone injury in grapevines. Sci Hortic (Amst) 8:137–142.

Ruark CA, Thomas CE, Bechtold WA, May DM (1991) Growth reductions in naturally regenerated southern pine stands in Alabama and Georgia. South J Appl For 15:73–79.
Reuss JO, Walthall PM (1988) Soil reaction and acidic deposition. In: Norton SA, Lindberg SE, Page AL (eds) *Acid Precipitation, Vol. 4, Soils, Aquatic Processes, and Lake Acidification.* Springer-Verlag, New York, pp 1–33.
Runeckles VC, Chevone BI (1992) Crop responses to ozone. In: Lefohn AS (ed) *Surface Level Ozone Exposures and Their Effects on Vegetation.* Lewis Publ, Chelsea, MI, pp 189–270.
Sasek TW, Richardson CJ, Fendick EA, Bevington SR, Kress LW (1991) Carry-over effects of acid rain and ozone on the physiology of multiple flushes of loblolly pine seedlings. For Sci 37:1078–1098.
Scherbatskoy T, Klein RM (1983) Responses of spruce and birch foliage to leaching by acidic mist. J Environ Quality 12:189–195.
Schier GA (1986) Seedling growth and nutrient relationships in a New Jersey Pine Barrens soil treated with "acid rain." Can J For Res 16:136–142.
Schier GA (1987) Germination and early growth of four pine species on soil treated with simulated acid rain. Can J For Res 17:1190–1196.
Schier GA, McQuattie CJ, Jensen KF (1990) Effects of ozone and aluminum on pitch pine (*Pinus rigida*) seedlings: growth and nutrient relations. Can J For Res 20:1714–1719.
Seiler JR, Paganelli DJ (1987) Photosynthesis and growth response of red spruce and loblolly pine to soil-applied lead and simulated rain. For Sci 33:668–675.
Senser M, Payer H-D (1989) The frost hardiness of young conifers exposed to different loads of ozone, sulfur dioxide or nitric oxide in environmental chambers. In: Bucher JB, Bucher-Wallin I (eds) *Proc 14th Int Meeting for Specialists in Air Pollut Effects on For Ecosystems*, Interlaken, Switzerland, October 1988. IUFRO P2.05, Birmensdorf, pp 523–526.
Shafer SR, Heagle AS (1989) Growth responses of field-grown loblolly pine to chronic doses of ozone during multiple growing seasons. Can J For Res 19:821–831.
Shafer SR, Heagle AS, Camberato DM (1987) Effects of chronic doses of ozone on field-grown loblolly pine: seedling responses in the first year. J Air Pollut Control Assoc 37:1179–1184.
Shafer SR, Reinert RA, Eason G, Spruill SE (1993) Analysis of ozone concentration—biomass response relationships among open-pollinated families of loblolly pine. Can J For Res 23:706–715.
Sheffield RM, Cost ND, Bechtold WA, McClure JP (1985) *Pine Growth Reductions in the Southeast.* Resource Bull SE-83. USDA Forest Service, Southeastern For Exp Stn, Asheville, NC.
Shelburne VB, Reardon JC, Paynter VA (1993) The effects of acid rain and ozone on biomass and leaf area parameters of shortleaf pine (*Pinus echinata* Mill.) Tree Physiol 12:163–172.
Sheu B-H, Chappelka AH (1993) *Effects of Long-Term Ozone Exposure on Superoxide Dismutase and Peroxidase Activities in Loblolly Pine (Pinus taeda L.).* Presented 86th Annu Meeting Air Waste Management Assoc, June 1993, paper no 93-TA-43.01. Denver, CO.
Shu-Wen Y (1984) Air pollution problems and research conducted on the effects of gaseous pollutants on plants in China. In: Koziol MJ, Whatley FR (eds) *Gaseous Air Pollutants and Plant Metabolism.* Butterworths, Boston, pp 49–57.
Simmons GL, Kelly JM (1989) Influence of O_3, rainfall acidity, and soil Mg status on growth and ectomycorrhizal colonization of loblolly pine roots. Water Air Soil Pollut 44:159–171.

Skarby L, Troeng E, Bostrom C (1987) Ozone uptake and effects on transpiration, net photosynthesis, and dark respiration in Scots pine. For Sci 33:801–808.

Skeffington RA, Roberts TM (1985) The effects of ozone and acid mist on Scots pine saplings. Oecologia (Berl) 65:201–206.

Skelly JM, Moore LD, Stone LL (1972) Symptoms of expression of eastern white pine located near sources of oxides of nitrogen and sulfur dioxide. Plant Dis Rep 56:3–6.

Smith WH (1991) Air pollution and forest damage. Chemical and Engineering News, November 11, pp 30–43.

Sommerville MC, Shadwick DS, Meldahl RS, Chappelka AH, Lockaby BG (1992) Use of a non-linear model in examining growth responses of loblolly pine to ozone and acid precipitation. Atmospheric Environ 26A:279–286.

Spence RD, Rykiel EJ Jr, Sharpe PJH (1990) Ozone alters carbon allocation in loblolly pine: Assessment with carbon-11 labeling. Environ Pollut 64:93–106.

Stark RW, Miller PR, Cobb FW Jr, Wood DL, Parmeter JR Jr (1968) Photochemical oxidant injury and bark beetle (Coleptera: Scolytidae) infestation of ponderosa pine. Hilgardia 39:121–126.

Stone LL, Skelly JM (1974) The growth of two forest tree species adjacent to a periodic source of air pollution. Phytopathology 64:773–778.

Stow TK, Allen HL, Kress LW (1992) Ozone impacts on seasonal foliage dynamics of young loblolly pine. For Sci 38:102–119.

Sutinen S (1989) Ultrastructure of conifer needles exposed to O_3 and $O_3 + SO_2$ compared to needle structure in the field in Finland, Sweden and in West Germany. Aquilo Ser Bot 27:29–34.

Taylor GE Jr (1994) Role of genotype in the response of loblolly pine to tropospheric ozone: effects at the whole-tree, stand, and regional level. J Environ Qual 23:63–82.

Taylor GE Jr, Murdy WH (1975) Population differentiations of an annual plant species, *Geranium carolinianum*, in response to sulfur dioxide. Bot Gaz 136:212–215.

Temple PJ (1986) Stomatal conductance and transpirational response of field-grown cotton to ozone. Plant Cell Environ 9:315–321.

Teskey RO, Fites JA, Samuelson LJ, Bongarten BC (1986) Stomatal and non-stomatal limitations to net photosynthesis in *Pinus taeda* L. under different environmental conditions. Tree Physiol 2:131–142.

Thompson AM (1992) The oxidizing capacity of the earth's atmosphere: Probable past and future changes. Science 256:1157–1165.

Tjoelker MG, Luxmoore RJ (1991) Soil nitrogen and chronic ozone stress influence physiology, growth and nutrient status of *Pinus taeda* L. and *Liriodendron tulipifera* L. seedlings. New Phytol 119:69–81.

Treshow M (1970) *Environment and Plant Response*. McGraw-Hill, New York.

Treshow M, Anderson FK (1982) Ecological assessment of potential fluoride effects on plants. In: Murray F (ed) *Fluoride Emissions*. Academic Press, Sydney, Australia, pp 177–189.

Turner D, Tingey DT (1989) Foliar leaching and root uptake of Ca, Mg, and K in relation to acid fog effects on Douglas-fir. Water Air Soil Pollut 49:205–214.

Turner RS, Olson RJ, Brandt CC (1986) *Areas Having Soil Characteristics that May Indicate Sensitivity to Acidic Deposition Under Alternative Forest Damage Hypotheses*. Environ Sci Division Publ No 2720, Oak Ridge Natl Lab, Oak Ridge, TN.

Turunen M, Huttunen S (1991) Effect of simulated acid rain on the epicuticular wax of Scots pine needles under northerly conditions. Can J Bot 69:412–419.

Tveite B (1980) Effects of acid precipitation on soil and forest. 8. Foliar nutrient concentrations in field experiments. In: Drablos D, Tollan A (eds) *Ecological*

Impact of Acid Precipitation. Proc Int Conf, Sandefjord, Norway. SNSF Project, Oslo-A, Norway, pp 204–205.
Van Deusen PC, Barnard JE (1989) *Indicator Plants in Forest Health Surveys*. Proc Advanced Forest Inventory Methods, August 1989, Syracuse, NY.
Vukovich FM, Fishman J, Browell EV (1985) The reservoir of ozone in the boundary layer of the eastern United States and its potential impact on the global tropospheric budget. J Geophys Res 90(D3):5687–5698.
Walker RF, McLaughlin SB (1991) Growth and root system development of white oak and loblolly pine as affected by simulated acidic precipitation and ectomycorrhizal inoculation. For Ecol and Management 46:123–133.
Walker RF, McLaughlin SB (1993) Growth and xylem water potential of white oak and loblolly pine seedlings as affected by simulated acidic rain. Am Midl Nat 129:26–34.
Weber JA, Clark CS, Hogsett WE (1993) Analysis of the relationships among O_3 uptake, conductance, and photosynthesis in needles of *Pinus ponderosa*. Tree Physiol 13:157–172.
Weinstein LH, Alscher–Herman R (1982) Physiological responses of plants to fluorine. In: Unsworth MH, (Goldstein RA (eds) *Effects of Gaseous Air Pollution in Agric and Hort*, Stanford University Press, Stanford, CA, pp 133–147.
Weir RJ (1977) *Genetic Variation in Loblolly Pine (Pinus taeda L) Tolerance to Ozone*. Ph.D. thesis, North Carolina State University, Raleigh.
Wellburn FAM, Wellburn AR (1994) Atmospheric ozone affects carbohydrate allocation and winter hardiness of *Pinus halepensis* (Mill.). J Exp Bot 45:607–614.
Wheeler GL (1972) *The Effects of Fluorine on the Cycling of Calcium, Magnesium, and Potassium in Pine Plantations of Eastern North Carolina*. Dissertation, North Carolina State University, Raleigh.
Winner WE, Cotter IS, Powers HR Jr, Skelly JM (1987) Screening loblolly pine seedling responses to SO_2 and O_3: Analysis of families differing in resistance to fusiform rust disease. Environ Pollut 47:205–230.
Winner WE, Mooney HA, Goldstein RA (eds) (1985) *Sulfur Dioxide and Vegetation: Physiology, Ecology, and Policy Issues*. Stanford University Press, Stanford, CA.
Wood T, Bormann FH (1975) Increases in foliar leaching caused by acidification of an artificial mist. Ambio 4:169–171.
Wood T, Bormann FH (1977) Short-term effects of simulated acid rain upon the growth and nutrient relations of *Pinus strobus* L. Water Air Soil Pollut 7:479–488.
Wright LM, Lockaby BG, Meldahl R, Thornton F, Chappelka AH (1990–1991) The influence of acid precipitation and ozone on nitrogen of young loblolly pine. Water Air Soil Pollut 54:135–142.
Yang Y-S, Skelly JM, Chevone BI (1983) Sensitivity of eastern white pine clones to acute doses of ozone, sulfur dioxide, or nitrogen dioxide. Phytopathology 73:1234–1237.
Zahner R, Saucier JR, Meyers RK (1989) Tree-ring model interprets growth declines in the southeastern United States. Can J For Res 19:612–621.
Zeide B (1992) Reevaluation of forest inventory data from loblolly pine stands in the Georgia piedmont and mountain areas. In: Flagler RB (ed) *The Response of Southern Commercial Forests to Air Pollution*. Air and Waste Management Assoc, Pittsburgh, PA, pp 18–35.
Zobel A, Nighswander JE (1991) Accumulation of phenolic compounds in the necrotic areas of Austrian and red pine needles after spraying with sulphuric acid: a possbile bioindicator of air pollution. New Phytol 117:565–574.

6. Biotic Agents of Stress in the South

James S. Meadows and John D. Hodges

Biotic agents affect millions of hectares of forests throughout the world each year. Of all the environmental stresses affecting trees, insects and diseases are by far the most widespread and most studied but, perhaps, the least understood. Their influence on forest productivity may be relatively small, as during low-level infestations that produce nagging drains on growth, or catastrophic, as during epidemic outbreaks that may destroy thousands of hectares in a short period of time. In contrast to air pollutants, whose effects are primarily regional in nature, biotic pests permeate nearly every forest in all parts of the world and constantly pose a threat to the health of these ecosystems. Biotic pests commonly interact with other agents of stress, both natural and anthropogenic, to produce complex problems for forest managers throughout the world. These complex interactions between biotic pests and other agents of stress are beyond the scope of this chapter and are discussed elsewhere in this book (Chapter 7). This chapter addresses the relationships between susceptible hosts and the major biotic pests in the southern United States.

Many biotic agents affect forest trees. The major insects that cause extensive damage to forest stands in the United States are the Douglas fir tussock moth, the gypsy moth, the various spruce budworms, the mountain pine beetle, and the southern pine beetle. The most important diseases in American forestry are dwarf mistletoes, fusiform rust, white pine blister rust, and various root rots (Loomis et al., 1985). This chapter concentrates on the four major biotic agents in the South: southern pine beetle, fusiform rust, littleleaf syndrome, and annosus root rot. Other insects and diseases affecting southern forests, although potentially important at the local level, include the *Ips* engraver beetles, black turpentine beetle, various pine sawflies, Nantucket pine tip moth, white pine weevil, pitch canker, comandra blister rust, and oak wilt.

Southern Pine Beetle

The southern pine beetle, *Dendroctonus frontalis* Zimmermann (Coleoptera: Scolytidae), is one of the most destructive forest pests in the United States. Unlike other bark beetles in the South, the southern pine beetle kills the host whenever attacks are successful and brood are produced. Therefore, it is an extremely important environmental stress on southern pine ecosystems. Because of its great biological and economic importance, the southern pine beetle has been intensively researched over the years. Numerous reviews have been published that deal with all aspects of this biotic agent of stress, including characteristics associated with the host tree, the site, climate, and the beetle itself (Belanger et al., 1986; Blanche et al., 1983; Christiansen et al., 1987; Coster and Searcy, 1981; Schowalter and Filip, 1993; Thatcher and Connor, 1985; Thatcher, 1980).

Extent and Magnitude in the South

The southern pine beetle attacks and kills all species of southern yellow pines, but its principal hosts are loblolly (*Pinus taeda* L.) and shortleaf (*P. echinata* Mill.) pines; slash (*P. elliottii* Engelm.), longleaf (*P. palustris* Mill.), and Virginia (*P. virginiana* Mill.) pines are also susceptible. The beetle's natural range extends well into many of the Midwestern states, but it is economically important only south of the Ozark Mountains and the Ohio River.

Regional outbreaks of the southern pine beetle have had devastating effects across the South. Two of the most severe outbreaks in recent history occurred in the Southeast from 1979 to 1980 and in Texas and Louisiana from 1982 to 1986. Hoffard (1985) reported that 6.1 million hectares across South Carolina, Georgia, Alabama, and Mississippi were infested with southern pine beetle outbreaks from 1979 to 1980, accounting for a loss of about 11.9 million cubic meters of timber across the South during those two years (Table 6.1). More recently, Carter et al. (1991) described the southern pine beetle epidemic that occurred in Texas and Louisiana from 1982 to 1986, with a peak year in 1985. The epidemic resulted in a loss of nearly 810 million board feet of timber during the five-year period in Texas alone, the worst outbreak in Texas

Table 6.1. Estimates of Timber Killed by Southern Pine Beetle in States with Outbreak Counties from 1979 to 1980[a]

State	1979	1980	Total
		Thousand Cubic Meters	
Alabama	2407	2614	5,021
Georgia	1841	1676	3,517
Mississippi	406	934	1,340
North Carolina	334	872	1,206
South Carolina	241	572	813
Tennessee	1	13	14
Total	5230	6681	11,911

[a] Adapted from Hoffard (1985).

history. Holmes (1991) estimated the total net welfare loss from this timber damage across the two-state area to range from $41.5 million to $102.6 million from 1984 to 1986.

Southern pine beetle population levels are generally sporadic (irregularly cyclic) and fluctuate rapidly from stable, endemic populations to unstable, epidemic populations. Hoffard (1985) reported that the area infested in any given year across the South may range from less than 0.4 million hectares to well over 4.0 million hectares. Despite a great deal of research on beetle behavior and ecology, our understanding of beetle population dynamics is limited. Consequently, our ability to predict beetle outbreaks, or the transition from an endemic to epidemic population level, is inadequate. Mawby et al. (1989) adapted a two-phase model, originally developed to predict mountain pine beetle outbreaks (Berryman, 1979; Raffa and Berryman, 1986), to southern pine beetle population dynamics at each of five hierarchical levels: the tree, the local infestation, the county, the state, and the region. The model consists of three points that represent a stable, endemic phase maintained by host-tree defensive capabilities; a transient, epidemic phase determined by the availability of host material; and a threshold between the two phases that depends on local environmental and biotic factors. Data from North Carolina and other parts of the Southeast were used to assess the utility of the model in predicting outbreaks of southern pine beetle. The model successfully predicted kilocords of damage from the beetle (r^2 ranged from 0.41 to 0.98) and number of infested counties (r^2 ranged from 0.52 to 0.83) at the regional and state hierarchical levels and number of beetle spots within several North Carolina counties (r^2 ranged from 0.61 to 0.83), but it was not tested at the individual tree and local infestation hierarchical levels (Mawby et al., 1989).

Another model to describe southern pine beetle population dynamics was recently proposed by Reeve and Turchin (1993). The authors suggest that natural enemies of the southern pine beetle may play a significant role in both the occurrence of regional outbreaks and the growth of local beetle infestations. This conceptual model divides the southern pine beetle system into two components: (1) movement of the beetle and its natural enemies between trees, and (2) interaction of the beetle and its enemies once they arrive on a tree under mass attack. Although untested, the conceptual model proposed by Reeve and Turchin (1993) identifies several underdeveloped avenues of experimentation needed to understand, describe, and predict southern pine beetle population dynamics.

Obviously, the southern pine beetle is one of, if not the most, devastating biotic agent of stress in the South, affecting hundreds of thousands of hectares and destroying millions of dollars of pine timber across the South each year. Although beetle population levels typically drop sharply after about three years of high activity during large outbreaks, the South is never completely free of beetle-infested areas or completely immune from the threat of future epidemics.

Characteristics of Susceptible Trees and Stands

Researchers often describe susceptible trees as being of low vigor and/or poor health. However, these terms have not been adequately defined. The concept of vigor, as it applies to resistance to bark beetle attack, is commonly associated with individual-tree radial growth. In general, low resistance to beetle attack may be related to very slow radial growth, and high resistance has been assumed to be related to rapid radial growth, but these relationships have not been adequately determined. In fact, resistance to beetle attack may decline as radial growth increases because a smaller proportion of photosynthate is allocated to the production of secondary compounds important to tree resistance (Lorio, 1993). The ambiguity in the use of the term *tree vigor* should be recognized in assessing the literature as presented in the following discussion.

Hicks (1980) listed three categories of factors that affect host susceptibility to southern pine beetle: climatic factors, site factors, and stand factors. Blanche et al. (1983) also discussed numerous physiological characteristics that are related to host susceptibility. These four categories of factors are summarized in Table 6.2.

In a similar fashion, Paine and Baker (1993) discussed three groups of factors that predispose conifers to bark beetle attack: abiotic, biotic, and anthropogenic. Abiotic predisposing factors include moisture stress, soil properties (especially soil texture and fertility), and damage from storms or fire. Biotic predisposition to bark beetle attack can be caused by fungal diseases, mistletoes, and by defoliators and other insects. Anthropogenic predisposing factors include exposure to air pollution and forest management practices, especially road construction, precommercial thinning, and logging damage to both the soil and to residual trees. Paine and Baker

Table 6.2. Factors that Affect Host Susceptibility to Southern Pine Beetle

Climatic factors:
1. rainfall
2. temperature
3. weather-related stand disturbances
Site factors:
1. landform
2. water regime
3. soil texture
4. soil chemical properties
5. site index
6. soil depth
Stand factors:
1. stand density
2. radial growth
3. species composition
4. average stand age, height, and diameter
5. presence of diseases and other insects
Physiological factors:
1. oleoresin exudation pressure
2. other oleoresin properties, such as total flow and viscosity
3. radial growth
4. water status
5. tree vigor

(1993) concluded that these abiotic, biotic, and anthropogenic factors all tend to reduce tree vigor and overall forest health and, thereby, increase susceptibility to invasion by bark beetles. These predisposing factors are probably much more important in increasing susceptibility to southern pine beetle attack at endemic population levels than at epidemic levels, when all trees can potentially succumb to beetle attack.

Climatic Factors

Although it has been assumed that the amount and timing of rainfall are related to southern pine beetle activity (Kalkstein, 1976; King, 1972), the mechanism through which this relationship may be manifested is poorly understood, and the evidence for a direct relationship is weak (Turchin et al., 1991). Hypothetically, rainfall influences the water status of the tree, which affects the host's resistance to beetle attack and, subsequently, affects beetle population levels. However, there is a lack of experimental evidence to support this theory. Hicks (1980) reported that infestations in the Western Gulf Coastal Plain were highest shortly after periods of abundant rainfall, while infestations in the Atlantic Coastal Plain occurred primarily during periods of drought.

Drought is an important agent of stress that may predispose otherwise healthy trees to invasion by insects. Mattson and Haack (1987) proposed six generalized mechanisms to explain the potential role of drought in outbreaks of phytophagous insects: (1) Drought may provide a more favorable thermal environment for growth of most insects; (2) drought-stressed plants may be behaviorally more attractive or acceptable for insects; (3) drought-stressed plants may be physiologically more suitable for most insects; (4) drought may enhance insect detoxication systems; (5) drought generally favors mutualistic microorganisms but not the insect's natural enemies; and (6) drought may induce genetic changes in insect populations through accelerated natural selection processes. In contrast, Martinat (1987) questioned the proposed role of drought and other climatic anomalies in contributing to insect outbreaks and concluded that the pattern of climatic variation in the temperate regions of North America is such that drought is unlikely to play a significant role in periodic insect outbreaks. In fact, Gagne et al. (1980) showed that drought, in association with high temperatures, may actually be detrimental to the survival of southern pine beetles. Thus, many of the mechanisms proposed by Mattson and Haack (1987) probably do not apply to outbreaks of southern pine beetle. More research is needed to assess accurately the possible role of drought in the ecology of the southern pine beetle.

In contrast, attack by bark beetles appears to be regulated primarily by the host tree's natural defense system. When this defense system is weakened, perhaps through drought or some other type of stress disturbance, the tree becomes susceptible to invasion by bark beetles. In this case, drought effects on insect growth, detoxication systems, and genetic structure, as proposed by Mattson and Haack (1987), are probably secondary to its debilitating effects on the host tree's natural defense system.

A second climatic factor, temperature, seems to be directly related to beetle survival (Beal, 1933; Gagne et al., 1980) but may also have an indirect effect on host susceptibility through its influence on the water status of the tree (Hicks, 1980). There have been numerous attempts to quantify the relationship between climate (primarily rainfall and temperature) and the intensity of southern pine beetle infestation (e.g., see Campbell and Smith, 1980; Kalkstein, 1981; Kroll and Reeves, 1978; Michaels, 1984; Michaels et al., 1985, 1986). However, these models are of limited use because they are applicable only to relatively small geographic areas and generally exhibit low rates of predictive success. Martinat (1987) examined the general relationship between the historical pattern of climatic variation and outbreaks of several forest insects and found only weak evidence that such a relationship exists. In contrast to these climate-driven models, Turchin et al. (1991) analyzed a 30-year history of southern pine beetle activity in east Texas and concluded that southern pine beetle outbreaks are not driven by random fluctuations of weather (a density-independent event) but by some unknown population process acting in a delayed density-dependent manner. Even though their analysis was never tested as a predictive model, regression analysis revealed that it accounted for 55% of the variation in southern pine beetle activity during that 30-year period. Additional research is needed to understand the interrelationships among all the abiotic and biotic factors that affect and regulate southern pine beetle population dynamics.

Weather-related stand disturbances, such as lightning strikes, wind, ice, and hail damage, serve to weaken the trees in the stand and render them more susceptible to initial beetle attack (Hicks, 1980). Numerous studies have confirmed the importance of lightning as a predisposing agent in southern pine beetle infestations (e.g., see Blanche et al., 1985; Flamm et al., 1993; Hodges and Pickard, 1971; Ku et al., 1980; Lorio and Bennett, 1974; Lovelady et al., 1991; Porterfield and Rowell, 1981; Thatcher and Pickard, 1964).

Site Factors

Hicks (1980) discussed several site factors that play a role in host susceptibility to southern pine beetle infestation (Table 6.2). Landform, or position in the landscape, apparently has an influence on beetle activity. Beetle infestations are generally most frequent on low-lying sites and on ridge sites. Apparently, the stress associated with excessive moisture on the low-lying sites and insufficient moisture on the ridge sites is key to the more frequent beetle activity. Other researchers have verified that the moisture regime of the site has a significant effect on the incidence of beetle infestation (Blanche et al., 1983; Ku et al., 1980; Porterfield and Rowell, 1981). The effect of soil texture on host susceptibility to beetle infestation varies geographically. In the Coastal Plain, infestations are highest on well-watered, productive, sandy and loamy soils; whereas in the Piedmont, clayey soils seem to be more conducive to beetle infestation (Hicks, 1980). Soil chemical properties also seem to be related to beetle activity. One study in east Texas found that beetle infestations were more frequent on soils deficient in potassium, calcium, or zinc (Hicks, 1980). A weak but positive correlation has been found between loblolly pine site index and beetle activity on the Gulf Coastal Plain (Rowell, 1978). Ku et al. (1980), however, found an inverse relationship between site index and beetle activity in Arkansas. Other site factors, such as soil depth, have a minor influence on beetle infestations.

Stand Factors

Various stand attributes exert a significant influence on occurrence of southern pine beetle infestations (Table 6.2). The most important of these is stand density. Numerous studies have found that high stand density is often associated with beetle infestation (e.g., see Belanger et al., 1979; Bennett, 1968; Cameron and Billings, 1988; Hicks et al., 1978; Ku et al., 1980; Lorio, 1980a; Lorio and Bennett, 1974; Porterfield and

Rowell, 1981; Schowalter and Turchin, 1993). High stand density causes a reduction in tree vigor, which then predisposes the tree to beetle attack (Hicks et al., 1978). Estimates of density thresholds for predisposition to beetle attack range from 100 square feet of basal area per acre (23.0 square meters per hectare) (Brown et al., 1987; Ku et al., 1980) to 140 square feet per acre (32.1 square meters per hectare) (Rowell, 1978). The reduction in tree vigor associated with high stand density (or with other stress factors) is manifested in very slow radial growth (Bennett, 1968). Slow radial growth has been consistently related to beetle attack all across the South (Belanger et al., 1979; Ku et al., 1980; Porterfield and Rowell, 1981). Thinning of slow-growing, overstocked stands has been shown to be an effective tool for reducing stand density, possibly increasing the resistance of residual trees and increasing intertree spacing and thus reducing stand susceptibility to southern pine beetle attack (Belanger et al., 1993; Brown et al., 1987; Hedden, 1983; Nebeker et al., 1985). Economic analyses have revealed that thinning can be expected to reduce losses from southern pine beetle attack on average and better sites and that, in general, benefits from this type of thinning increase with increasing site quality (Burkhart et al., 1986).

Species composition is also important in stand susceptibility to beetle infestation. Stands with high proportions of loblolly and/or shortleaf pines are more susceptible to beetle attack (Cameron and Billings, 1988; Hicks, 1980). Recent evidence indicates that the presence of hardwoods in southern pine stands does not increase susceptibility to beetle attack and may even interfere with infestation growth and spread (Schowalter and Turchin, 1993). Presence in the stand of diseases, particularly annosus root rot, or other insects, such as black turpentine beetle and *Ips*, is known to reduce resistance of individual trees and thus predispose them to attack by the southern pine beetle (Hicks, 1980). Kinn and Linit (1992) recently reported increased feeding activity of the cerambycid vector of the pinewood nematode [*Bursaphelenchus xylophilus* (Steiner and Buhrer) Nickel] on the periphery of southern pine beetle infestations in loblolly and shortleaf pine stands, possibly indicating that this insect vector may play a role in predisposing apparently healthy trees to beetle attack. In this way, the pinewood nematode may contribute to the spread of southern pine beetle infestations. Other stand attributes, such as average stand age, height, and diameter, have only minor effects on stand susceptibility to beetle infestation. However, overmature trees, independent of these site and stand factors, universally exhibit low resistance to southern pine beetle attack.

Physiological Factors

Blanche et al. (1983) discussed in great detail the physiological factors associated with the host that influence susceptibility to beetle attack. Because the oleoresin system in pines is the primary defense mechanism against beetle attack, any reduction in the amount of oleoresin present in the tree will result in increased susceptibility to attack. Early work indicated that low oleoresin exudation pressure is associated with successful beetle attack (Lorio and Hodges, 1968a, 1968b, 1977; Mason, 1971). Brown et al. (1987) showed, however, that thinning overstocked pine stands to basal areas of 23.0 square meters per hectare or less effectively increased oleoresin exudation pressure of residual trees for a period of two years after thinning. More recent research indicated that other properties of the oleoresin system, such as total resin flow, rate of flow, and relative viscosity, are also important in determining both host preference and host susceptibility (Blanche et al., 1983; Hodges et al., 1977, 1979, 1981). Total flow of oleoresin is particularly important in this respect and may be the single most important property of the oleoresin system when estimating the relative resistance of trees to beetle attack (Lorio, 1994). Nebeker et al. (1991, 1992) detected significant variation in total resin flow, rate of flow, and relative viscosity among 14 full-sib families of loblolly pine, indicating that it may be possible to incorporate potential bark beetle resistance into existing loblolly pine tree improvement programs. In addition to these physical properties of the oleoresin system, Bridges (1987) reported that the chemical composition of oleoresin may also be an important component of the host tree's natural defense system. Specifically, Bridges (1987) found that 4-allylanisole, a type of phenylpropanoid compound, was highly inhibitory to the growth of not only the blue-stain fungus but also the two mycangial fungi associated with southern pine beetles.

Under endemic population levels, stressed trees probably play an important role in maintaining southern pine beetle populations. The stress may have originated from any of several different sources: overstocking in high-density stands, structural and/or physiological damage from lightning strikes, excessive soil moisture, insufficient soil moisture, and many others. The most important diagnostic symptom of this stress is severely reduced radial growth. Consequently, individual trees susceptible to southern pine beetle attack are generally characterized by very slow radial growth, sparse crowns, and thin bark. Because these characteristics are more indicative of past performance than current vitality, Blanche et al. (1991) tested and recommended the use of stem electrical resistance as an effective, sensitive method of rating the current physiological vigor of individual loblolly pine trees. However, under epidemic population levels, the presence of these stressed trees may be of little consequence because, under these situations, the most important requirement for continued development of southern pine beetle outbreaks is an abundance of suitable food and habitat (mature and maturing southern pine trees) over a large land area.

Several site and stand attributes consistently contribute to increased attack susceptibility. Examples of susceptible stands are those that are overstocked, those that have been damaged through either natural events (such as lightning or wind) or forest management operations (such as logging), and those on sites with extreme moisture regimes (excessively wet or dry). All of these stand conditions act to impose a stress on the individual trees, weaken resistance, and increase their susceptibility to southern pine beetle attack.

A great many systems have been developed in recent years that attempt to rate individual stands and predict their susceptibility to southern pine beetle attack (e.g., see Billings et al., 1985; Hedden and Belanger, 1985; Hedden and Lorio, 1985; Hicks et al., 1980, 1981; Ku et al., 1980; Lorio, 1978, 1980b; Lorio and Sommers, 1985; Lorio et al., 1982; Mason et al., 1985; Reed et al., 1981, 1982; Sharpe and Wu, 1985; USDA Forest Service, 1987). Although it is beyond the scope of this chapter to review these rating schemes, it is important to note that most of them are useful and reasonably accurate, but only within the relatively small geographic area for which each was developed.

Visible Symptoms

Thatcher and Connor (1985) described the major visible symptoms associated with southern pine beetle attack: (1) foliage discoloration, (2) pitch tubes, (3) egg galleries, and (4) adult exit holes. Without close examination of individual stems, the most obvious indication of beetle attack is discolored foliage. The needles first fade from green to yellow and then, as the attack becomes more advanced, turn to red and finally to brown. Brown needles typically begin to appear within one to four months after the initial attack. Within the stand, beetle-killed trees may occur singly or in groups, ranging from a few trees to several thousand. Because several other environmental factors may cause foliar discoloration, it is necessary to ascertain the presence of other symptoms before establishing the beetle as the causal factor. Pitch tubes and boring dust are usually the first signs of bark beetle attack (Nebeker et al., 1993). Pitch tubes develop at the point of entry through the outer bark into the tree. They are small, whitish, popcorn-sized deposits of resin exuded by the tree in response to the beetle attack (USDA Forest Service, 1985). Boring dust is produced as the beetle tunnels into the bark of the tree and accumulates on the upper surfaces of understory plants and in spider webs near the base of the tree under attack. Removal of the outer bark of southern pine beetle-infested trees will reveal S-shaped tunnels, or egg galleries, on the surface of the inner bark. The galleries are constructed by the adult beetles and are generally filled with frass (a sawdust-like material) left behind by the feeding adults (Belanger et al., 1986). A final indication of southern pine beetle infestation is the presence of adult exit holes on the outer bark.

Effects on Growth and Physiological Processes

The natural defense system of southern pines against attack and invasion by the southern pine beetle has been studied extensively for many years. Excellent recent reviews have been published by Christiansen et al. (1987), Lorio (1993), and Nebeker et al. (1993). The defense system consists of two major components: (1) a preformed defensive system, in which oleoresin produced by the tree physically repels large numbers of attacking beetles; and (2) an induced defensive system, in which the invading beetles and their associated fungi are chemically and physically contained to a localized area and deprived of nutrients essential to their survival (Berryman, 1972; Nebeker et al., 1993).

The initial phase of any bark beetle attack involves identification by the beetle of a suitable host, followed by preliminary colonization attempts by the beetle. Colonization begins as the adult beetle bites the outer bark of the selected host tree and, if conditions are favorable, continues as the beetle tunnels through the outer bark. Host selection and the biting process may be stimulated by chemical and possibly by physical signals, but these processes are not well understood (Nebeker et al., 1993). Once the adult beetles have tunneled through the outer bark into the inner bark and cambium, the host tree's resin ducts are cut and resin flow usually begins. At this point, the invading beetles must cope with both the preformed and induced defensive systems to colonize the tree.

The preformed defensive system of southern yellow pines is the first line of defense against invasion by southern pine bark beetle. Southern pines are equipped with a well-formed resin duct system that consists of many radial resin ducts and their associated epithelial cells, which secrete the oleoresin into the ducts. This system serves two primary functions in the defense of the tree: (1) synthesis of new resin, and (2) delivery of resin to sites under attack (Nebeker et al., 1993). When the tree is attacked, these ducts exude large quantities of stored resin to the attack site, which physically repels or "pitches out" the invading beetles. Total resin flow to attack sites appears to be the most crucial component of this system in combating colonization by invading bark beetles (Hodges et al., 1977; Hodges et al., 1979; Lorio, 1994). Oleoresin exudation pressure was thought to be directly related to resin flow (Vité, 1961), but there is little experimental evidence to support this view. Rather, oleoresin exudation pressure apparently is an acceptable indicator of the relative water status of the tree, or of tree vigor in general, but does not necessarily indicate the tree's defensive capacity at any given time (Lorio, 1993, 1994). Because the oleoresin used by the tree to repel invading bark beetles is preformed, the storage capacity of the duct system and the ability of the system to deliver resin to attack sites are critical to successful resistance of the beetles' colonization efforts. DeAngelis et al. (1986b) identified four factors that influence the amount and rate of delivery of preformed resin in loblolly pine: (1) volume of the resin-producing system; (2) size of the stored resin pool at the time of attack; (3) physiological state of the resin duct system at the time of attack; and (4) resin viscosity. Resin chemistry may also be important in repelling the initial attack of southern pine beetles. Coyne and Lott (1976) demonstrated that several monoterpenes found in pine oleoresin were toxic to southern pine beetles. Even so, the preformed defensive system relies primarily on the physical rather than chemical properties of the oleoresin system to prevent bark beetle colonization.

If the flow of preformed oleoresin from the resin duct system fails to repel the beetle attack, the beetles infect the host tissues with propagules of a variety of associated fungi and other microorganisms (Christiansen et al., 1987). At this time, the invading beetles encounter the host tree's second line of defense, the induced defensive system. In general, the tree forms a necrotic area around the point of infection, physically and chemically containing the invading organisms within a localized area and depriving them of living tissues for food. Nebeker et al. (1993) described this hypersensitive response in terms of four related processes: (1) localized autolysis of parenchyma cells followed by rapid cellular desiccation; (2) tissue necrosis that results in the formation of lesions and the accumulation of toxic or allelochemic compounds; (3) secondary resinosis by adjacent epithelial and parenchyma cells; and (4) formation of wood periderm around the lesion to isolate it physically from normal metabolic and transport activities. The resinous and phenolic compounds found in the lesions prevent beetle gallery construction, are highly toxic to bark beetle eggs, and inhibit fungal growth (Bridges, 1987; Christiansen et al., 1987; Paine and Stephen, 1988). To be effective against bark beetle colonization efforts, this induced response must have a ready source of stored reserves that can be mobilized rapidly for use. Because these reserves are primarily carbohydrates, implementation of the induced defensive system creates a substantial drain on the host tree's energy supply. Containment of the invading fungi in the reaction zone depends on the tree's ability to accumulate resinous material. A tree under mass attack by bark beetles can rapidly deplete its stored reserves. Consequently, when the number of beetle attacks reaches some threshold level, the resin concentration in individual reaction zones decreases, the ability of the fungi to invade adjacent areas of living sapwood improves, and the induced defensive system fails (Christiansen et al., 1987).

Because of their well-developed resin duct system, southern pines utilize the preformed defensive system as their primary defense mechanism against attack by the southern pine bark beetle. Southern pines do possess the ability to utilize the induced defensive system as a secondary defense mechanism, and the production of hypersensitive reaction zones has been documented in loblolly pine (Paine et al., 1987; Paine and Stephen, 1987a, 1987b, 1988). However, because of the rapid mass-attack strategy employed by the southern pine beetle, Nebeker et al. (1993) hypothesized that the induced defensive system plays only a minor role in the defense of southern pines against beetle invasion.

Once the beetles, through the strategy of mass attack, have successfully overcome the host tree's preformed and induced defensive systems, unimpeded colonization of the tree begins. This process consists of two major events: (1) introduction of associated microorganisms into the phloem and xylem tissues; and (2) physical construction of egg galleries (Nebeker et al., 1993). Stated simply, the primary effect of a successful southern pine beetle attack is death of the host tree. However, the exact mechanism by which the beetles kill the tree is unknown. The beetles themselves may play a direct role in killing the tree, in that construction of egg galleries, by severing cells in the outer layers of xylem tissue, may interfere with downward translocation of materials and with upward movement of water. However, some evidence indicates that the microorganisms associated with the beetle may play a significant role in the eventual death of the tree, possibly through disruption of the tree's water relations (Blanche et al., 1983; Bridges et al., 1985; Hodges et al., 1985; Lorio and Hodges, 1985; Nebeker et al., 1993).

Anderson (1960) proposed four major mechanisms by which the associated fungi could cause death of the host tree: (1) production of toxic compounds; (2) plugging of the tracheids by fungal mycelia; (3) introduction of air bubbles into the tracheids; and (4) production of particles or compounds that block pit openings by causing tori aspiration. All of these events could occur, but none has been proven to be directly responsible for tree death.

Accumulation of additional evidence eventually led Hodges et al. (1985) to hypothesize that death of the host tree occurs as a result of a series of events initiated by both the beetle and its associated fungi: (1) suppression of initial tree resistance; (2) inoculation of fungi into beetle galleries; (3) production of fungal metabolites; (4) alteration of host physiological functions by the metabolites; and (5) death of the tree. This hypothesis has since served as a basis for further investigation into the roles played by the beetle's associated fungi in the death of host trees. It appears that fungal growth and production of metabolites are not responsible for the initial decline in tree resistance to beetle attack. Rather, the initial decrease in resistance is the result of a physical depletion of the oleoresin supply following a mass beetle attack.

The southern pine beetle is associated with a number of microorganisms, but three particular fungi have received most of the attention in recent research efforts. Two of these fungi, an unidentified basidiomycete (known as SJB 122) and *Ceratocystiopsis ranaculosus* Perry et Bridges [previously known as SJB 133 and as *Ceratocystis minor* (Hedgecock) Hunt var. *barrasii* Taylor], are carried in the mycangium of female beetles. A third associated fungus, *Ophiostoma minus* (Hedge.)

H. & P. Syd. [previously known as *Ceratocystis minor* (Hedgecock) Hunt], is a blue-stain fungus carried externally on both male and female beetles. *Ophiostoma minus* is not always present in southern pine beetle infestations and apparently is not required for successful beetle development (Bridges et al., 1985). The ecological relationships among these three fungi, the beetle, and the host tree are not well understood. Recent research indicates that the two mycangial fungi are important for optimal beetle reproduction, as evidenced by increased egg production and increased gallery length when both mycangial fungi are present, whereas the blue-stain fungus may interfere with southern pine beetle reproduction, as evidenced by decreased egg production in its presence (Goldhammer et al., 1990). Hodges et al. (1985) hypothesized that complete circumferential introduction of fungal inoculum into the living tree is necessary for successful beetle colonization and eventual death of the tree.

Sometime after the fungal associates of the southern pine beetle have been inoculated into the living tree, they begin producing toxic metabolites that may play a critical role in the eventual death of the tree (Hodges et al., 1985). Three toxic phenolic acids have been isolated from either liquid cultures of *O. minus* or from infected loblolly pine tissues (Hemingway et al., 1977; McGraw and Hemingway, 1977). Natural production of two of these metabolites, both isocoumarins, has been verified, and they have been shown to increase greatly the rate of transpiration in loblolly pine seedlings (DeAngelis et al., 1986a; Hodges et al., 1989). The increased transpiration observed in these infected seedlings may occur as a result of decreased stomatal control of water loss.

The final event leading to the death of the host tree, as hypothesized by Hodges et al. (1985), is the disruption of host physiological processes, especially water relations, which eventually leads to severe water stress and subsequent tree death. Secondary symptoms that appear in the host tree at this time include (1) a rapid drop in turgor pressure throughout living cells in the bole of the tree; (2) blockage of tracheids and a concomitant reduction in the volume of water transported to the crown; and (3) drying of the tissues in the outer bole as water is withheld from below and from within (Nebeker et al., 1993). Exact causal relationships for these events have not been proven, but the toxic metabolites produced by the beetle's associated fungi may play a major role in producing the severe water stress found in host trees just prior to death (DeAngelis et al., 1986a; Hodges et al., 1989; Nebeker et al., 1993).

Death of individual trees is only the first in a series of events associated with southern pine beetle infestations. Beetles typically kill trees in groups, or "spots," which enlarge as the infestation becomes more severe. Adult beetles emerging from infested trees apparently disperse in a random pattern (Thoeny et al., 1992) but may be attracted to new host trees by increased activity of the insect vector of the pinewood nematode, recently reported to be found on the periphery of active southern pine beetle infestations (Kinn and Linit, 1992). As the infestation continues, increasingly larger openings are created in the forest canopy. Species composition in both the overstory and understory may be significantly affected. Southern pine beetle infestations are also commonly associated with invasion by other biotic pathogens (Alexander et al., 1980, 1981). If the infestation is not controlled, the entire ecosystem may be drastically altered or destroyed.

In summary, the southern pine beetle greatly alters the water relations of individual trees and eventually kills them. Severe infestations may result in the death of many thousands of trees, thus substantially altering the species composition of the stand and the structure of the ecosystem.

Fusiform Rust

Fusiform rust, caused by the fungus *Cronartium quercuum* (Berk.) Miyabe ex Shirai f. sp. *fusiforme*, is the most important disease problem in loblolly and slash pines today. Anderson et al. (1986b) estimated annual losses due to fusiform rust to be over $35 million in the five Atlantic Coast states from Virginia to Florida. Annual losses throughout the South probably exceed $100 million (Anderson and Mistretta, 1982). The disease causes extensive mortality, especially in seedlings and young saplings, across the South. Infections in older trees usually do not directly kill the trees but do cause stem breakage and loss of quality. The biology, epidemiology, and control strategies associated with fusiform rust have been reviewed by Wells and Wakeley (1966), Czabator (1971), Powers et al. (1981, 1993), Anderson and Mistretta (1982), and Kuhlman (1982). The history of fusiform rust research in the South was reviewed by Verrall (1982).

Extent and Magnitude in the South

Fusiform rust primarily infects loblolly and slash pines, the most susceptible of all southern pine species. Longleaf pine, a resistant species, and pond pine (*Pinus serotina* Michx.) are occasionally infected; spruce pine (*P. glabra* Walt.), sand pine [*P. clausa* (Chapm. ex Engelm.) Vasey ex Sarg.], shortleaf pine, and Virginia pine are highly resistant and are only rarely infected. Oak (*Quercus* L. spp.), the alternate host of fusiform rust, is not severely damaged by the disease.

Fusiform rust occurs throughout the South but is most prominent in the southeastern states of Georgia, Florida, and South Carolina. In fact, Kuhlman (1982) reported that about 18% of all loblolly pines in Georgia and over 30% of planted loblolly pines in that state were infected with fusiform rust. Based on USDA Forest Service Forest Inventory and Analysis survey data, Anderson et al. (1988) developed fusiform rust hazard-zone maps for slash pine in Georgia and Florida and for loblolly pine in Georgia, South Carolina, North Carolina, and Virginia. Rust problems also exist in Alabama, Mississippi, and Louisiana but seem to be relatively minor in North Carolina and Virginia (Anderson, 1985). Incidence of fusiform rust has increased dramatically since 1969 in east Texas, particularly in slash pine plantations (Hunt and Lenhart, 1986; Lenhart et al., 1988, 1994).

Anderson (1985) outlined the extent of fusiform rust damage in the South from 1979 to 1983. Nearly 5.7 million hectares of pine timber across the South had at least 10% of the stems infected with fusiform rust, about 2.4 million hectares showed at least 30% infected, and there were 1.0 million hectares in which at least 50% of the trees were rust infected (Table 6.3). Powers et al. (1975) estimated that annual volume losses across the South were nearly 3.1 million cubic meters. Geron and Hafley (1988) simulated the impact of fusiform rust on product yields of loblolly pine plantations and found that yields of sawtimber and other solid wood products decreased sharply across plantations with increasing rust infection, whereas yields of pulpwood increased, but only because infected stems could only be utilized for that product.

Characteristics of Susceptible Trees and Stands

Fusiform rust can infect trees in all stages of development, from seedlings to mature trees. However, the effects of rust infection are more severe in seedlings, particularly in newly planted stands, than in older trees. Schmidt et al. (1974) estimated that rust incidence is increasing at a rate

Table 6.3. Slash and Loblolly Pine Stands in the South that Have at Least 10, 30, and 50% of the Trees Infected with Fusiform Rust on or Within 30 cm of the Main Stem, 1983[a]

State	Percent Infected		
	10	30	50
	Thousand Hectares		
Alabama	826	330	137
Arkansas	24	8	3
Florida	452	286	109
Georgia	1634	856	364
Louisiana	635	221	85
Mississippi	682	280	115
North Carolina	544	155	58
South Carolina	605	258	108
Texas	203	87	32
Virginia	3	2	1
Total	5608	2483	1012

[a] Adapted from Anderson (1985).

of 2 to 3% per year in slash pine; a similar rate of increase has also been reported in loblolly pine (Griggs and Schmidt, 1977). There are a number of factors that contribute to increased incidence of fusiform rust in southern pine forests: (1) high component of loblolly or slash pines; (2) increased existence of plantations in relation to natural stands; (3) increased oak component in pine stands; (4) intensive forest management practices; (5) soil factors; and (6) previous exposure to insects and other diseases.

Loblolly and slash pines are much more susceptible to fusiform rust than any of the other southern pines. The importance of these species in southern forestry has increased dramatically in relation to the other southern pines, particularly longleaf pine, a rust-resistant species. Consequently, the importance of fusiform rust as a destructive force in southern forest management has also increased dramatically.

Powers et al. (1975) provided convincing evidence that incidence of fusiform rust is significantly higher in planted stands than in natural stands. Their survey results showed that rust incidence in loblolly pine plantations was about 21%, whereas incidence in natural stands was only 12%. More recent data generally support the conclusion that rust incidence is higher in plantations than in natural stands, particularly in slash pine. For the five southeastern states along the Atlantic Coast, Anderson et al. (1986b) reported that the proportion of hectares in which at least 10% of the trees were rust infected was nearly two times greater in slash pine plantations as in natural slash pine stands (42 and 23%, respectively). Results from the same survey indicated, however, that the proportion of hectares affected by rust was about the same for both planted and natural stands of loblolly pine (35 and 41%, respectively). Nearly equal infection rates in planted and natural loblolly pine stands, as opposed to the greater rust-incidence found in plantations in the earlier survey by Powers et al. (1975), may be attributed to an increasing use of rust-resistant planting stock in most newer loblolly pine plantations. The rapid, vigorous growth of planted seedlings, particularly those planted on intensively site-prepared lands, apparently makes them more susceptible to rust infection than naturally established seedlings (Powers et al., 1981). Increased susceptibility of these fast-growing seedlings to rust infection may be due to lower allocation of photosynthate to defensive compounds during periods of rapid growth. Plantations established either outside or at the edges of the natural ranges of loblolly and slash pines showed particularly high levels of rust infection in the earlier survey (Powers et al., 1975).

A major factor influencing rust incidence is the number of oak stems within and adjacent to pine stands. Oaks, particularly red oaks, are the alternate hosts for the rust fungus. Prior to 1930, uncontrolled wildfires in southern forests kept the oak component in pine stands at a low level. However, with increased and more effective fire control efforts since then, the oak component has increased greatly (Kuhlman, 1982). Significant positive correlations have been found between the number of susceptible oak trees and fusiform rust incidence in pines (Froelich and Snow, 1986; Hollis and Schmidt, 1977; Schmidt et al., 1990; Squillace et al., 1978).

Intensive forest management practices, designed to increase growth rates of planted pines, have also been related to increased rust incidence. The vigorous growth produced through intensive site preparation generally leads to increased incidence of rust in both loblolly and slash pine plantations (May et al., 1973; Miller, 1972, 1977; Powers et al., 1975). Weed control (both mechanical and chemical) and fertilization have been shown to increase rust incidence in slash pine plantations (Dinus and Schmidtling, 1971; Haywood et al., 1994; Shoulders et al., 1991) but to have no effect on rust incidence in loblolly pine plantations (Dinus and Schmidtling, 1971; Kane, 1981). However, Zutter et al. (1987) reported increased rust incidence following herbaceous weed control in four out of seven loblolly pine plantations. In another study, fertilization with municipal sewage sludge on an intensively prepared, high-rust hazard site actually reduced the incidence of rust on loblolly pine (Stone and Powers, 1989). Obviously, the relationships between cultural practices and rust incidence in loblolly pine are much less well defined than for slash pine. In general, Miller (1972) concluded that rust incidence is positively correlated with seedling growth rate. Any forest management practice that promotes rapid seedling growth typically results in increased rust incidence. One possible explanation is that fast-growing seedlings probably utilize a greater proportion of photosynthate for growth, resulting in a smaller proportion available for production of defensive compounds necessary for rust resistance.

Some soil factors have also been reported to have an influence on rust incidence in southern pine forests. For example, Hollis et al. (1975) found that rust incidence of slash pine in Florida was highest on well-drained soils. Schmidt et al. (1990) confirmed this result and reported that rust incidence of slash pine plantations on Coastal Plain sites in Florida and Georgia was highest on moderately well- to well-drained soils (mostly Inceptisols and Ultisols) and was lowest on very poorly to moderately well-drained soils (mostly flatwoods Spodosols). In a separate study, Schmidt et al. (1988) found that rust incidence for both slash and loblolly pines was highest on sites with loamy sand or sandy loam soils and lowest on sites with sand soils. However, plantations of both species on high-hazard sites had a high incidence of rust infection regardless of soil drainage or texture class. Rust incidence of slash pine also increases when phosphorus-deficient soils are fertilized (Hollis et al., 1975; Schmidt et al., 1972). Increased rust incidence in fertilized seedlings may occur as a result of accelerated growth and subsequent changes in carbon partitioning, as discussed earlier.

Pine stands previously exposed to insects and other diseases may also be more susceptible to fusiform rust infection. In particular, attack by the Nantucket pine tip moth appears to predispose young loblolly pine trees to subsequent infection by the fusiform rust fungus (Hedden et al., 1991; Powers and Stone, 1988). The exact nature of this relationship is unknown at this time.

A number of hazard rating systems have been developed to predict the incidence of fusiform rust on various sites across the South. Squillace (1976) and Squillace and Wilhite (1977) presented isogram maps of the South showing regional differences and patterns of geographic variation in rust incidence in both loblolly and slash pine plantations. Anderson et al. (1986a) used these maps to develop a system—based on geographic location, species, and type of regeneration (natural or artificial)—to estimate the probability of different degrees of rust infection in newly established pine stands in the Southeast. Borders and Bailey (1986) developed a series of regression models to predict the proportion of infected trees in site-prepared loblolly and slash pine plantations from Virginia to Alabama; their system was based on geographic location, plantation age, and site index. Similar hazard rating systems and rust incidence prediction models have also been developed for Mississippi (Froelich and Snow, 1986) and east Texas (Arabatzis et al., 1991). Powers et al. (1993) recommended the use of these types of systems to aid the land manager in determining the intensity of site preparation, the selection of planting stock, and the choice of treatments in young stands.

Visible Symptoms

The earliest visible symptoms of fusiform rust infection are slight swelling and purplish epidermal discoloration of new shoots and needles. These protrusions are perennial and gradually enlarge to form spindle-shaped swellings, or galls, on the branches and/or main stem. Main-stem infections on older trees are often attacked by insects and other fungi, usually become flattened on one side, and form cankers, areas of diseased tissue that encircle the entire stem. In the spring, the fungus produces orange-yellow blisters on the surface of the galls. The blisters release the aeciospores, the spore stage of the fungus that infects the oak host. Later in the spring, or in early summer, basidiospores from the oak reinfect the pine host.

More detailed descriptions of symptomology may be found in Powers et al. (1981), Anderson and Mistretta (1982), Anderson (1985), USDA Forest Service (1985), and Kuhlman (1988).

Effects on Growth and Physiological Processes

The severity of the effects of fusiform rust on pine depends, to a large extent, on the age at which the tree is initially infected. Trees less than five years old are likely to be killed by fusiform rust infection (Campbell, 1965). In a study of the histology of fusiform rust in young loblolly pine seedlings, Kuhlman (1988) reported that rust symptoms generally began to appear within nine months after exposure. By that time, most symptomatic tissues had necrotic cells in the cortex. When infection was severe enough to cause depressed areas on the stem, the necrosis also included xylem and phloem cells. Seedlings with galls longer than 25 mm had very high mortality rates. In a histological study of young slash pine seedlings, Walkinshaw and Roland (1990) found that certain families previously thought to be rust resistant developed lethal stem-girdling galls shortly after exposure to the rust fungus. These stem-girdling galls were described as a bulbous swelling above a constriction on the stem and were found to disorganize the tissues of infected trees much more severely than typical spindle-shaped galls tapered on both ends. Slash pine seedlings infected with stem-girdling galls typically died within one year.

In contrast, older trees typically develop perennial cankers that severely weaken the tree and reduce its growth. Suppressed and intermediate trees are much more likely to die from fusiform rust cankers than are dominant trees. Damage to older trees may take the form of storm-caused stem breakage at the canker, apparently more common in slash pine than in loblolly pine (Belanger and Zarnoch, 1991), or a significant loss of quality for wood products at the time of harvest. Rust-infected trees are also highly susceptible to attack by other biotic pathogens.

Resistance to fusiform rust is apparently under a high degree of genetic control. Consequently, extensive tree breeding programs have developed over the past 25 years to select and breed rust-resistant strains of both loblolly and slash pines. The result of these efforts has been the production of highly rust-resistant planting stock for use in areas of high rust hazard. It is beyond the scope of this chapter to review the tree-breeding research performed over the past 25 years that has led to the development of rust-resistant loblolly and slash pines. More detailed accounts of various aspects of this effort can be found in Wells and Wakeley (1966), Powers et al. (1976), Powers and Kraus (1983), Sluder (1989), and Hodge et al. (1990).

Littleleaf Syndrome

Littleleaf syndrome is the most serious problem affecting shortleaf pine in the South. Loblolly pine may also be affected by it, but to a lesser extent. Littleleaf syndrome is caused by a complex of agents, including the fungus *Phytophthora cinnamomi* Rands, poor soil fertility (in particular, a deficiency of nitrogen), poor internal soil drainage (typically related to high clay content), and, in most cases, nematodes and another fungus, *Pythium* spp. (Anderson and Mistretta, 1982; USDA Forest Service,

1985). The biology, symptomology, consequences, and management strategies associated with littleleaf syndrome have been reviewed by Campbell et al. (1953), Campbell and Copeland (1954), Zak (1961), and, more recently, by Anderson and Mistretta (1982), USDA Forest Service (1985), and Belanger et al. (1986).

Extent and Magnitude in the South

Littleleaf syndrome affects about 6.1 million hectares of commercial shortleaf pine stands in the South, with annual losses exceeding about $15 million (Anderson and Mistretta, 1982). Its distribution forms a "littleleaf belt" from Virginia to Alabama. However, there is some evidence that a disorder resembling littleleaf syndrome has been found in shortleaf pine stands in southern Arkansas (Williams and Tainter, 1971). Symptoms similar to littleleaf syndrome and apparently caused by the same fungi have also been reported in loblolly pine stands in southwestern Louisiana (Lorio, 1966, 1973). Other than these seemingly isolated cases, there has been little or no geographic expansion of the distribution of littleleaf syndrome over the past 30 years (Anderson and Mistretta, 1982; Roth, 1954).

Characteristics of Susceptible Trees and Stands

A number of soil and site characteristics contribute heavily to stand susceptibility to littleleaf syndrome. In general, susceptibility increases as site quality decreases. More specifically, potentially high-hazard sites for littleleaf syndrome are characterized by shallow, severely eroded soil, with a high percentage of heavy, plastic clays, poor internal drainage (indicated by strong subsoil mottling), and low nitrogen availability (Anderson and Mistretta, 1982; Belanger et al., 1986; Oak, 1985). These conditions generally restrict root development and are conducive to the establishment of fungal diseases. For example, heavy clay soils limit aeration, tightly adsorb nutrients to soil particles, and contribute to soil moisture extremes. For the most part, these soils are intermittently waterlogged, an ideal condition for fungal spore dispersal and subsequent root infection. Recent research suggests that anaerobic conditions are necessary in the rhizosphere before pine root tips are predisposed to attack by *P. cinnamomi* (Fraedrich and Tainter, 1989).

Campbell and Copeland (1954) developed a hazard rating system to predict the incidence of littleleaf syndrome in shortleaf pine stands; the system was based on degree of soil erosion and internal soil drainage, the two most critical factors that control the incidence and severity of littleleaf syndrome. Internal soil drainage was further categorized into subsoil consistency, depth to a zone of greatly reduced permeability, and degree of subsoil mottling. This system, although time consuming to apply, is accurate and still in common use today. Oak (1985) developed a more rapid, but less precise, hazard rating system based on soil series and the presence of susceptible pine species. Computer-generated hazard maps may be produced from these ratings to aid in identification of high-hazard littleleaf sites. Both systems may be applicable to predict the risk of loblolly pine decline, a littleleaf-related disorder, on high-hazard littleleaf sites (Oak and Tainter, 1988).

Visible Symptoms

Early symptoms of littleleaf syndrome are very similar to those associated with nutrient deficiencies. During the initial stages of the infection, there is a slight yellowing of the foliage accompanied by slight reductions in needle production and elongation, so that new needles are shorter and fewer in number. Shoot growth is also reduced during the early stages of infection. The decrease in leaf area eventually leads to a reduction in diameter growth.

As the infection progresses to its later stages, the symptoms become more distinctive. The cumulative effects of the annual reductions in needle and twig growth produce sparse crowns with foliage "tufted" at the ends of the branches. Foliage is a pale yellow-green color, most easily recognized in the fall and winter. Increased branch dieback results in a thin, ragged crown. Trees in advanced stages of littleleaf syndrome produce a heavy crop of undersized, persistent cones with a high proportion of infertile seeds, typically within two to three years prior to tree death. By this stage, tree growth is negligible. Production of numerous epicormic branches along the bole of the tree is also a common response during the advanced stages of infection.

Detailed descriptions, accompanied by color photographs, of symptoms of littleleaf syndrome in shortleaf pine may be found in Anderson and Mistretta (1982), USDA Forest Service (1985), and Belanger et al. (1986).

Effects on Growth and Physiological Processes

The initial event in the infection by littleleaf syndrome is related to the characteristics of the soil in which the tree is growing. Poor soil aeration and lack of available nitrogen combine to limit root regeneration, the formation of new roots (Oak, 1985). At the same time, there is a gradual increase in mortality of the functional feeder roots, caused by infection by *Phytophthora cinnamomi* and *Pythium* spp. (USDA Forest Service, 1985). The diseased root system is unable to function properly, so nutrient and water uptake is reduced. The diseased root system imparts a physiological stress to the remainder of the tree, and the infection is well underway.

Littleleaf syndrome rarely infects trees less than 20 years old but becomes increasingly severe in older stands (Anderson and Mistretta, 1982). It develops slowly and causes a gradual reduction in growth over a long period of time. Infected trees seldom improve once littleleaf syndrome becomes established (Roth, 1954). In the later stages of infection, tree growth is severely impaired. For example, Jackson (1951) found that total annual radial growth of infected shortleaf pine trees was only 52% of that of symptomless trees. The cumulative effect of this gradual reduction in tree growth over the course of the infection is a large decrease in total volume production by the tree.

In many cases, infection by littleleaf syndrome predisposes the tree to attack by the southern pine beetle (Anderson and Mistretta, 1982; Belanger et al., 1986; Lorio, 1966, 1973). Susceptibility to beetle attack is highest during the early stages of the littleleaf infection. According to Belanger et al. (1986), trees in advanced stages of infection are rarely attacked by southern pine beetles because the internal moisture and nutrient supplies of the infected tree are unsuitable for brood production by the beetle.

On poor sites, trees infected with littleleaf syndrome usually die within six years after the initial appearance of visible symptoms; diseased trees on good sites may live 15 to 20 years after initial infection (USDA Forest Service, 1985). Development of the infection and subsequent tree mortality occur more rapidly in shortleaf pine than in loblolly pine (Belanger et al., 1986). A major ecological consequence of littleleaf syndrome is that shortleaf pine gradually dies out of infected stands and is replaced by other pine species or by hardwoods (Roth, 1954). In fact, loblolly pine is the recommended alternative for shortleaf pine on high-hazard littleleaf sites. However, loblolly pine has been reported to suffer from a growth decline disorder resembling littleleaf syndrome when planted on these sites (Jacobi et al., 1988).

Annosus Root Rot

Annosus root rot, caused by the fungus *Heterobasidion annosum* (Fr.) Bref. [previously known as *Fomes annosus* (Fr.) Karst], is a major disease problem of pines and other conifers throughout the northern temperate regions of the world. It is a serious problem in pine stands in the South and is the second most important disease of loblolly pine (Kuhlman, 1982). Annosus root rot causes long-term growth reductions and increased mortality in all species of southern pine. The problems associated with annosus root rot have been researched extensively and numerous review and summary articles have been written, such as those by Hodges (1969, 1974), Kuhlman et al. (1976), Froelich et al. (1977b), Anderson and Mistretta (1982), and Kuhlman (1982).

Extent and Magnitude in the South

Annosus root rot is a problem primarily in thinned pine stands on deep, well-drained sandy soils. These soils are found on about 20% of the land base in the South. Anderson and Mistretta (1982) presented a distribution map of high-hazard annosus root rot sites across the South.

Kuhlman et al. (1976) summarized the results of a damage survey of the South conducted in 1962. Although only 2.8% of loblolly pine trees in thinned plantations were dead or dying from annosus root rot, the disease was present in 59% of those stands. Even though reported mortality losses may be relatively small, long-term losses in growth may be large. Kuhlman (1982) reported that current levels of disease incidence and severity are similar to those found in 1962.

Characteristics of Susceptible Trees and Stands

Even though all species of southern pines are susceptible to annosus root rot, loblolly and slash pines are the most severely affected. Thinned stands are much more susceptible than unthinned stands because the stump surfaces and root wounds created during thinning operations act as points of entry for the fungus (Alexander et al., 1985).

There are certain soil characteristics that greatly increase the risk of annosus root rot infection. In general, hazard is greatest and damage is most severe on deep, well-drained, sandy soils. These soils have good internal drainage and do not have high seasonal water tables (Anderson and Mistretta, 1982). More specifically, high-hazard soils contain at least 65% sand in the surface horizon that is at least 30 cm thick above the clay subsoil (Alexander and Anderson, 1985; USDA Forest Service, 1985). Old-field sites and plantations also seem to be particularly susceptible to annosus root rot (Kuhlman, 1982).

One of the earliest and most successful soil hazard rating systems for annosus root rot was developed by Morris and Frazier (1966) for use in Virginia. In this system, high-hazard sites were defined as those having at least 70% sand at a depth of 20 to 30 cm and a water table level below 45 cm. Alexander et al. (1975) used this system to contrast in detail the soil characteristics associated with low- and high-hazard sites in thinned loblolly pine plantations in Virginia. The results of their study are presented in Table 6.4.

Visible Symptoms

In most cases, the earliest visible symptoms of annosus root rot are foliar discoloration from green to light yellow-green and crown thinning. Occasionally, however, the foliage simply turns red and the tree dies quickly. Crowns of infected trees may have a tufted appearance due to shorter-than-normal needles and internodes. Tree death generally occurs over an extended period of time. Infected trees may be scattered throughout the stand, usually adjacent to a stump from a previous thinning, or may occur in clumps of dead or dying trees, called "infection centers." Windthrown trees are commonly found in diseased stands.

Table 6.4. Soil Characteristics Averaged at 20 to 30 cm Depths and Percent Isfection of Loblolly Pine by *Heterobasidion annosum* for Low- and High-Hazard Sites[a]

Soil Characteristic	Low Hazard	High Hazard
Sand (%)	40	74
Silt (%)	30	20
Clay (%)	30	6
Noncapillary pores (%)	6	18
Capillary pores (%)	40	22
Field capacity (%)	27	10
Permanent wilting point (%)	14	3
Available moisture (%)	13	7
Organic matter (%)	1.7	0.8
Bulk density (g/cc)	1.44	1.66
pH	5.1	4.9
Infection (%)	8	33

[a] Values represent means from 10 stands. Adapted from Alexander et al. (1975).

Stands infected with annosus root rot are sometimes mistaken for stands attacked by southern pine beetles. However, the presence of conks (fruiting bodies) of the fungus at the base of affected trees will confirm the disease. When fresh, conks range in color from light gray or tan to dark grayish brown or reddish brown on the upper surface. The underside of the conk is creamy white but darkens with age and has many small pores. Conks are irregularly shaped and range in size from small buttons to brackets several inches wide. They are rubbery and difficult to tear. The conks may be difficult to find because they are usually buried beneath the layer of duff around the tree. Although the conks are perennial, they typically dry up and decay during hot, dry summers. It is easiest to locate conks between December and March. Consequently, absence of conks does not necessarily indicate that the tree is not infected with annosus root rot.

Further proof of annosus root rot infection may be obtained through an examination of the lateral roots of dead or dying trees. Infected roots during the early stages of the disease are resin soaked and brownish-red. In advanced stages, diseased roots are a white, stringy, or spongy mass of decayed tissue. Small, snowy-white or tan pustules (conks) may appear on the surface of diseased roots. Pustules are usually only about 0.3 cm in diameter.

More detailed descriptions, along with excellent color photographs, of the symptoms of annosus root rot may be found in Hodges (1974), Froelich et al. (1977b), Anderson and Mistretta (1982), Alexander and Anderson (1985), and USDA Forest Service (1985).

Effects on Growth and Physiological Processes

Froelich et al. (1977b) gave a brief, general description of the mode of infection and spread of annosus root rot. The agents responsible for the long-distance spread of the disease, from one stand to another, are the windblown spores produced on the underside of the conks, the fruiting structures of the fungus that are attached to the bark of diseased trees and stumps at the base of the main stem. These windblown spores readily germinate on the surfaces of fresh stumps and produce fungal mycelia that grow downward through the stump and its root system. Through root grafts or root contacts, the fungus spreads from the roots of diseased stumps to the roots of adjacent living trees. There is some evidence that, in sandy soils, spores deposited on the soil surface may wash down through the soil profile to penetrate the roots of healthy trees directly (Alexander et al., 1975), although this mechanism of infection is not common.

The most striking effect of annosus root rot in southern pine stands is tree mortality. Mortality usually begins about two to three years after a thinning operation and may continue for an additional five to seven years. Mortality generally occurs in small groups or pockets of trees, although the presence of individual dead trees scattered throughout the stand is not uncommon. Many diseased trees succumb to windthrow. A 1962 survey of the South, summarized by Kuhlman et al. (1976), found that total mortality in thinned loblolly pine plantations was about 3%. Kuhlman (1982) suggested that this percentage was still valid. Morris (1970) reported that annosus root rot caused a maximum volume loss, due to mortality, of 4.5 m^3/ha/yr in loblolly pine stands in Virginia. Cumulative volume loss nine years after thinning was about 5 to 8% of total stand volume.

In many cases, losses due to growth reductions are nearly as large as losses due to mortality alone. For example, Alexander and Skelly (1974) measured losses due to reductions in diameter growth that ranged from about 1.9 to 2.8 m^3/ha/yr in thinned loblolly pine plantations on high-hazard sites in Virginia. They also reported a growth loss of about 14.4 m^3/ha/yr on one low-hazard site, where disease-caused mortality was extremely low. Apparently, most of the total volume loss on high-hazard sites is due to mortality, whereas most of the total volume loss on low-hazard sites is attributed to reduced growth.

Reduced growth due to annosus root rot has also been reported in slash pine plantations (Froelich et al., 1977a). Although trees with less than half of their roots infected with *Heterobasidion annosum* suffered little or no growth reductions, those trees with more than half of their roots infected showed significantly reduced growth. In these trees, declines in stem cross-sectional area growth were detected within three years after thinning and averaged 20% in trees with symptomless crowns and 32% in trees with "tufted" crowns, six years after thinning. Height growth during this period was reduced about 40%.

Several studies have shown that annosus root rot predisposes trees to bark beetle attack (e.g., Alexander et al., 1980, 1981). Results indicate that loblolly pine trees attacked by southern pine beetle may be under stress induced by annosus root rot and that this stress is expressed as severely reduced radial growth. In these cases, separation of losses attributed to the disease from those due to the beetle is difficult.

At one time, annosus root rot was considered to be a serious threat to pine regeneration in second-generation plantations (Driver and Ginns, 1964; Hendrix et al., 1964). This belief was based primarily on research results from cool-climate, European stands. However, Kuhlman and Ross (1970) found little evidence of rot in young, second-generation pine plantations in the Southeast. Kuhlman (1986) verified these results in long-term measurements of pine plantations established on rot-infested sites and concluded that the earlier fears were unfounded and that annosus root rot is not a serious threat to pine regeneration on rot-infested sites in the Southeast. Apparently, rapid deterioration of *H. annosum*-infested stumps by the fungus depletes the inoculum from the sites. Consequently, there is insufficient food supply to maintain the fungus until the second-generation trees are large enough to serve as food bases for the long-term survival of the fungus.

Conclusions

The most important biotic agents of stress affecting southern pine forests are the southern pine beetle, fusiform rust, littleleaf syndrome, and annosus root rot. All of these agents may have devastating effects across large areas of the South. Reduced growth, tree mortality, species compositional shifts, and changes in ecosystem structure are but a few of the effects felt both locally and regionally by outbreaks of any of these biotic agents of stress. Economically, these outbreaks destroy millions of cubic feet of timber each year and result in millions of dollars lost in potential revenues. The South is never completely free from the effects of these infestations and is constantly under the threat of future epidemics.

The natural environment is a complex conglomerate of influencing factors and interactions to which trees must respond on a continuing basis. A great number of agents of stress are perpetually exerting their influence in various combinations and to various magnitudes. Long-term ecological studies of the interactions among these various agents of stress are required before the response of forest ecosystems to stress can be quantified. Of particular concern is the hypothesis that airborne chemicals may alter the susceptibility of affected trees to various biotic pests and pathogens. Consequently, we recommend that future research concentrate on the possible role of airborne chemicals in predisposing affected trees to subsequent attack by various insects and diseases. We suggest that these studies concentrate on the pests and pathogens most important to southern forestry: southern pine beetle, fusiform rust, littleleaf syndrome, and annosus root rot.

References

Alexander SA, Anderson RL (1985) *How To: Identify and Control Annosus Root Rot in the South*. For Bull R8-FB/P-19. USDA Forest Service, Southern Region, Atlanta, GA.

Alexander SA, Hokans RH, Fanelli ES, Kurdyla TM (1985) Methods for estimating annosus root rot in loblolly pine stands. In: Branham SJ, Thatcher RC (eds) *Proceedings of the Integrated Pest Management Research Symposium*, April 15–18, Asheville, NC. Gen Tech Rep SO-56. USDA Forest Service, Southern Forest Experiment Station, New Orleans, LA, pp 56–58.

Alexander SA, Skelly JM (1974) The effect of *Fomes annosus* infection on diameter growth of *Pinus taeda* L.: an initial report. In: Kuhlman EG (ed) *Proceedings of the 4th International Conference on Fomes annosus*. USDA Forest Service, Southeastern Forest Experiment Station, Asheville, NC, pp 30–34.

Alexander SA, Skelly JM, Morris CL (1975) Edaphic factors associated with the incidence and severity of disease caused by *Fomes annosus* in loblolly pine plantations in Virginia. Phytopathology 65:585–591.

Alexander SA, Skelly JM, Webb RS, Bardinelli TR, Bradford B (1980) Association of *Heterobasidion annosum* and the southern pine beetle on loblolly pine. Phytopathology 70:510–513.

Alexander SA, Skelly JM, Webb RS (1981) Effects of *Heterobasidion annosum* on radial growth in southern pine beetle-infested loblolly pine. Phytopathology 71:479–481.

Anderson RF (1960) *Forest and Shade Tree Entomology*. John Wiley and Sons, New York.

Anderson RL (1985) Fusiform rust: site-specific options promise to slow epidemic. In: Loomis RC, Tucker S, Hofacker TH (eds) *Insect and Disease Conditions in the United States, 1979–83*. Gen Tech Rep WO-46. USDA Forest Service, Washington, DC, pp 56–63.

Anderson RL, Cost ND, McClure JP, Ryan G (1986a) Predicting severity of fusiform rust in young loblolly and slash pine stands in Florida, Georgia, and the Carolinas. South J Appl For 10:38–41.

Anderson RL, McCartney TC, Cost ND, Devine H, Botkin M (1988) *Fusiform-Rust-Hazard Maps for Loblolly and Slash Pines*. Res Note SE-351. USDA Forest Service, Southeastern Forest Experiment Station, Asheville, NC.

Anderson RL, McClure JP, Cost N, Uhler RJ (1986b) Estimating fusiform rust losses in five southeast states. South J Appl For 10:237–240.

Anderson RL, Mistretta PA (1982) *Management Strategies for Reducing Losses Caused by Fusiform Rust, Annosus Root Rot, and Littleleaf Disease*. Agric Handb No 597. USDA Forest Service, Washington, DC.

Arabatzis AA, Gregoire TG, Lenhart JD (1991) Fusiform rust incidence in loblolly and slash pine plantations in east Texas. South J Appl For 15:79–84.

Beal JA (1933) Temperature extremes as a factor in the ecology of the southern pine beetle. J For 31:329–336.

Belanger RP, Hedden RL, Lorio PL Jr (1993) Management strategies to reduce losses from the southern pine beetle. South J Appl For 17:150–154.

Belanger RP, Hedden RL, Tainter FH (1986) *Managing Piedmont Forests to Reduce Losses from the Littleleaf Disease-Southern Pine Beetle Complex*. Agric Handb No 649. USDA Forest Service, Washington, DC.

Belanger RP, Osgood EA, Hatchell GE (1979) *Stand, Soil, and Site Characteristics Associated with Southern Pine Beetle Infestations in the Southern Appalachians*. Res Pap SE-198. USDA Forest Service, Southeastern Forest Experiment Station, Asheville, NC.

Belanger RP, Zarnoch SJ (1991) Evaluating and predicting tree mortality associated with fusiform rust in merchantable slash and loblolly pine plantations. In: Coleman, SS, Neary DG (eds) *Proceedings of the Sixth Biennial Southern Silvicultural Research Conference*; October 30–November 1 1990, Memphis, TN. Gen Tech Rep SE-70. USDA Forest Service, Southeastern Forest Experiment Station, Asheville, NC, pp 289–298.

Bennett WH (1968) Timber management and southern pine beetle research. For. Farmer 27(9):12–13.

Berryman AA (1972) Resistance of conifers to invasion by bark beetle-fungus association. BioScience 22:598–602.

Berryman AA (1979) Dynamics of bark beetle populations: analysis of dispersal and redistribution. Bull Swiss Entomol Soc 52:227–234.

Billings RF, Bryant CM, Wilson KH (1985) Development, implementation, and validation of a large area hazard- and risk-rating system for southern pine beetle. In: Branham SJ, Thatcher RC (eds) *Proceedings of the Integrated Pest Management Research Symposium*, April 15–18, Asheville, NC. Gen Tech Rep SO-56. USDA Forest Service, Southern Forest Experiment Station, New Orleans, LA, pp 226–232.

Blanche CA, Hodges JD, Nebeker TE (1985) Changes in bark beetle susceptibility indicators in a lightning-struck loblolly pine. Can J For Res 15:397–399.

Blanche CA, Hodges JD, Nebeker TE, Moehring DM (1983) *Southern Pine Beetle: the Host Dimension*. Bull 917, Mississippi Agricultural and Forestry Experiment Station.

Blanche CA, Nebeker TE, Hodges JD, Schmitt JJ, Honea CR (1991) The applicability of stem electrical resistance in rating loblolly pine tree vigor. In: Coleman SS, Neary DG, (eds) *Proceedings of the Sixth Biennial Southern Silvicultural Research Conference*; October 30–November 1, 1990, Memphis, TN. Gen Tech Rep SE-70. USDA Forest Service, Southeastern Forest Experiment Station, Asheville, NC, pp 314–322.

Borders BE, Bailey RL (1986) Fusiform rust prediction models for site-prepared slash and loblolly pine plantations in the Southeast. South J Appl For 10: 145–151.

Bridges JR (1987) Effects of terpenoid compounds on growth of symbiotic fungi associated with the southern pine beetle. Phytopathology 77:83–85.

Bridges JR, Nettleton WA, Connor MD (1985) Southern pine beetle (Coleoptera: Scolytidae) infestations without the bluestain fungus, *Ceratocystis minor*. J Econ Entomol 78:325–327.

Brown MW, Nebeker TE, Honea CR (1987) Thinning increases loblolly pine vigor and resistance to bark beetles. South J Appl For 11:28–31.

Burkhart HE, Haney HL Jr, Newberry JD, Leuschner WA, Morris CL, Reed DD (1986) Evaluation of thinning for reduction of losses from southern pine beetle attack in loblolly pine stands. South J Appl For 10:105–108.

Cameron RS, Billings RF (1988) Southern pine beetle: factors associated with spot occurrence and spread in young plantations. South J Appl For 12:208–214.

Campbell JB, Smith KE (1980) Climatological forecasts of southern pine beetle infestations. Southeast Geol 20:16–30.

Campbell WA (1965) Reducing losses from diseases in pine plantations. In: Wahlenberg WG (ed) *A Guide to Loblolly and Slash Pine Plantation Management in Southeastern USA*. Rep No 14. Georgia Forest Research Council, Macon, pp 146–155.
Campbell WA, Copeland OL Jr (1954) *Littleleaf Disease of Shortleaf and Loblolly Pines*. Circular No 940. US Department of Agriculture, Washington, DC.
Campbell WA, Copeland OL Jr, Hepting GH (1953) *Managing Shortleaf Pine in Littleleaf Disease Areas*. Stn Pap SE-25. USDA Forest Service, Southeastern Forest Experiment Station, Asheville, NC.
Carter DR, O'Laughlin J, McKinney C (1991) Southern pine beetle impacts and control policy in the 1982–1986 Texas epidemic. South J Appl For 15:145–153.
Christiansen E, Waring RH, Berryman AA (1987) Resistance of conifers to bark beetle attack: searching for general relationships. For Ecol Manage 22:89–106.
Coster JE, Searcy JL (eds) (1981) *Site, Stand, and Host Characteristics of Southern Pine Beetle Infestations*. Tech Bull No 1612. US Department of Agriculture, Washington, DC.
Coyne JF, Lott LH (1976) Toxicity of substances in pine oleoresin to southern pine beetles. J Ga Entomol Soc 11:297–301.
Czabator FJ (1971) *Fusiform Rust of Southern Pines—a Critical Review*. Res Pap SO-65. USDA Forest Service, Southern Forest Experiment Station, New Orleans, LA.
DeAngelis JD, Hodges JD, Nebeker TE (1986a) Phenolic metabolites of *Ceratocystis minor* from laboratory culture and their effects on transpiration in loblolly pine seedlings. Can J Bot 64:151–155.
DeAngelis JD, Nebeker TE, Hodges JD (1986b) Influence of tree age and growth rate on the radial resin duct system in loblolly pine (*Pinus taeda*). Can J Bot 64:1046–1049.
Dinus RJ, Schmidtling RC (1971) *Fusiform Rust in Loblolly and Slash Pines After Cultivation and Fertilization*. Res Pap SO-68. USDA Forest Service, Southern Forest Experiment Station, New Orleans, LA.
Driver CH, Ginns JH Jr (1964) Annosus root-rot in young southern pine planted on an infested site. Plant Dis Rep 48:803–807.
Flamm RO, Pulley PE, Coulson RN (1993) Colonization of disturbed trees by the southern pine bark beetle guild (Coleoptera: Scolytidae). Environ Entomol 22:62–70.
Fraedrich SW, Tainter FH (1989) Effect of dissolved oxygen concentration on the relative susceptibility of shortleaf and loblolly pine root tips to *Phytophthora cinnamomi*. Phytopathology 79:1114–1118.
Froelich RC, Cowling EB, Collicott LV, Dell TR (1977a) *Fomes annosus* reduces height and diameter growth of planted slash pine. For Sci 23:299–306.
Froelich RC, Kuhlman EG, Hodges CS, Weiss MJ, Nichols JD (1977b) *Fomes annosus* root rot in the South: guidelines for prevention. USDA Forest Service, Southeastern Area, State and Private Forestry, Atlanta, GA.
Froelich RC, Snow GA (1986) Predicting site hazard to fusiform rust. For Sci 32:21–35.
Gagne JA, Coulson RN, Foltz JL, Wagner TL, Edson L (1980) Attack and survival of *Dendroctonus frontalis* in relation to weather during three years in east Texas. Environ Entomol 9:222–229.
Geron CD, Hafley WL (1988) Impact of fusiform rust on product yields of loblolly pine plantations. South J Appl For 12:226–231.
Goldhammer DS, Stephen FM, Paine TD (1990) The effect of the fungi *Ceratocystis minor* (Hedgecock) Hunt, *Ceratocystis minor* (Hedgecock) Hunt var. *barrasii* Taylor, and SJB 122 on reproduction of the southern pine beetle,

Dendroctonus frontalis Zimmermann (Coleoptera: Scolytidae). Can Entomol 122:407–418.

Griggs MM, Schmidt RA (1977) Increase and spread of fusiform rust. In: Dinus RJ, Schmidt RA (eds) *Proceedings of the Symposium on Management of Fusiform Rust in Southern Pines*. University of Florida, Gainesville, pp 32–38.

Haywood JD, Tiarks AE, Snow GA (1994) Combinations of fungicide and cultural practices influence the incidence and impact of fusiform rust in slash pine plantations. South J Appl For 18:53–59.

Hedden RL (1983) Evaluation of loblolly pine thinning regimes for reduction of losses from southern pine beetle attack. In: Jones EP (ed) *Proceedings of the Second Biennial Southern Silvicultural Research Conference*. Gen Tech Rep SE-24. USDA Forest Service, Southeastern Forest Experiment Station, Asheville, NC, pp 371–375.

Hedden RL, Belanger RP (1985) Predicting susceptibility to southern pine beetle attack in the Coastal Plain, Piedmont, and southern Appalachians. In: Branham SJ, Thatcher RC (eds) *Proceedings of the Integrated Pest Management Research Symposium*, April 15–18, Asheville, NC. Gen Tech Rep SO-56. USDA Forest Service, Southern Forest Experiment Station, New Orleans, LA, pp 233–238.

Hedden RL, Belanger RP, Powers HR, Miller T (1991) Relation of Nantucket pine tip moth attack and fusiform rust infection in loblolly pine families. South J Appl For 15:204–208.

Hedden RL, Lorio PL Jr (1985) *Rating Stand Susceptibility to Southern Pine Beetle Attack on National Forests in the Gulf Coastal Plain*. Res Pap SO-221. USDA Forest Service, Southern Forest Experiment Station, New Orleans, LA.

Hemingway RW, McGraw GW, Barras S (1977) Polyphenols in *Ceratocystis minus* infected *Pinus taeda*: fungal metabolites, phloem and xylem phenols. Agric Food Chem 25:717–722.

Hendrix FF Jr, Kuhlman EG, Hodges CS Jr, Ross EW (1964) *Fomes annosus—a Serious Threat to Regeneration of Pine*. Res Note SE-24. USDA Forest Service, Southeastern Forest Experiment Station, Asheville, NC.

Hicks RR Jr (1980) Climate, site, and stand factors. In: Thatcher RC, Searcy JL, Coster JE, Hertel GD (eds) *The Southern Pine Beetle*. Tech Bull 1631. USDA Forest Service, Washington, DC, pp 55–68.

Hicks RR Jr, Howard JE, Coster JE, Watterston KG (1978) The role of tree vigor in susceptibility of loblolly pine to southern pine beetle. In: Hollis CA, Squillace AE (eds) *Proceedings of the 5th North American Forest Biology Workshop*. University of Florida, School of Forest Resources, Gainesville, pp 177–186.

Hicks RR Jr, Howard JE, Watterston KG, Coster JE (1980) Rating forest stand susceptibility to southern pine beetle in east Texas. For Ecol Manage 2: 269–283.

Hicks RR Jr, Howard JE, Watterston KG, Coster JE (1981) Rating east Texas stands for southern pine beetle susceptibility. South J Appl For 5:7–10.

Hodge GR, Schmidt RA, White TL (1990) Substantial realized gains from mass selection of fusiform rust-free trees in highly infected stands of slash pine. South J Appl For 14:143–146.

Hodges CS (1969) Modes of infection and spread of *Fomes annosus*. Annu Rev Phytopathol 7:247–266.

Hodges CS (1974) *Symptomatology and Spread of Fomes annosus in Southern Pine Plantations*. Res Pap SE-114. USDA Forest Service, Southeastern Forest Experiment Station, Asheville, NC.

Hodges JD, Elam WW, Bluhm DR (1981) *Influence of Resin Duct Size and Number on Oleoresin Flow in the Southern Pine*. Res Note SO-266. USDA Forest Service, Southern Forest Experiment Station, New Orleans, LA.

Hodges JD, Elam WW, Watson WF (1977) Physical properties of the oleoresin system of the four major southern pines. Can J For Res 7:520–525.

Hodges JD, Elam WW, Watson WF, Nebeker TE (1979) Oleoresin characteristics and susceptibility of four southern pines to southern pine beetle (Coleoptera: Scolytidae) attacks. Can Entomol 111:889–896.

Hodges JD, Nebeker TE, Blanche CA, Honea CR, Fisher TH, Schultz TP (1989) Southern pine beetle-microorganisms-host interactions: influence of compounds produced by *Ceratocystis minor*. In: Miller JH (ed) *Proceedings of the Fifth Biennial Southern Silvicultural Research Conference*. Gen Tech Rep SO-74. USDA Forest Service, Southern Forest Experiment Station, New Orleans, LA, pp 567–572.

Hodges JD, Nebeker TE, DeAngelis JD, Blanche CA (1985) Host/beetle interactions: influence of associated microorganisms, tree disturbance, and host vigor. In: Branham SJ, Thatcher RC (eds) *Proceedings of the Integrated Pest Management Research Symposium*, April 15–18, Asheville, NC. Gen Tech Rep SO-56. USDA Forest Service, Southern Forest Experiment Station, New Orleans, LA, pp 161–168.

Hodges JD, Pickard LS (1971) Lightning in the ecology of the southern pine beetle, *Dendroctonus frontalis* (Coleoptera: Scolytidae). Can Entomol 103: 44–51.

Hoffard WH (1985) Southern pine beetle: a would-be manager of southern forests. In: Loomis RC, Tucker S, Hofacker TH (eds) *Insect and Disease Conditions in the United States, 1979–83*. Gen Tech Rep WO-46. USDA Forest Service, Washington, DC, pp 32–37.

Hollis CA, Schmidt RA (1977) Site factors related to fusiform rust incidence in north Florida slash pine plantations. For Sci 23:69–77.

Hollis CA, Smith WH, Schmidt RA, Pritchett WL (1975) Soil and tissue nutrients, soil damage, fertilization and tree growth as related to fusiform rust incidence in slash pine. For Sci 21:141–148.

Holmes TP (1991) Price and welfare effects of catastrophic forest damage from southern pine beetle epidemics. For Sci 37:500–516.

Hunt EV Jr, Lenhart JD (1986) Fusiform rust trends in east Texas. South J Appl For 10:215–216.

Jackson, LWR (1951) Effect of littleleaf disease on growth of shortleaf pine. J For 49:778–779.

Jacobi JC, Tainter FH, Oak SW (1988) The effect of drought on growth decline of loblolly pine on littleleaf sites. Plant Dis 72:294–297.

Kalkstein LS (1976) Effects of climatic stress upon outbreaks of the southern pine beetle. Environ Entomol 5:653–658.

Kalkstein LS (1981) An improved technique to evaluate climate-southern pine beetle relationships. For Sci 27:579–589.

Kane MB (1981) Fertilization of juvenile loblolly pine plantations: impacts on fusiform rust incidence. In: Barnett JP (ed) *Proceedings of the First Biennial Southern Silvicultural Research Conference*. Gen Tech Rep SO-34. USDA Forest Service, Southern Forest Experiment Station, New Orleans, LA, pp 199–204.

King EW (1972) Rainfall and epidemics of the southern pine beetle. Environ Entomol 1:279–285.

Kinn DN, Linit MJ (1992) Temporal relationship between southern pine beetle (Coleoptera: Scolytidae) and pinewood nematode infestations in southern pines. J Entomol Sci 27:194–201.

Kroll JC, Reeves HC (1978) A simple model for predicting annual numbers of southern pine beetle infestations in east Texas. South J Appl For 2:62–64.

Ku TT, Sweeney JM, Shelburne VB (1980) Site and stand conditions associated with southern pine beetle outbreaks in Arkansas—a hazard-rating system. South J Appl For 4:103–106.

Kuhlman EG (1982) Diseases in stand management of loblolly pine. In: *Proceedings of the Symposium on the Loblolly Pine Ecosystem (East Region)*. North Carolina State University, School of Forest Resources, Raleigh, and USDA Forest Service, Southeastern Forest Experiment Station, Asheville, NC, pp 188–197.

Kuhlman EG (1986) Impact of annosus root rot minimal 22 years after planting pines on root rot infested sites. South J Appl For 10:96–98.

Kuhlman EG (1988) Histology and progression of fusiform rust symptoms on inoculated loblolly pine seedlings. Plant Dis 72:719–721.

Kuhlman EG, Hodges CS Jr, Froelich RC (1976) *Minimizing losses to Fomes annosus in the Southern United States*. Res Pap SE-151. USDA Forest Service, Southeastern Forest Experiment Station, Asheville, NC.

Kuhlman EG, Ross EW (1970) Regeneration of pine on *Fomes annosus*-infested sites in the southeastern United States. In: Hodges CS, Rishbeth J, Yde-Andersen A (eds) *Proceedings of the Third International Conference on Fomes annosus*. USDA Forest Service, Washington, DC, pp 71–76.

Lenhart JD, Gregoire TG, Kronrad GD, Holley AG (1994) Characterizing fusiform rust incidence and distribution in east Texas. South J Appl For 18:29–34.

Lenhart JD, McGrath WT, Hackett TL (1988) Fusiform rust trends in east Texas: 1969–1987. South J Appl For 12:259–261.

Loomis RC, Tucker S, Hofacker TH (eds) (1985) *Insect and Disease Conditions in the United States, 1979–83*. Gen Tech Rep WO-46. USDA Forest Service, Washington, DC.

Lorio PL Jr (1966) *Phytophthora cinnamomi* and *Pythium* species associated with loblolly pine decline in Louisiana. Plant Dis Rep 50:596–597.

Lorio PL Jr (1973) *Declining Pines Associated with Bark Beetle Activity in Allen Parish, LA*. Res Note SO-163. USDA Forest Service, Southern Forest Experiment Station, New Orleans, LA.

Lorio PL Jr (1978) *Developing Stand Risk Classes for the Southern Pine Beetle*. Res Pap SO-144. USDA Forest Service, Southern Forest Experiment Station, New Orleans, LA.

Lorio PL Jr (1980a) Loblolly pine stocking levels affect potential for southern pine beetle infestation. South J Appl For 4:162–165.

Lorio PL Jr (1980b) Rating stands for susceptibility to SPB. In: Thatcher RC, Searcy JL, Coster JE, Hertel GD (eds) *The Southern Pine Beetle*. Tech Bull 1631. USDA Forest Service, Washington, DC, pp 153–163.

Lorio PL Jr (1993) Environmental stress and whole-tree physiology. In: Schowalter RD, Filip GM (eds) *Beetle–Pathogen Interactions in Conifer Forests*. Academic Press, London, pp 81–101.

Lorio PL Jr (1994) The relationship of oleoresin exudation pressure (or lack thereof) to flow from wounds. J Sustainable For 1:81–93.

Lorio PL Jr, Bennett WH (1974) *Recurring Southern Pine Beetle Infestations near Oakdale, LA*. Res Pap SO-95. USDA Forest Service, Southern Forest Experiment Station, New Orleans, LA.

Lorio PL Jr, Hodges JD (1968a) Microsite effects on oleoresin exudation pressure of large loblolly pines. Ecology 49:1207–1210.

Lorio PL Jr, Hodges JD (1968b) Oleoresin exudation pressure and relative water content of inner bark as indicators of moisture stress in loblolly pines. For Sci 14:392–398.

Lorio PL Jr, Hodges JD (1977) *Tree Water Status Affects Induced Southern Pine Beetle Attack and Brood Production*. Res Pap SO-135. USDA Forest Service, Southern Forest Experiment Station, New Orleans, LA.
Lorio PL Jr, Hodges JD (1985) Theories of interactions among bark beetles, associated microorganisms, and host trees. In: Shoulders E (ed) *Proceedings of the Third Biennial Southern Silvicultural Research Conference*, November 7–8, 1984, Atlanta, GA. Gen Tech Rep SO-54. USDA Forest Service, Southern Forest Experiment Station, New Orleans, LA, pp 485–492.
Lorio PL Jr, Mason GN, Autry GL (1982) Stand risk rating for the southern pine beetle: integrating pest management with forest management. J For 80: 212–214.
Lorio PL Jr, Sommers RA (1985) Potential use of soil maps to estimate southern pine beetle risk. In: Branham SJ, Thatcher RC (eds) *Proceedings of the Integrated Pest Management Research Symposium*, April 15–18, Asheville, NC. Gen Tech Rep SO-56. USDA Forest Service, Southern Forest Experiment Station, New Orleans, LA, pp 239–245.
Lovelady CN, Pulley PE, Coulson RN, Flamm RO (1991) Relation of lightning to herbivory by the southern pine bark beetle guild (Coleoptera: Scolytidae). Environ. Entomol. 20:1279–1284.
Martinat PJ (1987) The role of climatic variation and weather in forest insect outbreaks. In: Barbosa P, Schultz JC (eds) *Insect Outbreaks*. Academic Press, San Diego, CA, pp 241–268.
Mason, GN, Lorio PL Jr, Belanger RP, Nettleton WA (1985) *Rating the Susceptibility of Stands to Southern Pine Beetle Attack*. Agric Handb No 645. USDA Forest Service, Washington, DC.
Mason RR (1971) Soil moisture and stand density affect oleoresin exudation flow in a loblolly pine plantation. For Sci 17:170–177.
Mattson WJ, Haack RA (1987) The role of drought in outbreaks of plant-eating insects. BioScience 37:110–118.
Mawby WD, Hain FP, Doggett CA (1989) Endemic and epidemic populations of southern pine beetle: implications of the two-phase model for forest managers. For Sci 35:1075–1087.
May JT, Rahman S, Worst RH (1973) Effects of site preparation and spacing on planted slash pine. J For 71:333–335.
McGraw GW, Hemingway RW (1977) 6,8-Dihydroxy-3-hydroxymethyl isocoumarin, and other phenolic metabolites of *Ceratocystis minus*. Phytochemistry (Oxford) 16:1315–1316.
Michaels PJ (1984) Climate and the southern pine beetle in Atlantic Coastal and Piedmont regions. For Sci 30:143–156.
Michaels PJ, Sappington DE, Stenger PJ (1985) Objective prediction of climate-related changes in the distribution of southern pine beetle. In: Branham SJ, Thatcher RC (eds) *Proceedings of the Integrated Pest Management Research Symposium*, April 15–18, Asheville, NC. Gen Tech Rep SO-56. USDA Forest Service, Southern Forest Experiment Station, New Orleans, LA, pp 41–52.
Michaels PJ, Sappington DE, Stenger PJ (1986) SPBCMP—a program to assess the likelihood of major changes in the distribution of the southern pine beetle. South J Appl For 10:158–161.
Miller T (1972) Fusiform rust in planted slash pines: influence of site preparation and spacing. For Sci 18:70–75.
Miller T (1977) Fusiform rust management strategies in concept: site preparation. In: Dinus RJ, Schmidt RA (eds) *Proceedings of the Symposium on Management of Fusiform Rust in Southern Pines*. University of Florida, Gainesville, pp 110–115.
Morris CL (1970) Volume losses from *Fomes annosus* in loblolly pine in Virginia. J For 68:283–284.

Morris CL, Frazier DH (1966) Development of a hazard rating for *Fomes annosus* in Virginia. Plant Dis Rep 50:510–512.

Nebeker TE, Hodges JD, Blanche CA (1993) Host response to bark beetle and pathogen colonization. In: Schowalter RD, Filip GM (eds) *Beetle–Pathogen Interactions in Conifer Forests*. Academic Press, London, pp 157–173.

Nebeker TE, Hodges JD, Blanche CA, Honea CR, Tisdale RA (1991) Exploring variation in the constitutive defensive system of woods run and full-sib families of loblolly pine in relation to bark beetle attack. In: Coleman SS, Neary DG (eds) *Proceedings of the Sixth Biennial Southern Silvicultural Research Conference*, October 30–November 1, 1990, Memphis, TN. Gen Tech Rep SE-70. USDA Forest Service, Southeastern Forest Experiment Station, Asheville, NC, pp 307–313.

Nebeker TE, Hodges JD, Blanche CA, Honea CR, Tisdale RA (1992) Variation in the constitutive defensive system of loblolly pine in relation to bark beetle attack. For Sci 38:457–466.

Nebeker TE, Hodges JD, Karr BK, Moehring DM (1985) Thinning practices in southern pines—with pest management recommendations. Tech Bull 1703. USDA Forest Service, Washington, DC.

Oak SW (1985) Adaptation of littleleaf disease hazard rating for use in forest management in South Carolina National Forests. In: Branham SJ, Thatcher RC (eds) *Proceedings of the Integrated Pest Management Research Symposium*, April 15–18, Asheville NC. Gen Tech Rep SO-56. USDA Forest Service, Southern Forest Experiment Station, New Orleans, LA, pp 246–251.

Oak SW, Tainter FH (1988) Risk prediction of loblolly pine decline on littleleaf disease sites in South Carolina. Plant Dis 72:289–293.

Paine TD, Baker FA (1993) Abiotic and biotic predisposition. In: Schowalter RD, Filip GM (eds) *Beetle–Pathogen Interactions in Conifer Forests*. Academic Press, London, pp 61–79.

Paine TD, Blanche CA, Nebeker TE, Stephen FM (1987) Composition of loblolly pine resin defenses: comparison of monoterpenes from induced lesion and sapwood resin. Can J For Res 17:1202–1206.

Paine TD, Stephen FM (1987a) Influence of tree stress and site quality on the induced defense system of loblolly pine. Can J For Res 17:569–571.

Paine TD, Stephen FM (1987b) The relationship of tree height and crown class to the induced plant defenses of loblolly pine. Can J Bot 65:2090–2092.

Paine TD, Stephen FM (1988) Induced defenses of loblolly pine, *Pinus taeda*: potential impact on *Dendroctonus frontalis* within-tree mortality. Entomol Exp Appl 46:39–46.

Porterfield RL, Rowell CE (1981) Characteristics of southern pine beetle infestations southwide. In: Coster JE, Searcy JL (eds) *Site, Stand, and Host Characteristics of Southern Pine Beetle Infestations*. Tech Bull No 1612. US Department of Agriculture, Washington, DC, pp 87–113.

Powers HR Jr, Kraus JF (1983) Developing fusiform rust-resistant loblolly and slash pines. Plant Dis 67:187–189.

Powers HR Jr, Kraus JF, Duncan HJ (1976) *Development of Rust Resistant Slash and Loblolly Pines in Georgia*. Res Pap 87. Georgia Forest Research Council, Macon.

Powers HR Jr, McClure JP, Knight HA, Dutrow GF (1975) *Fusiform Rust: Forest Survey Incidence Data and Financial Impact in the South*. Res Pap SE-127. USDA Forest Service, Southeastern Forest Experiment Station, Asheville, NC.

Powers HR, Miller T, Belanger RP (1993) Management strategies to reduce losses from fusiform rust. South J Appl For 17:146–149.

Powers HR Jr, Schmidt RA, Snow GA (1981) Current status and management of fusiform rust on southern pines. Annu Rev Phytopathol 19:353–371.

Powers HR Jr, Stone DM (1988) *Control of Tip Moth Reduces Fusiform Rust Infection on Loblolly Pine*. Res Pap SE-270. USDA Forest Service, Southeastern Forest Experiment Station, Asheville, NC.

Raffa KF, Berryman AA (1986) A mechanistic computer model of mountain pine beetle populations interacting with lodgepole pine stands and its implications for forest managers. For Sci 32:789–805.

Reed DD, Burkhart HE, Leuschner WA, Hedden RL (1981) A severity model for southern pine beetle infestations. For Sci 27:290–296.

Reed DD, Hedden RL, Daniels RF (1982) Estimating the annual probability of southern pine beetle outbreak. For Sci 28:202–206.

Reeve JD, Turchin P (1993) A mechanistic approach to understanding and predicting southern pine beetle dynamics. In: Liebhold AM, Barrett HR (eds) *Proceedings of the Symposium on Spatial Analysis and Forest Pest Management*. Gen Tech Rep NE-175. USDA Forest Service, Northeastern Forest Experiment Station, Radnor, PA, pp 100–110.

Roth ER (1954) Spread and intensification of the littleleaf disease of pine. J For 52:592–596.

Rowell CE (1978) *Describing and Predicting the Susceptibility of Gulf Coastal Plain Stands Associated with Southern Pine Beetle*. Unpubl. M.S. thesis, Mississippi State University.

Schmidt RA, Foxe MJ, Hollis CA, Smith WH (1972) Effect of N, P, and K on the incidence of fusiform rust galls of greenhouse-grown seedlings of slash pine. (Abstract) Phytopathology 62:788.

Schmidt RA, Goddard RE, Hollis CA (1974) *Incidence and Distribution of Fusiform Rust in Slash Pine Plantations in Florida and Georgia*. Tech Bull 763. Florida Agricultural Experiment Station, Gainesville, FL.

Schmidt RA, Jokela EJ, Allen JE, Belanger RP, Miller T (1990) Association between fusiform rust incidence and CRIFF soil classification for slash pine plantations in the Coastal Plain of Florida and Georgia. South J Appl For 14:39–43.

Schmidt RA, Miller T, Holley RC, Belanger RP, Allen JE (1988) Relation of site factors to fusiform rust incidence in young slash and loblolly pine plantations in the Coastal Plain of Florida and Georgia. Plant Dis 72:710–714.

Schowalter TD, Filip GM (eds) (1993) *Beetle–Pathogen Interactions in Conifer Forests*. Academic Press, London.

Schowalter TD, Turchin P (1993) Southern pine beetle infestation development: interaction between pine and hardwood basal areas. For Sci 39:201–210.

Sharpe PJH, Wu H (1985) A preliminary model of host susceptibility to bark beetle attack. In: *Proceedings of the IUFRO Host Insect Work Group*. Canadian Forestry Service, Victoria, British Columbia, pp 108–127.

Shoulders E, Scarborough JH Jr, Arnold RA (1991) Fusiform rust impact on slash pine under different cultural regimes. In: Coleman SS, Neary DG (eds) *Proceedings of the Sixth Biennial Southern Silvicultural Research Conference*; October 30–November 1, 1990, Memphis, TN. Gen Tech Rep SE-70. USDA Forest Service, Southeastern Forest Experiment Station, Asheville, NC, pp 282–288.

Sluder ER (1989) Fusiform rust in crosses among resistant and susceptible loblolly and slash pines. South J Appl For 13:174–177.

Squillace AE (1976) *Geographic Patterns of Fusiform Rust Infection in Loblolly and Slash Pine Plantations*. Res Note SE-232. USDA Forest Service, Southeastern Forest Experiment Station, Asheville, NC.

Squillace AE, Dinus RJ, Hollis CA, Schmidt RA (1978) *Relation of Oak Abundance, Seed Source, and Temperature to Geographic Patterns of Fusiform Rust Incidence*. Res Pap SE-186. USDA Forest Service, Southeastern Forest Experiment Station, Asheville, NC.

Squillace AE, Wilhite LP (1977) Influence of oak abundance and distribution on fusiform rust. In: Dinus RJ, Schmidt RS (eds) *Proceedings of the Symposium on Management of Fusiform Rust in Southern Pines*. University of Florida, Gainesville, pp 59–70.

Stone DM, Powers HR (1989) Sewage sludge increases early growth and decreases fusiform rust infection of nursery-run and rust-resistant loblolly pine. South J Appl For 13:68–71.

Thatcher RC, Connor MD (1985) *Identification and Biology of Southern Pine Bark Beetles*. Agric Handb No 634. USDA Forest Service, Washington, DC.

Thatcher RC, Pickard LS (1964) Seasonal variations in activity of the southern pine beetle in east Texas. J Econ Entomol 57:840–842.

Thatcher RC, Searcy JL, Coster JE, Hertel GD (eds) (1980) *The Southern Pine Beetle*. Tech Bull 1631. USDA Forest Service, Washington, DC.

Thoeny WT, Tiarks AE, Hayes JL, Bridges JR (1992) Marking the southern pine beetle (Coleoptera: Scolytidae) with rubidium within loblolly pine for dispersal studies. Environ Entomol 21:1377–1385.

Turchin P, Lorio PL Jr, Taylor AD, Billings RF (1991) Why do populations of southern pine beetles (Coleoptera: Scolytidae) fluctuate? Environ Entomol 20:401–409.

USDA Forest Service (1985) *Insects and Diseases of Trees in the South*. Gen Rep R8-GR-5. USDA Forest Service, Southern Region, Atlanta, GA.

USDA Forest Service (1987) *Piedmont Risk—a System to Calculate the Risks and Hazards of Pine Stand to Possible Attack by the Southern Pine Beetle*. Protection Rep R8-PR-2. Southern Pine Beetle Fact Sheet No 39. USDA Forest Service, Southern Region, Atlanta, GA.

Verrall AF (1982) *A History of Forest Pathology Research in the South and Southeast*. Gen Tech Rep SO-36. USDA Forest Service, Southern Forest Experiment Station, New Orleans, LA.

Vité JP (1961) The influence of water supply on oleoresin exudation pressure and resistance to bark beetle attack in *Pinus ponderosa*. Contrib Boyce Thompson Inst 21:37–66.

Walkinshaw CH, Roland TA (1990) Incidence and histology of stem-girdling galls caused by fusiform rust. Phytopathology 80:251–255.

Wells OO, Wakeley PC (1966) Geographic variation in survival, growth, and fusiform-rust infection of planted loblolly pine. For Sci Monogr 11:1–40.

Williams RA, Tainter FH (1971) Pine decline in southern Arkansas. Plant Dis Rep 55:1036.

Zak B (1961) *Aeration and Other Soil Factors Affecting Southern Pines as Related to Littleleaf Disease*. Tech Bull 1248. USDA Forest Service, Washington, DC.

Zutter BR, Gjerstad DH, Glover GR (1987) Fusiform rust incidence and severity in loblolly pine plantations following herbaceous weed control. For Sci 33: 790–800.

7. Interactive Effects of Air Pollutants with Abiotic and Biotic Factors on Southern Pine Forests

Jon D. Johnson, Arthur H. Chappelka, Fred P. Hain, and Allen S. Heagle

The potential interactions among abiotic and biotic factors in southern pine forests are extremely complex, particularly when one considers these interactions over a tree's lifetime. We know from experimental data that individual factors tend to impact trees either negatively or positively: Only factors that show a response tend to be published. Interactions among factors and their impact on trees can be antagonistic, neutral, additive, or synergistic, resulting in a continuum of tree response. When considering these complex relationships, four critical features must be considered (Table 7.1). The first consists of the abiotic aspects associated with the forest environment, whereas the other three encompass various levels of organization within the biotic features from the individual organism through the ecosystem level.

The objective of this chapter is to discuss air pollution and its interactions with insects, diseases, and abiotic factors and how they impact the processes of tree growth. It is hoped that this review will provide an understanding of how trees acclimate, adapt, and compensate in response to the interactions of abiotic and biotic stresses. Such an understanding could be used in developing amelioration strategies for southern pine forests.

Table 7.1. Four Classes of Critical Features to Be Considered in Air Pollution-Plant-Pest Interactions[a]

I. Physical and chemical characteristics of the environment
Climate (temperature, humidity, wind, insolation)
Site (soil, exposure)
Geomorphology (landforms, parent material)
Geochemistry (buffering capacity, hydrology, watershed)
Light penetration (in gaps and layers)
Pollutants (distribution in time and space)
II. Biological and chemical features of the individual organisms
Quality of host plant
Surface characteristics (especially of different organs of plants)
Structural characteristics (toughness, architecture, way in which different parts of plant are assembled)
Biochemical status (C, N, defense compounds)
Plant history (stress, age, size)
Vulnerability of host plant
Physiological status (carbon, nitrogen)
Resource availability (water, nutrients, light)
Phenology (age, juvenile, mature)
Vigor (crown rates, sapwood area, root starch, storage reserves)
Defense compounds (toxins, lignin, digestibilities)
Insects and pathogens
Dispersal and host-finding mechanisms
Initial establishment processes
Growth/utilization/biomass conversion
Reproductive output and mechanisms (sexual, asexual)
Detoxification/toxin production capacity (part of overall competitive abilities)
Parasite/disease/hyperparasite load
III. Population level features
Host
Age and size distribution
Spatial distribution and geographic extent
Genetic frequency of sensitive individuals, structure of sensitivity, diversity of gene pools
Infection status (history)
Infestation
Insect/pathogen
Sex ratio, i.e., number of females
Age and size distribution
Spatial distribution
Density (inoculum and survival characteristics)
Proportion virulent
Fecundity and survivorship
IV. Community and ecosystem level features
Interspecific interactions
Symbionts (e.g., mycorrhizae, N-fixers)
Vectors
Natural enemies of pest
Diffuse interspecific interactions
Species composition and distribution and interacting species
Community succession status and age of community
Microbial community dynamics (e.g., free-living N-fixers, decomposers)
Competitive interactions (e.g, pest–pest)

[a] After Bedford (1987), Hughes (1988).

Abiotic Factors

Trees growing in a natural environment are subjected to numerous stresses throughout their lifetime that are derived from the physical and chemical characteristics that constitute a tree's environment (Table 7.1). Abiotic stresses vary in duration and intensity. Depending on the phenological and ontological state of the tree, these stresses have effects ranging from positive to the death of the tree. Abiotic factors can elicit one to many physiological responses in the tree depending on the aforementioned conditions. With respect to southern pine's response to air pollutants, the factors of drought, flooding, nutrition, and low and high temperature have been reported in the literature that will be discussed in this section.

Due to the complexity of the interactions discussed here, there does not appear to be a single underlying physiological response, though changes in stomatal conductance may be considered a primary response (particularly changes in water and perhaps nutrition). Taking all of the information on abiotic stresses along with what is known about physiological responses of southern pines to air pollutants, claims of physiological acclimation, adaptation, and compensation can be better understood and perhaps explained, thus removing some of the speculation and conjecture associated with these claims.

Water

The importance of water to the growth and development of trees is well recognized (Kozlowski et al., 1991; Kramer, 1983). The water relations of southern pines have been studied extensively, so there is a large body of knowledge from which interactions with air pollutants can be discussed. There are two water conditions that can stress southern pines: drought and flood. Typically, drought resulting from sporadic and highly variable rainfall appears to be more widespread in the southeastern United States than flooding, except in limited areas along rivers and streams, where periodic flooding can occur.

Both of these water stresses cause stomatal closure. In response to drought, the stomata close as a result of loss of turgor in the guard cells; whereas with flooding, stomatal closure appears to be due to increased and/or decreased chemical (hormonal) signals coming from the flooded roots (Johnson, 1987). In either case, stomatal closure can have a positive effect on uptake because the primary path of pollutant uptake by the tree is through the stomata (Fig. 7.1). Hence, stomatal closure due to either type of water stress will necessarily diminish pollutant uptake and dose, reducing pollutant effects. Conversely, there is evidence that air pollutants can reduce root growth, and thus affect a tree's ability to withstand prolonged drought.

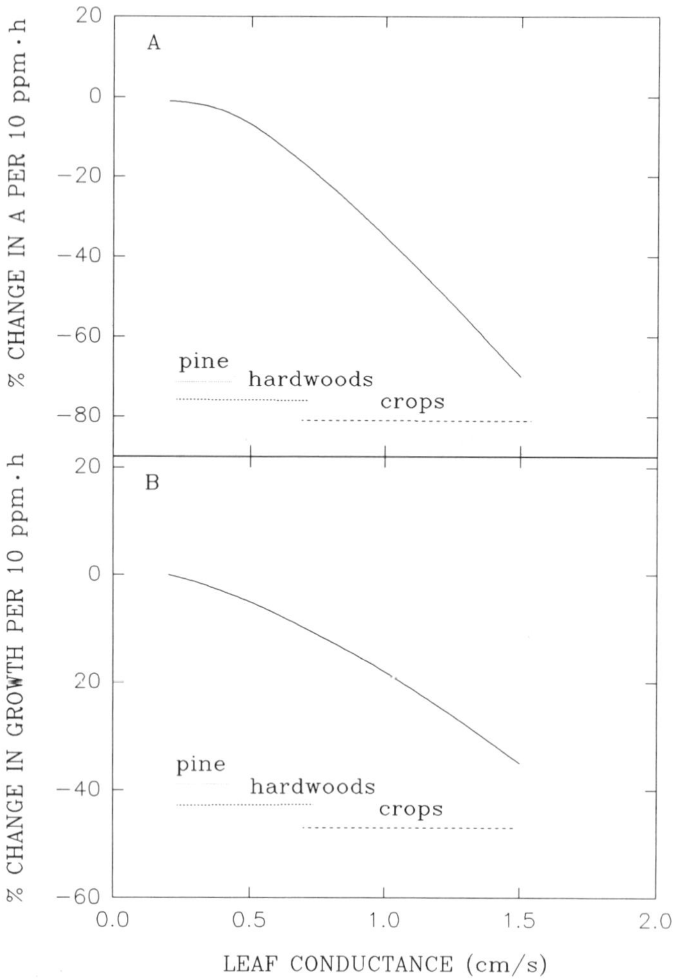

Figure 7.1. Percent reduction in net photosynthesis (A) and dry matter production due to equal ozone dose of pines, hardwoods, and agricultural crops in relation to leaf diffusive conductance of each species. The horizontal lines indicate the range of leaf conductance for each plant type (from Reich, 1987).

Drought

The interaction of drought stress with air pollutants can be complex. Air pollutants can affect trees in several ways, which may predispose them to drought by altering stomatal function and decreasing root growth. In contrast, air pollutants in some situations may improve drought avoidance by reducing canopy leaf area. Drought, on the other hand, can reduce gaseous air pollution effects on trees by decreasing stomatal aperture and thereby reduce pollutant uptake into the foliage.

The most commonly reported effect of air pollutants on stomata is closure as a direct result of pollutants either (1) decreasing photosynthesis, which, in turn, increases intercellular carbon dioxide concentration (C_i), which causes stomatal closure (Reich, 1987); or (2) directly affecting the guard cells (Maier–Maercker and Koch, 1992; Rozenkranz et al., 1989), leading to closure. Olszyk and Tibbitts (1981) suggested that stomatal closure was due to either ozone-induced solute efflux from the guard cells or ozone-stimulated abscisic acid accumulation in the foliage. Stomatal closure in response to ozone has been reported in loblolly pine (*Pinus taeda* L.) (Adams et al., 1990; Lee et al., 1990; Sasek and Richardson, 1989; Taylor et al., 1991), slash pine (*P. elliottii* Engelm.) (Byres et al., 1992; Johnson et al., 1995), and shortleaf pine (*P. echinata* Mill.) (Elsik et al., 1992).

More subtle and less obvious effects of ozone and other pollutants on stomatal functioning have been reported. Reich and Lassoie (1984) observed lower leaf conductance in poplar (*Populus* spp.) under high light and ozone exposure. Under low or no light, however, they found leaf conductance in ozone-treated plants to be higher. Barnes et al. (1990) and Keller and Hasler (1984, 1987) also found increased stomatal conductance in Norway spruce (*Picea abies* L.) exposed to ozone and observed that excised current-year needles dried more rapidly following either ozone or acid rain treatment (Barnes et al., 1990). Diurnal sampling of needle conductance of slash pine on three days in the spring of 1990 provided data suggesting that trees exposed to above ambient ozone (2× and 3× an ambient concentration of 38 ppb, based on a 12-hour seasonal mean) had more than double needle conductances at dusk than trees exposed to ambient and subambient ozone levels (Johnson et al., 1995). Additional circumstantial evidence for increased water loss during the night in this study included significantly lower soil moisture in the 3× ozone treatment even though these trees had significantly lower leaf area.

Increased water loss from ozone-treated foliage has been attributed to increased cuticular transpiration resulting from the degradation of the cuticle by pollutants and to impaired stomatal function leading to incomplete closure at night. Direct observations of leaf and needle cuticles, however, have provided little evidence for the former explanation (Keller, 1991; Maier–Maercker and Koch, 1992). Incomplete stomatal closure at night would have a subtle and long-term effect on a tree's water relations because water loss would be only slightly increased due to the low evaporative demand that exists at night. Additional water loss by trees exposed to pollutants may come from compensatory responses, particularly in those species that retain foliage for more than one year. In one slash pine family exposed to 3× ambient ozone, the newest needle age cohort was observed to have higher needle conductance, particularly in the late afternoon, than the same needle cohort in trees exposed to subambient ozone levels (Johnson et al., 1995). Such increased needle conductance could be due to enhanced photosynthesis reducing C_i. The additional water transpired by the needles in this cohort coupled with increased production of needles, as has been observed in ponderosa pine (*P. ponderosa* Laws) (Temple et al., 1993), could lead to increased water use by these trees, leading to depletion of soil water and the development of drought conditions.

Another process by which air pollutants may affect the ability of trees to respond to drought is through altered carbon allocation. McLaughlin (1985) reported reduced carbon transport to the roots of several pine species when exposed to ozone. The reduced transport to the roots appears to be a result of retention of carbon in the shoot associated with repair mechanisms in the foliage, including increased respiration (Skarby et al., 1987). Edwards (1991) observed root respiration in loblolly pine to decrease by 12% when exposed to ozone. It was found that both growth and maintenance respiration rates of roots were reduced, which was attributed to the lack of sucrose translocated from the foliage. This was supported by the work of Paynter et al. (1991, 1992), who found that ozone exposure of shortleaf pine (*Pinus echinata* mill.) lowered needle concentrations of sucrose, the primary transport sugar, while increasing total reducing sugars (including glucose). Meier et al. (1990) found a similar carbohydrate response in loblolly pine seedlings that resulted in a reduction in the number of mycorrhizal roots with increasing ozone exposure. Specific root length of beech (*Fagus sylvatica* L.) (root length per unit dry weight) was found to increase with ozone exposure (Taylor and Davies, 1990). Proliferation of long, thin roots occurred in the lower half of the top 1 m of soil during July, but this had no effect on drought susceptibility when water was subsequently withheld from these trees. Slash pine exposed to 3× ambient ozone levels exhibited significantly greater water loss from 75 to 100 cm soil depth in north Florida (Johnson et al., 1995), but there were no observable differences in root length or density at these depths as measured by a root periscope (Johnson, unpublished data), nor were there any differences in xylem water potential among ozone treatments (Johnson et al., 1995).

Reduced carbon allocation to roots as a result of air pollution exposure apparently can have variable effects depending on species and soil conditions. If a species has a tap root and its growth at lower soil depths is stimulated by pollution exposure, it may actually be less susceptible to drought if the tap root is able to extract water from those lower soil depths (Johnson et al., 1995). However, the growth of the tap root would probably be at the expense of fine root growth (if the overall carbon to the roots is decreased), which would be unimportant during a drought but could severely impair nutrient uptake (see the subsequent section on nutrition) upon relief from the drought. Further study on this topic is required to understand ecological implications.

The most widely reported interaction between drought and air pollution effects on trees has been the lessening of pollutant injury due to water stress. This response is attributed to the water stress–induced reduction of stomatal aperture decreasing the uptake of pollutant into the foliage. Drought-stressed ponderosa pine exposed to ozone exhibited less foliar injury than their watered counterparts (Temple et al., 1992), and after the second year the stressed trees showed no response to ozone. At the end of the three-year study, water-stressed trees receiving 1.5× ambient ozone had lost half the needle biomass that the watered trees did in the same ozone treatment (Temple et al., 1993). Similar reports have been published for loblolly pine. Meier et al. (1990) reported that droughted seedlings exhibited less ozone symptoms, which was attributed to stomatal closure. Adams et al. (1990) concluded that a drought during the second and third year of a three-year study made both ozone and acid rain treatments nonsignificant with respect to growth parameters. Based on carbon isotope discrimination in tissues from these same trees, Taylor et al. (1991) inferred that during the wet year (first of the study), the significant ozone × genotype interaction was a result of genetic differences in ozone-induced stomatal closure and associated increases in water use efficiency in some of the families. During the subsequent two-year drought, water stress reduced stomatal aperture in all genotypes, lessening ozone uptake and masking any genetic differential ozone effect.

Flooding

Little has been published on this topic with respect to interaction with air pollutants. Norby and Kozlowski (1983) and Shanklin and Kozlowski (1985) reported the interaction of flooding and SO_2 fumigation in two *Betula* species and *Taxodium distichum* (L.) Rich., respectively. As with drought, flooding caused stomatal closure, lessening the impact of pollution. The only study with southern pines was reported by Elsik et al. (1992), and their flooded treatment was actually their watered control. Flooded (control) loblolly and shortleaf pine showed decreased photosynthesis and chlorophyll concentration compared to the treatment receiving less water. There were no significant ozone × water interactions in the gas exchange parameters, but biomass was consistently lower in the flooded treatment across ozone treatments for both species. As ozone level increased, biomass decreased in both water treatments. Several interactions were present in the biomass data. In the 1.7× ambient ozone treatment (78 ppb 12-hour seasonal mean), foliage (and total) biomass increased in the less watered treatment compared to the flooded seedlings for both species. Leaf area ratio (leaf area to total seedling biomass) and specific leaf area (leaf area to leaf biomass) in loblolly pine increased

150% and 140%, respectively, with increasing ozone in the flooded treatment, whereas it remained virtually unchanged in the less watered treatment. Shortleaf pine leaf area ratio and specific leaf area were unaffected by ozone and water treatment. It is apparent that there is much more information required before we understand the response of southern pine species to the interaction between air pollutants and flooding.

Humidity

Another environmental factor that interacts with pollutants and is associated with a tree's water balance is humidity. McLaughlin and Taylor (1981) first reported the importance of humidity on ozone and sulfur dioxide uptake, which was three to four times and two to three times greater, respectively, at high humidity. They attributed this increased uptake to reduced internal leaf resistance and not to stomatal regulation and suggested that vegetation growing in humid areas like the southeastern United States may experience a greater uptake of pollutants. Jensen and Roberts (1986) found that ozone had no effect on yellow-poplar (*Liriodendron tulipifera* L.) stomata at 40% relative humidity, but at high humidity the ozone treatment resulted in opening of the stomata. In the review by Darrall (1989), several studies were cited that indicated that stomatal closure occurs faster in the presence of ozone at lower humidities. The response of slash pine to vapor pressure deficit (VPD) showed greater closure at high VPD in current-year needles exposed to 3× ambient ozone (114 ppb 12-hour seasonal mean) (Johnson et al., 1995). If humidity tends to decrease during drought and increase during flooding, then there is the potential for a pollutant × water × humidity interaction affecting the response of southern pines.

Nutrition

The general low fertility of southeastern U.S. soils impacts not only the growth of the pines in this region but also tree response to pollutants (Binkley et al., 1989). The often reported fertilizer effect of acid rain on southern pines growing in native soils has been attributed to the deposition of nitrogen (see Chapter 6; also see Johnson et al., 1992b; Wright et al., 1991). This, however, can be considered a direct pollutant effect and is outside the purview of this chapter.

The interaction of tree nutrition and pollutants, while not as extensively studied as water pollutant interactions, has been documented in several studies, including some on southern pines. One of the earlier hypothesized interactions between nutrients and pollutants was that deterioration of foliage cuticle by either ozone or acid corrosion would increase leaching of nutrients, resulting in either nutrient deficiencies and/or nutrient imbalances, which would lead to growth reduction. There is little evidence, however, to support this hypothesis, at least in pine species (Edwards et al., 1991; Skeffington and Roberts, 1985).

A commonly reported observation of ozone-treated southern pines has been an increase in foliage nutrient concentrations, particularly nitrogen (Fig. 7.2) (Baker et al., 1994; Johnson et al., 1992b; Skeffington and Roberts, 1985; Temple and Reichers, 1992; Wright et al., 1991). Tjoelker and Luxmoore (1991) reported increased needle concentrations of nitrogen, sulfur, and phosphorus in loblolly pine exposed to ozone, whereas Baker et al. (1994) also found increases in nitrogen, phosphorus, potassium, magnesium, and calcium in all components of the shoot: stem, branches, and foliage. Associated with these foliar nitrogen increases were observations of increased photosynthesis in current-year needles (Manderscheid et al., 1992; Temple and Reichers, 1992; Tjoelker and Luxmoore, 1991), which may explain some of the observed physiological compensation in pines exposed to ozone. Edwards et al. (1990b) reported 23 and 30% increases in chlorophyll a and b, respectively, and 21% more carotenes in needles of loblolly pine trees treated with 2× ambient ozone (7-hour annual mean = 90 ppb). The increased nitrogen in needles of ozone-

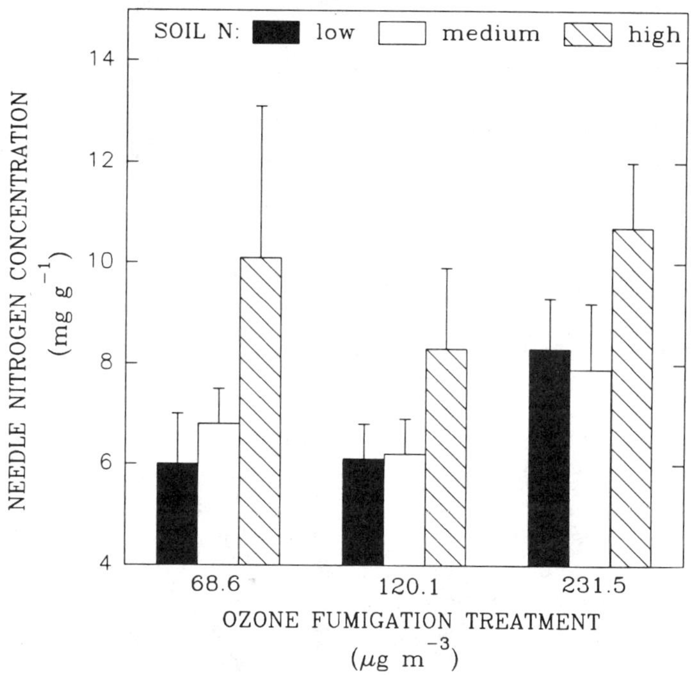

Figure 7.2. Current-year loblolly pine needle nitrogen concentration in response to ozone exposure and three soil nitrogen treatments (redrawn from Tjoelker and Luxmore, 1991).

treated trees apparently originates in older needles that prematurely senesce and drop from the trees in response to ozone. Temple and Reichers (1992) found 33% higher nitrogen in current-year needles and 14% higher in the previous-year needles of ponderosa pine exposed to elevated ozone and that the ozone treated trees reabsorbed 57% of the nitrogen from the senescing foliage while the control trees only reabsorbed 52%.

In a study of loblolly pine, Manderscheid et al. (1992) suggest that the aforementioned observations are due to a strategy for reallocating, and thus conserving, essential nutrients, particularly nitrogen, from senescing foliage damaged by ozone. The nitrogen is imported to remaining foliage, typically the most recent needle cohort, where it is metabolized into proteins and chlorophyll, thus increasing photosynthesis. This additional metabolism conserves nitrogen but also utilizes photosynthate, which would normally have been exported to the root system. Thus, the reported decrease in root biomass and respiration (Edwards, 1991; Edwards et al., 1991; see previous section) resulting from ozone treatment may, in part, be attributed to nitrogen conservation by the tree. The results of Baker et al. (1994) with loblolly pine support the retranslocation of nutrients from older, senescing needle cohorts into the needles of late-season flushes.

The stimulation of photosynthesis under these circumstances, however, rarely leads to increased or even maintained growth rates. In a study of ozone-treated slash pine, Johnson et al. (1992b) found that the nitrogen productivity (g shoot dry weight/g needle nitrogen/unit time) relationship to foliage increment decreased as ozone level increased to 3× ambient (114 ppb 12-hour seasonal mean) over a three-year period. In the last year of that study, all treatments had similar nitrogen productivity values of 0.07 g shoot/g nitrogen/month, but the foliage increment in the highest ozone treatment was 700 g while the other three treatments —2× ambient, ambient, and subambient—all exhibited foliage increment of around 1100 g.

Nutrient status as well as type of the soil has been shown to mitigate pollutant effects in a few studies. Reich et al. (1987) found the response of eastern white pine (*Pinus strobus* L.) to acid rain to be inversely related to available soil nitrogen. This is not surprising since this acid rain response is the previously mentioned "fertilizer response." In a study with loblolly pine, an interaction between soil nitrogen and ozone was observed (Tjoelker and Luxmoore, 1991). Under ambient ozone levels, photosynthesis increased with increasing soil nitrogen while needle conductance remained unchanged resulting in an increase in instantaneous water use efficiency (IWUE). In the elevated ozone treatment, photosynthesis increased with soil nitrogen as in the ambient treatment, but needle conductance was about a third lower, leading to further increases in IWUE. In the latter treatment, reduced needle conductance drove the increase in IWUE, which would also reduce ozone uptake into the foliage.

Needle nitrogen concentration in this study increased as ozone treatment increased except at the highest soil nitrogen level, where there was little difference among ozone treatments. Loblolly pine seedlings grown in a nitrogen-poor soil (a vermiculite–peat mixture; needle nitrogen = 1.10%) showed a greater reduction in root biomass with increasing exposure to ozone than seedlings grown in a mineral soil–peat mixture (needle nitrogen = 1.93%) (Horton et al., 1990).

Nutrition has such an important, and sometimes major, effect on the growth and physiology of southern pines that manipulating soil nutrients may be one means to overcome growth reductions caused by air pollutants. Additional research is needed to understand the physiological ramifications of improved nutrition and to conduct research that will help determine the interactive effects of nutrition and air pollutants.

Temperature

Exposure to air pollutants has been demonstrated to affect tree response to low temperature, particularly freezing temperatures. Keller (1991) summarized the potential physiological effects pollutants may have on frost hardiness. These include loss or delay in development of the cuticle, cell membrane injury, prolonged fall growth due to increased nitrogen availability, and low potassium contents leading to the delay of carbohydrate conversion to osmotica. Although it is not usually considered a problem in the southeastern United States, there are instances of early fall and late spring freezes associated with rapidly moving cold fronts that have resulted in foliar damage. The processes involved in late fall versus early spring frost damage are different, however. In the former instance, pollutants apparently postpone the hardening process, with damage occurring primarily to the older needle age cohorts. In the latter case, newly elongated, succulent foliage is damaged while the older needle cohorts show little damage.

The older needles of loblolly pine had reduced frost hardiness in fall (1°C reduction in [lethal temperature] LT_{50}) and spring (3°C reduction) when exposed to 2× ambient ozone levels (Edwards et al., 1990a). These changes were attributed to changes in solute transport and osmoregulation resulting from ozone treatment and suggested that latent damage to the older needles could result in early senescence of these needles during the subsequent growing season. This idea is consistent with the observations of accelerated loss of the older needle cohorts during their second year (see Chapter 6).

The only report of ozone effects on spring hardiness of newly elongating needles was that of Chappelka et al. (1990) as a result of an unusually late April frost (Fig. 7.3). In this case, no visible damage to the older needles was observed, but the newly emerged needles showed increasing tip photo bleaching as ozone level increased. These damaged needles became

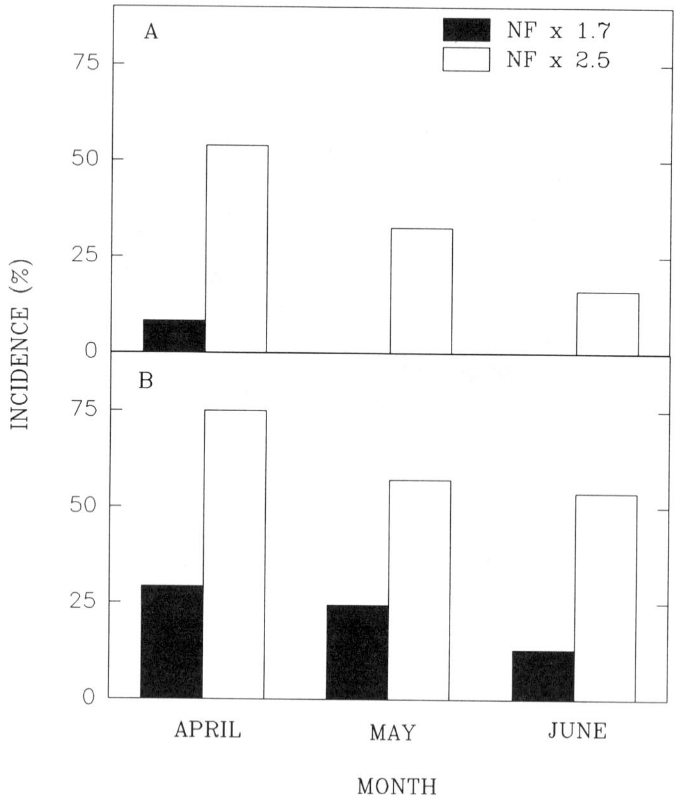

Figure 7.3. Incidence of frost-induced tip die-back in two families of loblolly pine: (A) family GAKR 15-23, and (B) family GAKR 15-91. The seedlings were exposed to ozone beginning in April, when the frost occurred. No damage was observed in the subambient and ambient (NF: nonfiltered) treatments (from Chappelka et al., 1990).

necrotic within a week of the frost. The ozone-sensitive family showed greater damage than the ozone-resistant one, and this was attributed to a genetic effect. This effect, however, could result from phenological differences in bud break and elongation and may not necessarily be related to ozone sensitivity.

Biotic Factors

Biotic factors of insects and disease can influence tree responses to air pollutants such as ozone and acid rain. Southern pines are exposed to numerous disease and insect pests throughout their lifetime (see Table 1.6 in Chapter 1). A major disease is fusiform rust (*Cronartium quercuum* f.sp. *fusiforme*) and a major insect is the southern pine beetle (*Dendroctonus frontalis*), both of which cause annual monetary losses in the millions of dollars (Hoffard, 1985). Other serious insect and disease problems include Ips beetles (*Ips* spp.), littleleaf disease (*Phytophthora cinnamomi*), annosus root rot (*Heterobasidium annosum*), pitch canker (*Fusarium subglutinans*), and pine sawflies (*Neodiprion, Diprion* spp.). Occurrence of these pests is often more severe on weakened or stressed trees, which may be due to air pollution.

Pollutants can reduce host tree defense mechanisms, predisposing them to insect attacks and disease infection. Ozone and acid rain have been demonstrated to alter plant response to both insect herbivores and plant pathogens (Hughes and Laurence, 1984). To understand these interactions, causal relationships need to be established, and the biochemical and physiological mechanisms controlling these host–pest interactions must be identified. Some of the critical features of air pollution–host–pest interactions are shown in Table 7.1.

Insects

A number of authors have reviewed the correlative evidence that air pollution influences outbreak patterns of forest insect species (Alstad et al., 1982; Baltensweiler, 1985; Führer, 1985; Hain, 1987; Hay, 1977; Heagle, 1973; Hughes, 1988; Morrison, 1984; Sierpiński, 1984, 1985; Skrzypcaynska 1984; Smith, 1990, 1984a, 1984b). Prior to 1980, there were few reliable studies on the causality or mode of action of pollutants on insects either directly or indirectly. Since then, more emphasis has been placed on developing causal relationships and on understanding basic mechanisms. However, most of this work has been in agricultural and other nonforest systems; very few studies have been in coniferous systems, including the southern pine system. This short review will concentrate on studies of pollutants in forest ecosystems since 1987. For a more comprehensive review of insect populations on host plants subjected to air pollution, see Hughes (1988) and Heliövaara and Väisänen (1993).

Pollutants may have a negative or positive impact on insect populations (Führer, 1985). Negative correlations between pollutant and insect populations have generally occurred in areas of very high pollutant concentrations. The insect species are either sensitive to high levels of pollution, or the host substrate has been altered to render it unsuitable for insect development. In the latter case, the pollution injury to the host may result in a more susceptible but less suitable host. Attacks are more likely to be successful, but development and survival are poor (Cobb et al., 1968a, 1968b; Miller et al., 1968; Stark et al., 1968). Positive correlations between pollutant and insect populations have generally occurred in areas of moderate or low pollutant concentrations. The population may increase due to a reduction in competitors and/or natural enemies by the pollutant, or due to increased susceptibility and suitability of the host as a result of the pollution, or both.

Our current knoweldge of the impact of air pollutants on insects remains sketchy. For example, a number of studies have correlated high aphid populations with exposure to elevated levels of pollutants. Exposure of host plants to different pollutants has resulted in high levels of aphid attack (Heliövaara and Väisänen, 1993), suggesting increased susceptibility and suitability of the host plant. However, a recent study in Finland (Holopainen et al., 1993) following two species of aphids on pine and spruce seedlings along a pollution gradient of mainly sulphur dioxide concluded "that aphid species show diverse responses to air-pollutant emissions from a paper mill, and, therefore, it is very difficult to generalize about the effect of air pollution on aphid populations on conifers" (p. 190). Earlier work (Villemant, 1981) had also reported differing responses for different aphid species.

Effects of Pollutants on Insects

The direct effects of pollutants on invertebrate herbivores has been examined in only a few cases (see Hughes, 1988, and Heliövaara and Väisänen, 1993, for a complete review). The direct effects can be toxic, metabolically stimulating, or behaviorally altering. Generally, direct toxicity occurs only at very high concentrations of either gases or particulates. At lower levels of a toxin, arthropod metabolism may be stimulated (Feir and Hale, 1983). Alteration of insect behavior by pollutants, either through repellency or attraction, has not yet been studied.

Beneficial insects may be affected negatively by pollutants, thus resulting in a higher population of detrimental insects. Parasites and predators might be reduced by pollutants (Alstad et al., 1982), especially SO_2 (Freitag and Hastings, 1973; Freitag et al., 1973; Hillmann, 1972; Hillmann and Benton, 1972; Przybylski, 1968) and inert dusts (Bartlett, 1951; Callenbach, 1940; Edmunds, 1973; Fye, 1983). Although it is generally assumed that toxic pollutants accumulate through the food chain from herbivore to predator (Buse, 1986; Dewey, 1973; Kettle et al., 1983), the biological consequences of this have not been determined (Hughes, 1988).

Pollution might increase the susceptibility of insects to natural enemies. In the case of industrial melanism (Bishop, 1972; Kettlewell, 1955, 1956, 1958; Mikkola, 1975; Owen, 1961), moths exposed to high amounts of pollution change color forms, due to selective predation, from a pale form to a darker form containing heavy melanism. The pale forms became apparent to predators when resting on tree trunks because SO_2 pollution reduced the occurrence of light-colored lichens on the tree bark. The moths had depended on cryptic coloration to avoid predation.

Effects of Pollutants on Host–Insect Interactions

Air pollution can affect insects indirectly by altering certain properties of the host plant that are important to the insect's survival, such as nutrition and sites for oviposition and shelter. The many possible mechanisms by which pollutants may alter the plant–insect interactions have been reviewed in detail (Hughes, 1988; Hughes and Laurence, 1984; and Laurence et al., 1983). To summarize, pollutants may alter plant–insect relationships in the following ways. First, the host's vulnerability to discovery may be changed by changing the host density, altering both the chemical and physical behavioral cues of the host (color, surface morphology, surface chemicals, volatile chemicals, nonvolatile primary and secondary metabolites) or by modifying the host species diversity or community structure. Second, the host's nutritional quality may be changed by altering plant nutrition, primary and secondary metabolite levels, plant water balance, metabolic activity, or plant hormones. Third, both the constitutive (surface morphology, toughness, or secondary metabolites) or induced compounds involved in host resistance or tolerance of insect herbivory may be altered.

Trees in the family *Pinaceae* are particularly susceptible to attacks by bark beetles and their vectored fungi. In North America, the two most aggressive beetles, the southern pine beetle (SPB), *Dendroctonus frontalis* Zimmermann, and the mountain pine beetle (MPB), *D. ponderosae* Hopkins, alone are responsible for considerable mortality of pine species. The destructive power of the symbiotic relationship between bark beetles (*Scolytidae*) and the pathogenic fungi they vector (Basham, 1970) is largely associated with the fungi's ability to kill the tree and improve the suitability of the tree for further beetle development and continued attack.

Beetles initiate attack by boring into phloem tissue, where they encounter the tree's preformed resin system. Oleoresin flows out of the wound and acts as a physical barrier. The total flow, flow rate, viscosity, and rate of crystallization differ between the more SPB resistant (slash, *Pinus elliottii* Engelm. and longleaf pine, *Pinus palustris* mill.) and susceptible southern pines (loblolly, *Pinus taeda* L. and shortleaf, *Pinus echinata* mill.) (Hodges et al., 1979). Oleoresin also acts as a chemical barrier since monoterpene fractions of oleoresin are toxic to both beetle and fungi (Bridges, 1987; Raffa et al., 1985). Bark beetle outbreaks are most often associated with environmental stresses such as drought, lightning, and flooding, which drastically reduce the quantity of, and the pressure with which preformed resin is exuded from wounds (Blanche et al., 1985; Hodges and Pickard, 1971; Lorio and Hodges, 1977). Diminished flow of the preformed resin eliminates both the physical and chemical barriers to successful beetle colonization. Although not specifically identified as a stressor, air pollution—particularly ozone—could have a significant effect on the ability of southern pines to resist an SPB attack. Stark and Cobb (1969) provided evidence that exposure of *Pinus ponderosa* Laws to oxidants in the San Bernadino mountains of California increased incidence of tree mortality attributed to bark beetle attack (Fig. 7.4).

The monoterpenes found within induced lesions are a combination of those from the preformed resins, which immediately inundate the wound, and the induced resins, which are synthesized shortly after wounding. All terpenes are produced by the mevalonic acid pathway (Porter and Spurgeon, 1981), and the primary substrates for this pathway are reducing sugars. As discussed earlier, retention of reducing sugars in ozone-treated needles could significantly reduce substrate availability for terpene production in the stem, contributing to lower resin exudation pressure and resistance to beetle attack.

Although considerable research has demonstrated that pollutant-induced changes in the host plant have affected particular insects, little attention has been given to the specific mechanisms involved. The effects of SO_2 on Mexican bean beetle attacking soybeans is probably the most thoroughly studied system (Hughes, 1988; Hughes et al., 1982, 1983, 1985). Mexican bean beetles prefer feeding on soybean foliage exposed for 24 hours to SO_2 and increase in growth, rate of development, and fecundity. Pupal

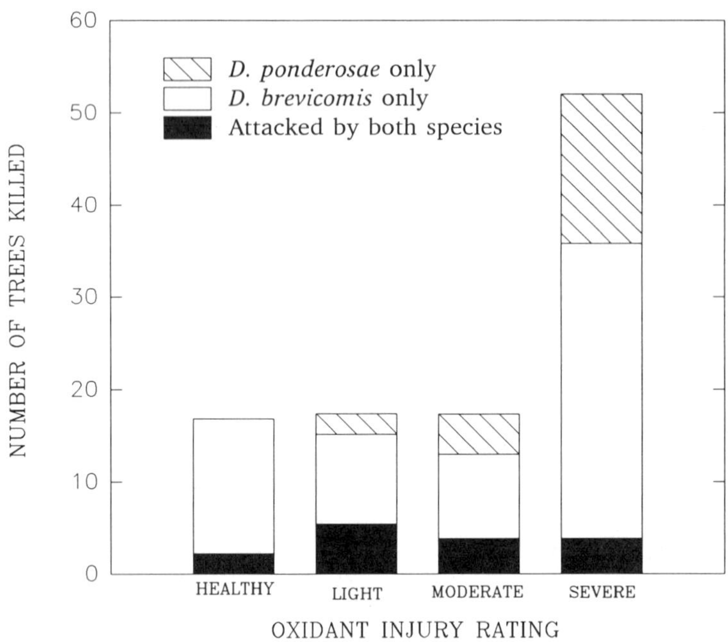

Figure 7.4. Relationship between the degree of photochemical air pollution injury and mortality due to bark beetle attack on ponderosa pine in the San Bernadino Mountains of California (from Smith, 1990).

weight increases with increased concentration of the pollutant up to 0.30 ppm SO_2. Pupal weights declined thereafter and were no different from the controls at 0.75 ppm SO_2. When removed from a polluted environment, plants showed some degree of recovery after 24 hours, but at a much slower rate than the response time to the pollutant.

Foliar concentrations of reduced glutathione (GSH) were highly correlated with the Mexican bean beetle's growth, rate of development, fecundity, and feeding preferences (Hughes, 1988). Foliar concentrations of GSH increased to a maximum in soybean leaves exposed to concentrations of SO_2 up to 0.30 ppm and returned to control levels at a concentration of 0.75 ppm (Chiment et al., 1986). Clearly there is a strong correlation, if not a causal relationship, between GSH and Mexican bean beetle success.

An increase in foliar concentrations of GSH is a common response to stresses such as drought, salinity, cold stress, and oxidative pollutants (Gamble and Burke, 1984; Hughes, 1988; Levitt, 1972; Shevyakova and Loshadkina, 1965). SO_2-exposed foliage of Norway spruce, Scots pine (*Pinus sylvestris* L.) European larch (*Larix decidua* Mill.), and European white birch (*Betula pendula* Roth) respond with increased GSH. A greater increase in GSH occurred in spruce needles exposed to a mixture of SO_2 and O_3 than SO_2 alone (Smidt, 1984).

Although elevated foliar concentrations of GSH are correlated with plant stress, they are not a general predictor of insect success on such plants. Some insects are little affected by increases in foliar GSH (Hughes, 1988). Even Mexican bean beetle showed little response to elevated GSH when feeding on bean plants. Thus, precise mechanisms that explain the interrelationship between herbivore, host, and pollutant are still largely unknown.

The studies on the effects of SO_2 on the Mexican bean beetle/soybean system (Hughes, 1988; Hughes et al., 1982, 1983, 1985) are a good example of the type of research required for understanding host–herbivore–pollutant interactions. Although the interactive mechanisms are still unknown, it is clear that pollutants can cause plant stress and physiological and biochemical changes, which can result in increased herbivory. Pollutants may also interact with the herbivore's natural enemies. Increased pollution levels may be analogous to the scenario that occurred when synthetic pesticides became widely used after World War II. Many insignificant herbivores, especially spider mites, suddenly attained the status of pest. Pollutants could have the same influence on as yet nonpest herbivores, through alterations in the herbivore–host relationship or through a reduction in the natural enemy complex. Indeed, a spider mite, *Oligonychus milleri* (McGregor), was found infesting loblolly pine exposed to various concentrations of O_3 (Forcum et al., unpublished data). The higher the O_3 concentration, the higher the spider mite population. Although *O. milleri* is an occasional pest of young, even-aged stands of pine in the South, the correlation with O_3 suggests that its status as a pest of pine may increase if O_3 increases.

Exotic Pests and Pollutants

It is important that we distinguish between native insects and their potential to become pests, and introduced (exotic) insects and their pest potential. Native insects and hosts have coevolved and, in an undisturbed state, have established life strategies that allow for the continued existence of both. On the other hand, exotic insects in their new environment have not co-evolved with their new host. Thus, with exotic insects, the complexity of the host–insect–pollutant interactions may be an order of magnitude greater. In addition to the stress created by the introduced herbivore, for which the host may have little or no defense, the stress created by the pollutant may prevent or inhibit the development of any life strategies that would allow for a coexistence of host and herbivore.

The balsam woolly adelgid (*Adelges piceae*)/North American fir system is an example of a forest system under stress from pollutants and an exotic pest. To date, research on this system has concentrated on the interactions between the host and insect (Hollingsworth and Hain, 1991). The logical next step is to evaluate the impact of pollutants on the host–insect interactions. For example, some of the effects of balsam woolly adelgid on fir are the formation of abnormal wood that is a poor conductor of water, lower xylem water potentials, and increased areas of heartwood. Although the causal mechanisms for the decline of certain forests in North America and Europe are poorly understood, there is a general consensus that it may be pollution related (Pitelka and Raynal, 1989), resulting in reduced annual increment. Of course, narrower growth rings mean reduced sapwood area. Trees with less than 25% basal area of sapwood become highly vulnerable to death from secondary pathogens and insects because less sapwood is available for defense, food storage, and water conduction (Shortle and Bauch, 1986; Shortle and Smith, 1988). A reduced proportion of sapwood area results from a reduced rate of sapwood production and an accelerated rate of heartwood production (Zimmermann and Brown, 1971). Thus, the effects on fir trees of both the balsam woolly adelgid and pollutants may be very similar. The combined effects of both may mean that the North American hosts are less capable of tolerating an infestation of this exotic insect, which is not considered a pest in its native range of central Europe. Indeed, the incidence of Fraser fir [*Abies frazeri* (Pursh) Poir.] mortality is highly correlated with the presence of the adelgid (Zedaker et al., 1987) and with high inputs of atmospheric pollutants at the highest elevations of the southern Appalachians, where fir dominates over spruce. Higher pollution levels at the higher elevations are due to higher precipitation and frequent cloud immersion (Dull et al., 1987). Thus, it has been hypothesized that atmospheric pollutants may either lower the tolerance of Fraser fir to the insect or add a final lethal stress to trees already heavily stressed by the insect (Hain and Arthur, 1985).

The balsam woolly adelgid/North American fir/pollutants scenario is not an isolated case. Since about 1860, there has been a dramatic increase in the rate of establishment of exotic species, presumably as a result of intercontinental mobility of humans and their materials, resulting in the breakdown of natural geographic barriers that once limited the movement of organisms. Many of these exotics become pests. Of the 70 major insect pests of U.S. forests, 19 were introduced (Pimentel, 1986). Introductions such as the balsam woolly adelgid, gypsy moth (*Lymantria dispar*), Dutch elm disease (*Ceratocystis ulmi*), and chestnut blight (*Endothia parasitica*) have caused catastrophic tree mortality and in some cases caused the virtual elimination of previously dominant tree species. The larval feeding preference of gypsy moth on white oak (*Quercus alba* L.) was apparently affected by ozone, although the dose–response relationships were unclear (Jeffords and Endress, 1984).

Introductions are occurring to this day, perhaps at an increasing rate. In the last few years, both the pine shoot beetle and the Asian gypsy moth have been discovered in North America. Both have the potential of becoming serious, if not devastating pests. As our economy truly becomes a world economy, we can assume that the type of exotic introductions described earlier will continue, if not escalate. We can only speculate about the impact these exotics will have on our native tree species. One thing is clear, however: Minimizing the amount of stress our native trees are subjected to will improve their chances for adjusting to exotic pest species. There are indications, for example, that some North American firs, like the European species, can tolerate the balsam woolly adelgid if they are growing in a healthy environment. The challenge for research foresters and entomologists is to expand on the type of research described here for the Mexican bean beetle and the balsam woolly adelgid. Understanding the mechanisms involved in host–insect interactions is critical to our understanding of the impact pollutants may have on those interactions.

Fungi

The occurrence of fungal diseases on southern pines is widespread. These diseases infect and colonize foliage, stems, branches, and roots (see Table 1.6 in Chapter 1). The effects of air pollutants on diseases caused by fungal plant pathogens can range from stimulatory to inhibitory (Heagle, 1973; Smith, 1990; Treshow, 1975). Pollutants can affect host–fungi interactions in many ways, as shown in Table 7.1, including alteration of host tissue susceptibility, plant resistance, fungal virulence, and inoculum density (Bruck and Shafer, 1984).

Effects of Pollutants on Fungi

Several air pollutants have been investigated with respect to their fungicidal potential, including ozone and sulfur dioxide (Hartman, 1924; Treshow, 1975). Fungicidal effects, however, were found to occur only at very high concentrations. For O_3, effective concentrations were reported at above 250 ppb, and they were at least 900 ppm for SO_2, though fungal species did show differential resistance, with some requiring over 1 ppm (O_3) and 2500 ppm (SO_2) (Smith, 1990). Such concentrations are not currently possible in a forested area in the southeastern United States, and therefore pollutants do not act as fungicides. It is unknown, however, if long-term chronic exposure directly affects the fungal pathogens of southern pines. The literature on other fungi reported varied responses (Smith, 1990). Presumably, only those fungi that would be exposed to the air (i.e., foliage, branch, and stem diseases) would be directly affected by air pollutants such as ozone.

Effects of Pollutants on Host–Fungi Interactions

The majority of reports concerning plant–pathogen–air pollutant relationships are correlative in nature (Manning and Keane, 1988) and pertain to herbaceous crop systems. Only a few studies involve the interaction among air pollutants and the southern pine–pathogen system.

Ozone has been reported to weaken trees in natural stands and increase susceptibility to invasion by plant pathogens, such as *Heterobasdion annosum*, the causal agent of annosum root rot (Fig. 7.5) (James et al., 1980a and 1980b). Lackner and Alexander (1983), in a correlative study, investigated the association of "procerum root disease," *Leptographium procerum* (Kendr.) Wingf., with oxidant-stressed *Pinus strobus* L. growing along the Blue Ridge Parkway in Virginia. Occurrence of *L. procerum* was greater on severely stressed trees, and there was more insect feeding on these trees. In another study, root disease incidence associated with oxidant-stressed eastern white pines was surveyed in the Coweeta Basin of North Carolina, but no relationship between oxidant injury and root disease incidence was observed in these stands (Leininger et al., 1990).

Ozone may also enhance disease development of normally saprophytic fungi such as *Lophodermium pinastri* and *Pullalaria pullulans*, which were associated with foliar ozone injury of eastern white pine (*Pinus strobus* L.) (Costonis and Sinclair, 1972). Weidensaul and Darling (1979) found that ozone, alone or in combination with sulfur dioxide, increased infection and colonization of *Pinus sylvestris* L. needles compared with controls, but only when inoculation occurred before pollutant treatment.

Observed increases in a needle blight (*Lophodermella cerina*) on slash pine (*Pinus elliottii* Engelm.) were attributed to air pollutants, including ozone and acidic deposition (Snow, 1986). Van Deusen and Snow (1991) correlated this disease complex with growth loss and found that the most

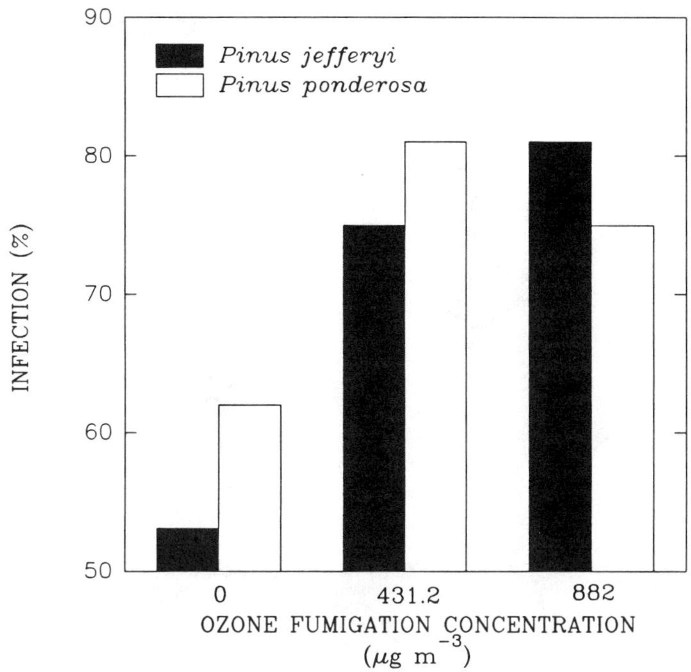

Figure 7.5. Infection by *Heterobasdion annosum* of Jeffrey and ponderosa pine exposed to various ozone treatments for a period of 58 to 87 days (redrawn from Smith, 1990).

severe occurrence of the blight occurs near large metropolitan areas along the Gulf Coast.

In one of the few cause-and-effect studies pertaining to ozone–plant pathogen interaction with southern pine, Carey and Kelley (1994) observed the growth and disease development of loblolly pine (*P. taeda*) seedlings exposed to ozone and inoculated with *Fusarium subglutinans*, the causal agent of pitch canker disease. Seedlings were exposed to subambient, ambient, and elevated levels of ozone in open-top chambers for approximately three months; then one half of the seedlings were inoculated with the fungus and the pines were fumigated an additional 50 days (Fig. 7.6). Canker dimensions increased with increasing ozone concentrations (i.e., the cankers were longer and wider at higher concentrations).

Little research exists on the effects of simulated acid rain on southern pine pathosystems. Shriner (1978) found that simulated acid rain applied to oak leaves resulted in an 86% reduction of telia produced by *Cronartium quercuum* f. sp. *fusiforme*. Anderson et al. (1993) reported that dogwood anthracnose (*Discula distructiva* sp. nov.) disease development on *Cornus florida* L. increased with increasing rain acidity. Although not a pine,

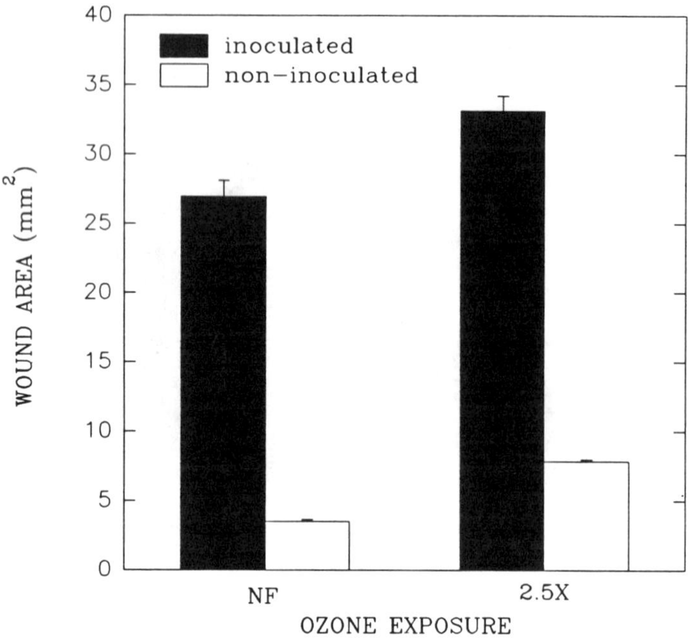

Figure 7.6. Area of cankers inoculated with *Fusarium subglutinans* or wound scars of noninoculated seedlings of loblolly pine. The seedlings were exposed to ozone treatments of NF (nonfiltered ambient) and 2.5 × NF for 105 days before treatments and then were assessed 50 days later. The areas were calculated by multiplying lengths and widths of the cankers/wounds and represent the means and standard errors of two separate experiments (redrawn from Carey and Kelley, 1994).

flowering dogwood is commonly associated with southern pine, and these results indicate the need for future research in this ecosystem.

Host Defense Mechanisms

Trees possess both chemical and structural defenses against pathogens and herbivores (McLaughlin and Shriner, 1980). However, only a few studies have addressed the interactions among southern pines, insects and diseases, and air pollution. Jordan et al. (1991) investigated the response of total needle tannins and phenolics of loblolly pine saplings growing in open-top chambers exposed to ozone and acid rain. Both tannin concentration (quantity per gram) and content (quantity per fascicle) were increased in trees exposed to elevated ozone but were unaffected by acid rain. Total phenolics were unaffected by either pollutant at any level.

Ozone apparently affects distribution of chemicals within the phenolic group but not total needle phenolic level.

Increased foliar nitrogen of trees exposed to elevated ozone and simulated acidic precipitation has been reported for loblolly and slash pine (Johnson et al., 1992a; Wright et al., 1991). While the increase in N with acid rain was believed to result from a fertilizer effect, increased N in response to ozone was attributed to N retranslocation out of premature senescing foliage. Increased foliar N may increase susceptibility of southern pines to insects and diseases through enhanced performance of these pests (Hughes, 1988; Huges and Laurence, 1984; Manning and Keane, 1988).

Carey and Kelley (1994) found that increasing ozone concentrations inhibited the wound-healing process such that inoculation wounds did not heal as fast in ozone-treated trees (Fig. 7.6). While the cause of this phenomenon was unknown, changes in host defense mechanisms, perhaps due to reduced stem carbohydrates, may be involved. Reduced wound healing could provide infection courts for wound pathogens such as *F. subglutinans* and wood-boring insects.

The ability of slash pine to defend against blue-stain fungi (*Ophiostoma minus*) was affected by ozone exposure and family (Johnson et al., 1992c). A major defense mechanism against the fungus is the ability to produce terpenes rapidly, which may be fungicidal and/or provide a physical barrier to further colonization. Seedlings were pulsed-labeled with $^{14}CO_2$ and inoculated with fungal spores or water, and then ethanol-soluble sugars and pentane-soluble terpenes were extracted 2, 5, and 10 days after inoculation and analyzed for radioactivity in the fractions. One family, selected for high gum yield (greater terpene production), exhibited increased ^{14}C in sugars and terpene under elevated ozone. In contrast, the other family, a low gum yielder, showed the opposite response: lower sugars and terpenes under elevated ozone. The interaction among host genetics, fungi, and air pollution is obviously complex and awaits further research.

It has been suggested that mycorrhizae can provide protection from root pathogens. Recently, Bonello et al. (1993) reported that ozone–root disease susceptibility decreased by the presence of mycorrhizae in Scots pine seedlings (*Pinus sylvestris* Linn.). This response was attributed to the presence of mycorrhizae decreasing pathogen-induced secondary compound production. The response of mycorrhizae to air pollution is discussed in Chapter 9.

It is obvious that our understanding of how air pollution interacts with a complex process like host defense is limited. Additional research is needed to elucidate mechanisms of host responses to fungi and how air pollutants modify the host tree's ability to defend itself from infection and colonization.

Conclusions

This chapter has attempted to summarize existing literature on the effect of air pollution on the interactions among host tree, environmental stresses, insects, and disease in the southeastern United States. Collectively, this topic encompasses the processes that constitute forest health and productivity. Understanding these processes is necessary to manage our forests properly in a rapidly changing environment driven by anthropogenic influences of air pollution and greenhouse gases.

Abiotic and biotic factors interact to affect the growth and development of southern pines. The potential magnitude and direction of the responses, however, mostly remain unknown. Most studies have involved either environmental stress effects on the host tree or the effect of a pest on the tree. Only a few have considered the interaction of environment, pest, and host tree. Hughes (1988) summarized characteristics of an ideal system for studying these interactions (Table 7.2), many of which fit the

Table 7.2. Characteristics of an ideal experimental system for studying effects of air pollutants on plant-pest interactions.[a]

- The system should by long term in nature.
- Storage mechanisms should be involved in the pest–host-pollutant.
- Plants in the long-term model system should have genetic diversity that influences demographic changes in the population.
- There should be intraspecific competition in the model system.
- The system should lend itself to study of resource allocation in the host.
- The host in the system should have a spectrum of well-described foliar and soil-borne pests and parasitoids.
- The host should have some symbiotic relationships with other organisms.
- The host should be insect pollinated to allow pollinator studies.
- The host plant should be economically and/or ecologically important.
- The plant-pest system should be amenable to controlled experimentation.
- The host should have more than one use, such that the different uses cause different ecological relationships to develop.
- There should be enough experience with the system that experimental design criteria are available.
- There should be diversity in the host for resistance and sensitivity to pollutants and pests.
- There should be documented responses in the system to changes in mineral nutrient status.
- The system should have well-known phenological characteristics.
- There should be diversity in the host for resistance and sensitivity to pollutants and pests.
- There should be documented responses in the system to changes in mineral nutrient status.
- The system should have well-known phenological characteristics.
- The model system should be amenable to both laboratory and field studies, short- and long-term studies, and mechanistic and correlative studies.

[a] Adapted from Huges (1988).

southern pine ecosystem. It is clear that as more factors are added, it will become more difficult to maintain experimental integrity and to interpret the subsequent results. Another aspect that is rarely addressed experimentally is time, particularly given the longevity of trees. Unfortunately, most research is now tied to short funding cycles of two to three years, which often lead to the publication of preliminary and/or premature results, at least with respect to long-term tree response. The level of research funding also inhibits the necessary collaboration among ecologists, entomologists, pathologists, and physiologists required to address these complex research questions.

There are many challenges and opportunities for forest scientists to begin developing an understanding of the interactions discussed in this chapter. Multi disciplinary teams employing holistic approaches will be necessary to fill many of the questions raised in this and other chapters in this book.

References

Adams MB, Kelly JM, Taylor GE, Edwards NT (1990) Growth of five families of *Pinus taeda* L. during three years of ozone exposure. New Phytol 116:689–694.

Alstad DN, Edmunds GF Jr, Weinstein LH (1982) Effects of air pollutants on insect populations. Ann Rev Entomol 27:369–384.

Anderson RL, Berrang P, Knighten J, Lawton KA, Britton KO (1993) Pretreating dogwood seedlings with simulated acidic precipitation increases dogwood anthracnose symptoms in greenhouse-laboratory trials. Can J For Res 23:55–58.

Baker TR, Allen HL, Schoeneberger MM, Kress LW (1994) Nutritional response of loblolly pine exposed to ozone and simulated acid rain. Can J For Res 24:453–461.

Baltensweiler W (1985) "Waldsterben": forest pests and air pollution. Z Ang Ent 99:77–85.

Barnes JD, Eamus D, Brown KA (1990) The influence of ozone, acid mist and soil nutrient status on Norway spruce [*Picea abies* (L.) Karst.]. New Phytol 115:149–156.

Bartlett BR (1951) The action of certain "inert" dust materials on parasitic Hymenoptera. J Econ Entomol 44:891–896.

Basham HG (1970) Wilt of loblolly pine inoculated with blue stain fungi of the genus *Ceratocystis*. Phytopath 60:750–754.

Binkley D, Driscoll CT, Allen HL, Schoeneberger P, McAvoy D (1989) *Acidic Deposition and Forest Soils: Context and Case Studies in the Southeastern United States*. Springer-Verlag, New York.

Bishop JA (1972) An experimental study of the decline of industrial melanism in Biston betularia (L.) (Lepidoptera) between urban Liverpool and rural North Wales. J Anim Ecol 41:209–243.

Blanche CA, Hodges JD, Nebeker TE (1985) Changes in bark beetle susceptibility indicators in a lightning-struck loblolly pine. Can J For Res 15:397–399.

Bonello P, Heller W, Sandermann H Jr (1993) Ozone effects on root-disease susceptibility and defense responses in mycorrhizal and non-mycorrhizal seedlings of Scots pine (*Pinus sylvestris* L.). New Phytol 124:653–663.

Bridges JR (1987) Effects of terpenoid compounds on growth of symbiotic fungi associated with the southern pine beetle. Phytopath 77:83–85.

Bruck RI, Shafer SR (1984) Effects of acid precipitation on plant diseases. In: Linthurst RA (ed) *Direct and Indirect Effects of Acidic Precipitation on Vegetation*. Butterworths, Stonehaven MA, pp 19–32.

Buse A (1986) Fluoride accumulation in invertebrates near an aluminum reduction plant in Wales. Environ Pollut 41:199–217.

Byres DP, Johnson JD, Dean TJ (1992) Seasonal response of slash pine (*Pinus elliottii* var. *elliottii* Engelm.) photosynthesis to long-term exposure to ozone and acidic precipitation. New Phytol 122:91–96.

Callenbach JA (1940) Influences of road dust upon codling moth control. J Econ Entomol 33:803–807.

Carey WA, Kelley WD (1994) Effects of ozone exposure and *Fusarium subglutinans* inoculation on growth and disease development of loblolly pine seedlings. Environ Pollut 84:35–43.

Chappelka A, Kush JS, Meldahl RS, Lockaby BG (1990) An ozone–low temperature interaction in loblolly pine (*Pinus taeda* L.). New Phytol 114:713–720.

Chiment JJ, Alscher R, Hughes PR (1986) Glutathione as an indicator of SO_2-induced stress in soybean. Environ Expl Bot 26:147–152.

Cobb FW Jr, Wood DL, Stark RW, Miller PR (1968b) Effect of injury upon physical properties of oleoresin, moisture content, and phloem thickness. Hilgardia 39:127–134.

Cobb FW Jr, Wood DL, Stark RW, Parmeter JR Jr (1968b) Theory on the relationships between oxidant injury and bark beetle infestation. Hilgardia 39:141–152.

Costonis AC, Sinclair WA (1972) Relationship of atmospheric ozone to needle blight of eastern white pine. Phytopathology 59:1566–1574.

Darrall NM (1989) The effects of air pollutants on physiological processes in plants. Plant, Cell and Environ 12:1–30.

Dewey JE (1973) Accumulations of fluorides by insects near an emission source in western Montana. Environ Entomol 2:179–182.

Dull CW, Ward JD, Brown HD, Ryan GW (1987) Evaluation of tree mortality in the spruce-fir forest of the southeastern United States. In: *Effects of Atmospheric Pollutants on the Spruce-Fir Forests of the Eastern United States and the Federal Republic of Germany*. Proc US/FRG Research Symp, Burlington, VT, October 19–23. US For Serv General Technical Report NE-120, pp 107–110.

Edmunds GF Jr (1973) Ecology of black pineleaf scale (*Homoptera: Diaspididae*). Environ Entomol 2:765–777.

Edwards GS, Edwards NT, Kelly JM, Mays AP (1991) Ozone, acidic precipitation and soil magnesium effects on growth and nutrition of loblolly pine seedlings. Environ Exper Bot 31:67–78.

Edwards GS, Pier PA, Kelly JM (1990) Influence of ozone and soil magnesium status on the cold hardiness of loblolly pine (*Pinus taeda* L.) seedlings. New Phytol 115:157–164.

Edwards NT (1991) Root and soil respiration response to ozone in *Pinus taeda* L. seedlings. New Phytol 118:315–321.

Edwards NT, Taylor GE, Adams MB, Simmons GL, Kelly JM (1990b) Ozone, acidic rain and soil magnesium effects on growth and foliar pigments of *Pinus taeda* L. Tree Physiol 6:95–104.

Elsik CG, Flagler RB, Boutton TW (1992) Effects of ozone and water deficit on growth and physiology of *Pinus taeda* and *Pinus echinata*. In: Flagler RB (ed) *The Response of Southern Commercial Forests to Air Pollution*. 1992 Transactions of A&WMA, No 21. Pittsburgh, PA, pp 225–245.

Feir D, Hale R (1983) Responses of the large milkweed bug, *Oncopeltus fasciatus* (*Hemiptera: Lygaeidae*) to high levels of air pollutants. Int J Environ Studies 20:269–273.

Freitag R, Hastings L (1973) Kraft mill fallout and ground beetle populations. Atmospheric Environ 7:587–588.

Freitag R, Hastings L, Mercer WR, Smith A (1973) Ground beetle populations near a kraft mill. Can Entomol 105:299–310.

Führer E (1985) Air pollution and the incidence of forest insect problems. Z Ang Entomol 99:371–377.

Fye RE (1983) Impact of volcanic ash on pear psylla (*Homoptera: Psyllidae*) and associated predators. Environ Entomol 12:222–226.

Gamble PE, Burke JJ (1984) Effect of water stress on chloroplast antioxidant system. Plant Physiol 76:615–621.

Hain FP (1987) Interactions of insects, trees and air pollutants. Tree Physiol 3:93–102.

Hain FP, Arthur FH (1985) The role of atmospheric deposition in the latitudinal variation of Fraser fir mortality caused by the balsam woolly adelgid, *Adelges piceae* (Ratz.) (Hemipt.: Adelgidae): A hypothesis. Z Angew Entomol 99: 145–152.

Hartman FE (1924) The industrial application of ozone. J Am Soc Heat Vent Engin 30:711–727.

Hay JC (1977) *Bibliography on Arthropoda and Air Pollution*. Forest Service Gen Tech Rep NE-24, pp 1–16.

Heagle AS (1973) Interactions between air pollutants and plant parasites. Ann Rev Phytopathol 11:305–388.

Heliövaara K, Väisänen R (1993) *Insects and Pollution*. CRC Press, Boca Raton, FL.

Hillmann RC (1972) *Biological Effects of Air Pollution on Insects, Emphasizing the Reactions of the Honey Bee (Apis mellifera L.) to Sulfur Dioxide*. Ph.D. dissertation, Pennsylvania State University, University Park, PA.

Hillman RC, Benton AW (1972) Biological effects of air pollution on insects, emphasizing the reaction of the honey bee (*Apis mellifera* L.). Elisha Mitchell Sci Soc 88:195.

Hodges JD, Pickard RL (1971) Lightning in the ecology of the southern pine beetle, *Dendroctonus frontalis* (*Coleoptera: Scolytidae*). Can Ent 103:44–51.

Hodges JD, Elam WW, Watson WF, Nebeker TE (1979) Oleoresin characteristics and susceptibility of four southern pines to southern pine beetle (*Coleptera: Scolytidae*) attacks. Can Ent 111:889–896.

Hoffard WH (1985) A would-be manager of southern forests. In: *Insect and Disease Conditions in the United States, 1979–1983*. USDA For Serv, St and Priv For, FPM, GTR WO-46, pp 32–37.

Hollingsworth RG, Hain FP (1991) Balsam woolly adelgid (Homoptera: Adelgidae) and spruce-fir decline in the southern Appalachians: Assessing pest relevance in a damaged ecosystem. Florida Entomol 74:179–187.

Holopainen JK, Mustaniemi A, Kainulainen P, Satka H, Oksanen J (1993) Conifer aphids in an air-polluted environment. I. Aphid density, growth and accumulation of sulfur and nitrogen by Scots pine and Norway spruce seedlings. Environ Pollut 80:185–191.

Horton SJ, Reinert RA, Heck WW (1990) Effects of ozone on three open-pollinated families of *Pinus taeda* L. grown in two substrates. Environ Pollut 65:279–292.

Hughes PR (1988) Insect populations on host plants subjected to air pollution. In: Heinricks EA (ed) *Plant–Stress–Insect Interactions*. John Wiley and Sons, New York, pp 249–319.

Hughes PR, Chiment JJ, Dickie AI (1985) Effect of pollutant dose on the response of Mexican bean beetle (Coleoptera: Coccinellidae) to SO_2-induced changes in soybean. Environ Entomol 14:718–721.

Hughes PR, Dickie AI, Penton MA (1983) Increased success of Mexican bean beetle on field-grown soybeans exposed to sulfur dioxide. J Environ Qual 12:565–568.

Hughes PR, Laurence JA (1984) Relationship of biochemical effects of air pollutants on plants to environmental problems: insects and microbial interactions. In: Koziol MK,Whatley FR (eds) *Gaseous Air Pollutants and Plant Metabolism*. Butterworths, London, pp 361–377.

Hughes PR, Potter JE, Weinstein LH (1982) Effects of air pollution on plant–insect interactions: increased susceptibility of greenhouse grown soybeans to the Mexican bean beetle after plant exposure to SO_2. Environ Entomol 11: 173–176.

James RL, Cobb FW Jr, Wilcox WW, Rowney DL (1980a) Effects of oxidant air pollution of pine roots to *Fomes annosus*. Phytopath 70:560–563.

James RL, Cobb FW Jr, Wilcox WW, Rowney DL (1980b) Effects of photochemical oxidant injury of ponderosa and Jeffrey pines on susceptibility of sapwood and freshly cut stumps to *Fomes annosus*. Phytopath 70:704–708.

Jeffords MR, Endress AG (1984) A possible role of ozone in tree defoliation by the gypsy moth (*Lepidoptera: Lymantriidae*). Environ Entomol 13:1249–1252.

Jensen KF, Roberts BR (1986) Changes in yellow-poplar stomatal resistance with SO_2 and O_3 fumigation. Environ Pollut 41:235–245.

Johnson JD (1987) Stress physiology of forest trees: the role of plant growth regulators, In: Kossuth SV, Ross SD (eds) *The Role of Plant Hormones in Forest Trees*. Martinus Nijhoff, Boston, pp 193–215.

Johnson JD, Byres D, Chappelka AH, Dean TJ (1992a) The role of homeostatic growth in the response of slash and loblolly pine to ozone and acid rain. In: Flagler RB (ed) *The Response of Southern Commercial Forests to Air Pollution*. Transactions, AWMA, Pittsburgh, PA, pp 273–286.

Johnson JD, Byres DP, Dean TJ (1992b) *Nitrogen Productivity Response of Slash Pine (Pinus elliottii) to Ozone and Acid Rain*. Resource Capture by Crops, 52nd Easter School in Agricultural Science, University of Nottingham.

Johnson JD, Byres DP, Dean TJ (1995) Diurnal water relations and gas exchange of two slash pine (*Pinus elliottii* Engelm.) families exposed to chronic ozone levels. New Phytol.

Johnson JD, Michelozzi M, Byres D, Osmond J (1992c) The effect of ozone on the induced host resistance of slash pine. In: Flagler RB (ed) *The Response of Southern Commercial Forests to Air Pollution*. Transactions, AWMA, Pittsburgh, PA, p 323.

Jordan DN, Green TH, Chappelka AH, Lockaby BG, Meldahl RS, Gjerstad DH (1991) Response of total tannins and phenolics in loblolly pine foliage exposed to ozone and acid rain. J Chem Ecol 17:505–513.

Keller T (1991) Air pollutants. In: Raghavendra AS (ed) *Physiology of Trees*. John Wiley and Sons, New York, pp 337–357.

Keller T, Hasler R (1984) The influence of fall fumigation with ozone on the stomatal behavior of spruce and fir. Oecologia 64:284–286.

Keller T, Hasler R (1987) Some effects of long-term ozone fumigations on Norway spruce. I. Gas exchange and stomatal response. Trees 1:129–133.

Kettle A, Port G, Davison A (1983) The impact of fluoride pollution on invertebrates. Bull Brit Ecol Soc 14:110–111.

Kettlewell HBD (1955) Selection experiments on industrial melanism in the *Lepidoptera*. Heredity 9:323–342.

Kettlewell HBD (1956) Further selection experiments on industrial melanism in the *Lepidoptera*. Heredity 10:287–301.

Kettlewell HBD (1958) A survey of the frequencies of *Biston betularia* (L.) (Lep.) and its melanic forms in Great Britain. Heredity 12:51–73.

Kozlowski TT, Kramer PJ, Pallardy SG (1991) The physiological ecology of woody plants. Academic Press, San Diego, p 657.

Kramer PJ (1983) *Water Relations of Plants*. Academic Press, New York.

Lackner AL, Alexander SA (1983) Root disease and insect infestations on air-pollution sensitive *Pinus taeda* and studies of pathogenicity of *Verticicladiella procera*. Plant Dis 67:679–681.

Laurence JA, Hughes PR, Weinstein LH, Geballe GT, Smith WH (1983) *Impact of Air Pollution on Plant–Pest Interactions: Implications of Current Research and Strategy for Future Studies*. ERC Report No 20. Ecosystem Research Center, Ithaca, NY.

Lee WS, Chevone BI, Seiler JR (1990) Growth and gas exchange of loblolly pine seedlings as influenced by drought and air pollutants. Water, Air and Soil Pollut 51:105–116.

Leininger TD, Winner WE, Alexander SA (1990) Root disease incidence in eastern white pine plantations with and without symptoms of ozone injury in the Coweeta Basin of North Carolina. Plant Dis 74:552–554.

Levitt J (1972) *Responses of Plants to Environmental Stresses*. Academic Press, New York.

Lorio PL Jr, Hodges JD (1977) Oleoresin exudation pressure and relative water content of inner bark as indicators of moisture stress in loblolly pines. For Sci 14:392–398.

Maier–Maercker U, Koch W (1992) Histological examination of spruce needles from a long-term gas exchange experiment in pure and polluted air in the field. Trees 6:186–194.

Manderscheid R, Jager HJ, Kress L (1992) Effects of ozone on foliar nitrogen metabolism of *Pinus taeda* L. and implications for carbohydrate metabolism. New Phytol 121:623–633.

Manning WJ, Keane KD (1988) Effects of air pollutants on interactions between plants, insects, and pathogens. In: Heck WW, Taylor OC, Tingey DT (eds) *Assessment of Crop Loss from Air Pollutants*. Elsevier, London, pp 365–386.

McLaughlin SB (1985) Effects of air pollution on forests: a critical review. J Air Pollution Control Assoc 35:512–534.

McLaughlin SB, Shriner DS (1980) Allocation of resources to defense and repair. In: Horsfall JG, Cowling EB (eds) *Plant Disease: An Advanced Treatise*, Vol. V. Academic Press, New York, pp 407–431.

McLaughlin SB, Taylor GE (1981) Relative humidity: Important modifier of pollutant uptake by plants. Science 211:167–168.

Meier S, Grand LF, Schoeneberger MM, Reinert RA, Bruck RI (1990) Growth, ectomycorrhizae, nonstructural carbohydrates of loblolly pine seedlings exposed to ozone and soil water deficit. Environ Pollut 64:11–27.

Mikkola K (1975) Frequencies of melanic forms of *Oligia* moths (*Lepidoptera, Noctuidae*) as a measure of atmospheric pollution in Finland. Ann Zool Fennici 12:197–204.

Miller PR, Cobb FW Jr, Zavarin E (1968) Effect of injury upon oleoresin composition, phloem carbohydrates, and phloem pH. Hilgardia 39:135–140.

Morrison JK (1984) Acid rain a review of literature on acid deposition effects in forest ecosystems. For Abstr 45:483–506.

Olszyk DM, Tibbitts TW (1981) Stomatal response and leaf injury of *Pisum sativum* L. with SO_2 and O_3 exposures. II. Influence of moisture stress and time of exposure. Plant Physiol 67:545–549.

Norby RJ, Kozlowski TT (1983) Flooding and SO_2 stress interaction in *Betula papyrifera* and *B. nigra* seedlings. For Sci 29:739–750.

Owen DF (1961) Industrial melanism in North American moths. Am Nat 95:227–233.

Paynter VA, Reardon JC, Shelburne VB (1991) Carbohydrate changes in shortleaf pine (*Pinus echinata*) needles exposed to acid rain and ozone. Can J For Res 21:666–671.

Paynter VA, Reardon JC, Shelburne VB (1992) Changing carbohydrate profiles in shortleaf pine (*Pinus echinata*) after prolonged exposure to acid rain and ozone. Can J For Res 22:1556–1561.

Pimentel D (1984) Biological invasions of plants and animals in agriculture and forestry. In: Mooney HA, Drake JA (eds) *Ecology of Biological Invasions of North America and Hawaii*. Ecological Studies 58. Springer-Verlag, New York, pp 149–162.

Pitelka LG, Raynal DJ (1989) Forest decline and acidic deposition. Ecology 70:2–10.

Porter JW, Spurgeon K (1981) *Biosynthesis of Isoprenoid Compounds*. Vol 1. John Wiley and Sons, New York.

Przybylski Z (1968) Results of consecutive observations of effect of SO_2, SO_3 and H_2SO_4 gases and vapors on trees, shrubs, and entomofauna of orchards in the vicinity of sulfur mines and sulfur processing plant in Machow. Postepy Nauk Rolm 15:131–138.

Raffa KF, Berryman AA, Simasko J, Teal W, Wong BL (1985) Effects of grand fir monoterpenes on the fir engraver, *Scolytus ventralis* (*Coleoptera: Scolytidae*), and its symbiotic fungus. Environ Entomol 14:552–556.

Reich PB (1987) Quantifying plant response to ozone: a unifying theory. Tree Physiol 3:63–91.

Reich PB, Lassoie JP (1984) Effects of low O_3 exposure on leaf diffusive conductance and water-use efficiency in hybrid poplar. Plant, Cell and Environ 7:661–668.

Reich PB, Schoettle AW, Stroo HF, Troiano J, Amundson RB (1987) Effects of O_3 and acid rain on white pine seedlings grown in five soils. I. Net photosytnhesis and growth. Can J Bot 65:977–987.

Rozenkranz T, Bareis C, Braun J, Wild A (1989) Investigation on the water status of spruce trees in relation to the occurrence of novel forest decline. Plant Physiol Biochem 27:333–340.

Sasek TW, Richardson CJ (1989) Effects of chronic doses of ozone on loblolly pine: photosynthetic characteristics in the third growing season. For Sci 35: 745–755.

Shanklin J, Kozlowski TT (1985) Effect of flooding of soil on growth and subsequent responses of *Taxodium distichum* seedlings to SO_2. Environ Poll Ser A 38:199–212.

Shevyakova NI, Loshadkina AP (1965) Variation of the sulfhydryl group content in plants under conditions of salinization. Soviet Plant Physiol 12:280–286.

Shortle WC, Bauch J (1986) Wood characterization of *Abies balsamea* in the New England states compared to *Abies alba* from sites in Europe with decline problems. Int Assoc Wood Anatomists Bull 7:375–387.

Shortle WC, Smith KT (1988) Aluminum-induced calcium deficiency syndrome in declining red spruce. Science 240:1017–1018.

Shriner DS (1978) Effects of simulated acidic rain on host–parasite interactions in plant diseases. Phytopath 68:213–218.

Sierpiński Z (1984) On the effect of air pollution on harmful insects in stands of coniferous trees in Poland. Forst Centralblatt 103:83–91.

Sierpiński Z (1985) Air pollution and forest pests. Z Ang Entomol 99:1–6.

Skarby L, Troeng E, Botstrom CA (1987) Ozone uptake and effects on transpiration, net photosynthesis, and dark respiration in Scots pine. For Sci 33:801–808.

Skeffington RA, Roberts TM (1985) The effects of ozone and acid mists on Scots pine saplings. Oecologia 65:201–206.
Skrzypcaynska M (1984) Preliminary studies on entomofauna of cones of *Abies alba* in Ojcowski and Tatrzanski national parks in Poland. Z Ang Entomol 98:375–379.
Smidt VS (1984) Begasungsversuche mit SO_2 und Ozon an jungen Fichten. Eur J For Pathol 14:241–248.
Smith WH (1984a) Ecosystem pathology: a new perspective for phytopathology. For Ecol Manag 9:193–219.
Smith WH (1984b) Effects of regional air pollutants on forests in the USA. Forst Centralblatt 103:48–61.
Smith WH (1990) Air pollution and forest interactions between air contaminants and forest ecosystems. 2nd ed. Springer-Verlag, New York.
Snow GA (1986) A needle blight of slash and loblolly pines in south Mississippi. In: Peterson GW (ed) *Recent Research on Conifer Needle diseases*. Gen Tech Rep WO US For Serv WO-50, pp 20–21.
Stark RW, Cobb FW Jr (1969) Smog injury, root diseases and beetle damage in ponderosa pine. Calif Agric pp 13–15.
Stark RW, Miller PR, Cobb FW Jr, Wood DL, Parmeter JR Jr (1968) Incidence of bark beetle infestation in injured trees. Hilgardia 39:121–126.
Taylor GE, Davies WJ (1990) Root growth of *Fagus sylvatica*: impact of air quality and drought at a site in southern Britain. New Phytol 116:457–464.
Taylor GE, Edwards NT, Adams MB, Friend AL, Dickson RE, Grizzard T, Gunderson CA, O'Neill EG, McEvers JA, Tomlinson PT (1991) *Influence of Ozone, Acidic Precipitation and Soil Magnesium Status on the Physiology and Growth of Pinus taeda L. (Loblolly Pine) Under Field Conditions*. Final report of the Physiological Ecology Research Task of the Southern Component of the EPRI Program, RP-2799-2.
Temple PJ, Riechers GH (1992) Internal nitrogen cycling in ozone-injured ponderosa pine. Bull Ecol Soc Amer 73:363–364.
Temple PJ, Riechers GH, Miller PR (1992) Foliar injury responses of ponderosa pine seedlings to ozone, wet and dry acidic deposition, and drought. Environ Exper Bot 32:101–113.
Temple PJ, Riechers GH, Miller PR, Lennox RW (1993) Growth responses of ponderosa pine to long-term exposure to ozone, wet and dry acidic deposition, and drought. Can J For Res 23:59–66.
Tjoelker MG, Luxmoore RJ (1991) Soil nitrogen and chronic ozone stress influence physiology, growth and nutrient status of *Pinus taeda* L. and *Liriodendron tulipifera* L. seedlings. New Phytol 119:69–81.
Treshow M (1975) Interaction of air pollutants and plant diseases. In: Mudd JB, Kozlowski TT (eds) *Responses of Plants to Air Pollution*. Academic Press, New York, 307–334.
Van Deusen PC, Snow GA (1991) Paired-tree study suggests 20–year recurrent slash pine blight. Can J For Res 21:1145–1148.
Villemant C (1981) Influence de la pollution atmospherique sue les populations d'aphides du pin sylvestre en Forêt de Roumaire (Seine Maritime). Environ Pollut Ser A 24:245.
Weidensaul TC, Darling SL (1979) Effects of ozone and sulfur dioxide on the host–pathogen relationship of Scotch pine and *Scirrhia acicola*. Phytopath 69:939–941.
Wright LM, Lockaby BG, Meldahl R, Thornton F, Chappelka AH (1991) Influence of acid precipitation and ozone on nitrogen nutrition of young loblolly pine. Water, Air and Soil Pollut 54:135–142.

Zedaker SM, Nicholas NS, Eager C, White PS, Burk TE (1987) Stand characteristics associated with potential decline of spruce-fir forests in the southern Appalachians. In: *Effects of Atmospheric Pollutants on the Spruce-Fir Forests of the Eastern United States and the Federal Republic of Germany*. Proc US/FRG Research Symp, Burlington, VT, October 19–23. US For Serv General Technical Report NE-120, pp 123–131.

Zimmermann MH, Brown CL (1971) *Trees Structure and Function*. Springer-Verlag, New York.

3. Growth and Physiological Effects of Regional Air Pollutants on Southern Pines

8. Atmospheric Deposition and Soil Resources of the Southern Pine Forest

Daniel D. Richter and Daniel Markewitz

Soils are open systems. As such, atmospheric deposition affects soil properties, processes, and genesis. Windblown loess deposits influence physical and chemical properties in many soils on all continents. Andisols, one of 11 soil orders of the world, owe their existence mainly to atmospheric depositions of volcanically ejected particulates (Soil Survey Staff, 1992). Near marine coastlines, sea-spray aerosols entrained in the atmosphere are deposited to soils, elevating concentrations of sodium, chloride, magnesium, and sulfate (Låg, 1968). Similarly, industrial emissions of acid- and base-forming compounds are transported through the atmosphere and are eventually deposited to soils as gases, aerosols, and as solutes in precipitation, sometimes up to thousands of kilometers from their points of emission.

Atmospheric deposition of industrial emissions interacts with soil biogeochemical processes over a wide range of time scales and with great complexity. Although these phenomena have been studied intensively over the last two decades in Europe and North America, we have remarkably little direct, quantitative evidence of interactive effects of acidic deposition on forests, soils, and soil biota. It is therefore wise to maintain some skepticism about the rate at which regional air pollutants can affect forest ecosystems (Tabatabai, 1985).

The objective of this chapter is to evaluate the chemical reactions and ecological consequences of atmospheric deposition of acids and bases on the soils that support southern pine forests. These soils are mainly found on the Atlantic and Gulf Coastal Plains, the Piedmont, and the Ozark Plateau. Based on predominant ecosystem and soil conditions in this region (Binkley et al., 1989; Johnson and Lindberg, 1992; Ralston, 1978; Switzer and Nelson, 1972; Turner et al., 1986; Wells and Jorgensen, 1975), atmospheric deposition appears to be most significant to these soils by its potential alteration of (1) nitrogen status via reactions with soil organic matter, plant roots, and microbes; and (2) soil acidity and nutrient cations via the reactions affected by deposition of sulfuric and nitric acids and base cations.

These two soil problems of nitrogen and acidity are a part of a larger set of inherent nutritional problems associated with the soils that support southern pine forests. Soil factors and management control forest productivity because much of this soil resources has a low cation exchange capacity (CEC) and relatively few weatherable primary minerals, and because the organic matter in most of this soil resource has been seriously impacted by past agricultural use.

These issues were addressed by Binkley et al. (1989) in a book entitled *Acid Deposition and Forest Soils*, which used soil conditions of the southeastern United States as a case study because many southeastern soils are susceptible to chemical change, especially those with low CEC and low weathering release of nutrient cations. Since this book by Binkley et al. (1989) was written, the biogeochemistry of five southern pine stands has been particularly well documented (Johnson and Lindberg, 1992; Richter et al., 1994a), one of which includes an unusually detailed record of soil acidification over the last three decades. These recent studies provide such significant data that a reconsideration of the impacts of atmospheric deposition on southern forest soils is warranted.

Nitrogen Deposition

Most southern pine forests grow in soils that have been used intensively for agriculture, some for up to 250 years. Soil fertility and nutrient cycling have been greatly altered not only by the cutting over of the original forest but by the cultivation, fertilization, liming, and crop harvests that followed. On many sites, the combination of intense rainfall, erodible soils, and exploitive agricultural practices has led to severe soil erosion problems that are widely known throughout the world. In the twentieth century, farm abandonment and conversion to pine forest have become widespread, at least in part due to low crop productivity and what has been called "soil exhaustion." This history of agricultural soil disturbance in the Piedmont is chronicled in detail by Trimble (1974).

As a result of this land-use history, pine forested soils in the southeastern United States are currently in a stage of recovery from the effects of agriculture (Richter et al., 1994, 1995; Switzer and Schelton, 1979). Many upland soils that support pine forests in the southeastern United States have relatively low concentrations of nitrogen that is available for plant uptake. Elevated atmospheric nitrogen deposition is a significant ecological issue because the productivity of many southeastern pine forests is so commonly limited by low soil nitrogen. The deposition and uptake of atmospheric nitrate and ammonium by plants and soils thus directly contributes to the recovery of the soil nitrogen capital. It is reasonable to hypothesize that plant-available nitrogen will eventually be elevated by cumulative effects of atmospheric deposition, although the rate at which soil nitrogen mineralization will respond to relatively low-level annual inputs is not quantified.

Southern pine forests typically develop a mor-like forest floor that has a high C/N ratio and a large capacity to immobilize mineral nitrogen, whether this nitrogen is added in fertilizer or deposited from the atmosphere. Soil solution concentrations of mineral nitrogen (ammonium plus nitrate) in precipitation and soil solutions in five pine forest ecosystems of the southern United States illustrate that immobilization results in most nitrogen becoming retained by the ecosystem (Table 8.1). Strong immobilization in upper horizons is indicated in four of five systems by the sharp decreases in mineral nitrogen concentrations as rainwater moves through the upper horizon of the soil (i.e., even within the surficial O and A horizons; Table 8.1). Leaching of nitrogen from these systems is relatively

Table 8.1. Total Mineral Nitrogen (Nitrate plus Ammonium) in Solutions of Five Southern Pine Ecosystems

Component	Calhoun Exp. For., SC (*P. taeda*)	Grant Exp. For., GA (*P. taeda*)	Duke Exp. For., NC (*P. taeda*)	Oak Ridge Nat Lab, TN (*P. taeda*)	Univ. Florida Exp. For. (*P. elliottii*)
Mineral N Concentration (umolc L^{-1})					
Precipitation[c]	22.2[d]	74.4	88.2	60.8	38.5
Throughfall	21.2	77.2	65.6	43.1	23.0
O horizon	17.7	—	7.3	26.7	15.0
A horizon	5.8	1.6	5.3	13.0	37.7
B horizon	3.3	7.5	4.5	4.7	27.6
B/C horizon	4.5	7.4	6.8	25.0	2.8
Mineral N Flux (kg ha^{-1} yr^{-1})					
Atmosphere	4.47[d]	9.06	14.0	9.79	6.03
Throughfall	3.68	8.11	9.30	6.58	3.19
O horizon	3.07	—	0.41	1.61	1.18
A horizon	0.96	0.10	0.29	0.78	3.91
B horizon	0.47	0.25	0.25	0.28	1.43
B/C horizon	0.35	0.24	0.62	1.02	0.14

Adapted from Markewitz, unpublished, and Johnson and Lindberg, 1992. Soil taxonomic subgroups and C/N ratios are listed in table footnotes a and b. Strong immobilization of nitrogen is evident in four of five systems.

[a] Soil subgroup, respectively, Typic Kanhapludult, Typic Hapludult, Typic Hapludult, Fluventic Dystrochrept, and Typic Haplaqod.

[b] C/N of O horizons, respectively, 48, 53, 70, 26, and 65.

[c] Calculated concentration with total atmospheric deposition dissolved in the annual volume of precipitation (Johnson and Lindberg, 1992).

[d] Wet-only precipitation.

small, ranging from about 0.24 to 1.4 kg ha^{-1} per year through B and C horizons, in contrast with annual atmospheric inputs of up to 14 kg ha^{-1}. Over a typical rotation length for southern pine of 30 years, nearly 300 kg N ha^{-1} may be accumulated in the forest with a total deposition of 10 kg N ha^{-1} yr^{-1}. Provided that pine ecosystems immobilize atmospheric nitrogen such as in the Calhoun, Grant, Duke, and Oak Ridge forests in Table 8.1, atmospheric deposition contributes directly to long-term nitrogen capital.

In managed pine forests that are periodically harvested for fiber products and also sometimes burned, inputs of nitrogen from the atmosphere help resupply losses incurred by management practices. Prescribed burning, even if at low intensity, results in losses of nitrogen that are comparable to current depositions from the atmosphere. Low-intensity prescribed fires on a four-year fire cycle in the South Carolina Coastal Plain volatilized from 10 to 40 kg ha^{-1}, equivalent to about 2.5 to 10 kg ha^{-1} yr^{-1} (Richter et al., 1982). In comparison, nitrogen removals by conventional harvests of stems remove about 2.5 to 5.8 kg ha^{-1} yr^{-1} of nitrogen from the pine ecosystems discussed in Table 8.1, whereas nitrogen removals in whole-tree methods of harvesting (wholetree harvests of boles, branches, bark, and foliage) amount to about 4.6 to 8.9 kg ha^{-1} yr^{-1}. Atmospheric deposition of nitrogen at current regional rates (Table 8.2) compensates for an important fraction of nitrogen losses associated with forest management practices such as harvesting and burning. It might also be argued, however, that given generally low plant availability of soil nitrogen, nitrogen fertilization will continue to boost productivity of southeastern soils under intensive management, even with elevated atmospheric deposition of N.

On the other hand, unmanaged pine forests are also often nitrogen deficient, and atmospheric deposition of nitrogen may well relieve nitrogen limitations over a period of decades. Atmospheric nitrogen input appears biologically significant for ecosystem nitrogen requirements, especially when considered over several decades (Table 8.2). In unharvested systems, even the deposited nitrogen that is immobilized by microbes in soil organic matter should eventually lead to increased nitrogen mineralization and plant nutrient availability. It is not too speculative to suggest ameliorative effects on forest productivity in these currently N-limited systems.

Acid- and Base-Cation Deposition and Soil Leaching

Acid Deposition

The major detrimental effects of acid deposition on terrestrial ecosystems are often suggested to be a gradual acidification of soils and depletion of

Table 8.2. Mean Annual Nitrogen Fluxes of Five Southern Pine Stands

Nitrogen Flux	Calhoun Exp. For., SC (*P. taeda*)	Grant Exp. For., GA (*P. taeda*)	Duke Exp. For., NC (*P. taeda*)	ORNL, TN (*P. taeda*)	U. Florida Exp. For. (*P. elliottii*)
Mean Nitrogen Flux ($kg\,ha^{-1}\,yr^{-1}$)					
Wood/bark increment	5.80	—	3.74	2.54	5.41
Total aboveground biomass increment[a]	7.17	—	8.9	4.60	7.91
O horizon mass increment	21.5	18.2	19.6	7.91	13.0
Total aboveground stand increment[b]	28.7	—	29.7	11.9	26.4
Root uptake for aboveground biomass[c]	35.0	—	25.5	21.4	24.1
Stand age (yr)	34	17	23	35	22

Adapted from Urrego, 1993, and Johnson and Lindberg, 1992.
[a] Wood, bark, plus foliage. Mean annual increment.
[b] Mean annual increment in aboveground biomass plus O horizon.
[c] Current foliar litterfall, new wood, new bark, and net canopy exchange (Johnson and Lindberg, 1992).

exchangeable base cations. The deposition of sulfuric and nitric acids alters the chemistry of soil solutions and cation exchange reactions, generally by increasing the concentration of polyvalent cations in soil solutions and accelerating the rate of soil cation leaching. Since the predominant soil taxa in the southeastern United States are Ultisols, Inceptisols, and Spodosols, many soils in the region possess relatively little acidneutralizing capacity. Acid deposition may have cumulative "erosional" effects on soil exchangeable nutrients, leaching and potentially depleting the divalent cations calcium and magnesium (Binkey et al., 1989; Johnson and Lindberg, 1992; Richter et al., 1992).

Many Ultisols, Inceptisols, and Spodosols in the South have low contents of primary minerals and consequently low rates of cation resupply from the weathering of these minerals. In addition, several million hectares of forested soils in the southeastern United States are extremely acidic (e.g., Paleudults, Umbrepts, and Aquods) and may exchange aluminum into soil solutions that are elevated in electrolyte concentration due to atmospheric deposition of sulfate. Although acid deposition does not acidify soils rapidly, many soils in the region appear to be susceptible over decades to acidification by atmospheric deposition.

Numerous studies in the southeastern United States demonstrate that sulfate is a dominant anion in solutions of many forest soils, an observation that indicates that sulfate deposition is associated with much of the leaching of cations from soils (Johnson and Lindberg, 1992; Johnson et al., 1985; Richter et al., 1983). Table 8.3 summarizes anion distributions of the five southern pine stands whose nitrogen dynamics were previously examined. Sulfate balances a large fraction of the cations in A horizons and a widely variable fraction of cations in subsoils, depending on deposition rates and on the ability of subsoils to adsorb sulfate and remove it from solution.

Table 8.3. Concentration of Sulfate and Sum of Cations (H, Na, Ca, Mg, and K) in Soil Solutions of A and B Horizons in Five Southern Pine Stands

Horizon	Calhoun Exp. For., SC (*P. taeda*)	Grant Exp. For., GA (*P. taeda*)	Duke Exp. For., NC (*P. taeda*)	ORNL, TN (*P. taeda*)	U. Florida Exp. For. (*P. elliottii*)
Sulfate Concentration ($\mu mol\,l^{-1}$)					
A	72	73	174	194	136
B	97	16	312	294	156
B/C	16	—	79	—	—
Total Measured Cations Concentration ($\mu mol\,l^{-1}$)					
A	197	377	279	433	552
B	157	367	376	413	376
B/C	98	—	173	—	—

Adapted from Markewitz, unpublished data, and Johnson and Lindberg, 1992.

Table 8.4. Wet-Only Precipitation Fluxes of Ions to 27 NADP (National Atmospheric Deposition Program) Collection Stations Located Within the Southern Pine-Occupied Southeastern United States

Statistic	Ca	Mg	K	NH_4–N	SO_4–S	NO_3–N	H
				kg ha^{-1} yr^{-1}			
Southeast United States							
Mean	1.29	0.42	0.38	1.58	5.20	2.14	0.25
Coef. var. %	37.4	50.8	21.9	28.8	23.6	20.7	39.7
East of Mississippi River							
Mean	1.02	0.44	0.37	1.37	5.59	2.15	0.30
Coef. var.%	23.8	58.0	26.7	26.7	21.6	22.9	28.7
West of Mississippi River							
Mean	1.71	0.40	0.38	1.88	4.63	2.13	0.18
Coef. var. %	26.7	35.7	13.7	21.9	22.9	18.0	42.3
Probability of East vs. West Difference	0.002	0.26	0.43	0.002	0.02	0.44	0.007

Data accumulated only from stations with greater than five years of quality-controlled analysis (Luther Smith, personal communication). Years 1979 to 1992.
[a] Student's t test with unequal variances, n = 16 for east of Mississippi River, 11 for west.

Base-Cation Deposition

Increasingly it is clear that the deposition of acids and bases has important effects on the nutrient cations of a soil (Table 8.4). The atmosphere is an important source of nutrient cations, especially Ca and Mg, to soils of the southeast United States. Across the Southeast, Ca, Mg, and K deposited as wet-only precipitation amount to 1.29, 0.42, and 0.38 kg ha^{-1}, respectively, as estimated by the National Atmospheric Deposition Program (NADP) precipitation chemistry monitoring system (Luther Smith, personal communication). Geographic variation in deposition appears relatively great across the region, with wet-only Ca deposition averaging about 70% greater in the somewhat drier region west of the Mississippi River as compared to the East (Table 8.4). This is presumably due to higher Ca-enriched soil particles entrained in the atmosphere to the west of the Mississippi River.

The total deposition of nutrient cations from the atmosphere is actually much higher than is indicated by estimates of cations dissolved in precipitation (wet-only precipitation). For nutrient cations, deposition of coarse particles can be especially substantial. Table 8.5 summarizes total atmospheric deposition of nutrient cations compared with wet-only estimates for four loblolly pine plantations that were part of the IFS, or Intergrated Forest Study (Johnson and Lindberg, 1992). These studies included special measurements of dry deposition rates for several years. At the

Table 8.5. Total Cation Deposition Estimates of Four Southern Pine Forest Stands (Johnson and Lindberg, 1992) in Relation to Wet-Only Precipitation Collections

Cation Deposition	Calhoun Exp. For., SC (*P. taeda*)	Duke Forest, NC (*P. taeda*)	Grant Exp. For., Univ. GA (*P. taeda*)	Oak Ridge Nat. Lab, TN (*P. taeda*)	Univ. Florida Exp. For. (*P. elliottii*)	NADP Southeast Mean (n = 27)
Calcium						
Wet-only precipitation	72	202	45	91	243	64.5 (37.4)[a]
Total deposition	—	344	100	284	402	—
Magnesium						
Wet-only precipitation	28	38	23	26	131	34.7 (50.8)
Total deposition	—	75	41	53	205	—
Potassium						
Wet-only precipitation	12	—	13	7	12	9.7 (21.9)
Total deposition	—	—	33	20	26	—

Units: $mol_c\,ha^{-1}\,yr^{-1}$.
[a] Coefficient of variation (% in parentheses) among the southeastern NADP wet-only collectors.

four southern pine stands, total deposition of Ca ranged from 1.7- to 3-fold greater than wet-only deposition, as estimated by measurements described in Johnson and Lindberg (1992). Annual rates ranged from 100 to 402 $mol_c\,ha^{-1}$ (mole charge per hectare). Similar patterns for Mg were observed at these stands, with total deposition for Mg exceeding wet-only deposition by 1.5- to 2.0-fold. Annual rates of total atmospheric deposition of Mg ranged from 41 to 205 $mol_c\,ha^{-1}$, with the higher rates at sites near the marine coastline. Total deposition of K exceeded wet-only deposition by 2.2- to 2.9-fold. Annual rates of total K deposition ranged from 20 to 33 $mol_c\,ha^{-1}$. As will be indicated later in this chapter, these rates of cation deposition are biologically significant as judged by rates of forest nutrient uptake and soil leaching.

In North America and Europe, the deposition of nutrient cations have been decreasing over a decade or more (Driscoll et al., 1989; Hedin et al., 1994). In northwestern Europe and the northeastern and southeastern United States, base-cation deposition has decreased since the late 1970s baed on measurements by wet-only deposition collections. These negative time trends are statistically significant and are apparently associated with pollution controls over fly-ash emissions and reduced wind-erosion events in recent years. Between 1979 and 1990, the decreased deposition of base cations (Ca, Mg, K, and Na, corrected for sea salts) averaged a rate of 7.4 $mol_c\,ha^{-1}\,yr^{-1}$ (mole charge per hectare per year) at eight NADP wetonly precipitation-gaging stations in the Southeast (Hedin et al., 1994). This rate of decrease closely parallels the decrease in sulfate in the same wet-only samples, so the deposition of wet-only acidity has remained approximately the same during the period in which nutrient cation input has decreased substantially. Unfortunately, there are no time-trend estimates for the total deposition of cations (wet plus dry deposition) over this period, although the total deposits are expected to have decreased substantially as well.

Short-Term Changes in Soil Leaching

In the short term, acid atmospheric deposition affects the distribution of cationic species in the soil solution and accelerates soil cation leaching. Acid atmospheric deposition increases ionic strength of soil solution by up to several hundreds of $\mu mol_c\,l^{-1}$ in the southeastern United States (Johnson and Lindberg, 1992), an increase that is sufficient to deprease the pH of soil solutions and alter the distribution of cations in solution (Dai, 1994; Reuss, 1983; Richter et al., 1988, 1992). In the laboratory, it is readily demonstrated that increases in ionic strength of several hundreds of μmol/l reduces soil solution pH by several tenths of a unit via the salt effect (Richter et al., 1988) and increases the proportions of polyvalent cations in solution, such as Ca, Mg, and Al (Dai, 1994; Richter et al., 1992).

Such increases in cation activity can be described by the activity ratio law (Bohn et al., 1985; Dai, 1994; Reuss and Johnson, 1986; Richter et al., 1992). Figure 8.1 illustrates how cation activity ratios may be relatively constant over at least a 15-fold range in solution ionic strength (Dai, 1994; Richter et al., 1992). There are, however, remarkably few laboratory or field data to evaluate theoretical calculations or model simulations of soil solution chemistry at low ionic stength, especially with variable-charged soils (Bohn et al., 1985; Richter et al., 1988). Nonetheless, laboratory and theoretical data (Dai, 1994; Reuss, 1983, Richter et al.,

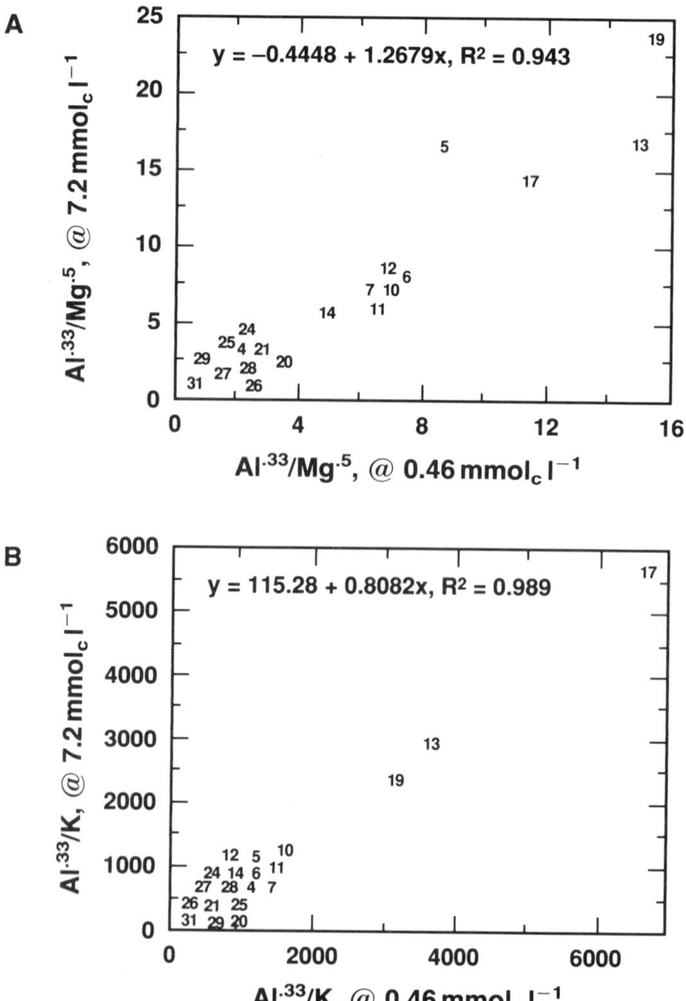

Figure 8.1. Schofield's activity ratios of (a) Al/Mg exchange and (b) Al/K exchange at 0.46 and 7.2 $mmol_c l^{-1}$ $CaCl_2$ (Johnson and Lindberg, 1992).

1992) suggest that as ionic strength increases in very dilute soil solutions in contact with acidic soils, solution activities of trivalent cations increase by the cube, and solution activities of divalent cations increase by the square of the solution activity of monovalent cations.

In the short term, soil exchangeable aluminum may thus enter the soil solution of extremely acid soils that are leached by acidic deposition (Reuss, 1983). The exchange properties of the soil will determine the adsorption energy with which aluminum is bound to exchange sites; but at a relatively high aluminum saturation of the CEC, exchangeable aluminum will be displaced and enter soil solutions. Richter et al. (1992) and Dai (1994) estimated that increases in soil solution electrolytes of several hundreds of $\mu mol_c l^{-1}$ might bring exchangeable aluminum into solution when KCl-exchangeable aluminum saturated about 70% or more of the soil's ECEC (effective cation exchange capacity, or the CEC as determined at soil pH; Richter, 1986). Although these studies indicated much soil-to-soil variation, in many extremely acid soils (KCl-acidity saturation of >85%) aluminum was displaced into solution with relatively small increases in soil solution ionic strength. One the basis of model simulation, Binkely et al. (1989) estimated that Al would be displaced into soil solution only in soils with >80% acid saturation.

The extent to which such elevated aluminum activity in soil solution affects toxicity to roots of pine forests in not well appreciated. Compared to many plant species, however, roots of loblolly pine are tolerant of relatively high concentrations of aluminum (Raynal et al., 1990).

Long-Term Changes as a Consequence of Leaching

Over the long term, the most serious potential problems resulting from cumulative and relatively high deposition of atmospheric acids are likely to be the gradual displacement of exchangeable cations into soil solution, the leaching loss of cations into drainage waters, and the depletion of exchange sites of nutrient cations. Especially in low CEC Ultisols, Inceptisols, and Spodosols, soil-exchangeable nutrient cations may gradually be depleted because of the scarcity of primary minerals to buffer acidification and because of low atmospheric inputs.

In soils subject to solutions with elevated ionic strength from acidic deposition, polyvalent cations such as calcium and magnesium have much greater probability of being exchanged into solution compared to monovalent cations such as potassium (Dai, 1994; Richter et al., 1992). Under such conditions, soil leaching appears most likely to deplete low CEC soils of calcium and magnesium rather than of exchangeable potassium. These dynamics are supported by theory, laboratory experimentation, and, most recently, by a long-term field study that helps verify the fundamental differences in the exchange and cycling behavior of polyvalent and monovalent cations (Richter et al., 1994).

Soil Sulfate Adsorption

In many soils in the southeastern United States, cation leaching is limited by the soil's ability to adsorb sulfate. Adsorption of sulfate reduces the ionic strength of soil solutions and specifically the leaching of polyvalent nutrient cations from soils. Clay mineral suites in many southeastern soils are dominated by kaolinite and associated with gibbsite, illite, vermiculite, and chlorite intergradational clays. Free iron and aluminum hydroxides (sesquioxides) are prominent in the B horizons of many soils. Such surfaces often have the potential to remove sulfate from solution by several adsorption reaction mechanisms.

The surfaces of sesquioxide-rich materials possess a variable surface charge that is dependent on the pH and ionic strength of solution and that often has a strong ability to adsorb anions such as sulfate (Harrison and Johnson, 1992; Johnson and Todd, 1983) and phosphate. Elevated ionic strength and depressed pH thus promote anion adsorption in variable charged soils, which can greatly reduce ion leaching (Harrison and Johnson, 1992; Richter et al., 1995). Both laboratory and field studies support the significance of sulfate retention in many soils in the southeastern US (Harrison and Johnson, 1992; Hodges and Johnson, 1987; Johnson and Henderson, 1979; Johnson and Lindberg, 1992; Richter et al., 1983, 1995). Watershed input–output studies also indicate that retention of deposited sulfate varies widely among soils in the southeastern region (Cosby et al., 1985; Johnson and Henderson, 1979; Johnson and Lindberg, 1992; Johnson and Todd, 1987; Richter, 1980; Swank et al., 1984) and that some watersheds retain more than half the sulfate deposited from the atmosphere. Sulfate may also be incorporated into soil organic matter (Swank et al., 1984), although much less is known about the biological significance of this soil process (Fitzsgerald and Autry, 1992).

Soil Nutrient Cation Depletion

Accelerated cation leaching from soils with low weathering potential is thus a special ecosystem problem for many southeastern soils (Binkley et al., 1989; Johnson and Lindberg, 1992; Turner et al., 1986). Many soils in the southeastern United States—but especially acidic Ultisols, Inceptisols, and Spodosols—are highly weathered, are inherited from materials with low weatherable minerals, or are composed of materials that have gone through several cycles of weathering, geological erosion, and deposition. Much of the Piedmont is pedogenically aged, and many soils of the Atlantic Coastal Plain are composed of materials previously weathered and derived from geological erosion of the Peidmont and the Appalachian Mountains. The contents of exchangeable nutrient cations and weatherable minerals in such soils are often low. On the other hand, to help satisfy nutrient requirements under these soil conditions, forests may be rooted in a large soil depth and volume (Stone and Kalisz, 1991) are rely on recirculated nutrient cations that are in relatively short supply.

The extent to which biogeochemical recycling (Switzer and Nelson, 1972) successfully counters leaching depends on the balance of leaching, root uptake, recirculation, and the CEC of soils. McCracken et al. (1989) concluded that Piedmont Ultisols (similar to Cecil series) without agricultural fertilization may be extremely infertile throughout the upper meter of the soil profile, despite the hardwood forest that takes up relatively large contents of nutrient cations and recirculates them to O horizons. Soils under native hardwood stands examined by McCracken et al. (1989) were acid throughout the profile because biological immobilization and leaching outpaced weathering resupply and nutrient recirculation.

Two approaches are used to evaluate potential effects of acidic deposition on depletion rates of exchangeable nutrient cations in the forested soils of the southeastern United States. The first and by far the best approach is direct observations of soil depletion and acidification through time. The second approach is to compare rates of the major processes that remove nutrient cations from soils: pine-forest harvesting, sulfate leaching, and carbonic acid leaching. We can gain a useful perspective about the potential problem of soil cation depletion from a better knowledge of the rates of these removal processes.

Direct Observations of Soil Change and Depletion

The ability of forest soils to sustain their supply of nutrient cations to a growing forest is controlled by a complex of biogeochemical processes. Nonetheless, the rates and magnitudes of forest-soil chemical change are poorly quantified, and forest-soil data are notably absent that describe sustained nutrient supply or nutrient depletion in specific ecosystems over decades. One well-documented case of soil acidification and depletion of exchangeable cations is located in the southeastern United States and is the result of a remarkable ecosystem sampling program initiated by C.G. Wells in 1962 (Binkley et al., 1989; Richter et al., 1994, 1995). The soil is a low-CEC Piedmont Ultisol in the Calhoun Experimental Forest of South Carolina that has been resampled on six occasions since 1962 with eight permanent plots arrayed in a rigorous statistical design. The explicit purpose is to quantify how low-CEC soils with low or marginal weathering resupply rates are able to sustain their supply of nutrients to a rapidly growing pine forest.

During 28 years of forest growth (1962 to 1990), increases in soil acidity were relatively rapid (Fig. 8.2), and the acidification has continued through the most recent sampling date in 1990 (Richter et al., 1994). In conjunction with increased soil acidity, exchangeable nutrient cations have decreased markedly throughout the three decades of forest development (Table

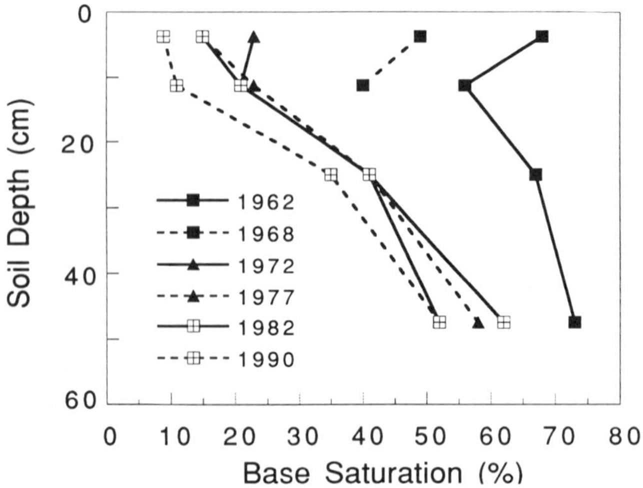

Figure 8.2. Three-decade changes in soil base saturation, effective cation exchange capacity (ECEC [= CEC at ambient soil pH]), pH in dilute $CaCl_2$, and exchangeable KCl acidity at four soil depths at the Calhoun Experimental Forest, South Carolina.

8.6). The observed soil acidification at the Calhoun Forest is estimated to amount to 1.59 (kilomole charge per hectare per year) $kmol_c\,ha^{-1}\,yr^{-1}$ from 1962 and 1990 and is attributed to base-cation accumulation in biomass and forest floor and to soil leaching loss. Perhaps the most

Table 8.6. Concentrations of Soil Exchangeable Cations in 1962 and 1990 in the Calhoun Experimental Forest in South Carolina

Soil Layer (m)	Soil Exchangeable			
	Ca	Mg	K	KCl-acidity
1962 Concentrations ($cmol_c\,kg^{-1}$)				
0–0.075	0.500	0.098	0.056	0.31
0.075–0.15	0.298	0.068	0.042	0.33
0.15–0.35	0.366	0.105	0.050	0.27
0.35–0.60	1.750	0.520	0.127	0.89
1990 Concentrations ($cmol_c\,kg^{-1}$)				
0–0.075	0.060****	0.023****	0.030***	1.13****
0.075–0.15	0.047****	0.018****	0.024***	0.74****
0.15–0.35	0.227†	0.072*	0.049	0.66****
0.35–0.60	1.024***	0.315**	0.110	1.34*

Adapted from Richter et al., 1994a.
*, †, **, ***, **** Significant at the 0.05, 0.02, 0.01 0.001, 0.0001 probability levels, respectively, for obtaining a greater *F* for the statistical contrast between 1962 and 1990.

fundamentally important observation at the Calhoun Forest is that cation removals greatly outpaced the resupply of nutrient cations to surface mineral soils via mineral weathering, atmospheric deposition, and recycling. The acidification patterns in Fig. 8.2 illustrated how base saturation in the upper 15 cm of soil decreased from 65% to about 10% in three decades. It is significant, therefore, that acid deposition inputs in many forests of the southeastern United States and elsewhere are higher than that at the Calhoun (Johnson and Lindberg, 1992; Lindberg et al., 1986). Many sites in North America and Europe have inputs up to and exceeding 1 $kmol_c\,ha^{-1}\,yr^{-1}$. When these high rates of acid input are to rapidly growing forests supporting soils with properties such as those of the Calhoun Ultisol (with low CEC and weathering releases), depletion of exchangeable base cations may be the rule rather than the exception.

Soil acidification during this long-term study was attributed to the accumulation of nutrient cations in biomass and forest floor (55%) and to leaching loss (45%). Two-year collections of soil solutions (1992 to 1994) indicated that sulfate balanced between 36 and 62% of the total cations in soil solutions in the A and B horizons, respectively (Table 8.3). Although sulfur cycling and sulfate leaching are important to forest nutrient dynamics in the absence of acidic deposition, we conclude that a maximum of 28% of the observed soil acidification (from the product of 0.45 and 0.62) was attributable to acid deposition and sulfate leaching from 60 cm.

The Calhoun Forest soil is clearly readjusting to a condition of preagricultural acidity characteristic of an unfertilized Ultisol. Liming and fertilization of the Calhoun soil prior to forest planting in 1957 substantially increased base saturation to a level above which the system could sustain without refertilization. In fact, the Calhoun soils are readjusting toward the soil conditions common to unfertilized, acid soils in the region that remain under hardwood cover and that have not been agriculturally fertilized (McCracken et al., 1989).

Comparisons of Cation Removal Rates by Leaching and Harvest

Although the Calhoun Experimental Forest data are significant, they are only from one forest site. A more regional evaluation of soil acidification and cation depletion was based on data from sites that are dispersed across the region. Although these data lack the direct observations of soil acidification that are demonstrated in the Calhoun Experimental Forest (e.g., Fig. 8.2), comparison of the cation removal processes of sulfate leaching, bicarbonate leaching, and forest harvesting gives a good perspective of the relative role of acid deposition in contributing acidity to the southern pine forest.

Soil leaching of cations varies widely among soils in the Southeast (Tables 8.3 and 8.7), but in many soils sulfate appears to balance a relatively large fraction of the cations in soil solutions (Table 8.3). In fact,

Table 8.7. Nutrient Cation Removals in Two Harvest Intensities Compared with Soil Leaching Rates of Sulfate and Bicarbonate Anions Within B Horizons

Southern Pine Forest	Bole Harvest	Whole-Tree Harvest	Sulfate Leaching	Bicarbonate Leaching	Reference[a]
		$kmol\ ha^{-1}\ yr^{-1}$			
Tennessee and South Carolina	—	0.28 to 0.46	0.28 to 1.21	0.08 to 1.01	1
Grant Exp. Forest, GA	1.09	1.29	0.038	0.41	2
Duke Forest, NC	0.257	0.534	0.77	0.32	2
Univ. Florida Exp. Forest, FL	0.336	0.456	0.58	NA	2
Oak Ridge National Lab, TN	0.194	0.270	1.32	0.037	2
Calhoun Exp. For., SC	0.420	0.540	0.78	0.28	3
Mississippi State Univ., MS	—	0.45	—	—	4
IBP, Saxpahaw, NC	—	1.10	—	—	5
Slash pine, FL	—	0.62	—	—	6
Median	0.336	0.534	0.78	0.30	

[a] References as follows: 1 = Johnson et al. (1988); 2 = Johnson and Lindberg (1992); 3 = Richter et al. (1994a, 1994b); 4 = Switzer and Nelson (1972); 5 = Wells and Jorgensen (1975); 6 = Gholz et al. (1985).

sulfate leaching has been observed to be on the same order of magnitude as that of bicarbonate (Table 8.7), a natural product of soil respiration and carbonic acid leaching (Binkley and Richter, 1987; Johnson and Lindberg, 1992; Richter, 1986; Richter et al., 1983, 1994). Johnson et al. (1988) estimated sulfate leaching from young loblolly pine stands in Tennesses and South Carolina to range from about 0.28 to 1.21 $kmol\ ha^{-1}$, compared to annual bicarbonate leaching of between 0.08 and 1.01 $kmol\ ha^{-1}$. Over a two-year period in eastern Tennessee, atmospheric deposition was estimated to effect a cation leaching loss from soils of between about 0.5 to 1.0 $kmol\ ha^{-1}$ per year (Lindberg et al., 1986; Richter, 1986), depending on location; whereas carbonic acid leaching in three Ultisols with low CEC averaged between 0.30 and 0.65 $kmol\ ha^{-1}$ per year. Such results indicate that atmospheric deposition can significantly alter cation leaching and that, over time, soil base saturation may be decreased in forest soils with low CEC and few primary minerals that can resupply nutrient cations lost by leaching and biomass accumulation.

These leaching losses can be compared to cation accumulations by aggrading forests and their removals by harvest operations since both processes remove at least nutrient cations from the soil and result in acidification (Table 8.7). A number of studies indicate that aggrading southern pine stands accumulate nutrient cations in aboveground biomass between 0.26 and 1.1 $kmol\ ha^{-1}$ per year (Table 8.7). These rates of nutrient removal approximate rates of removal from whole-tree harvesting operations. Conventional bolewood removals may amount to 50 to 70%

of those associated with whole-tree harvests. These rates are about the same order of magnitude as either sulfate or bicarbonate leaching (Table 8.7).

Although rates of soil leaching and harvest removals are on the same order of magnitude in soils supporting pine stands, these two removal processes affect individual cations very differently in soil systems (Johnson and Todd, 1987; Richter et al., 1994; Switzer and Nelson, 1972; Wells and Jorgensen, 1975). Potassium and calcium, for example, are taken up by roots to satisfy plant-nutrient requirements in much greater quantities than magnesium. In contrast, calcium and magnesium, as polyvalent cations, are much more susceptible than potassium to accelerated leaching caused by elevated concentrations of solution electrolytes (Dai, 1994; Richter et al., 1992). In low-CEC soils with little weathering potential, soil exchangeable calcium may be susceptible to depletions from a combination of leaching and root uptake and harvest removals, whereas potassium is most specifically affected by harvest removal. These broad generalizations are supported by long-term depletion data collected at the Calhoun Forest (Richter et al., 1994).

Extent of Poorly Buffered Soils in the Southeast

A variety of regional studies have been conducted of soil properties in the southeastern United States and have estimated the extent of soils having properties that may be altered by acidic air pollutants (Binkley et al., 1989; Klopatek et al., 1980; McFee, 1980; Olson et al., 1982). Two types of soil chemical conditions were of most interest in these regional analyses: (1) soils that were poorly buffered and that can acidify over decades to centuries; and (2) extremely acid soils that may exchange Al into soil solutions and drainage waters as a response to acid deposition inputs. The analyses of Turner et al. (1986) and Binkley et al. (1989) are by far the most sophisticated attempts at evaluating regional sensitivity of soil resources to acid deposition.

The soil criteria of Turner et al. (1986) for poorly buffered acidifiable forested soils were soil water pH >4.5, exchangeable nutrient cations <90 mmol/kg, and exchangeable Al saturation >40% of pH 7 CEC. About 59% (15,930,000 ha) of the pine resource in the Southeast is supported by soils with these properties (i.e., they are considered to have A and E horizons with CECs low enough to be susceptible to acidification from accelerated cation leaching). These soils are mainly Hapludults and Paleudults and are located throughout the southeastern region on the Piedmont and Atlantic and Gulf Coastal Plains. The Appling series soils in the Calhoun Experimental Forest of South Carolina, described in Fig. 8.2 and Table 8.6, are examples of such acidifiable soils with low CEC.

The criteria of Turner et al. (1986) for extremely acid soils that may exchange Al to soil solutions in response to additions of atmospheric sulfate are >80% exchangeable Al saturation of pH 7 CEC, soil water pH of <4.5, and soil organic carbon of <5%. These soils are estimated to support about 10% (2,690,000 ha) of pine forests in the southeastern United States and are mainly classified as Paleaquults in soil taxonomy. These soils are primarily located on the Atlantic Coastal Plain of the Carolinas and Georgia (Turner et al., 1986).

The foregoing criteria are broad and are based on relatively small-scale soil mapping of the Southeast but are generally similar to those proposed by others, including Reuss and Johnson (1986), Binkley et al. (1989), and Richter et al. (1992). Perhaps the most important result of these regional analyses lies not in the precision of areal estimates of poorly buffered southeastern soils but rather in the general and clear message that the southeastern region has widespread edaphic problems associated with the sustainability of its soil nutrients.

Conclusions: The Need for Long-Term Forest-Soil Biogeochemical Research

Atmospheric deposition of air pollutants is having cumulative effects on the nutrient status of soils that support the pine forest resource of the southeastern United States. Immobilization of atmospheric ammonium and nitrate is an important process in O and A horizons of many southeastern pine soils. Since availability of nitrogen in southern pine stands is characteristically low, the cumulative additions of atmospheric nitrogen in forms compatible for plant root or microbial uptake are resulting in a slow accretion of ecosystem nitrogen capital. Although this nitrogen accretion is considered to be biologically significant, the rate at which this input of nitrogen will manifest itself in enhanced productivity is unknown.

Sustaining nutrient availability, however, represents a special ecological problem for the region's forest soils, which are dominated by Ultisols, Inceptisols, and Spodosols. Accelerated leaching of soils with low contents of primary minerals may preferentially deplete exchangeable cations due to the balance or relatively high removal rates compared to inputs. The apparent reductions in atmospheric depositions of nutrient cations throughout the last several decades have also reduced the content of exchangeable cations in circulation. Without weathering release from primary minerals, maintaining exchangeable cations in low-CEC soils depends on atmospheric deposition of nutrient cations, sulfate adsorption to retard cation leaching rates, nutrient uptake and recycling of nutrient cations by deep tree roots, and weathering releases of nutrient cations from secondary minerals. Cycling processes that contribute to the renewal or the inputs of nutrient cations deserve special study, including deep root uptake (Stone and Kalisz, 1991), weathering releases from primary and secondary minerals (Johnson and Lindberg, 1992), and long-term atmospheric deposition of base cations (Hedin et al., 1994).

In the southeastern United States, the wood products industries generate products from the forest worth many billions of dollars per year; other valuable commodities and services are affected by the southeastern forest. At the same time, the southern pine forest is supported by a soil resource that is often typified by low CEC, inherent infertility, low soil organic matter due to long-term cultivation, and a history of long-term agricultural disturbance. Especially because of these properties, there continues to be a great need to understand the soil processes that sustain the productivity of southern pine ecosystems.

Like other forest scientists in the past, we strongly recommend that an efficient network of southern pine research sites be organized to evaluate long-term biogeochemical changes in forest and soil systems. A network that is well designed, efficiently operated, and carefully resampled could provide data needed for better understanding how land-use management, natural soil processes, acid deposition, and fertilizers alter the dynamic nutrient chemistry of southeastern soils and forest ecosystems. The Calhoun Experimental Forest demonstrates the feasibility for such long-term forest-soil studies. Considering that much of the forest-soil resource in the southeast United States is in a dynamic state of recovery from long-term agricultural inpacts and that many low-CEC soils are nearly devoid of primary minerals, a network of long-term forest-soil study plots can provide invaluable information about decadal changes in soil resource that supports the productivity and health of these remarkable ecosystems.

References

Alexander M (1977) *Introduction to Soil Microbiology*. John Wiley, New York.

Binkley D, Driscoll CT, Allen HL, Schoenenberger P, McAvoy D (1989) *Acid Deposition and Forest Soils: context and Case Studies in the Southeastern United States*. Springer-Verlag Ecological Studies 77, New York.

Binkley D, Richter DD (1987) Nutrient cycles and H^+ budgets of forest ecosystems. Adv Ecol Res 16:1–51.

Binkley D, Valentine D, Wells C, Valentine U (1989) An empirical analysis of the factors contributing to 20-yr decrease in soil pH in an old-field plantation of loblolly pine. Biochemistry 8:39–54.

Bohn H, McNeal BL, O'Connor GA (1985) *Soil Chemistry*. John Wiley, New York.

Cosby BJ, Hornberger GM, Galloway JN (1985) Modelling the effects of acid deposition: assessment of a lumped parameter model of soil water and streamwater chemistry. Water Resources Research 21:51–63.

Dai KH (1994) *Soil Aluminum Chemistry in Response to Low-Ionic Strength Solutions and Ligands*. Ph.D. dissertation, Duke University, Durham, NC.

Driscoll CT, Likens GE, Hedin LO, Eaton HS, Bormann FH (1989) Changes in the chemistry of surface waters. Environ Sci Technol 23:137–143.

Fitzgerald JW, Autry AR (1992) Organic sulfur dynamics including mineralization and immobilization of various organic fractions. in: Johnson DW, Lindberg SE (eds) *Atmospheric Deposition and Nutrient Cycling in Forest Ecosystems*. Springer-Verlag, New York, pp 118–129.

Gholz HI, Perry CS, Cropper WP, Hendry LC (1985) Litterfall, decomposition, and nitrogen and phosphorus dynamics in a chronosequence of slash pine (*Pinus elliottii*) plantations. For Sci 31:463–478.

Harrison RB, Johnson DW (1992) Inorganic sulfate dynamics. In: Johnson DW, Lindberg SE (eds) *Atmospheric Deposition and Nutrient Cycling in Forest Ecosystems*. Springer-Verlag, New York, pp 104–118.

Hedin LO, Granat L, Linkens GE, Buishand TA, Galloway JN, Butler TJ, Rodhe H (1994) Steep declines in atmospheric base cations in regions of Europe and North America. Nature 367:351–354.

Hodges SC, Johnson GC (1987) Kinetics of sulfate adsorption and desorption by Cecil soil using miscible displacement. Soil Sci Soc Am J 51:323–331.

Johnson DW, Henderson GS (1979) Sulfate adsorption and sulfur fractions in a highly weathered soil under a mixed deciduous forest. Soil Sci 128:34–40.

Johnson DW, Kelly JM, Swank WT, Cole DW, Van Miegroet H, Hornbeck JW, Pierce RS, Van Lear D (1988) The effects of leaching and whole-tree harvest on cation budgets of several forests. J Environ Qual 17:418–424.

Johnson DW, Lindberg SE (1992) *Atmospheric Deposition and Nutrient Cycling in Forest Ecosystems*. Springer-Verlag, New York.

Johnson DW, Richter DD, Lovett GM, Lindberg SE (1985) Effects of atmospheric deposition on potassium, calcium, and magnesium cycling in two deciduous forests. Can J For Res 15:773–782.

Johnson DW, Todd DE (1983) Relationships among iron, aluminum, carbon, and sulfate in a variety of forest soils. Soil Sci Soc Am J 47:792–800.

Johnson DW, Todd DE, (1987) Nutrient export by leaching and whole-tree harvesting in a loblolly pine and mixed oak forest. Plant and Soil 102:99–109.

Klopatek JM, Harris WF, Olson RJ (1980) A regional ecological assessment approach to atmospheric deposition: effects on soil systems. In: Shriner DS, Richmond CR, Lindberg SE (eds) *Atmospheric Sulfur Deposition: Environmental Impact and Health Effects*. Ann Arbor Science Publishers, Ann Arbor, MI, pp 539–553.

Låg (1968) Relationships between the chemical composition of the precipitation and the contents of exchangeable ions in the humus layer of natural soils. Acta Agric Scand 18:148–152.

Lindberg SE, Lovett GM, Richter DD, Johnson DW (1986) Atmospheric deposition and canopy interactions of major ions in a forest. Science 231:141–145.

McCracken RJ, Daniels RB, Fulcher WE (1989) Undisturbed soils, landscapes, and vegetation in a North Carolina Piedmont virgin forest. Soil Sci Soc Am J 53:1146–1152.

McFee W (1980) Sensitivity of soil regions to long-term acid precipitation. In: Shriner DS, Richmond CR, Lindberg SE (eds) *Atmospheric Sulfur Deposition: Environmental Impact and Health Effects*. Ann Arbor Science Publsihers, Ann Arbor, MI, pp 495–505.

Olson RJ, Johnson DW, Shriner DS (1982) *Regional Assessments of Potential Sensitivity of Soils in the Eastern United States to Acid Precipitation*. Oak Ridge National Laboratory Technical Manual 9374, Oak Ridge, TN.

Ralston CW (1978) *The Southern Pinery*. Forests, physiography, and soils. In: Tom Tippin (ed) A symposium on principles of maintaining productivity on prepared sites. Proceedings USFS and Southern Region of the Association of State College and University Forestry Research Organizations, Mississippi State University, Starkville, MS. USFS, New Orleans, LA, pp 6–13.

Raynal D (1990) Sensitivity of tree seedlings to Al: III. Red spruce and loblolly pine. J Environ Qual 19:180–187.

Raynal D, Joslin JD, Thornton FC, Schaedle M, Henderson GS (1990) Sensitivity of tree seedlings to Aluminum: III. Red spruce and loblolly pine. J Environ Qual 19:180–187.

Reuss JO (1983) Implications of the calcium–aluminum exchange system for the effect of acid precipitation on soils. J Environ Qual 12:591–595.

Reuss JO, Johnson DW (1986) *Acid Deposition and the Acidification of Soils*. Springer-Verlag, New York.

Richter DD (1980) *Prescribed Fire: Effects on Forest Nutrient Cycling and Water Quality on the Santee Experimental Forest, South Carolina*. Ph.D. dissertation, Duke University, Durham, NC.

Richter DD (1986) Sources of acidity in some forested Udults. Soil Sci Soc Am J 50:1584–1589.

Richter DD, Comer PJ, King KS, Sawin HS, Wright DW (1988) Effects of low ionic strength solutions on pH of acid forested soils. Soil Sci Soc Am J 52:261–264.

Richter DD, Johnson DW, Dai KH (1992) Cation exchange reactions in acid forested soils: effects of atmospheric pollutant deposition. In: Johnson DW, Lindberg SE (eds) *Atmospheric Deposition and Nutrient Cycling in Forest Ecosystems*. Spring-Verlag, New York, pp 339–358.

Richter DD, Johnson DW, Todd DE (1983) Atmospheric deposition, neutralization, and leaching in two forest ecosystems. J Environ Qual 12:263–270.

Richter DD, Markewitz D, Wells CG, Allen HL, April R, Heine PR, Urrego B (1994) Soil chemical change during three decades in a old-field loblolly pine (*Pinus taeda* L.) ecosystem. Ecology 75:1463–1473.

Richter DD, Markewitz D, Wells CG, Allen HL, Dunscomb J, Harrison K, Heine PR, Stuanes A, Urrego B, Bonani G (1995) Carbon cycling in a loblolly pine forest: Implications for the missing carbon sink and for the concept of soil. In: Kelly JM, McFee WW (eds) Proc 8th N Amer For Soils Conf, Gainesville, FL, pp 233–251. Soil Sci Soc Am Publ, Madisor, WI.

Richter DD, Ralston CW, Harms WR (1982) Prescribed fire: Effects on forest nutrient cycling and water quality. Science.

Soil Survey Staff (1992) *Keys to Soil Taxonomy*. Cornell University, Ithaca, NY.

Stone EL, Kalisz PJ (1991) On the maximum extent of tree roots. For Ecol and Management 46:59–102.

Swank WT, Fitzgerald JW, Ash JT (1984) Microbial transformation of sulfate in forest soils. Science 223:182–184.

Switzer GL, Nelson LE (1972) Nutrient accumulation and cycling in loblolly pine (*Pinus taeda* L.) plantation ecosystems: the first twenty years. Soil Sci Soc Am Proc 36:143–147.

Switzer GL, Shelton MG (1979) Successional development of the forest floor and soil surface on upland sites of the east gulf coastal plain. Ecology 60:1162–1171.

Tabatabai MA (1985) Effect of acid rain on soils. Crit Rev Environ Control 15:65–110.

Trimble SW (1974) *Man-Induced Soil Erosion on the Southern Piedmont 1700–1970*. Soil Water Conservation Society, Ankeny, IA.

Turner RS, Olson RJ, Brandt CC (1986) *Areas Having Soil Characteristics that May Indicate Sensitivity to Acid Deposition Under Alternative Forest Damage Hypotheses*. Oak Ridge National Laboratory Technical Manual 9917, Oak Ridge, TN.

Urrego JB (1993) *Nutrient Accumulation in Biomass and Forest Floor of a 34-Year-Old Loblolly Pine Plantation*. M.S. thesis, North Carolina State University Dept of For, Raleigh.

Wells CG, Jorgensen JR (1975) Nutrient cycling in loblolly pine plantations. In: Bernier B, Winget CH (eds) *Forest Soils and Forest Land Management*. Fourth North American Forest Soils Conference. Laval University Press, Quebec, Canada, pp 137–158.

9. Atmospheric Nitrogen Deposition and the Mycorrhizae of Southern Commercial Forest Trees

Michael L. Cline and Donald H. Marx

Ecosystem responses to increased nitrogen (N) deposition are primarily soil mediated. Concentrations of nitrogen oxides in the ambient atmosphere rarely, if ever, reach levels that cause visible injury to plants (Garner, 1994). Soil borne mycorrhizal fungi, because of their essential role in the growth and development of forest trees, must be considered when studying factors that influence the overall health of forest ecosystems. To understand forest stresses caused by atmospheric deposition of nitrogen, this chapter will consider, both qualitatively and quantitatively, the potentially toxic effects of excess N on mycorrhizae of forest trees.

Toxicity, as defined in this book, occurs when growth-inhibiting factors disrupt an organism's normal functioning, eventually causing its death. In the context of this definition, a substance required for normal growth and functioning (e.g., N) may become toxic at high concentrations. In Fig. 9.1, three cardinal points describe an organism's response to environmental parameters: the minimum, the optimum, and the maximum (Salisbury and Ross, 1978). The curve between the minimum and optimum points illustrates a growth response to an increasing supply of a previously growth-limiting factor. The descending portion of the curve depends on the specific factor being considered. For example, if high temperature is involved, enzyme denaturation may cause rapid growth inhibition.

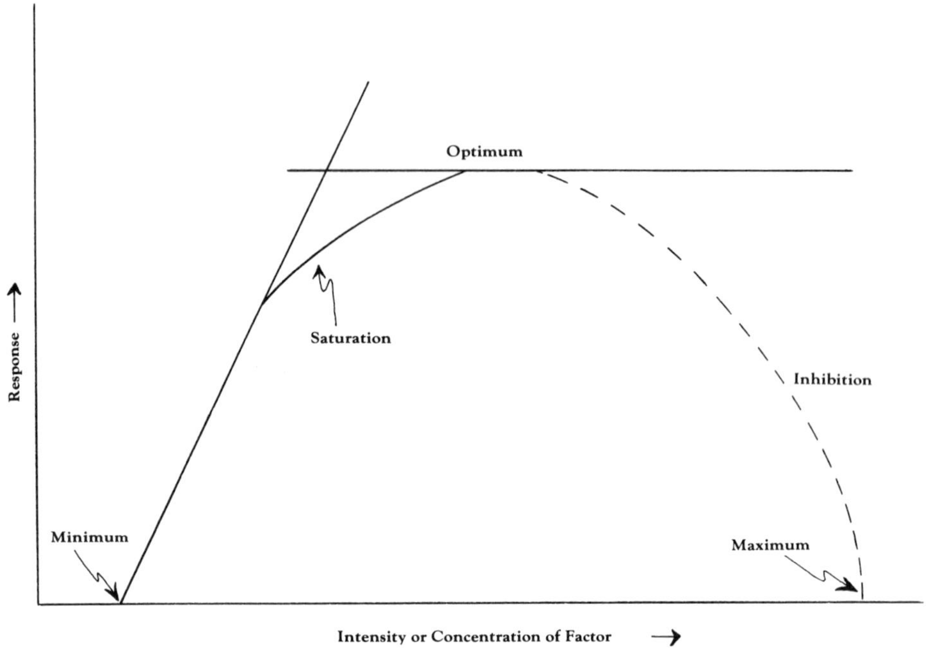

Figure 9.1. A generalized curve showing organism response to an environmental parameter. Minimum, optimum, and maximum are called cardinal points (from Salisbury and Ross, 1978).

Figure 9.2 represents organismic growth responses to mineral nutrient concentration. Applying nutrients to an organism in the deficient zone produces a dramatic growth increase. After the optimal concentration is attained, additional nutrients do not appreciably affect growth. The adequate, or luxury, consumption zone is narrower for micronutrients (e.g., zinc [Zn], copper [Cu], and iron [Fe]) than for macronutrients (e.g., N, phosphorus [P] and potassium [K]) (Epstein, 1972; Salisbury and Ross, 1978). At a certain point, continued increases in mineral nutrient concentrations usually cause reduced plant growth rates and toxicity over time.

Information is scarce relating atmospheric deposition to mycorrhizal development. To date, studies have dealt primarily with acidic deposition's impact on mycorrhizae, limited to general observation of field sites and dose–response studies in greenhouses. Few have studied the potential negative impact of atmospheric N deposition on mycorrhizal development; although results of numerous N fertilization experiments suggest that high soil N inhibits mycorrhizae (Bjorkman, 1949; Kormanik et al., 1977; Marx et al., 1977b; Richards and Wilson, 1963).

These studies and many others were reviewed to determine whether N application can be toxic to mycorrhizae and how experimental N fertilization rates compared to current N deposition and indigenous N levels in soil. The lack of research addressing directly the impacts of atmospherically deposited N on mycorrhizae required a circumstantial approach, drawing on analogies with many N fertilization studies.

Although this chapter is intended to concentrate findings on impacts to the southeastern commercial forest, supporting information is presented for commercial tree species from temperate regions worldwide. The forest N cycle and the role of mycorrhizae are reviewed, an estimate of atmospheric N deposition for the Southeastern U.S. forests is developed, and the impacts of N on mycorrhizal development and function examined. Limitations in interpreting past work due to experimental design differences, nonstandard treatments, and variable assessment techniques are also discussed and improvements recommended.

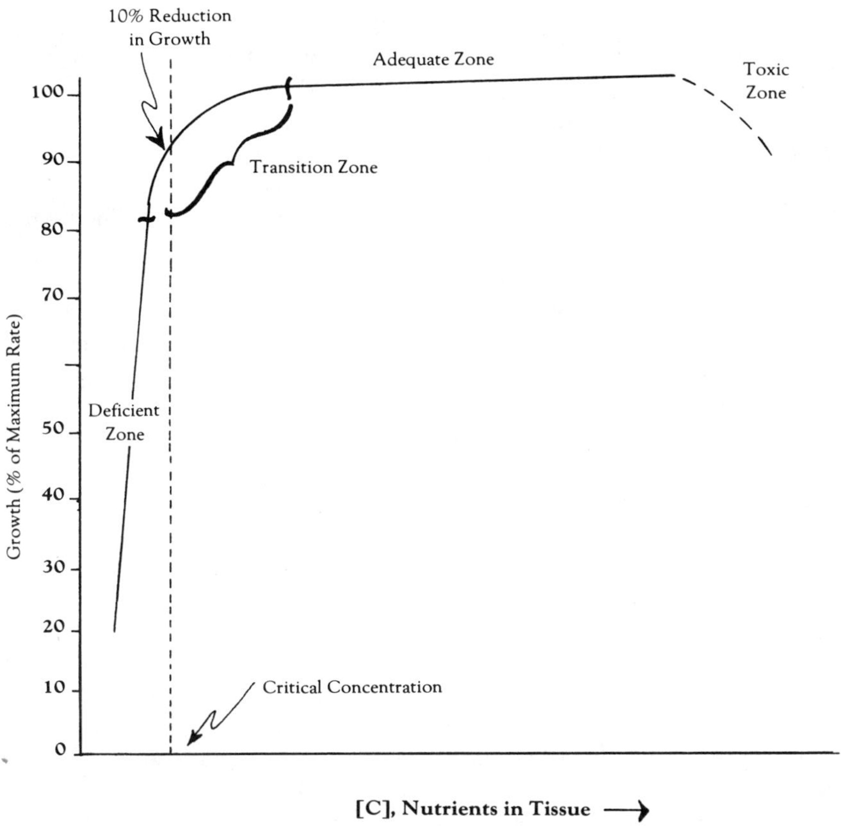

Figure 9.2. Generalized plot of growth as a function of the concentration of a nutrient in plant tissue (from Salisbury and Ross, 1978).

Nitrogen Cycling in Forest Ecosystems

Nitrogen demand in natural ecosystems exceeds the supply of available N; it is the single element that most often limits natural productivity (Stewart et al., 1981). The nitrogen cycle is extremely complex (Fig. 9.3), involving N fixation, nitrification, denitrification, mineralization, immobilization, leaching, and volatilization (Clark, 1981).

Forest systems maintain a tight N cycle in which timber harvesting causes the greatest N loss. Physiological processes of trees change in response to N availability (Gosz, 1981), so if additional N is present, trees increase N requirements and N is cycled rapidly. The microbial population, which varies widely with environmental conditions, is quick to immobilize inorganic N, making it unavailable to vegetation. Therefore, only a portion of deposited N reaches the rhizosphere in a form that forest trees can use.

The N cycle includes dynamic processes that adjust to varying atmospheric and soil conditions. Deposited N moves through the system without necessarily being retained in soil. Undisturbed forests generally maintain a closed N cycle, but both leaching and denitrification rates increase rapidly when N is added (Allen, 1987; Hauck, 1981). In most forests, N additions generally benefit forest productivity as long as nutrients remain balanced. For these reasons, N accumulation in soil should not occur in most southeastern U.S. forests.

Some atmospheric N, both wet and dry, is intercepted by foliage and affects the soil system only indirectly. Other natural inputs of N to soil are biological fixation of N_2 and decomposition of organic matter. These processes provide fluxes of N to the forest system, depending on region, season, and climate. In areas largely unaffected by industrialization, about $2\,kg\,N\,ha^{-1}$ are deposited annually (Lovett et al., 1982). The present amount of resident soil N is only a small fraction of what has been added over the years (E. Stone, personal communication). If $2\,kg\,N\,ha^{-1}\,yr^{-1}$ approximates a historic deposition rate for atmospheric N, maximum deposition rates may have increased in some areas as much as 10- to 15-fold in the past 50 years. Although southeastern U.S. forests seem to incorporate current amounts of deposited N into their nutrient cycles, continuing increases in N emissions may eventually disturb forest ecosystems.

The operational N fertilization level in southeastern U.S. forests is generally 168 to $224\,kg\,N\,ha^{-1}$. Even at this concentration, single applications do not substantially increase the N capital of a site, and forest response is short lived (Allen, 1987). Applying a single element at high concentrations may disrupt the nutrient balance. Without imbalances, annual applications of 50 to $60\,kg\,N\,ha^{-1}$ would improve many, but not all, southern U.S. forest ecosystems (L. Allen, personal communication). Therefore, the current level of N deposition may benefit tree growth in southern forests.

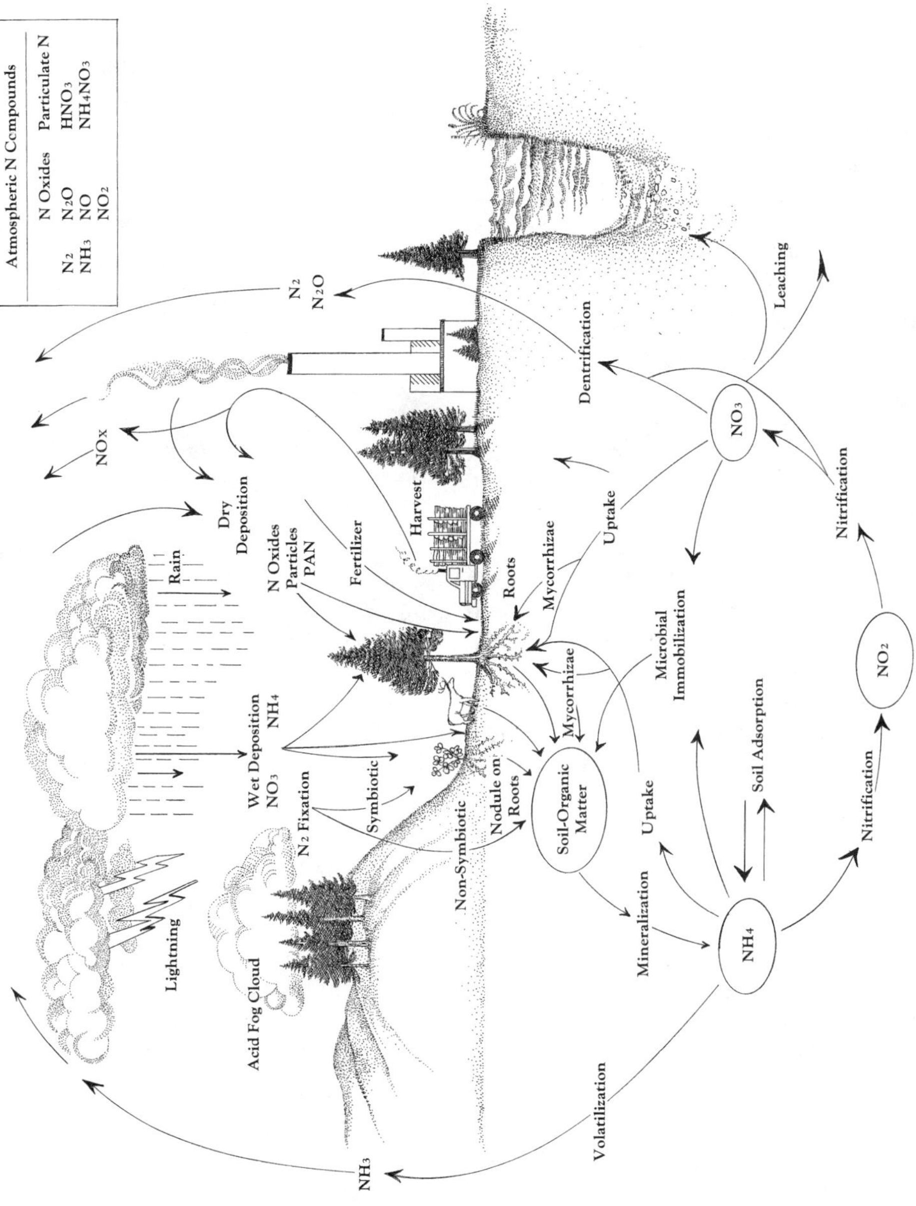

Figure 9.3. The nitrogen cycle in a forest ecosystem.

Nitrogen deposition and N fertilization affect forests differently since the amount of N added and duration of exposure are not comparable. Atmospheric deposition varies with season and amount of precipitation but generally adds consistently low N concentrations over long periods of time. Operational N fertilization of forests usually consists of one high-concentration application of N. Following fertilization, the microbial population captures much of the added N. Depending on precipitation, leaching then removes additional N from the soil. Immobilization and leaching also act on deposited N, but on smaller scales. The N cycle returns to equilibrium even after fertilization, with smaller N additions incorporated into the N cycle more rapidly. One potential benefit of atmospheric N deposition is to increase the ecosystem's ability to support biomass. Greater productivity, however, depends on sufficient supplies of other essential elements as well (e.g., P, K, Mg, and available soil water).

One major concern related to N deposition's effect on mycorrhizae is the potential for N accumulation in soil. In Germany, N saturation of soil is accepted by some as a contributing factor in forest decline (Meyer, 1986), but other scientists are skeptical that N could reach saturation (Fernandez, 1985; Galloway and Dillon, 1983) or even accumulate in forest soils (L. Nelson and G. Switzer, personal communication).

Deposition rates in Europe are reported to be higher than in the southeastern United States Hinrichsen (1986) estimates an average of $30\,kg\,N\,ha^{-1}\,yr^{-1}$. Zottl (1990) cites annual figures varying from $0.8\,kg\,N\,ha^{-1}$ in northern Finland to $35\,kg\,N\,ha^{-1}$ in southern Germany, with a mean in central Europe between 10 and $20\,kg\,N\,ha^{-1}$. According to Meyer (1986), traffic alone emits $40\,kg\,N\,ha^{-1}\,yr^{-1}$; some N is exported, but a substantial amount enters the forest ecosystem. Nitrogen saturation of forest soils is apparently occurring on some sites in Germany (Meyer, 1986). A forest stand naturally accumulates N in biomass as it matures; however, inorganic N is generally thought to become less available (Switzer and Nelson, 1972).

Aber et al. (1989) suggest that the ability of temperate forest ecosystems to accumulate N from atmospheric deposition is limited. They suggest that high-elevation northern forest ecosystems have reached N saturation, in which cumulative N deposition inputs exceed the capacity of these systems to accumulate N based on elevated levels of nitrate leaching in these systems. Schulze (1989) studied spruce decline in German forests and suggested that canopy uptake of atmospheric N, in addition to root uptake, stimulated growth and caused an N:cation imbalance that resulted in decline symptoms.

Role of Mycorrhizae in Nitrogen Cycling

The prevalence of mycorrhizal associations on plant roots throughout the world is so widespread that a nonmycorrhizal plant is a notable exception

(Gerdemann, 1975; Marx and Bryan, 1973). However, mycorrhizae generally have not been included in nutrient-cycling studies of forest ecosystems, probably because their study is labor intensive and researchers often make no functional distinctions between mycorrhizae and roots (Fogel, 1980). Data specific to mycorrhizae are not available on nutrient content and transfers in production, senescence, and decomposition, but studies demonstrate that mycorrhizae contribute to nutrient cycling.

Intense mycorrhizal development is expected in organic horizons in which little nitrogen is available but nitrogen mineralization occurs. Generally, root infection is greater in soils containing low or unbalanced levels of available nutrients (Alexander, 1981). Evidence suggests that ectomycorrhizae efficiently exploit and compete for N. Release of nutrients into the available pool is likely to be periodic; therefore, plants must fully exploit flushes of nutrient inputs (Harley, 1978). In natural conditions, mycorrhizae may intercept 60 to 70% of available nutrients (Vitousek, 1977). Ectomycorrhizal and vesicular arbuscular mycorrhizal (VAM) fungi do not appear to be directly capable of fixing N, but the mycorrhizal symbiosis may stimulate N fixation (Gerdemann, 1975; Voigt, 1971).

Fine root decomposition returns a substantial amount of organic N to the soil (Keeney, 1980). Annual production of fine roots and mycorrhizae may be double the amount of N in annual litterfall (Fogel, 1980). Other studies show mycorrhizae turnover contributing 30 to 50% as much N as litterfall (Adams et al., 1980) and combined loblolly pine fine roots and mycorrhizae turnover adding 25% of the total annual N increment (Ralston, 1978). Variations occur, depending on tree and fungal species and season, but fungal components generally seem to have a great potential for accumulating N (Alexander, 1981).

In a two-year study by Fogel and Hunt (1983), the allocation of biomass and nutrients was measured in various stand components of a young *Pseudotsuga menziesii* (Mirb.) Franco stand. Fine roots and mycorrhizae accounted for over 70% of annual organic matter additions to the soil. Mycorrhizae return of nitrogen was 83 to 87% of total tree return, a figure representing four to five times the return of N by nonmycorrhizal roots. Evidence suggests that a substantial portion of the nutrients present in active mycorrhizae are withdrawn by the tree before senescence of mycorrhizae.

The potential contribution of mycorrhizal fungi to total ecosystem biomass and turnover and to nutrient distribution and turnover was examined in a 23-year-old and a 180-year-old *Abies amabilis* (Dougl.) Forbes stand (Vogt et al., 1982). Mycorrhizae accounted for about 1% of the ecosystem's biomass but contributed 14% of net primary production in the young stand and 15% in the mature stand. Fine roots, including mycorrhizae, cycle at least 60 kg ha^{-1} yr^{-1} of nitrogen in the young stand and 110 kg ha^{-1} yr^{-1} in the mature stand.

Aber et al. (1985) examined sequential changes in root biomass in forest stands with a variety of characteristics in Massachusetts and Wisconsin. They reported patterns of fine root biomass production and turnover related to the quantity and form of N availability. Specifically, total fine root biomas declined with increasing nitrification in deciduous stands, seasonal changes in root biomass were greater in ammonium-dominated stands, and the fine root turnover rate appeared to increase exponentially with increased N availability.

Haines and Best (1976) studied the effect of the VAM fungi, *Glomus mosseae* (Nic. & Gerd.) Gerd. & Trappe, in association with *Liquidambar styraciflua* L. seedlings on N movement through soil profiles. Seedlings were watered with solutions containing ammonium and nitrate and percolates were collected and analyzed. The endomycorrhizal symbiosis reduced loss of both ammonium and nitrate. Nitrate concentration in percolates collected at a depth of 25 cm was always greater than the concentration in the applied solution, so the role of mycorrhizae as N interceptors was particularly important.

Alexander (1981) suggests that mycorrhizal fungi have a competitive advantage over other microorganisms for N uptake because of carbohydrates supplied by their host. He cites a study by Turner (1977), in which carbon, supplied as sucrose and sawdust, applied to a nitrogen-limited *P. menziesii* stand reduced N uptake into the trees by 80%. Apparently, available N was immobilized in microbial tissue. According to Alexander, under conditions of N stress, mycorrhizae, with their extensive hyphal network and host carbon source, will preferentially absorb available ammonium, thus inhibiting the nitrifier population. The presence of mycorrhizal roots also has been shown to reduce the rate of litter decomposition (Gadgil and Gadgil, 1975). Intense competition for N produces a closed system in which N uptake is rapid, little free N exists, and only plants capable of forming mycorrhizal associations can break into the cycle and become established.

In a series of three articles, Abuzinadah and Read (1986) and Abuzinadah et al. (1986a, 1986b) propose that mycorrhizal competition for N includes organic N. The potential proteolytic capabilities of mycorrhizal fungi are vitally important in the N cycle because they provide an alternative sink for organic N. Mycorrhizal use of organic N would bypass the less efficient mineralization pathways that require a separate decomposer population. The transfer of organic N directly to the host, before reimmobilization, would produce a much tighter N cycle.

Atmospheric Nitrogen Deposition

NO_x emissions are produced during industrial combustion processes, in which high temperatures oxidize atmospheric nitrogen, and by the oxidation of N compounds during fossil fuel combustion, especially coal and petroleum. Sources emitting NO_x include power plants, industrial boilers, industrial processes, commercial and residential heaters, natural gas pipelines, highway vehicles, and off-highway diesel engines. The current NO_x emission rate over the United States is between 18.2 and 20.7 Teragrams (Tg) N yr^{-1} from anthropogenic sources (U.S. Environmental Protection Agency, 1993).

Much of the information on N deposition is limited to nitrate and ammonium deposition in rainfall. Dry deposition of nitrate may constitute a significant portion of atmospheric N inputs and should be included in ecosystem nitrogen cycling studies (Lovett and Lindberg, 1986). Devices and methods for measuring dry deposition need improvement; little quantitative data exist for point sources, and there is even less regional data. According to Husar and Holloway (1983), only about 20% of the anthropogenic NO_x emissions in the eastern United States is accounted for by wet deposition. The remainder is evidently deposited in dry form or exported from the continent. Fay et al. (1987) estimated that 29% of the N in NO_x emissions in the eastern United States is deposited in wet form, 58% is deposited in dry form within the region, and about 13% flows outside the continent. Lovett (1992) summarized research data for a number of forested sites in North America and Norway and concluded that annual dry deposition of N was approximately equal to N deposited in precipitation.

Annual wetfall deposition of nitrate and ammonium in each of the 13 southern states was estimated by the Atlas project of the National Vegetation Survey (Marx et al., 1987). The data were obtained mainly from the National Atmospheric Deposition Program (NADP), with additional data from the Multistate Power Production Pollution Study (MAP3S) and the Utility Acid Precipitation Study Program (UAPSP). The periods of data availability, sampling frequencies, and numbers of sampling stations for the three networks are shown in Table 9.1.

Wetfall deposition and rainfall recorded in the 13 southern states are presented in Table 9.2. The range of N deposition throughout the southern region varies considerably. Total N deposition is generally highest in the northern part (Kentucky, Tennessee, Arkansas, North

Table 9.1. Description of Data From Annual Wetfall Deposition Studies

Year	Network	Sampling Frequency	#Southeast Stations
1980–1984	NADP	Weekly	45
1978–1983	UAPSP	Event	7
1978–1982	MAP3S	Event	1

Table 9.2. Annual Wetfall Concentrations and Deposition of Nitrogen and Rainfall During 1978 Through 1984 for Each of the Thirteen Southern States of the United States

State		$\bar{X}$ Yearly Rainfall (mm)	Concentration (mg/l)			Deposition (kg/ha)		
			NO_3	NH_4	Total N	NO_3	NH_4	Total N
Alabama	$\bar{X}$	1416	0.672	0.170	0.284	9.300	2.430	3.992
	high	1716	0.806	0.178	0.321	10.044	2.934	4.553
	low	1178	0.589	0.160	0.265	8.370	2.178	3.586
Arkansas	$\bar{X}$	1308	0.959	0.273	0.429	10.471	3.034	4.727
	high	1759	1.122	0.410	0.563	14.942	4.068	6.177
	low	966	0.756	0.158	0.294	8.246	2.196	3.628
Florida	$\bar{X}$	1307	0.558	0.118	0.218	6.668	1.235	2.468
	high	1617	0.880	0.225	0.374	8.680	1.800	3.040
	low	855	0.366	0.052	0.130	3.410	0.486	1.415
Georgia	$\bar{X}$	1167	0.754	0.187	0.316	8.046	2.054	3.416
	high	1390	1.184	0.304	0.420	10.602	4.320	4.916
	low	894	0.527	0.108	0.203	6.138	1.260	2.368
Kentucky	$\bar{X}$	1075	1.255	0.253	0.481	12.094	2.783	5.091
	high	1207	1.445	0.434	0.664	15.376	4.554	6.976
	low	934	1.035	0.157	0.356	8.494	1.242	2.886
Louisiana	$\bar{X}$	1515	0.871	0.256	0.396	11.108	3.216	5.013
	high	1705	1.023	0.362	0.445	12.338	4.716	6.275
	low	1240	0.719	0.203	0.338	8.184	2.322	3.656
Mississippi	$\bar{X}$	1538	0.738	0.183	0.309	9.848	2.595	4.245
	high	2013	0.868	0.223	0.333	13.082	4.482	6.444
	low	984	0558	0.128	0.226	6.572	1.692	2.802
North Carolina	$\bar{X}$	1187	1.017	0.218	0.399	10.449	2.246	4.101
	high	2004	1.506	0.520	0.703	14.756	4.266	6.416
	low	719	0.614	0.106	0.231	6.696	1.133	2.206
Oklahoma	$\bar{X}$	921	1.277	0.308	0.528	8.949	2.208	3.740
	high	1623	1.451	0.353	0.602	12.462	3.024	5.169
	low	457	0.942	0.227	0.389	6.634	1.620	2.760
South Carolina	$\bar{X}$	1149	1.194	0.240	0.457	11.358	2.624	4.329
	high	1387	1.680	0.392	0.685	15.190	3.132	5.870
	low	674	1.029	0.166	0.389	8.680	1.818	3.558
Tennessee	$\bar{X}$	1282	0.986	0.214	0.389	11.532	2.624	4.653
	high	1526	1.370	0.443	0.532	14.384	4.554	5.856
	low	936	0.738	0.140	0.276	8.556	1.728	3.740
Texas	$\bar{X}$	842	0.883	0.268	0.408	6.115	1.967	2.912
	high	1406	1.364	0.515	0.619	8.990	4.806	5.449
	low	241	0.620	0.187	0.294	1.860	0.738	1.009
Virginia	$\bar{X}$	1146	1.124	0.214	0.421	11.935	2.390	4.557
	high	1755	1.705	0.301	0.615	18.848	3.420	6.430
	low	796	0.620	0.103	0.230	4.278	0.720	1.527
All South	$\bar{X}$	1189	0.939	0.220	0.383	9.660	2.289	3.961
	high	2013	1.705	0.520	0.703	18.848	4.806	6.975
	low	241	0.366	0.052	0.130	1.860	0.486	1.009

Carolina, and Virginia) of the southern region and lowest in the southern part (Georgia, Florida, Alabama, and Texas). Average total N deposition ranges from about 2.5 to 5.1 $kg\,ha^{-1}\,yr^{-1}$; NO_3 represents nearly 75% of this amount. Total N wetfall concentrations varied from a low of 0.130 $mg\,l^{-1}$ to a high of 0.703 $mg\,l^{-1}$. The states furthest from the industrialized North (i.e., Florida, Georgia, Alabama, Mississippi, Texas, and Louisiana) recorded the lowest concentrations of NO_3, NH_4, and total N. Four of these states (Alabama, Florida, Louisiana, and Mississippi) also had rainfall higher than the regional average.

Average annual N concentrations and depositions for the entire South, shown by year, are listed in Table 9.3. Annual concentration of total N varies from 0.367 to 0.421 $mg\,l^{-1}$, and total N deposition ranges from 3.436 to 4.228 $kg\,ha^{-1}$. Deposition appears to be related to rainfall in that the wetter years (e.g., 1983 with 1340 mm of rainfall) had the highest deposition of total N and the driest year (i.e., 1981 with 942 mm of rainfall) had the lowest deposition. Alhough total N deposition is lower in drier years, N concentration in the rain is high due to washout.

Other deposition studies verify these deposition level ranges in the Southeast. Brezonik et al. (1983) sampled bulk and wet-only precipitation in Florida and reported that average annual total N deposited ranged from 3 to 13 $kg\,ha^{-1}\,yr^{-1}$, with a statewide mean of 7.2 $kg\,ha^{-1}\,yr^{-1}$. The National Atmospheric Deposition Program (1988) collected mean annual wet N deposition data in nine of the 13 southern states. Deposition rates ranged from 4 to 8 $kg\,N\,ha^{-1}$, with the highest levels reported for Georgia, North Carolina, and Tennessee. Lovett and Lindberg (1986) estimated wet and dry deposition of N and reported 10.1 $kg\,N\,ha^{-1}\,yr^{-1}$ total deposition at their eastern Tennessee oak–hickory study site. On specific high-elevation sites in New England, N deposition levels have reached 44 $kg\,N\,ha^{-1}\,yr^{-1}$ (Lovett et al., 1982); however, in the Southeast, deposition rates that high are not common.

Husar (1986) and Fay et al. (1987) estimate that wet deposition contributes 20 and 29%, respectively, of total N deposition. Assuming that wet deposition accounts for 20% and adjusting the highest recorded

Table 9.3. Annual Average Nitrogen Concentration/Deposition and Rainfall from 1980 Through 1984 Averaged for All Stations in the Southern United States

Year	No. Sites	$\bar{X}$ Yearly Rainfall (mm)	Concentration (mg/l)			Deposition (kg/ha)		
			NO_3	NH_4	Total N	NO_3	NH_4	Total N
1980	10	1058	1.115	0.200	0.408	10.242	1.859	3.761
1981	20	942	1.075	0.237	0.428	8.655	1.918	3.436
1982	28	1227	0.905	0.208	0.367	10.097	2.298	4.070
1983	29	1340	0.865	0.238	0.380	9.621	2.638	4.228
1984	38	1212	0.896	0.211	0.367	9.742	2.324	4.006

N deposition level from the Atlas project, a maximum working deposition level for the Southeast is 35 kg N $ha^{-1}yr^{-1}$. Since emissions increase over time, this level surpasses maximum amounts deposited in earlier years. When considering the effect of N deposition on mycorrhizae, 35 kg N ha^{-1} yr^{-1} will be used as the maximum deposition level.

The Role of Mycorrhizae in Nitrogen Nutrition of Trees

Hacskaylo (1958) thoroughly reviews early work on the role of mycorrhizae in the host's mineral nutrition. The study of mycorrhizae and nutrition received much attention after Melin (1921) devised techniques allowing controlled studies in aseptic conditions. Kramer and Wilber (1949) pioneered the use of radioactive isotopes in conjunction with mycorrhizae, using P32 to study uptake of excised mycorrhizal and nonmycorrhizal pine roots. Most past research has been concentrated in the nutrient uptake phase, with less emphasis on the use of absorbed nutrients (Bowen, 1973). The relationship between vigorous tree growth and abundant mycorrhizal development has not been demonstrated convincingly (Bowen, 1973), but mycorrhizal contributions to nutrient uptake are well documented (Gerdemann, 1975; Melin and Nilsson, 1952; Mosse, 1973; Rygiewicz et al., 1984).

The sheath of the ectomycorrhizae may be expensive for the host, requiring carbohydrates that might otherwise go to timber production. Odum and Biever (1984) estimated that 15 to 25% of net primary production flows to ectomycorrhizae of pines on stressed sites. Harley (1978) and Meyer (1984) conclude that there must be some compensating selective advantages for the trees to support the carbon expense of mycorrhizae. Enhanced absorption of nutrients is commonly accepted as a mycorrhizal function, but other possible benefits are still matters of speculation. If absorption of nutrients was the only function involved, the fungal sheath or closely grouped strands of mycelia would suffer from the same disadvantages in absorbing nutrients from solution as do closely grouped root hairs (Nye, 1966). This would be relevant only to the absorption rate factor, not other factors related to distribution of absorbing surface area in soil. Mycorrhizal plants usually contain higher quantities of essential nutrients than nonmycorrhizal plants, but the presence of nutrients does not prove greater efficiency in uptake. If increased growth accompanied increased nutrient concentration, this pattern would confirm that mycorrhizae were directly responsible for increased uptake (Gerdemann, 1975).

Harley (1978) suggested several mechanisms by which mycorrhizae may benefit a host plant. Hyphae branch and may grow to cross nutrient depletion zones, with less expense per unit of absorbing area than would be required for root growth. Flushes of nutrients following seasonal changes in soil and leaf fall may be exploited more readily by mycorrhizal roots. Khan (1972) suggests that mycorrhizae may produce growth-promoting substances such as cytokinin, which results in a larger root system that can exploit more soil. Khan also suggests that salt absorption depends on metabolic processes, which occur at a greater rate in mycorrhizae (Khan, 1972).

Mycorrhizae can ameliorate deficiencies of other nutrients, thus increasing N absorption (Bowen and Smith, 1981). A striking feature of mycorrhizae is their sustained absorbing ability in a fixed location; this longevity allows them to exploit larger volumes of soil than more transient uninfected roots (Bowen, 1973). Bowen also suggests that mycorrhizal plants may simply use nutrients more efficiently. Characteristics important in efficient nutrient uptake are the absorbing capacity for ions and water and the abundance, distribution, and effective radius of mycorrhizae. Mycorrhizal fungi have many roles, and efficiency or effectiveness of a species in one area may not mean high performance in all roles.

Voigt (1971) proposes that increased solubility of minerals in the presence of mycorrhizae may result from the higher metabolic activity of mycorrhizal roots over nonmycorrhizal roots and may involve byproducts of fungal metabolism. The weathering mechanism has not been explained fully, but evidence exists that mycorrhizal plants mobilize nutrient ions from soil minerals. In infertile soils, the nutrient capital exists primarily as soil mineral components, so mycorrhizae's ability to dissolve silicates, in addition to a greater absorbing surface, would be beneficial.

Numerous studies have dealt with the role of ectomycorrhizae in nutrient uptake and of vesicular–arbuscular mycorrhizae in phosphorus uptake, but little literature exists on interactions between N and vesicular–arbuscular mycorrhizae. Ho and Trappe (1975) found that *G. mosseae* and *Glomus macrocarpus* (Tul. & Tul.) are capable of reducing nitrate to nitrite; therefore, presumably they possess nitrate reductase. The ability to use nitrates enhances the symbiotic effectiveness of the host–fungus association in terms of N assimilation and translocation. Ames et al. (1983) studied hyphal uptake and transport of N by *G. mosseae* mycorrhizal with celery (*Apium graveolens* L.). Nitrogen was added organically as ground sorghum leaves and inorganically as ammonium sulfate. Mycorrhizal celery plants derived significantly more radioactive N (15_N) from the two sources of N than did nonmycorrhizal plants. Ammonium was absorbed faster than organic N, apparently because the organic N had to be mineralized before it became available.

In 1973, Bowen suggested that less emphasis should be placed on the importance of N mineralization for higher plant nutrition. Abuzinadah and Read (1986) broke new ground studying N nutrition of mycorrhizal fungi by supplying protein in the form of bovine serum albumen (BSA) as the nitrogen source. The ability of mycorrhizae to use protein would be of particular significance for mycorrhizal symbiosis, because in addition to obtaining N from the organic polymer, the assimilation of the carbon skeleton would reduce the drain on the host photosynthate. *Pinus contorta* Dougl. in ectomycorrhizal association with *P. involutus*, *Rhizopogon roseolus* Fries or *Suillus bovinus* (Fries) O. Kuntze readily obtained nitrogen from protein, but this was not the case in association with *Pisolithus tinctorius* (Abuzinadah et al., 1986a). Previous experiments with more complex N sources have usually used humic compounds, but the fact that mor–humus naturally accumulates in thick horizons demonstrates its inherent resistance to microbial attack. Mycorrhizal roots proliferate above the humified layer, where they have access to proteins derived from plant, animal, and microbial tissue in the decomposition horizon.

Abuzinadah and Read, (1986) expanded the previous experiment to include spruce and birch and found that with protein as the N source, birch tissue N content was unusually high (5% N). Mycorrhizal seedlings used 50% of the nitrogen supplied. Dependence on mycorrhizae for N was related to seed size, which controls the N reserve. Smaller seeded tree species were more dependent.

A number of ecologically significant ectomycorrhizal fungi can use peptides and proteins as N sources, and the difference in proteolytic capability between fungi is likely to be important in their distribution in time and space (Abuzinadah and Read, 1986). Access to protein nitrogen is particularly important for tree species in arctic–alpine, boreal, and temperate zones, where long-term or seasonal climatic factors inhibit N mineralization and cause organic matter to accumulate. Possible use of proteins, however, should not be overlooked in southeastern United States commercial forests.

Extrapolating results of nitrogen uptake experiments into broad generalizations would be inadvisable. Host and fungal species variation, host age, soil properties, and root system structure all influence the functioning of the symbiosis. Harley (1978) expressed concern about the use of seedlings in nutrition experiments. Comparing nonmycorrhizal to mycorrhizal seedlings can be problematic because uninfected rootlets of seedlings grow rapidly and exploit the soil widely, whereas mycorrhizae are short, branched, and slow growing and exploit the substrate intensively. The proportion of mycorrhizal to uninfected roots is generally large and stable in a mature tree but can be small and variable in a seedling. Harley also noted that seedling root systems depend more directly on current photosynthesis for carbon; thus ambient conditions influence root development in seedlings more than in mature trees.

Caution should also be used in predicting ion uptake from soil, using results from solution uptake studies (Bowen, 1973). Nitrate diffuses through the soil at a rate an order of magnitude higher than ammonium, so conducting experiments with ammonium in solution does not adequately represent a field situation. Mycorrhizal uptake of ammonium ions is aided by increased life span of the root and hyphal growth into the soil. In soil, distances between roots and movement of ions from soil to roots must be considered since the supply of ions, rather than the absorbing capacity of the mycorrhizae, may limit N uptake. Despite these problems and the need for further research on organic nitrogenous compounds, there is ample evidence that mycorrhizal uptake and transfer of various N forms significantly influences N cycling in forest stands.

Nitrogen Effects on Mycorrhizae

The effect of N application on mycorrhizal infection has been studied extensively to produce vigorous mycorrhizal seedlings of acceptable size for reforestation use (Brown et al., 1981; Castellano et al., 1985; Molina and Chamard, 1983; Ruehle and Marx, 1977; Schultz et al., 1981) and to explore reasons for inconsistent response to forest fertilization (Gill and Lavender, 1983; Menge et al., 1977; Tetreault et al., 1978). Pollution's effect on mycorrhizae and its potential role in forest decline have been examined more recently (Bell, 1986; Gulden and Hoiland, 1985; Meyer, 1985). Experimental designs for these studies vary greatly, rendering comparison difficult—the form, amount, and means of applying N fertilizer as well as the number of applications and the expression of solution N concentration (e.g., $mg\,l^{-1}$, ppm, $kg\,ha^{-1}$) all differ among experiments.

Other variables include the experiment length, fungal species, host species, growth medium, environment (e.g., greenhouse, chamber, field), and method of expressing mycorrhizal infection. Most experiments involve ectomycorrhizal fungi and seedlings in a controlled environment. After adding N, researchers report that mycorrhizal infection declines, remains stable, increases, or temporarily declines and then recovers. Given the variety of experiments and the individuality of each host–fungus interaction, diversity of results is to be expected.

Studies of Nitrogen Effects on Ectomycorrhizae in Controlled Environments

In an early study of N effects on mycorrhizae, Bjorkman (1940) grew pine and spruce seedlings in different soils and exposed them to varying light conditions and N fertilization treatments. Pots were watered with ammonium nitrate solutions at concentrations of 1, 3, 9, 27, and 81 g to 7 liters of water. In a rich mull with added nitrate, the available N content rose over 1000 mg per liter of soil. Following N fertilization, mycorrhizal development decreased but was not eradicated. In the unfertilized but nutrient-rich soils, mycorrhizal infection reached high percentages and was well developed.

In 1959, Hacskaylo and Snow studied nutritional influences on mycorrhizal frequency and on photoperiodic responses in *Pinus virginiana* Mill., *P. taeda*, and *P. strobus*. Seedlings were grown outdoors in pots for five months and treated with solutions containing 310 or 930 ppm N. Nutrient solution additions reduced the prevalence of mycorrhizae. Mycorrhizal development was best on unfertilized seedlings grown in full sunlight with natural daylengths.

Richards and Wilson (1963) reported changes in mycorrhizal development of *Pinus caribaea* Morelet seedlings that received 0 or 336 lb/acre of ammonium nitrate. In the greenhouse study, increased soil nitrate levels decreased mycorrhizal infection. In a later greenhouse study, Richards (1965) investigated the relative influence of soil pH and nitrate supply on mycorrhizal development of *P. taeda*. Seedlings were raised in surface soil collected from a young *P. taeda* plantation and treated either with lime or one of four sodium nitrate concentrations (0, 20.7, 41, or 83 lb/acre). As N concentration increased, mycorrhizal development temporarily declined (22 weeks after sowing); but at 45 weeks, no significant differences were apparent and plants' N contents were similar in all treatment groups. Richards hypothesized that root carbohydrate–N balance predicts mycorrhizal development.

Dumbroff (1968) observed an inverse relationship between the supply of N and P and the intensity of mycorrhizae development. However, he noted that some mycorrhizae persisted even with the best nutritional conditions. *Pinus clausa* (Chapm.) Vasey and *P. elliottii* seedlings received 210 ppm N in one experiment and 112 kg N ha^{-1} in another. Ectomycorrhizal fungi were not artifically introduced, but the seedlings consistently became infected naturally.

Theodorou and Bowen (1969) studied effects of pH and high nitrate on mycorrhizal fungal growth in a laboratory medium and in the rhizosphere of seedlings. They observed decreases in fungal growth at the highest sodium nitrate levels and with equivalent sodium concentrations in the absence of nitrate. However, Richards (1965) found that sodium nitrate, but not sodium carbonate, reduced both yield and mycorrhizae, and he concluded that the nitrate ion caused the reductions. In another experiment, Theodorou and Bowen (1969) inhibited fungal colonization by applying 115 ppm nitrate. They concluded that high soil nitrate at acid pH inhibits the mycorrhizal infection process, not rhizosphere colonization.

Marx et al. (1977b) studied N fertilization–mycorrhizae interactions using 10 nutrient solutions combinations applied to *P. taeda* seedlings: complete low fertility (10 ppm each of N, P, K, and Ca), complete high fertility (300 ppm N, 75 ppm P, 100 ppm K, and 80 ppm Ca), and complete treatments minus each element. After growing in the aforementioned conditions, seedlings were inoculated with *P. tinctorius*, transplanted to a rooting medium of moderate soil fertility, and assessed for mycorrhizal development after 20 days. Nonmycorrhizal short roots from other seedlings grown under each of the 10 fertility regimes were assayed for soluble carbohydrates. Less sucrose was found in short roots from the high N and P than lower fertility regimes. Significantly more mycorrhizae formed on seedlings receiving the low-fertility regime than those receiving the high regime. High N and P levels in soil decreased sucrose content in short roots and were correlated with decreased susceptibility of roots to colonization by *P. tinctorius*.

In the past decade, this area of research has received much attention. High-fertility treatments have reduced ectomycorrhizal infection for the following host–fungus combinations: *Quercus rubra* L. with *P. tinctorius* (Beckjord et al., 1985; Pope and Chaney, 1985); *P. taeda* with *P. tinctorius* (Ruehle and Wells, 1984; Torbert et al., 1986); *Quercus velutina* Lam. and *Quercus alba* L. with *P. tinctorius* (Beckjord et al., 1985; Dixon et al., 1981, 1985); *P. contorta* with *P. tinctorius* or *S. granulatus* (Ekwebelam and Reid, 1983); *Pinus rigida* Mill. and *P. virginiana* with *P. tinctorius* (Maronek et al., 1982); *P. sylvestris* with *S. bovinus* and *Paxillus involutus* (Batsch.: Fr.) Fr. (Termorshuizen and Ket, 1991); *Picea abies* and *Lactarius rufus* (Scop.: Fr.) Fr. (Jentschke et al., 1991); and *P. menziesii* and *L. rufus* (Gorissen, 1991). Experimental conditions of these studies are outlined in Table 9.4.

Dixon et al. (1981) found that foliar-fertilized, inoculated *Q. velutina* seedlings consistently produced more mycorrhizal short roots, and a greater percentage of mycorrhizal lateral roots than soil-fertilized, inoculated seedlings. Fructose content of short roots was significantly related to mycorrhizal development. Other factors affecting mycorrhizal infection include light intensity (Ekwebelam and Reid, 1983), form of N applied (Beckjord et al., 1983; Rudawska, 1986), and time of fertilization (Beckjord et al., 1980). High light intensity enhances photosynthesis, providing more carbohydrates to the roots and increasing mycorrhizal infection (Ekwebelam and Reid, 1983). Both Rudawska (1986) and Beckjord et al. (1983) found that nitrate reduced mycorrhizae more than ammonium. Rudawska observed that sucrose levels in seedling roots were altered by the form and concentration of applied N and were significantly correlated with mycorrhizal development. Beckjord et al. (1980) fertilized seedlings either 15 or 40 days after planting. The percentage of mycorrhizae declined with increased N concentrations when seedlings were fertilized at 15 days but rose with increased N concentration when seedlings were fertilized at 40 days.

Maronek et al. (1982) and Crowley et al. (1981) studied the influence of slow-release fertilizers on ectomycorrhizae. In both studies, higher fertilizer rates proportionately reduced plant growth, percent of mycorrhizal development, and the number of mycorrhizal seedlings. In Maronek's study, soluble fertilizer applied during each watering at a rate of 200 ppm N eliminated mycorrhizae.

Alexander and Fairley (1986) studied growth and N uptake rates per unit of root and mycorrhizae for *P. sitchensis* seedlings. Mycorrhizal plants were produced by incorporating humus from a mature *P. sitchensis*

Table 9.4. Experimental Conditions for Controlled Environment Studies of Nitrogen Effects on Ectomycorrhizae of Seedlings

Inoculum	Host	Nitrogen Source	Application Rate	Application Frequency	Mycorrhizal Sampling Schedule	Mycorrhizal Response	Reference
vegetative	*Pinus sylvestris* *Picea excelsa*	NH_4NO_3	0, 51, 157, 471, 1414, 4229 ppm	applied at watering	end of 1 summer and 2 summers	decreased at high N but not eliminated	Bjorkman, 1940
vegetative	*P. virginiana* *P. taeda* *P. strobus*	NO_3	310, 930 ppm	once a week	5 months	reduced at 310 ppm eliminated at 930 ppm	Hacskaylo and Snow, 1959
vegetative	*P. caribea*	NH_4NO_3	0, 150 ppm	one treatment	5 months	reduced at 150 ppm	Richards and Wilson, 1963
vegetative	*P. taeda*	$NaNO_3$	0, 8.3, 16.5, 33 ppm N	one treatment	21 weeks; 45 weeks	reduced to 2.5% at 33 ppm N; no response (at 45 wks.)	Richards, 1965
(natural)	*P. clausa*	—	0, 25, 44.6 ppm N	one treatment	8 months	reduced with increased N	Dumbroff, 1968
P. tinctorius vegetative	*P. taeda*	NH_4NO_3	0, 10, 300 ppm N	3 times	16 weeks	decreased at 300 ppm N	Marx et al., 1977b
P. tinctorius vegetative	*Q. rubra*	NH_4NO_3	0, 100, 200 ppm N	every 37 days	17 weeks	reduced at 200 ppm	Pope and Chaney, 1985
P. tinctorius *S. auranteum* spores	*Q. alba* *Q. rubra*	NH_4NO_3	1, 10, 100, 200, 400 ppm N	3 times/week	100 days	reduced at 100 ppm N	Beckjord et al., 1985

P. tinctorius vegetative	*P. taeda*	NH_4NO_3	25, 100, 400 ppm N or 100, 250, 500 ppm N	once a week once a month	22 weeks	reduced at 400 and 500 ppm	Ruehle and Wells, 1984
P. tinctorius spores	*P. taeda*	KNO_4 NH_4SO_4	.25, .50, 1, 2, 4 mg N	every 2 weeks	20 weeks	reduced from 30% at low N to 10% at highest N	Torbert et al., 1986
P. tinctorius vegetative	*Q. velutina*	NH_4NO_3	0, 750, 1, 150 ppm N	once a week to medium or twice a week to foliage	18 weeks	lower with medium-fert. than foliar fert.	Dixon et al., 1981
P. tinctorius *S. granulatus*	*P. contorta*	NH_4NO_3	3, 62, 248 ppm N	daily	6 weeks after inoculation	decreased at 248 ppm N	Ekwebelam and Reid, 1983
P. tinctorius vegetative	*P. rigida* *P. virginiana*	osmocote or soluble fert.	0, 1.1, 2.3, 4.5 9.0, 18.0 kg/m 200 ppm N	mixed with medium or at each watering	5 months and 8 months	reduced at 9 kg/m^3 osmocote eliminated with soluble fert.	Maronek et al., 1982
unidentified ECM	*P. sylvestris*	NH_4 NO_3 urea	10, 20, 50, 100, 200 ppm 20, 100 ppm 20, 100 ppm	100 ml 2 times/week	6 months	reduced at 100 ppm; nitrate caused greatest reduction	Rudawska, 1986

(continued)

Table 9.4. *Continued*

Inoculum	Host	Nitrogen Source	Application Rate	Application Frequency	Mycorrhizal Sampling Schedule	Mycorrhizal Response	Reference
P. tinctorius vegetative	*Q. rubra*	NH_4 NO_3	0, 13.3, 26.6, 53.2 mg N/seedling	once at 15 days or 40 days	100 days	reduced at 53.2 mg N at early application	Beckjord et al., 1980
Spruce humus	*P. sitchensis*	NH_4SO_2	0, 37.5, 75.0 ppm N	2 times/week	28 weeks	reduced at 75 ppm	Alexander and Fairley, 1986
Spruce humus	*Picea* sp.	NH_4NO_3	60 kg N/ha	2 times (at 6 wks. and 10 wks.)	—	reduced to 3% or eliminated	Meyer, 1985
P. tinctorius vegetative	*Q. rubra*	NH_4NO_3	12.5, 25, 50, 100 ppm N	6 times	20 weeks	greatest at 100 ppm	Ruehle, 1980
P. tinctorius vegetative	*Q. palustris*	14-6-11.6 osmocote	0, .11 g, .45 g/l	mixed with medium	8 weeks 14 weeks 18 weeks	greater at .11 g/l at 8 weeks, but no sig. diff. at 18 wks.	Maronek et al., 1981
L. laccata vegetative	*P. menziesii* *P. ponderosa*	20-19-18 NPK soluble or 18-6-12 osmocote	7, 14, 28 mg/seedling or 3.5, 7.0 g/l	bi-weekly or mixed in medium	5 months	94.5% mycorrhizal roots over all fertility treatments	Molina and Chamard, 1983
R. vinicolor *R. colossus*	*P. menziesii*	20-19-18 NPK soluble or 18-6-12 osmocote	3.5, 9, 7, 18 mg/seedling or 7.1 g/l	bi-weekly or mixed in medium	20 weeks	no response to soluble fert. reduced by osmocote	Castellano et al., 1985

S. bovinus *P. involutus* vegetative	*P. sylvestris*	NH_4SO_4 $NaNO_3$	0, 10, 50, 100, 400 kg N/ha/yr	once a week	26 weeks	reduced above 50 kg N/ha	Termorshuizen and Ket, 1991
(natural)	*P. sylvestris*	NH_4	0, 48, 480 kg N/ha/yr		7 months	reduced at 480 kg N/ha/yr	Van Dijk et al., 1990
L. rufus vegetative	*Picea abies*	NH_4NO_3 $(NH_4)_2SO_4$ NH_4Cl $Ca(NO_3)_2$ $Mg(NO_3)_2$ KNO_3	C, 1.5 mM NH_4, 1.5 mM NO_3, 2.7 mM NH_4, 2.7 mM NO_3	flow system	13 weeks	reduced at 2.7 mM NH_4	Jentschke et al., 1991
(natural)	*P. menziesii*	$(NH_4)_2SO_4$ NH_3 (air)	0, 50, 250, 500, 2500 mM $(NH_4)_2SO_4$/l	weekly	14 weeks	reduced at all NH_4 and NH_3 levels	Van der Eerden et al., 1992
R. vincolor *L. rufus* vegetative	*P. menziesii*	$(NH_4)SO_4$	5, 50, 200 kg N/ha/yr	weekly	28 days	increased at 50 kg N/ha/yr	Gorissen et al., 1991
(natural)	*P. menziesii*	$(NH_4)_2SO_4$	5, 50, 200 kg N/ha/yr	weekly	23 months	reduced at 200 kg N/ha/yr	Gorissen, 1991
P. tinctorius	*P. taeda*	NH_4NO_3 $(NH_4)_2SO_4$	150, 300, 450 kg N/ha/yr		1 growing season	no effect	Marx, unpublished
P. tinctorius	*P. taeda*	NH_4NO_3	150, 300, 450 kg N/ha/yr	10, 15, or 30-day intervals	10 months	increased with increased N	Marx, 1990

plantation in the germination medium. A nutrient solution containing either 0 (N_0), 37.5 (N_1), or 75.0 (N_2) mg N l^{-1} as ammonium sulfate was applied twice each week. At the beginning of the experiment, 42% of the root tips on inoculated seedlings were mycorrhizal. Infection level rose to 83% at 28 weeks for the N_0 application, remained constant for the N_1 treatment, and fell to around 35% at level N_2. At level N_2, there was a marked decrease in root tip number on both mycorrhizal and nonmycorrhizal seedlings. Results also suggested that in the field, mycorrhizae absorb ammonium more efficiently than uninfected roots.

Meyer (1985) studied nitrogen's influence on mycorrhizal infection of spruce seedlings grown in humus from a damaged forest area. In one experiment, humus was mixed with 0, 3, 6, 12, 24, or 48% sawdust by volume. Sawdust was added to increase the C/N ratio in the soil; at 48% sawdust, the C/N ratio reached 400:1. Under these conditions, scarce available N is immediately immobilized by microorganisms. Mycorrhizal frequency rose with increasing sawdust (decreasing available N) to an optimum level of 12% sawdust, at which infection reached 70%. Mycorrhizal response to limited N supply followed a curve often observed in biological processes.

In a second experiment, two applications of 60 kg N ha^{-1} as ammonium nitrate were added to seedlings from each sawdust mixture. Mycorrhizal development was reduced visibly or completely eliminated with N fertilization; the highest infection level (3%) was observed in the 48% sawdust mixture (Meyer, 1985).

The mechanism of nitrogen's effect on mycorrhizae is often attributed in part to root carbohydrate availability (Bjorkman, 1940; Ekwebelam and Reid, 1983; Hacskaylo and Snow, 1959; Marx et al., 1977b; Meyer, 1985; Richards and Wilson, 1963). Meyer (1985) observed that following N fertilization, roots broke through the fungal mantle and grew so rapidly that fungi were unable to infect the new roots.

Melhuish and Hacskaylo (1980) examined the metabolism of *P. tinctorius* and its response to N and C source ratios. Fungal dry weight, total lipid content, and total fatty acid content increased as concentrations of ammonium tartrate increased from 0.1 g l^{-1} to 1.0 g l^{-1} and then dropped off sharply after 5.0 g l^{-1} High levels of available N may suppress mycorrhizae development by modifying fatty-acid content, membrane permeability, and internal pH.

Fertilization treatments from 50 kg upward negatively affected *P. sylvestris* root length and mycorrhizal infection by *S. bovinus* and *P. involutus* (Termorshuizen and Ket, 1991). Mycorrhizae were more inhibited by ammonium than nitrate. The investigators concluded that inhibition of mycorrhizal formation occurs at lower nitrogen levels than does the inhibition of tree growth. The N level at which inhibition occurs depends on availability and form of mineral N, the fungal species, tree species, and the carbohydrate economy of the plant.

Three-year-old *P. menziesii* with naturally occurring mycorrhizae were grown in chambers and fumigated with filtered air and filtered air enriched with ammonium for 14 weeks (Van der Eerden et al., 1992). Exposure levels were high relative to ambient in the experiment, but not high enough to cause visible injury to needles. Seedlings in the ammonia and ammonium treatments were poorly infected with mycorrhizae compared to controls. Mycorrhizal infection was weakly correlated with N content in needles. Results support the notion that infection rate is related to nitrogen metabolism in the tree and not direct changes in the soil.

Van Dijk et al. (1990) reported on studies in which *Pinus sylvestris* seedlings were planted into pots and grown in the greenhouse at three N levels (0, 48, 480 $kg\,ha^{-1}\,yr^{-1}$) for seven months in different soils. Fruiting body production by *Thelephora* spp. was reduced at 48 and totally inhibited at 480 $kg\,ha^{-1}\,yr^{-1}$, and biomass of fine roots and mycorrhizae development were inhibited by the highest N level.

Several experiments demonstrated no mycorrhizal suppression following N applications. Ruehle (1980) suppressed *Q. rubra*/*P. tinctorius* mycorrhizae using a commercial fertilizer formulation (Rapidgro), but at the highest level of ammonium nitrate applied (100 µg/ml) every three weeks, mycorrhizae were abundant. Marx (1990) studied five *P. taeda* families in field microplots inoculated with *P. tinctorius* in soils with varying N and pH levels. Mycorrhizal infection was strongly suppressed by pH values above 6.3; however, increased N generally resulted in increased mycorrhizal development. No N treatment had any significant effect on fungal inoculum survival in the soil. Different inoculation and fertilization techniques and different fungal isolates may explain discrepancies between experiments.

Two-year-old nursery-grown Douglas fir trees were inoculated with *R. vincolor* and *L. rufus* and grown for four months outdoors (Gorissen et al., 1991). Subsequently, they were placed in a phytotron and received ozone and ammonium sulphate treatments. The N treatments were equivalent to 5, 50, and 200 $kg\,N\,ha^{-1}yr^{-1}$. The 50-N treatment stimulated both root growth activity and mycorrhizal frequency.

Maronek et al. (1981) applied a slow-release fertilizer to *Quercus palustris* Muenchh. and found that the best mycorrhizal development occurred at the lower fertilizer rate during the first 14 weeks. After 18 weeks, the researchers found no significant differences in infection between fertilizer treatment groups. Experimenting with a slow-release fertilizer and inoculated seedlings, Molina and Chamard (1983) observed that different nutrient treatments did not significantly alter mycorrhizal formation for *P. menziesii* or *Pinus ponderosa* Laws. seedlings. The latter had an average of 94.5% mycorrhizal short roots over all fertility treatments. In contrast, Castellano et al. (1985) fertilized mycorrhizal *P. menziesii* seedlings with soluble and that slow-release fertilizers and reported that slow-release fertilizer suppressed mycorrhizal formation, but abundant mycorrhizae remained on more than 54% of feeder roots under both high and low soluble fertilizer regimes. The differing results might be attributed to the fungal symbiont, since Molina and Chamard (1983) inoculated with *L. laccata* whereas Castellano et al. (1985) experimented with *R. vinicolor* and *Rhizopogon colossus* Smith.

Danielson et al. (1984) tested soluble fertilizer effects at three levels on *Pinus banksiana* Lamb. seedling growth and studied the mycorrhizae development by four symbionts. The reaction of the four symbionts ranged from suppression at all but the lowest fertilizer level to 95% infection in all treatments, demonstrating the importance of fungal species' influence on N fertilization response.

It is currently recommended (May et al., 1985) that southern pine seedlings in bareroot nurseries receive the minimum equivalent of 1120 $kg\,ha^{-1}$ of 10-10-10 fertilizer plus 336 $kg\,ha^{-1}$ of ammonium nitrate each crop year. This N, totaling 225 $kg\,ha^{-1}$, is applied at seeding in the spring and at several times during the initial three to five months of the seven- to eight-month growing season. In practice, the actual amount of N applied may vary from half to over twice these amounts (Marx et al., 1984). The amount of N actually absorbed and used by the seedlings averages about 50 to 70% of the amount applied. The remainder of the N is leached through the root zone, volatilized, or absorbed by soil microorganisms.

In an investigation of mycorrhizae of tree seedlings in operational bareroot nurseries, correlations were not found between ectomycorrhizal development and varying N applications (i.e., totaling 100 to 450 kg $N\,ha^{-1}$) in 54 discrete tests undertaken on 13 tree species in 33 conventional bareroot nurseries in the United States (Marx et al., 1984).

The various components of the aforementioned experiments are outlined in Table 9.4. Many of the experimental treatments are typical of standard tree nursery procedures, and most were carried through one growing season. When experimental conditions (including host and fungal species) were similar between studies, results were fairly consistent and could be used to guide nursery management. There are problems, however, with translating study results to a mature forest stand exposed to atmospheric N deposition. Soil type, competition between plants, variation among fungal species, competition with microorganisms for available N, physiological differences between seedlings and mature trees, and amount, frequency, and duration of N addition all complicate predictions of N deposition's effect on a mature stand.

Field Studies of Nitrogen Effects on Ectomycorrhizae

Due in part to the aforementioned complications, very few field studies have examined N effects on mycorrhizae. Menge et al. (1977) related ectomycorrhizae frequency and type to both foliar nutrient concentrations and growth of N-fertilized *P. taeda*. Seven fertilizer treatments were applied to two 11-year-old *P. taeda* plantations, with N additions ranging from 0 to 112 kg ha^{-1} in various combinations with P and K. For two years, root samples were removed at two-month intervals during the growing season and assessed for mycorrhizae. Nitrogen fertilization appeared to be responsible for most changes in mycorrhizal tip frequency. In the six months following fertilization, 40% fewer mycorrhizal tips were found in N-fertilized plots than in prefertilization plots. After two years, however, the number of mycorrhizal tips between plots did not differ significantly. Fertilization substantially altered relative populations of the 11 ectomycorrhizae types in the two plantations; types that had adapted well to N-deficient sites were most adversely affected by N fertilization.

Basidiocarp production also has been monitored as a potential indicator of the extent of symbiosis, at least for ectomycorrhizal fungi that fruit regularly (Menge and Grand, 1977). In addition to the treatments previously described (Menge et al., 1977), a 21-year-old *P. taeda* plantation was fertilized several times throughout one growing season with 336 kg N/ha as ammonium nitrate. The largest numbers of basidiocarps were found on unfertilized plots or plots fertilized only with P. The number of living mycorrhizae was not well correlated with epigeous basidiocarp numbers on plots receiving similar treatments.

Termorshuizen and Schaffers (1991) studied carpophore production of ectomycorrhizal fungi in young stands (5 to 10 years old) and old stands (50 to 80 years old) of *P. sylvestris* in response to air pollution over a three-year period. Young stands had higher numbers and more species than old stands. The deposition of all air pollutants, but especially N, was correlated with declines in carpophores.

Tetreault et al. (1978) examined short-term and long-term N fertilization effects on *Abies balsamea* (L.) Mill. ectomycorrhizae and long roots in natural stands. Five- and six-year-old seedlings were fertilized with granular urea in concentrations up to 336 kg N ha^{-1}. Intensifying N fertilization initially reduced the average number of mycorrhizae per seedling, but after three years, no significant difference was apparent in mycorrhizae numbers between fertilized seedlings and controls. Mycorrhizal fungi were not differentiated by species, so the type and efficiency of the mycorrhizae that recolonized after fertilization may have changed.

Studying a 35-year-old *P. sitchensis* plantation, Alexander and Fairley (1983) reported lower mycorrhizal infection on fertilized plots and that sampling time influenced results, with infection markedly reduced in winter. Gill and Lavender (1983) studied urea fertilization in young *T. heterophylla* stands and found that it significantly reduced the total number of mycorrhizae in all but one coastal stand. Mycorrhizae were suppressed through nine months, but the magnitude of difference between mycorrhizal numbers in the control and fertilizer plots diminished steadily as time progressed. Reductions in mycorrhizae were attributed to root mortality rather than to decreased infection of new roots. The overall decrease in total mycorrhizae masked the individual species data; however, in urea fertilization treatments the numbers of some mycorrhizae appeared to increase. Moreover, the change in relative populations of mycorrhizal types persisted longer than the reduction in total mycorrhizae.

In most of these studies, the number of mycorrhizal roots gradually increased over time after N fertilization and eventually matched prefertilization numbers (Table 9.5). Only Gill and Lavender (1983) differentiated between species of mycorrhizal fungi, but other researchers suggested that absolute numbers are deceiving because N fertilization may alter the type and efficiency of mycorrhizae. Another similarity among the experiments involved N fertilizer being applied at high concentrations once or through one growing season. Atmospheric N deposition generally occurs at low concentrations for prolonged periods of time.

Studies of Nitrogen Effects on Endomycorrhizae

Hayman (1982) reviewed the influence of soils and fertility on vesicular–arbuscular mycorrhizae (VAM) activity and survival. To predict effects of N fertilization on VAM populations requires knowledge of soil fertility. In infertile soil, spore production is limited by the small amount of total root infection that accompanies poor plant growth, so percent infection would not adequately describe mycorrhizal frequency. Low soil fertility is not always a prerequisite for extensive mycorrhizal development; the relationship between mycorrhizae and soil fertility at various sites is not consistent. Vesicular–arbuscular mycorrhizal fungi in a nutrient-poor soil may be intolerant of added nutrients, whereas other endophyte species are adapted to fertile soil. Hayman (1982) proposes that in some soils, high fertility can preclude mycorrhizal benefits to the extent that the host–endophyte balance may shift from mutualistic symbiosis to parasitism.

There are few studies that examine N effects on vesicular–arbuscular endomycorrhizae of tree species. Most research on VAM focuses on agricultural crops and P fertilization. Among N fertilization experiments reported, methods and results vary as with the ectomycorrhizae studies.

Kormanik et al. (1977) experimented with the interaction of soil fertility, VAM, and growth and development of *L. styraciflua* seedlings. A 10-10-10 fertilizer was applied once at 140, 280, 560, or 1120 kg/ha, and ammonium nitrate was applied three times throughout the growing season at 560 kg ha^{-1}. All inoculated seedlings were mycorrhizal, and high soil fertility did not alter the VAM beneficial effects for the *L. styraciflua* seedlings. In a similar study but with a higher fertility level (2240 kg N ha^{-1}), VAM infection percentages decreased on seedlings receiving the highest N applications and on those receiving no supplemental N (Brown

Table 9.5. Experimental Conditions for Field Studies of N Fertilization Effects on Ectomycorrhizae

Inoculum	Host	Nitrogen Source	Application Rate	Application Frequency	Mycorrhizal Sampling Schedule	Mycorrhizal Response	Reference
indigenous	*P. taeda* (11-yr-old plantation)	NH_4NO_3	0, 56, 112 kg N/ha	once	bimonthly for 2 years	40% fewer at 6 months; then recovered	Menge et al., 1977
indigenous	*A. balsamea* (natural stand)	granular urea	112, 224, 336 kg N/ha	2–5 times	1 year or 3 years	reduced by 50% at 336 kg N/ha after 1 year, recovered after 3 yrs.	Tetreault et al., 1978
indigenous	*T. heterophylla* (natural stand)	prilled urea	224 kg N/ha	once	bi-monthly for 9 months; once at 18 months	reduced thru 9 months, recovered at 18 mo.	Gill and Lavender, 1983

et al., 1981). Highest infection intensities occurred on seedlings grown at N application levels optimum for tree growth (i.e., 560 kg N ha^{-1}).

Kiernan et al. (1983) also studied fertilizer effects on VAM development of *L. styraciflua* seedlings, comparing development with indigenous stripmine fungi to that of *G. fasciculatus*. When treated with 18-6-12 slow-release fertilizer at the manufacturer's recommended rate (4.5 g N l^{-1} of soil) and at twice that rate, stripmine isolates inhibited plant growth. High fertilization tended to reduce spore production and root colonization of stripmine isolates. *Glomus fasciculatus* usually stimulated early seedling development. Kiernan et al. (1983) recommended that fertilization practices be evaluated for their short-term effects on seedling growth to ensure survival of the ecologically adapted fungal symbiont.

Thapar et al. (1984) established a pot experiment to study the effects of mineral fertilizer and compost on *Agathis robusta* C. Moore, R.M. Bailey (Kauri pine) growth, Endogone spore population, and *G. macrocarpus* development. Fertilizer was applied to deliver 0, 452, 807, 897, 1614, and 3230 kg N ha^{-1}. The highest fertilizer level was toxic to all seedlings. Compost significantly increased plant growth, enhanced VAM development, and increased the number of red-brown laminate spores. Mineral fertilizers significantly decreased plant growth and tended to decrease VAM development. In a study of woody ornamentals, Johnson et al. (1980) reported that increased N application rates reduced mycorrhizal infection. At the highest N fertilization rate of 1250 kg N/ha, infection percent dropped from 58 to 44%. Vesicular–arbuscular mycorrhizae apparently adapt to a wide range of N fertilization regimes.

Schultz et al. (1981) studied N fertilization effects on eight hardwood species in association with *G. mosseae* or *G. etunicatus* (Becker & Gerd). Percentage of VAM infection varied between species from a high of about 80% to a low of about 40%, indicating a difference in host preference for the *Glomus* spp. symbionts. Only *Acer saccharum* Marsh. and *L. styraciflua* showed an infection response to N fertilizer levels—infection increased with higher N fertilizer application. High soil N concentrations did not reduce VAM infection in any treatment for any of the symbionts.

In another study, seedlings of *Liriodendron tulipifera* L. were inoculated with *G. fasciculatus* and fertilized at three levels with 19-6-12 slow-release fertilizer (0, 2, and 4 g N l^{-1} of soil) (Verkade and Hamilton, 1983). Inoculated plants that received no supplemental fertilizer showed limited VAM development. Percentage of infection was greater in fertilized treatments but did not differ significantly between the moderate and high fertilizer treatments. Ponder (1984) studied *Fraxinus americana* L. and *Juglans nigra* L., inoculated with *G. fasciculatus* and fertilized with 1200 ppm N (applied to the soil or foliage). The percentage of mycorrhizae-infected root segments was unaffected by the method of fertilization or fertilizer type, but no account was given of infected root length or infection intensity.

These experiments are outlined in Table 9.6. Different fertilization treatments make direct comparisons difficult. However, it is evident from all the results that the choice of host–fungus association and prefertilization conditions are important determinants of VAM response to fertilization.

Discussion

Although N effects cannot be separated entirely from other factors influencing mycorrhizal infection, researchers have demonstrated that some species of mycorrhizal fungi, in association with certain host trees, respond adversely to significant additions of N fertilizer. In most studies, experiments were conducted in greenhouses with fumigated soil and controlled inoculation of single symbionts. These procedures allow observations of individual host–fungus responses without complications from microbial competition or influence from other mycorrhizal fungi, which may be adapted to different soil conditions. Other factors regulated in some experiments included nutrient balance, soil type, light intensity, and day length.

Most studies were terminated within one growing season (five months was a typical length). Seedling root systems were generally analyzed once, at the termination date. Ectomycorrhizal infection was often assessed visually or expressed as a percentage of mycorrhizae to total root numbers.

Nitrogen treatments are unique to each experiment. Without exception, however, in studies that show a mycorrhizal response, moderate treatment levels (e.g., 60 to 80 kg N ha^{-1}) produce seedlings with good mycorrhizal development, whereas no fertilization or high fertilization (e.g., 200 to 300 kg N ha^{-1}) produce seedlings with poor mycorrhizal development. Mycorrhizal response to fertilization seems to depend largely on host and fungal species and on the host–fungus association. Both ectomycorrhizae and endomycorrhizae apparently adapt to a wide range of fertility regimes.

Researchers conducting fertilization studies must adopt a standardized method of expressing application rates in a biologically meaningful manner that considers the amount and contact time of each element in the root zone. Although it is necessary to report fertilization amounts in conventional terms for reference, some form of dose computation is needed. This might express fertilization as μg N/day (or week or month) for the growing season, or length of test. This information would be even more useful if it were augmented with plant tissue N content measurements. Acceptance and use of a standard procedure of N fertilization application would facilitate valid comparisons of N fertilization studies.

Table 9.6. Experimental Conditions for Controlled Environment Studies of N Fertilization Effects on Endomycorrhizae of Seedlings

Inoculum	Host	Nitrogen Source	Application Rate	Application Frequency	Mycorrhizal Sampling Schedule	Mycorrhizal Response	Reference
Clomus etunicatus spores	*Liquidambar styraciflua*	ammonium sulfate ammonium nitrate potassium nitrate	140, 280, 560, 1, 120, 2, 240 kg N/ha	8 equal treatments	5 months	decreased at 1, 120	Brown et al., 1981
C. etunicatus spores	*L. styraciflua*	ammonium nitrate	560 kg N/ha	3 times	5 months	no response	Kormanik et al., 1977
		10-10-10 fartilizer	140, 280, 560, 1, 120 kg N/ha	one time			
C. fasciculatus spores and root pieces	*L. styraciflus*	18-6-12 osmocote (8–9 month release)	0, 1.1, 2.2, 4.5, 9.0 g/l	mixed with medium	14 weeks	reduced at 9.0 g/l	Kiernan et al., 1983
C. macrocarpus roots of hoop pine	*A. robusta*	ammonium sulfate	0, 452, 807, 897, 1614, 3230 kg N/ha	1 treatment	one growing season	reduced at 67.5 g	Thapar et al., 1984

C. fasciculatus and *C. mosseae*	woody ornamentals	ammonium nitrate potassium nitrate	250, 750, 1, 250 kg N/ha	each month	9 months	reduced by 8–14% at 1, 250	Johnson et al., 1980
C. etunicatus *C. mosseae* sorghum roots	8 hardwood species	ammonium nitrate 10-10-10 fertilizer	168 kg/ha 140, 560, 1, 120 kg/ha	10 times 3 times	5 months	increased at 1, 120 for *A. sacharum* and *L. styraciflua*, otherwise no response	Schultz et al., 1981
C. Fasciculatus spores and root pieces	*L. tulipifera*	19-6-12 osmocote 3–4 month release	0, 2, 4 g N/l	mixed with medium	14 weeks	no response	Verkade and Hamilton, 1983
C. fasciculatus infected roots	*F. americana* *J. nigra*	ammonium nitrate	0.1200 ppm N (total 7.14 g N/ seedling)	2 times/ week	20 weeks	no response	Ponder, 1984

Speculation abounds about how N affects mycorrhizae, but no conclusions have been reached. Theodorou and Bowen (1969) propose that the ectomycorrhizal infection process is inhibited, whereas Gill and Lavender (1983) observed mortality of existing mycorrhizae. The importance of the N/P balance is emphasized (Ekwebelam and Reid, 1983; Richards and Wilson, 1963) along with the indirect effect of pH, which influences host nutrition (Richards, 1965). Saturating the growing medium daily with water decreases the number of seedlings with ectomycorrhizae. Increased temperature apparently stimulates ectomycorrhizal infection (Dixon et al., 1985). Light intensity, and hence the photosynthetic rate, also seems to influence mycorrhizal development (Beckjord et al., 1983; Bjorkman, 1940; Ekwebelam and Reid, 1983; Hacskaylo and Snow, 1959; Marx et al., 1977b). Melhuish and Hacskaylo (1980) suspected that fatty acid content modifications, influenced by N levels, change membrane permeability, which affects ECM development.

Sampling time also appears to influence results. Schultz et al. (1981), following the sequential development pattern of endomycorrhizae from hyphae to arbuscles to vesicles, noted that the percent of infections was complicated by seasonal influences on fungal morphology. Alexander and Fairley (1983) observed pronounced differences depending on sampling time (infection was reduced during winter).

Results of studies that ran from nine months to two years generally showed mycorrhizal infection levels recovering after an initial suppression. When individual fungal species are identified in field studies, different species are found to recolonize following fertilization. This response has serious implications for a natural stand because species that may be adapted to existing soil conditions may be replaced by species that tolerate fertilized soils. Ectomycorrhizal fungi demonstrate species preference for various N forms and may facilitate nutrient uptake with varying degrees of efficiency. Fungal species differentiation and the method of mycorrhizal infection assessment are especially important in determining mycorrhizal responses to N application.

Two Major Factors Limiting the Applicability of Past Research to Atmospheric Deposition of Nitrogen

The ecological niche of mycorrhizal fungi and forest succession are usually overlooked in discussion of discrepancies in research results. Findings that appear contradictory may reflect adaptive ecological differences in fungal species, which make one-to-one comparisons inappropriate.

The ectomycorrhizae development assessment methods used in past studies are accurate for a given set of conditions, but researchers may often mistakenly assume a level of homogeniety in experimental conditions that does not exist. These two factors may serve to confound research results on the influence of mycorrhizae on tree growth, especially plant–nitrogen relations. The following discussion sheds light on the importance of these two considerations to future mycorrhizal research.

Forest Succession and Mycorrhizae

Consideration of factors that influence the growth, development, and function of mycorrhizal fungi and associated roots must include the succession of fungal species that accompanies changes in soil characteristics, forest type, and forest age. As forest succession occurs and new tree species invade a site, mycorrhizal fungal populations appear to change accordingly. The impact of cultural treatments or environmental changes (e.g., N deposition) on mycorrhizal fungi populations in forest stands varies with stand characteristics.

According to Meyer (1973), competition between different trees and fungi on particular sites creates a relatively stable equilibrium, enabling the best-adapted fungi and trees to persist. Humans often influence this balance by harvesting, planting, and employing silvicultural practices that create even-aged forest stands. Although mycorrhizal fungi are widespread in managed forests, quantitative and qualitative fungal population differences have been observed in response to cultural treatments (Meyer, 1973). Romell (1930) observed that sporophores present prior to clearcutting were absent following harvest. However, absence of fruiting bodies does not indicate the absence of fungi, which can remain dormant in soil for many decades (Mikola, 1973) or on roots of minor forest plants. Site preparation techniques that often follow clearcutting (e.g., burning of slash) may destroy fungal symbionts directly or modify soil characteristics (e.g., pH, nutrient relations), which indirectly affects fungi. McAfee and Fortin (1986) compared ectomycorrhizal fungi populations on cleared and burned sites with those in mature jack pine stands. The indigenous and inoculated fungi on jack pine seedlings planted in the disturbed sites infected fewer roots, which researchers attributed to changes in the physical and chemical soil properties.

Aging of a forest stand also alters the population composition of mycorrhizal fungi. Chu–Chou and Grace (1981, 1982) and Chu–Chou (1979) found that nursery seedlings of *Psuedotsuga menziesii* and *Pinus radiata* were infected with *Hebeloma crustuliniforme* (Bull. ex St. Amans) Quelet, *Laccaria laccata* (Scop ex Fr.) Cooke, *Rhizopogon luteolus* Fr., and *Rhizopogon rubescens* Tul.; however, *H. crustuliniforme* disappeared shortly after outplanting in New Zealand. Adams (1951) studied *P. radiata* in South Australia and showed that ectomycorrhizae formed by *Rhizopogon* spp. in the nursery were replaced by *Boletus* spp. following outplanting.

Little is known regarding specific changes in mycorrhizal fungi populations during the life of a forest stand. However, studies suggest that as forest stands age, fungal species change in composition and their diversity increases. Miller (1981) suggests three nutritional stages that occur as a forest stand matures and that are relevant to fungal populations. To colonize tree roots of a young stand, early stage mycorrhizal fungi are ecologically adapted to an open environment, severe competition for nutrients, and the physiology of the host tree. In the second stage, smaller soil temperature fluctuations, accumulating organic matter, and decreased demand from the host tree for nutrients create a more habitable environment. The function and balance of the ectomycorrhizal symbiosis may be different in these two stages of stand development. The third stage is characterized by continued immobilization of N in humus over time and may lead to N deficiencies on N-poor sites, but this third stage is uncommon on intensively managed sites.

Many studies of fungal populations rely on fruiting body production as an indirect measure of indigenous ectomycorrhizal populations. Laiho (1970) and Tominaga (1975) demonstrate a close relationship between fruiting body production and the abundance of different types of ectomycorrhizae. Certain fungi (e.g., imperfect fungi) are not included in this type of survey. An assessment of *Pinus monticola* Dougl. stands in Idaho revealed increased diversity of mycorrhizal fungi with greater stand age (Miller, 1983). Five fungal species were found in 15-year-old stands, 34 to 37 species in 30- to 40-year-old stands, and 98 species in 175- to 215-year-old stands.

In Europe, Mason et al. (1982) studied mycorrhizal fungi population changes in the production of predominantly agaric fruiting bodies associated with outplantings of *Betula pendula* Roth. and *Betula pubescens* Ehrh. Although weather varied appreciably from year to year, average fruiting bodies per tree increased progressively from zero the first year after outplanting to 0.4, 28, and 170 in the second, fourth, and sixth years, respectively. Collectively, *Hebeloma* fruiting bodies increased from 173 in the second year to 8224 fruiting bodies in the sixth year. In year 2, virtually all fruiting bodies were *Hebeloma*, but by year 6 this decreased to 79% of the total. *Betula pubescens* supported only fruiting bodies of *Hebeloma* through year 5.

Introduced ectomycorrhizae of outplanted seedlings disrupt natural populations of ectomycorrhizae fungi associated with early successional forest stands (Last and Flemming, 1985). To avoid this disruption, Last and Flemming planted germinated *Betula* spp. seed directly in the field, to approximate natural conditions, and observed the progression of ectomycorrhizae fungi. *Hebeloma crustuliniforme* and *Laccaria torteles* [(Bolt.) S.F. Gray] Cke. produced fruiting bodies first, followed by *T. terrestris* in year 3, *Inocybe lanuginella* and *L. pubescens* in year 4, *Continarius* sp. and *Leccinum* sp. in year 6, and *Russula* spp. in year 10. Although qualitative differences were evident in the comparison of nursery transplants and directly seeded seedlings, the progression of fungal species over time was similar.

Dighton and Mason (1985) used population behavior terminology to characterize ectomycorrhizal fungi associated with forest stands of various ages. Early stage fungi are similar to r-species, which spend most of their time colonizing, and give way to others in natural succession. Conversely, late-stage fungi are similar to k-species, which specialize in resource-limited environments restricted by intense interference from neighbors. Early stage fungi, which produce lighter fruiting bodies, may be less demanding of host resources than late-stage fungi. However, the greater number of smaller fruiting bodies produced by early stage fungi may partially offset this weight difference (Deacon et al., 1983).

Fleming (1984) suggests that trees regenerating in or near mature woodlands may display different patterns of ectomycorrhizal fungal population development than those discussed previously. He found that *L. pubescens*, a late-stage fungus, developed on *Betula* sp. seedlings in soils permeated with intact roots of older trees infected by *L. pubescens* but did not develop when older tree roots were severed. Thus, it appeared that *L. pubescens* and probably other late-stage fungi depend, to a degree, on host-derived assimilates for sufficient inoculum potential to colonize seedling roots. Although both late- and early stage fungi can infect seedlings under axenic conditions, only early stage fungi could colonize seedlings under nonsterile conditions from spores (Last and Flemming, 1985).

Mycorrhizal fungal populations are also influenced by soil type (Last et al., 1984; Mason et al., 1986). Relating soil factors that influence mycorrhizal development to forest age-related successional patterns is complex. Edaphic factors (e.g., soil temperature, moisture, fertility, texture, reaction) determine, to a great extent, which fungal species are early or late-stage symbionts. In the southeastern United States, both *T. terrestris* and *P. tinctorius* are considered early stage fungi; however, *P. tinctorius* is tolerant of extreme soil conditions (e.g., acidity, temperature) and ecologically adapted to poor sites (Marx et al., 1977a). Conversely, *T. terrestris* is ecologically adapted to the excellent tilth, fertility, and moisture of nursery soil but often fails to adapt to the harsh soil conditions of many outplanting sites (Marx and Artman, 1979).

Abuzinadah and Read (1986) suggest that the changing N status, as soil organics increase, causes population changes in mycorrhizal fungi occurring on roots as stands mature; however, high soil organic matter does not always imply good N status. Grier et al. (1981) demonstrated that mineral soils supporting a young stand of *Abies amabilis* will be progressively transformed to an organic soil as detritus additions increase with time. A shift in the intensive mycorrhizal exploitation zone from mineral to organic soil layers accompanied changes in soil conditions. Fungi with proteolytic capabilities may have a selective advantage on organic versus mineral soils (Abuzinadah and Read, 1986).

In addition to influence from forest stand age and soil conditions, mycorrhizal fungi populations may be influenced by the host tree. Trappe

(1962) studied patterns of ectomycorrhizae host specificity among Pacific Northwest conifers and grouped fungi according to their ability to form mycorrhizae with diverse host trees. Molina and Trappe (1982) formed three groups of fungi based on their host specificity: (1) fungi with wide host potential and low specificity forming sporocarps in association with diverse host trees, (2) fungi with intermediate host potential and specificity with limited sporocarp–host associations, and (3) fungi with narrow host specificity.

Much of the existing information on VAM fungal populations is derived from crop plants. Fungi that form arbuscular, vesicular–arbuscular, or other endomycorrhizae are members of the orders Endogonales or Glomales (Morton and Benny, 1990) and are referred to collectively as VAM fungi in this chapter. The few reports of VAM fungal population changes suggest that less species diversity exists in natural communities than in disturbed or cultivated soils (Nicolson, 1967). Nicolson (1960) studied natural populations of fungi on sand dunes and found that both frequency and extent of VAM infection were related to host plant successional stages. In later stages of succession, VAM infection increased, but it decreased with the closing of the community at the most mature stage.

Barnard (1977) surveyed a 36-year-old *L. styraciflua* plantation in the southern United States and reported three predominant spore types: *G. macrocarpus*, *G. fasciculatus*, and *Glomus* spp. Unlike tree species with ectomycorrhizae associations, VAM-forming tree species appear to lack diversity of fungal symbionts.

Kiernan et al. (1983) sampled abandoned strip mines in Kentucky and identified 38 spore types, which they estimate is comparable to the number in highly productive agricultural soils. The most common were *G. fasciculatus* found on three sites, *G. macrocarpus* var. *geosporus* found on two sites, and *Glomus constrictus* found frequently only on one site. These differences may be ecological adaptations of certain species to particular edaphic conditions or they may be merely random occurrences.

According to available literature, VAM fungi appear to have a broad host range. Mosse (1972) concluded that specificity in VAM fungi may be more a function of interactions between fungal strains and the soil than between the fungus and host plant. Marx and Bryan (1973) maintain that *Endogone (Glomus) mosseae* will form mycorrhizae with cotton, corn, pepper, soybeans, sorghum, and other agricultural crops as well as sycamore, sweetgum, citrus, peach, black locust, and other trees, predominantly angiosperms. Molina et al. (1978) examined the ability of six species of *Festuca* to form mycorrhizae with 11 species of VAM fungi in western U.S. grasslands and reported no signs of host specificity.

Mycorrhizal fungi population dynamics associated with forest developmental stages, host specificity, and edaphic factors are poorly understood. Pioneering work in this area questions the validity of broad generalizations concerning mycorrhizal development and function in different environments. Population shifts are caused by environmental modifications that occur as a stand ages and by changes in the host tree physiology. Mycorrhizal fungi, which are ecologically adapted to site conditions at specific stages of forest stand development, react differently to environmental changes.

Determination of Mycorrhizal Infection

Determining the impact of experimental treatments on mycorrhizal fungal growth and development, as well as the host plant response, requires accurate assessment of root colonization. Consistent assessment techniques applied in individual studies provide relative comparisons, but the lack of an accurate method for absolute quantification of mycorrhizal infection hampers comparisons among different studies. Over the years, researchers have developed numerous methods for determining ectomycorrhizal and VAM infection; however, none have combined accuracy with the simplicity required to handle large root systems and sample sizes. Current assessment techniques also fail to consider extramatrical biomass associated with surrounding soil. Studies that attempt to quantify extramatrical fungal biomass are laborious and have met with only limited success (Pacovsky and Bethlenfalvay, 1982; Sanders and Tinker, 1973).

Quantifying ectomycorrhizal infection is further complicated by several factors: the diverse morphology it assumes on different host plants; the branching patterns of infected roots ranging from simple, monopodial structures to complex, corrolloid ectomycorrhizae; the dichotomous branching of noninfected roots of *Pinus*; and the unavoidable subjectivity of different examiners. Richards and Wilson (1963) maintained that since mycorrhizae function as nutrient-absorbing organs, development should be described in terms of the nutrient uptake process. Hence, if prolific forking increases nutrient-uptake efficiency, the extent of forking should be considered when assessing treatment effects. This concept was originally proposed by Hatch (1937), who attempted several volumetric and colorimetric determinations that proved impractical. Hatch developed a compromise method that considered the number of ectomycorrhizae separately given different amounts of branching, extent of total infection, and number of noninfected root tips. Combining these values, he obtained "mycorrhizal points." His method has been modified over time but remains the basic method for expressing infection.

The degree of infection is typically expressed as a numeric ratio of mycorrhizal short roots to total short roots (both mycorrhizal and nonmycorrhizal) present on the root system. Even this simple method is confounded because some researchers report infection percentage based on total numbers of infected root tips while others record numbers of ectomycorrhizal structures, which may have multiple tips. Marx and Bryan

(1975) modified the counting procedure to a simple visual estimate now used by many researchers faced with evaluating numerous root systems.

Quantifying VAM infection is also problematic and has additional methodological difficulties. Vesicular–arbuscular mycorrhizal infection is usually examined microscopically on roots or root segments that have been cleared and stained using techniques similar to Phillips and Hayman (1970). Infection estimates include counting infected roots or segments; measuring lengths of infected roots; counting vesicles, arbuscles, and internal spores; and recording infection intensities. Giovannetti and Mosse (1980) introduced a technique of grading infection microscopically using a gridline-intersect method that considers both the infected root length and infection intensity. Other methods for quantifying VAM infection are based on the colorimetric determinations of chitin, a constituent of fungi but not plants (Haselwandter, 1978; Hepper, 1977), and on the intensity of yellow pigmentation accompanying mycorrhizal infection (Becker and Gerdemann, 1977). Regardless of the quantifying method used, comparison of results among studies is hampered by lack of a uniform assessment technique.

Interpretations and comparisons of studies examining N fertilization effects on mycorrhizal development often neglect treatment, temporal, and seasonal factors that directly or indirectly influence root growth and mycorrhizal infection. Wilcox (1968) suggested that frequently encountered variable distribution of mycorrhizae along roots may be caused by the vigorous growth of emerging lateral roots. Roots that emerge slowly allow sufficient time for slow-growing fungi to "capture" short roots and form mycorrhizae, while vigorously growing roots may escape infection.

Regardless of the inoculum form in soil or on root systems, roots may rapidly outgrow associated mycorrhizal fungi under average growing conditions. According to Marks and Foster (1967), the mycorrhizal infection zone will be restricted in size or exposed to infection for a much shorter duration on roots that have stopped growing or grow rapidly than on slow-growing roots and those with slow secondary differentiation. For southern pines, secondary differentiation appears to occur slowly enough that mycorrhizal fungi "catch up" and infect short roots once root growth has slowed.

Percentages of mycorrhizal infection can be different throughout the course of an experiment. For example, time may be construed as inhibiting mycorrhizal infection in a treatment that stimulates root growth (Fig. 9.4). In the presence of a root growth stimulator, mycorrhizal infection lags behind root growth and short root proliferation; therefore, based on percentage of short roots converted to mycorrhizae, total mycorrhizal infection decreases (1X and 2X). Over time, new root growth eventually slows and mycorrhizal infection may "catch up" and represent a greater percentage of total short roots. If a study was sampled during time 1X or 2X, results would indicate that the treatment suppressed mycorrhizal

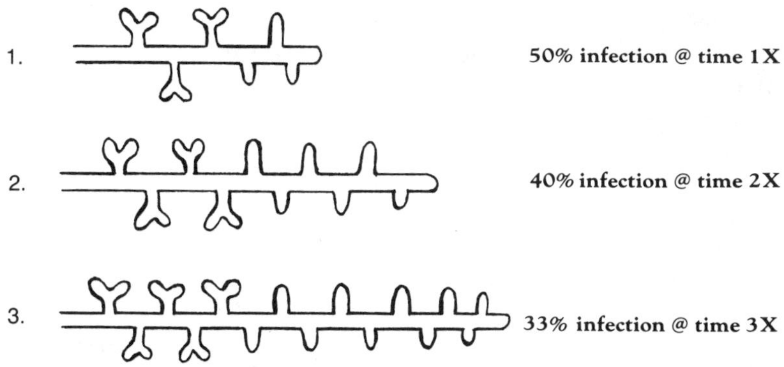

Figure 9.4. Illustrations (1) through (3) show rapid root elongation and formation of short roots at a rate faster than fungi can form new ectomycorrhizae. Although more ectomycorrhizae are being formed, new short roots are forming faster resulting in more ectomycorrhizae but lower percentage representation. Such root growth activity normally occurs in the spring of each year and following soil fertilization during the growing season.

infection. In reality, the absolute rate of infection and fungal growth did not decrease but only slowed relative to the stimulated root growth rate. In a second example, the time factor might be interpreted as a mycorrhizae stimulator in the presence of a root growth inhibitor (Fig. 9.5). If root growth is inhibited, growth rates of fungi and roots will be more closely matched with ectomycorrhizae developing soon after short root emergence

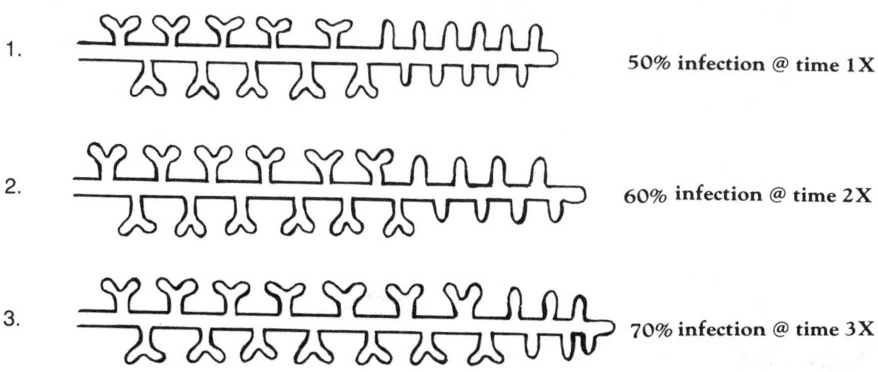

Figure 9.5. Illustrations (1) through (3) show lateral elongation and short root formation rates that permit fungi to increase the speed of conversion to ectomycorrhizae without continued short root development. This results in a higher percentage of ectomycorrhizae but no change in the total number of feeder roots (short roots and ectomycorrhizae) and lateral roots. Examples of (1) through (3) occur during mild water deficits, during winter months, and during periods of depleted soil fertility.

(Bowen and Theodorou, 1973). At times 1X, 2X, and 3X (Fig. 9.5), for example, root growth inhibition would be interpreted as a mycorrhizae stimulator.

Another influence on mycorrhizal development is tree growth patterns in roots and shoots, which vary over a growing season (Drew, 1982; Drew and Ledig, 1980). During a period of rapid root or rapid shoot growth, mycorrhizal infection could be influenced. Because internal and external factors influence root growth rate, repeated sampling over time is important. Studies must ensure that treatment effects are consistent and valid over time.

Treatment effects on root growth and mycorrhizal development can hinder attempts to interpret fertilization experiments. Marks and Foster (1967) suggested that mycorrhizal formation depends on the rate of root growth and factors that affect it directly or indirectly. In his critique of Bjorkman's carbohydrate hypothesis for mycorrhizal formation, Robertson (1954) suggested that light intensity and N fertilization may differentially affect fungal and root growth and that for any set of environmental conditions, a balance is achieved between the number of short roots produced and the ability of fungi to reach, infect, and form mycorrhizae.

Enhanced root growth in response to N fertilization is well documented (Epstein, 1972; Russell, 1977); however, in most studies that examine the impact of N on mycorrhizal infection, it is difficult to determine whether mycorrhizal fungal growth and infection have been affected directly or only relative to root growth. It may be that the relative rate of fungal growth and the infection rates remained constant while the relative root growth rate increased. This situation would imply a decreased percentage of short roots converted to mycorrhizae without a change in mycorrhizal fungal growth in response to fertilization. A mycorrhizal assessment technique that is coupled to the relative growth rate of the root system (i.e., biomass or length) might alleviate this confounding factor. Moreover, during short-term tests (i.e., less than one year), roots should be sampled at least monthly to follow the root growth relationship with mycorrhizal progression or suppression (Figs. 9.4 and 9.5).

Conclusions

According to wetfall deposition data obtained from the Atlas Project of the National Vegetation Survey (Marx et al., 1987), modified to include dry deposition, maximum annual N deposition in the southeastern United States could theoretically be 35 kg/ha/yr. This figure is small compared to a moderate forest fertilization rate of 168 kg N/ha (Allen, personal communication) and similar to reported N deposition rates in Europe (Hinrichsen, 1986). Nevertheless, even at relatively low rates, N deposition may influence N cycle processes.

Arnold (1991) reported that recent changes in species diversity of ectomycorrhizal fungi in different regions of Europe were most likely due to indirect effects of air pollution, in particular to increased available N, and/or to decreased tree vitality with decreased transport of tree assimilates to roots and mycorrhizae. Gulden and Hoiland (1985) propose that N inhibition of mycorrhizae may contribute to forest decline through two mechanisms: reduced root carbohydrate availability and direct damage to mycorrhizae. Foliar uptake and absorption through the roots may make additional N available to the tree. The additional N in the tree then ties up photosynthate in the production of proteins, reducing the carbohydrate supply to the mycorrhizal fungi. Any circumstance that limits photosynthesis (e.g., shading, water stress, and disease) also reduces available carbohydrates. Most mycorrhizal fungi depend completely on the host plant for carbon, so any decrease in soluble sugars in the roots will be detrimental to the mycorrhizae.

Gulden and Hoiland's hypothesis that N directly damages mycorrhizal fungi evolved from many of the N fertilization studies cited in this chapter. However, even at high levels of N fertilization, mycorrhizae were seldom eradicated. Soil solution N concentrations in natural forest systems are much lower than the "low" N values used in many of these studies. In predicting the level at which N will suppress mycorrhizae, one must consider fungal species, tree species, N form, fungal succession, infection assessment method, and initial soil fertility. Moreover, the difference between a one-time high application of N fertilizer and N deposition should be noted. The N concentration in soil solution, not the total quantity of N, affects mycorrhizae. Therefore, one large dose of N might adversely affect mycorrhizal infection more than a long-term low dose.

In field studies in which fungal species were differentiated, changes in fungal populations occurred, but mycorrhizae were not eliminated with N applications. Throughout the life of a forest stand, mycorrhizal fungi follow a natural pattern of population dynamics. Although N fertilization may cause a temporary decline in occurrence of mycorrhizae, they usually recover to prefertilization infection levels. Moreover, experiments using more than one tree species demonstrate that host species influences the mycorrhizal response to N addition.

Most mycorrhizal fungi use inorganic N and prefer ammonium, although others possess nitrate reductase and thrive on nitrate (Lundeberg, 1970; Trappe, 1967). Smith (1974) suggested that increased efficiency of nutrient absorption leading to higher nutrient content of mycorrhizal plants could be brought about by increased rate of uptake and by increased surface area over which absorption takes place. However, in mycorrhizae formed by non-nitrate-using fungi, the capacity of the fungal sheath to absorb nitrate may be limited to diffussion through the intercellular spaces and cell walls of the sheath to reach root cells. Conversely, ammonium absorption by mycorrhizae may occur also by fungal cells from external solutions and intercellular spaces, with direct transfer from fungus to root cells (Smith, 1974).

Abuzinadah et al. (1986a, 1986b) found that some mycorrhizal fungi were able to exploit organic N sources, such as protein. Other work has shown that certain amino acids are readily used by mycorrhizal fungi (Lundeberg, 1970). Mycorrhizae's ability to absorb amino acid and protein N from natural sources means that more N is already available to them than was previously recognized.

The schedule and method of sampling mycorrhizal infection critically influence an experiment's results. Studies carried out over more than one growing season showed mycorrhizal populations recovering after N fertilization effects subsided. Because seasonal fluctuations in mycorrhizal populations occur, sampling time influences results.

Many researchers have reported suppression of mycorrhizae by N fertilization, but none found signs of toxicity as defined in this chapter. Because various fertilizer forms and application methods are used in these studies, estimating N concentrations that consistently suppress mycorrhizae is difficult. In general, "moderate" levels of N (i.e., approximately 150 kg N ha^{-1}) have no detrimental effect on mycorrhizae.

There is no basis to conclude that even the highest level of N deposition in the South, approximately 35 kg ha^{-1} yr^{-1}, is likely to be toxic to mycorrhizae. Field, nursery, greenhouse, and pure culture studies, in which N has been applied to mycorrhizal plants at many orders of magnitude greater than atmospheric deposition rates, have shown no toxic effects on the mycorrhizae. Host physiology would be more readily influenced by atmospheric pollutants than the associated mycorrhizal fungi. Any disruption of the host physiology will indirectly affect the mycorrhizal fungi that depend on the host supply of carbon. However, an increase in N deposition rates may alter nutrient cycling in general. Agren (1983) used a theoretical C/N model to demonstrate that N deposition of 30 kg ha^{-1} yr^{-1} would cause N saturation of fertile soils within 25 years and would saturate poor soils within 50 years. Whereas N-saturated soils may profoundly affect plant growth and may alter mycorrhizal fungi species composition, there is no evidence to suggest that mycorrhizae would be eradicated.

Nitrogen applications do not appear to damage mycorrhizae directly. Because the fungal infection rate lags behind the root growth rate, the infection process seems to be the point at which excess N may inhibit mycorrhizae's development. Mycorrhizal fungi have been grown in pure cultures with nutrient solutions containing higher N concentrations than normally encountered in soil, so N alone does not seem to harm fungi.

The dual nature of the host–fungus relationship complicates understanding of its functioning. The trees and fungi, individually and in association as mycorrhizae, adapt to the environmental conditions of the site they inhabit. Nitrogen deposition, which has increased significantly with greater industrialization, potentially pollutes forest ecosystems, perhaps contributing to forest decline. The mechanism of nitrogen's effect on forest decline is still in question. Mycorrhizae may respond to excess N through the functioning of the host, the fungus, or the interaction between host and fungus. At the current N deposition rate in the southeastern United States, however, N should not be toxic to mycorrhizae. In case the current situation changes, continued monitoring of N deposition levels is recommended for North American and European studies of forest decline.

References

Aber JD, Melillo JM, Nadelhoffer KJ, McClaugherty CA, Pastor J (1985) Fine root turnover in forest ecosystems in relation to quanitity and form of nitrogen availability: a comparison of two methods. Oecologia 66:317–321.

Aber JD, Nadelhoffer KJ, Steudler P, Melillo JM (1989) Nitrogen saturation in northern forest ecosystems. Bioscience 39:378–386.

Abuzinadah RA, Read DJ (1986) The role of proteins in the nitrogen nutrition of ectomycorrhizal plants. I. Utilization of peptides and proteins by ectomycorrhizal fungi. New Phytol 103:481–493.

Abuzinadah RA, Finlay RD, Read DJ (1986a) The role of proteins in the nitrogen nutrition of ectomycorrhizal plants. II. Utilization of protein by mycorrhizal plants of *Pinus contorta*. New Phytol 103:495–505.

Abuzinadah RA, Finlay RD, Read DJ (1986b) The role of proteins in the nitrogen nutrition of ectomycorrhizal plants. III. Protein utilization by *Betula, Picea*, and *Pinus* in mycorrhizal association with *Hebeloma crustuliniforme*. New Phytol 103:507–514.

Adams AJS (1951) A forest nursery for *Pinus radiata* at Mt. Burr in the southeast of South Australia. Aust Forest 15:47.

Adams SN, Dickson EL, Quinn C (1980) The amount and nutrient content of litterfall under Sitka spruce on poorly drained soils. J Forest 53:65–70.

Agren G (1983) Model analysis of some consequences of acid precipitation on forest growth. In: *Ecological Effects of Acid Deposition*. Natl Swedish Environ Protect Board, Report PM 1636, pp 233–244.

Alexander IJ (1981) The significance of ectomycorrhizas in the nitrogen cycle. In: Lee JA, McNeill S, Rorison IH (eds) *Nitrogen as an Ecological Factor*. 22nd Symp Brit Ecol Soc, Oxford, pp 69–93.

Alexander IJ, Fairley RI (1983) Effects of N-fertilization on populations of fine roots and mycorrhizas in spruce humus. Plant and Soil 71:49–53.

Alexander IJ, Fairley RI (1986) Growth and nitrogen uptake rates of ectomycorrhizal spruce seedlings. In: Gianinazzi V, Gianinazzi S (eds) *Physiological and Genetical Aspects of Mycorrhizae*. Proc 1st Eur Symp Mycorrhizae, Dijon, pp 377–382.

Allen HL (1987) Forest fertilizers: nutrient amendment, stand productivity, and environmental impact. J Forest 85:37–46.

Ames RN, Reid CPP, Porter LK, Cambardella C (1983) Hyphal uptake and transport of nitrogen from two 15N-labelled sources by *Glomus mosseae*, a vesicular–arbuscular mycorrhizal fungus. New Phytol 95:381–396.

Arnold E (1991) Decline of ectomycorrhizal fungi in Europe. Agric; Ecosystem.s and Environ 35:209–244.

Barnard EL (1977) *Endoganaceae and "Endoganaceae-Like" Spore Associates of Yellow-poplar in Some Southeastern Forest Tree Nurseries and Selected Field Soils*. Internat Paper Co Res Note.
Becker WN, Gerdemann JW (1977) Colorimetric quantification of vesicular–arbuscular mycorrhizal infection in onions. New Phytol 78:289–295.
Beckjord PR, Adams RE, Smith DW (1980) Effects of nitrogen fertilization on growth and ectomycorrhizal formation of red oak. For Sci 26:529–536.
Beckjord PR, Melhuish JH Jr, McIntosh MS (1985) Effects of nitrogen and phosphorus fertilization on growth and formation of ectomycorrhizae of *Quercus alba* and *Q. rubra* seedlings by *Pisolithus tinctorius* and *Scleroderma auranteum*. Can J Bot 63:1677–1680.
Beckjord PR, Melhuish JM Jr, McIntosh MS, Hacskaylo E (1983) Effects of nitrogen fertilization on growth and ectomycorrhizal formation of *Quercus alba*, Q. *rubra*, Q. *falcata*, and Q. *falcata* var. *pagodifolia*. Can J Bot 61: 2507–2514.
Bell JNB (1986) Effects of acid deposition on crops and forests. Experientia 42:363–371.
Bjorkman E (1940) Om mykorrhizans utbildning hos tall och granplantor, odlade i naringsrika jordar vid olika kvavetillforsel ock ljustillgang. Meadelanden fran States Skogsforsoksanstalt 32:23–74.
Bjorkman E (1949) The ecological significance of the ectotrophic mycorrhizal association in forest trees. Sven Bot Tids 43:223–262.
Bowen GD (1973) Mineral nutrition of ectomycorrhizae. In: Marks GC, Kozlowski TT (eds) *Ectomycorrhizae, Their Ecology and Physiology*. Academic Press, New York, pp 151–205.
Bowen GD, Smith SE (1981) The effects of mycorrhizas on nitrogen uptake by plants. In: Clark FE, Rosswall R (eds) *Terrestrial Nitrogen Cycles: Processes, Ecosystem Strategies and Management Impacts*. Stockholm, Ecol Bull 33, pp 237–247.
Bowen GD, Theodorou C (1973) Growth of ectomycorrhizal fungi around seeds and roots. In: Marks GC, Kozlowski TT (eds) *Ectomycorrhizae, Their Ecology and Physiology*. New York, Academic Press; New York, pp 107–150.
Brezonik PL, Hendry CD Jr, Edgerton ES, Schulze RL, Crisman TL (1983) *Acidity, Nutrients and Minerals in Atmospheric Precipitation over Florida: Deposition Patterns, Mechanisms and Ecological Effects*. US EPA Project Summary EPA600/S3-83-004.
Brown RW, Schultz RC, Kormanik PP (1981) Response of vesicular–arbuscular endomycorrhizal sweetgum seedlings to three nitrogen fertilizers. For Sci 27:413–420.
Castellano MA, Trappe JM, Molina R (1985) Inoculation of container-grown Douglas-fir seedlings with basidiospores of *Rhizopogon vinicolor* and *R. colossus*: effects of fertility and spore application rate. Can J For Res 15:10–13.
Chu-Chou M (1979) Mycorrhizal fungi of *Pinus radiata* in New Zealand. Soil Biol Biochem 11:557–562.
Chu-Chou M, Grace LJ (1981) Mycorrhizal fungi of *Pseudotsuga menziesii* in the north island of New Zealand. Soil Biol Biochem 13:247–249.
Chu-Chou M, Grace LJ (1982) Mycorrhizal fungi of Eucalyptus in the north island of New Zealand. Soil Biol Biochem 14:133–137.
Clark RE (1981) The nitrogen cycle, viewed with poetic license. In: Clark FE, Rosswall T (eds) *Terrestrial Nitrogen Cycles Processes, Ecosystem Strategies and Management Impacts*. Stockholm: Ecol Bull 33, pp 13–24.
Crowley DE, Maronek DM, Hendrix JW (1981) Fertility–mycorrhizal interactions on container-grown *Pinus echinata*. Hortscience 16:66.

Danielson RM, Griffiths DL, Parkinson D (1984) Effects of fertilization on the growth and mycorrhizal development of container-grown Jack pine seedlings. For Sci 30:828–835.

Deacon JW, Donaldson SD, Last FT (1983) Sequences and interactions of mycorrhizal fungi on birch. Plant and Soil 71:257–262.

Dighton J, Mason PA (1985) Mycorrhizal dynamics during forest tree developmento In: Moore D, Casselton LA, Woods DA, Frankland JC (eds) *Developmental Biology of Higher Fungi*. Cambridge University Press, Cambridge, pp 117–139.

Dixon RK, Behrns GT, Garrett HE, Cox GS, Sander IL (1985) Synthesis of ectomycorrhizae on container-grown oak seedlings. South J Appl For 9:95–99.

Dixon RK, Garrett HE, Bixby JA, Cox GS, Tompson JG (1981) Growth, ectomycorrhizal development and root soluble carbohydrates of black oak seedlings fertilized by two methods. For Sci 27:617–624.

Drew AP (1982) Shoot-root plasticity and episodic growth in red pine seedlings. Ann Bot 49:347–357.

Drew AP, Ledig FT (1980) Episodic growth and relative shoot-root balance in loblolly pine seedlings. Ann Bot 45:143–148.

Dumbroff EB (1968) Some observations on the effects of nutrient supply on mycorrhizal development in pine. Plant and Soil 28:463–466.

Ekwebelam SA, Reid CPP (1983) Effect of light, nitrogen fertilization, and mycorrhizal fungi on growth and photosynthesis of lodgepole pine seedlings. Can J For Res 13:1099–1106.

Epstein E (1972) *Mineral Nutrition of Plants: Principles and Perspectives*. John Wiley & Sons, New York.

Fay JA, Golomb D, Kumar S (1987) Anthropogenic nitrogenoxide transport and deposition in Eastern North America. Atmospheric Environment 21:61–68.

Fernandez IJ (1985) Potential effects of atmospheric deposition on forest soils. In: *Proc Symp on Effects of Air Pollutants on Forest Ecosystems*. Acid Rain Foundation, St. Paul, pp 237–250.

Fleming LV (1984) Effects of soil trenching and coring on the formation of ectomycorrhizas of birch seedlings grown around mature trees. New Phytol 98:143–153.

Fogel R (1980) Mycorrhizae and nutrient cycling in natural forest ecosystems. New Phytol 86:199–212.

Fogel R, Hunt G (1983) Contribution of mycorrhizae and soil fungi to nutrient cycling in a Douglas-fir ecosystem. Can J For Res 13:219–232.

Gadgil RL, Gadgil PD (1975) Supression of litter decomposition by mycorrhizal roots of *Pinus radiata*. New Zealand J For Sci 5:33–41.

Galloway JN, Dillon PJ (1983) *Effects of Acid Deposition: The Importance of Nitrogen. Ecological Effects of Acid Deposition*. Natl Swedish Environ Protect Board Report PM 1636, pp 145–160.

Garner JHB (1994) Nitrogen oxides, plant metabolism and forest ecosystem response. In: Alscher RG, Wellburn AR (eds) *Plant Responses to the Gaseous Environment*. Chapman and Hall, London, pp 301–314.

Gerdemann JW (1975) Vesicular–arbuscular mycorrhizae. In: Torrey JG, Clarkson DT (eds) *The Development and Function of Roots*. Academic Press, London, pp 575–591.

Gill R, Lavender DP (1983) Urea fertilization: effects on primary root mortality and mycorrhizal development of young-growth western hemlock. For Sci 29:751–760.

Giovannetti M, Mosse B (1980) An evaluation of techniques for measuring vesicular-arbuscular mycorrhizal infection in roots. New Phytol 84:489–500.

Gorissen A (1991) *Effects of Air Pollutants in the Rhizosphere of Douglas-Fir.* Ph.D. thesis, University of Amsterdam.

Gorissen A, Joosten NN, Jansen AE (1991) Effects of ozone and ammonium sulphate on carbon partitioning to mycorrhizal roots of juvenile Douglas-fir. New Phytol 119:243–250.

Gosz JR (1981) Nitrogen cycling in coniferous ecosystems. In: Clark FE, Rosswall T (eds) *Terrestrial Nitrogen Cycles: Processes, Ecosystem Strategies and Management Impacts*. Stockholm, Ecol Bull 33, pp 405–426.

Grier CC, Bogt KA, Keyes MR, Edmonds RL (1981) Biomass distribution and above- and below-ground production in young and mature *Abies amabilis* zone ecosystems of the Washington Cascades. Can J For Res 11:155–167.

Gulden G, Hoiland K (1985) The role of ectomycorrhizae in a situation of air pollution and forest death. Agarica 6:341–357.

Hacskaylo E (1958) The role of mycorrhizae in the mineral nutrition of trees. Duke Univ School of Forest Bul 15:111–115.

Hacskaylo E, Snow AG Jr (1959) *Relation of Soil Nutrients and Light to Prevalence of Mycorrhizae on Pine Seedlings*. USDA Forest Service, NEFES Paper 125.

Haines BL, Best GR (1976) *Glomus mosseae* endomycorrhizal with *Liquidambar styraciflua* L. seedlings retards NO3, NO2, and NH4 nitrogen loss from a temperate forest soil. Plant and Soil 45:257–261.

Harley JL (1978) Ectomycorrhizas as nutrient absorbing organs. Proc Res Soc London B Biol Sci 203:1–22.

Haselwandter K (1978) Quantification of mycorrhizal infection in Ericaceae. In: Riedacker A, Gagnaire–Michard J (eds) *Root Physiology and Symbiosis*. Proc IUFRO Symp, Nancy, France, pp 477–485.

Hatch AB (1937) *The Physical Basis of Mycotrophy in Pinus*. Black Rock Forest Bulletin 6.

Hauck RD (1981) Nitrogen fertilizer effects on nitrogen cycle processes. In: Clark FE, Rosswall T (eds) *Terrestrial Nitrogen Cycles: Processes, Ecosystem Strategies and Management Impacts*. Stockholm, Ecol Bull 33, pp 551–562.

Hayman DS (1982) Influence of soils and fertility on activity and survival of vesicular–arbuscular mycorrhizal fungi. Phytopath 72:1119–1125.

Hepper CM (1977) A Colorimetric method for estimating vesicular–arbuscular mycorrhizal infection in roots. Soil Biology and Biochemistry 9:15–18.

Hinrichsen D (1986) Multiple pollutants and forest decline. Ambio 14:258–265.

Ho I, Trappe JM (1975) Nitrate reducing capacity of two vesicul–ararbuscular mycorrhizal fungi. Mycologia 67:886–888.

Husar RB (1986) *Emissions of Sulfur Dioxide and Nitrogen Oxides and Trends for Eastern North America. Acid Deposition Long-Term Trends*. National Academy Press, Washington, DC, pp 48–92.

Husar RB, Holloway JM (1983) *Sulfur and Nitrogen over North America. Ecological Effects of Acid Deposition*. Natl Swedish Environ Protect Board Report PM 1636, pp 95–115.

Jentschke G, Godbold DL, Huttermann A (1991) Culture of mycorrhizal tree seedlings under controlled conditions: effects of nitrogen and aluminium. Physiol Plant 81:408–416.

Johnson CR, Joiner JN, Crews CE (1980) Effects of N, K, and Mg on growth and leaf nutrient composition of three container-grown woody ornamentals inoculated with mycorrhizae. J Amer Soc Hort Sci 105:286–288.

Keeney DR (1980) Prediction of soil nitrogen availability in forest ecosystems: a literature review. For Sci 26:159–171.

Khan AG (1972) Mycorrhizae and their significance in plant nutrition. Biologia April:42–78.

Kiernan JM, Hendrix JW, Maronek DM (1983) Fertilizer-induced pathogenicity of mycorrhizal fungi to sweetgum seedlings. Soil Biol Biochem 15:257–262.
Kormanik PP, Bryan WC, Schultz RC (1977) Influence of endomycorrhizae on growth of sweetgum seedlings from eight mother trees. For Sci 23:500–505.
Kramer PJ, Wilber KM (1949) Absorption of radioactive phosphorus by mycorrhizal roots of pine. Science 110:8–9.
Laiho O (1970) *Paxillus involutus* as a mycorrhizal symbiont of forest trees. Acta For Fenn 106:1–72.
Last FT, Fleming LV (1985) Factors affecting the occurrence of fruitbodies of fungi forming sheathing (ecto-) mycorrhizas with roots of trees. Proc Indian Acad Sci 94:111–127.
Last FT, Mason PA, Ingleby K, Fleming LV (1984) Succession of fruitbodies of sheathing mycorrhizal fungi associated with *Betula pendula*. For Ecol Manage 9:229–234.
Lovett GM (1992) Atmospheric deposition and canopy interactions of nitrogen. In: Johnson DW, Lindberg SE (eds) *Atmospheric Deposition and Nutrient Cycling: A Synthesis of the Intergrated Forest Study*. Springer-Verlag, New York, pp 152–166.
Lovett GM, Lindberg SE (1986) Dry deposition of nitrate to a deciduous forest. Biogeochemistry 2:137–148.
Lovett GM, Reiners WA, Olsen RK (1982) Cloud droplet deposition in subalpine balsam fir forests: Hydrological and chemical inputs. Science 218:1303–1304.
Lundeberg G (1970) Utilization of various nitrogen sources, in particular bound soil nitrogen. Stud For Suec 79:1–95.
Marks GC, Foster RC (1967) Succession of mycorrhizal associations on individual roots of radiata pine. Aust J Forest 31:193–201.
Maronek DM, Hendrix JW, Kiernan JM (1982) Adjusting nursery practices for production of mycorrhizal seedlings during propagation. Int Plant Prop Soc Proc 31:461–471.
Maronek DM, Hendrix JW, Stevens CD (1981) Fertility–mycorrhizal–isolate interactions in production of containerized pin oak seedlings. Scientia Hort 15:283–289.
Marx DH (1990) Soil pH and nitrogen influence *Pisolithus* ectomycorrhizal deveopment and growth of loblolly pine seedlings. For Sci 36:224–245.
Marx DH, Artman JD (1979) *Pisolithus tinctorius* ectomycorrhizae improve survival and growth of pine seedlings on acid coal spoils in Kentucky and Virginia. Reclam Rev 2:23–31.
Marx DH, Bryan WC (1973) The significance of mycorrhizae to forest trees, forest soils and forest land management. In: Bernier E, Winger CH (eds) Proc N 4th Amer For Soils Conf, Laval Univ, Quebec, pp 107–117.
Marx DH, Bryan WC (1975) Growth and ectomycorrhizal development of loblolly pine seedlings in fumigated soil infested with the fungal symbiont *Pisolithus tinctorius*. For Sci 22:245–254.
Marx DH, Bryan WC, Cordell CE (1977a) Survival and growthof pine seedlings with *Pisolithus* ectomycorrhizae after two years on reforestation sites in North Carolina and Florida. For Sci 23:363–373.
Marx DH, Clerke WH, Lefohn AS, Paul JT, Beltz RC, Jorgensen JR (1987) *Design and Pilot Test of a Long-term Monitoring Study of Eastern United States Forest for Response to Atmospheric Deposition*. Atlanta, GA, Terrestrial Effects Peer Review.
Marx DH, Cordell CE, Kenney DS, Mexal JG, Artman JD, Riffle JW, Molina RJ (1984) Commercial vegetative inoculum of *Pisolithus tinctorius* and inocula-

tion techniques for development of ectomycorrhizae on bare-root tree seedlings. Forest Service Monograph 25.
Marx DH, Hatch AB, Mendicino JF (1977b) High soil fertility decreases sucrose content and susceptibility of loblolly pine roots to ectomycorrhizal infection by *Pisolithus tinctorius*. Can J Bot 55:1569–1574.
Mason PA, Last FT, Pelham J, Ingleby K (1982) Ecology of some fungi associated with aging stands of birches (*Betula pendula* and *B. pubescens*). For Ecol Manage 4:19–39.
Mason PA, Last FT, Wilson J (1986) Effects of different soils on the establishment and influence of sheathing mycorrhizas. In: Gianinazzi V, Gianinazzi S (eds) *Physiological and Genetical Aspects of Mycorrhizae*. Dijon, France, Proc 1st Eur Symp Mycorrhizae, pp 767–772.
May JT, Belcher EW Jr, Cordell CE, Filer TH Jr, South D, Lantz CW (1985) *Southern Pine Nursery Handbook*. USDA Forest Service, Southern Region Cooperative Forestry, Atlanta, GA.
McAfee BJ, Fortin JA (1986) Competitive interactions of ectomycorrhizal mycobionts under field conditions. Can J Bot 64:848–852.
Melhuish JH Jr, Hacskaylo E (1980) Fatty-acid content of *Pisolithus tinctorius* in response to changing ratios of nitrogen and carbon source. Mycologia 72: 1041–1044.
Melin E (1921) Uber die Mydorrhizenpilze von *Pinus Silvestris* L. und *Picea Abies* (L.) Karst Sv Bot Tidskr 15:192–203.
Melin E, Nilsson H (1952) Transport of labelled nitrogen from an ammonium source to pine seedlings through mycorrhizal mycelium. Sven Bot Tids 46: 281–285.
Menge JA, Grand LF (1977) Effect of fertilization on production of epigeous basidiocarps by mycorrhizal fungi in loblolly pine plantations. Can J Bot 56:2357–2362.
Menge JA, Grand LF, Haines LW (1977) The effect of fertilization on growth and mycorrhizae numbers in 11-year-old loblolly pine plantations. For Sci 23:37–44.
Meyer FH (1973) Distribution of ectomycorrhizae in native and man-made forests. In: Marks GC, Kozlowski TT (eds) *Ectomycorrhizae, Their Ecology and Physiology*. Academic Press, New York, pp 79–105.
Meyer FH (1984) *Mykologische Beobachtungen zum Baumsterben*. Allgemeine Forst Zeitshrift, pp 212–228.
Meyer FH (1985) *Einfluss des Stickstoff-faktors auf den Mykorrhizabesatz fon Fichtensamlingen im Humus einer Waldschadensflache*. Allgemeine Forst Zeitschrift, pp 208–219.
Meyer FH (1986) Das Waldsterben aus okologischer Sicht. Automobil-industrie 86:473–483.
Mikola P (1973) Application of mycorrhizal symbiosis in forestry practice. In: Marks GC, Kozlowski TT (eds) *Ectomycorrhizae, Their Ecology and Physiology*. Academic Press, New York, pp 383–411.
Miller HG (1981) *Nutrient Cycles in Forest Plantations, Their Change with Age and the Consequence for Fertilizer Practice*. Canberra, Australia, Proc Aust For Nutrition Workshop: Productivity in Perpetuity, pp 187–199.
Molina R, Chamard J (1983) Use of the ectomycorrhizal fungus *Laccaria laccata* in forestry. II. Effects of fertilizer forms and levels on ectomycorrhizal development and growth of container-grown Douglas-fir and ponderosa pine seedlings. Can J For Res 13:89–95.
Molina R, Trappe JM (1982) Patterns of ectomycorrhizal host specificity and potential among Pacific Northwest conifers and fungi. For Sci 28:423–458.

Molina R, Trappe JM, Strickler GS (1978) Mycorrhizal fungi associated with *Festuca* in the western United States and Canada. Can J Bot 56:1691–1695.

Morton JB, Benny GL (1990) Revised classification of arbuscular mycorrhizal fungi (zygomycetes): A new order, Glomales, two new suborders, Glomineae and Gigasporineae, and two new families, Acaulosporaceae and Gigasporaceae, with an emendation of Glomaceae. Mycotaxon 37:471–491.

Mosse B (1972) The influence of soil type and Endogone strain on the growth of mycorrhizal plants in phosphate-deficient soils. Rev Ecol Biol Sol 9:529–537.

Mosse B (1973) Advances in the study of vesicular–arbuscular mycorrhiza. A Rev Phytopathol 11:171–196.

National Atmospheric Deposition Program (1988) *NADP/NTN Annual Data Summary*. National Atmospheric Deposition Program/National Trends Network Coordination Office, Fort Collins, Co.

Nicolson TH (1960) Mycorrhiza in the Graminaceae. II. Development in different habitats, particularly sand-dunes. Trans Brit Mycol Soc 43:132–145.

Nicolson TH (1967) Vesicular–arbuscular mycorrhiza—a universal plant symbiosis. Sci Prog (Oxford) 55:561–581.

Nye PH (1966) The effect of the nutrient intensity and buffering power of a soil, and the absorbing power, size and root hairs of a root, on nutrient absorption by diffusion. Plant and Soil 25:81–86.

Pacovsky RS, Bethlenfalvay GJ (1982) Measurement of the extraradical mycelium of a vesicular-arbuscular mycorrhizal fungus in soil by chitin determination. Plant and Soil 68:143–147.

Odum EP, Biever LJ (1984) Resource quality, mutualism, and energy partitioning in food chains. Am Natur 124:360–376.

Phillips JM, Hayman DS (1970) Improved procedures for clearing roots and staining parasitic and vesicular–arbuscular mycorrhizal fungi for rapid assessment of infection. Trans Brit Mycol Soc 55:158–161.

Ponder F Jr (1984) Growth and mycorrhizal development of potted white ash and black walnut fertilized by two methods. Can J Bot 62:509–512.

Pope PE, Chaney WR (1985) Influence of *Pisolithus tinctorius* and fertilization on the development of container grown red oak seedlings. In: Shoulder E (ed) *Proc Third Biennial South Silvic Res Conf*, Gen Tech Rept S-54, Athens, GA, pp 403–409.

Ralston CW (1978) Mineral cycling in temperate forest ecosystem. In: Youngberg CT (ed) *Forest Soils and Land Use*. Proc 5th N Amer For Soils Conf, Fort Collins, CO, pp 320–340.

Richards BN (1965) Mycorrhiza development of loblolly pine seedlings in relation to soil reaction and the supply of nitrate. Plant and Soil 22:187–199.

Richards BN, Wilson GL (1963) Nutritional supply and mycorrhizal development in Carribean pine. For Sci 9:405–412.

Robertson NF (1954) Studies on the mycorrhiza of *Pinus sylvestris*. I. The pattern of development of mycorrhizal roots and its significance for experimental studies. New Phytol 53:253–283.

Romell LG (1930) Blodriskan, en granens foljesvamp. Sv Bot Tidskr 24:524.

Rudawska M (1986) Sugar metabolism of ectomycorrhizal Scots pine seedlings as influenced by different nitrogen froms and levels. In: Gianinazzi V, Gianinazzi S (eds) *Physiological and Genetical Aspects of Mycorrhizae*. Dijon, France, Proc 1st Eur Symp Mycorrhizae, pp 389–394.

Ruehle JL (1980) Ectomycorrhizal colonization of container-grown northern red oak as affected by fertility. USDA For Serv Res Note SE-297.

Ruehle JL, Marx DH (1977) *Developing Ectomycorrhizae on Containerized Pine Seedlings*. USDA For Serv Res Note SE-242.

Ruehle JL, Wells CG (1984) Development of *Pisolithus tinctorius* ectomycorrhizae on container-grown pine seedlings as affected by fertility. For Sci 30:1010–1016.
Russell RS (1977) *Plant Root Systems: Their Function and Interaction with the Soil*. McGraw-Hill, London.
Rygiewicz PT, Bledsoe CS, Zasoski RJ (1984) Effects of ectomycorrhizae and solution pH on [15N] nitrate uptake by coniferous seedlings. Can J For Res 14:885–892.
Salisbury RB, Ross CW (1978) *Plant Physiology*. Wadsworth Publ Co, Belmont, CA.
Sanders FE, Tinker PB (1973) Phosphate flow into mycorrhizal roots. Pestic Sci 4:385–395.
Schulze ED (1989) Air pollution and forest decline in spruce (*Picea abies*) forest. Science 244:776–783.
Schultz RC, Kormanik PP, Bryan WC (1981) Effects of fertilization and vesicular–arbuscular mycorrhizal inoculation on growth of hardwood seedlings. Soil Sci Amer J 45:961–965.
Smith SE (1974) *Mycorrhizal Fungi*. CRC Critical Reviews in Microbiol. CRC Press, Boca Raton, FL, pp 275–311.
Stewart WDP, Preston T, Rai AN, Rowell P (1981) Nitrogen cycling. In: Lee JA, McNeill S, Rorison IH (eds) *Nitrogen as an Ecological Factor*. Blackwell Sci Publ, Oxford, pp 1–27.
Switzer GL, Nelson LE (1972) Nutrient accumulation and cycling in loblolly pine (*Pinus taeda* L.) plantation ecosystems: the first twenty years. Soil Sci Amer Proc 36:143–147.
Termorshuizen AJ, Ket PC (1991) Effects of ammonium and nitrate on mycorrhizal seedlings of *Pinus sylvestris*. Eur J For Path 21:404–413.
Termorshuizen A, Schaffers A (1991) The decline of carpophores of ectomycorrhizal fungi in stands of *Pinus sylvestri* L. in The Netherlands: possible casues. Nova Hedwigia 53:267–289.
Tetreault JP, Bernier R, Fortin JA (1978) Nitrogen fertilization and mycorrhizae of balsam fir seedlings in natural stands. Natural Can 105:461–466.
Thapar HS, Khan SN, Ram MAR, Bakshi BK (1984) Effect of fertilizers on development of endomycorrhiza in Kauri pine (*Agathis robusta*). Indian Forester 110:1093–1097.
Theodorou C, Bowen GD (1969) The influence of pH and nitrate on mycorrhizal associations of *Pinus radiata* D. Don. Aust J Bot 17:59–67.
Tominaga Y (1975) Studies on the mycorrhiza of "fairy ring" of *Tricholoma matsutake* (S. Ito et Imai). Sing. VI. on the mycorrhiza of *Pinus densiflora*. Bull Hiroshima Agric Coll 5:159–163.
Torbert JL, Burger JA, Kreh RE (1986) Nutrient concentration effects on *Pisolithus tinctorius* development on containerized loblolly pine (*Pinus taeda* L.) seedlings. Tree Planter's Notes 37:17–22.
Trappe JM (1967) *Principles of Classifying Ectotrophic Mycorrhizae for Identification of Fungal Symbionts*. Munich, Proc 14th Int Union Forest Organ Kongress.
Trappe JM (1962) Fungus associates of ectotrophic mycorrhizae. Bot Rev 38: 538–606.
Turner J (1977) Effect of nitrogen availability on nitrogen cycling in a Douglas-fir stand. For Sci 23:307–316.
U.S. Environmental Protection Agency (1993) *Air Quality Criteria of Nitrogen*. Office of Res Develop, EPA/600/8-91/049bF, Washington, DC.
Van der Eerden LJ, Lekkerkerk LJA, Smeulders SM (1992) Effects of atmospheric ammonia and ammonium sulphate on Douglas-fir (*Pseudotsuga menziesii*). Envirn Pollut 76:1–9.

Van Dijk HFG, de Louw MHJ, Roelofs JGM, Verburgh JJ (1990) Impact of artificial, ammonium-enriched rainwater on soils and young coniferous trees in a greenhouse. Part II—Effects on the trees. Environ Pollut 63:41–59.

Verkade SD, Hamilton DF (1983) Effects of soil fertility on growth, nutrient concentration and mycorrhizal development of *Liriodendron tulipifera* seedlings inoculated with the vesicular-arbuscular fungus, *Glomus fasciculatus*. Scientia Hort 21:243–252.

Vitousek PM (1977) The regulation of element concentrations in mountain streams in the Northeastern United States. Ecological Monographs 47:65–87.

Vogt KA, Grier CC, Meiir CE, Edmonds RL (1982) Mycorrhizal role in net primary production and nutrient cycling in *Abies amabilis* ecosystems in western Washington. Ecology 63:370–380.

Voigt GK (1971) Mycorrhizae and nutrient mobilization. In: Hacskaylo E (ed) *Mycorrhizae*. USDA For Serv Misc Publ 1189. US Govt Printing Office, Washington, DC, pp 122–131.

Wilcox HE (1968) Morphological studies of the roots of red pine, *Pinus resinosa*. II. Fungal colonization of roots and the development of mycorrhizae. Amer J Bot 55:686–700.

Zottl HW (1990) Remarks on the effects of nitrogen deposition to forest ecosystems. Plant Soil 128:83–89.

10. Growth Response of Southern Pines to Acidic Precipitation and Ozone

Richard B. Flagler and Arthur H. Chappelka

Data from 10-year remeasurements of Forest Inventory and Analysis plots in the South indicated a recent, unexplained decline in radial growth of natural pine stands in some portions of the region in the mid-1980s (Sheffield and Cost, 1987; Zahner et al., 1989). These observations, along with reports of damage in high-elevation forests of eastern North America (Johnson and Siccama, 1983), southern California (Miller, 1983), and Europe (Blank, 1985) led to an increase in public and industry concern. Although no conclusive evidence had emerged on a specific cause of these declines, air pollutants—in particular, acid deposition and ozone—were implicated as possible primary or contributing factors (McLaughlin, 1985).

Ozone is the most phytotoxic air pollutant occurring in North America (USEPA, 1986). It causes reductions in growth of agricultural crops and forest tree species, even in the absence of macroscopic, foliar symptoms. In addition, awareness of possible increases in rain acidity has developed due to the increased input of industrial pollutants into the atmosphere and potential impacts on forested ecosystems (McLaughlin, 1985).

In 1986, the Southern Commercial Forest Research Cooperative (SCFRC) was established under the Forest Response Program (FRP), a component of the National Acid Precipitation Assessment Program (Fox et al., 1992; see also Introduction to this book). The FRP goals were to determine the answers to three policy questions. In abbreviated form, these questions are as follows: (1) Is there a problem of forest damage that might be caused by acid rain and associated pollutants? (2) What are the causal relationships between forest damage and these pollutants? and (3) What are the dose–response relationships? Within the framework of the FRP, the purpose of the SCFRC was to determine if acidic precipitation and ozone, alone or in combination, were causing, or have the potential to cause, changes in the health and productivity of commercially important southern pines—specifically, loblolly (*Pinus taeda* L.), shortleaf (*P. echinata* Mill.), and slash pine (*P. elliottii* Engelm.).

To achieve this goal, several scientific questions, from which hypotheses could be formulated, were devised (see Introduction). All of the studies discussed in this chapter addressed policy question 2. Priorities were based on the use of critical reviews (Binkley et al., 1989; Garner et al., 1989; Medows et al., 1987). Conclusions regarding the probability of each individual mechanism having a significant impact on southern forest health and productivity were then made and priorities were assigned. The highest priority was placed on questions concerning mechanisms involving tree physiological processes directly related to growth. Ranking then progressed downward to questions about mechanisms having an indirect impact through soils and soil microorganisms (2.1). Of lowest priority was the question involving nitrogen deposition and winter injury (2.4). Research sites that participated in the SCFRC are listed in Table 1 of the Introduction.

The purpose of this chapter is to review and summarize the data collected by SCFRC scientists and relate this work to other studies conducted with southern pines. Where insufficient SCFRC data are available, a summary of pertinent data for the relevant pollutants and species is provided.

Methodologies Employed

Controlled Laboratory Studies

The first studies funded by the SCFRC included four projects (eight studies) to test the variability in sensitivity to ozone among a large proportion of commercially available half-sibling families of loblolly pine (100 open-pollinated families). These studies were intended to be relatively short-term (two to four months) exposures to relatively high ozone concentrations (Table 10.1). Data from these studies were used to select families for exposure to these pollutants in long-term field studies (see the next section).

No single institution possessed facilities large enough to expose all of the selected families, so this effort was divided among three institutions: North Carolina State University, Oak Ridge National Laboratory, and Texas A&M University. Each site was assigned a subgroup of half-sib

Table 10.1. Experimental Conditions for SCFRC Controlled Laboratory/Greenhouse Ozone Exposures of Loblolly Pine Seedlings

Exposure System Type	Ozone Exposure Levels (ppb)[c]	Exposure Duration	# Families	Reference
Laminar Flow[a]	0, 160, 320	8 h day^{-1}, 4 days $week^{-1}$, 9 weeks	30	Wiselogel et al., 1991
CSTR[a]	0, 160, 320	6 h day^{-1}, 4 days $week^{-1}$, 22 weeks	8	McLaughlin et al., 1988
CSTR[b]	0, 80, 160, 240, 320	6 h day^{-1}, 4 days $week^{-1}$, 12 weeks	30	Shafer et al., 1993
CSTR[b]	0, 80, 160, 240, 320	6 h day^{-1}, 4 days $week^{-1}$, 12 weeks	12	Reinert et al., 1988
CSTR[b]	0, 160, 320	6 h day^{-1}, 4 days $week^{-1}$, 8 weeks	3	Horton et al., 1990
CSTR[b]	0, 80, 160, 240	8 h day^{-1}, 7 days $week^{-1}$, 12 weeks	21	Flaglter and McKinley, 1992

[a] Laboratory study.
[b] Greenhouse study.
[c] Refers to ozone concentration added to charcoal-filtered air.

Table 10.2. Experimental Conditions for SCFRC Controlled Field Studies Using Open-Top Chambers to Determine Pollutant Effects on Growth of Southern Pine Seedlings/Saplings

Species	Ozone[b]	Acid Rain (pH)	Exposure Duration	# Families	Reference
Loblolly[a]	CF, A, A + 40, A + 80, A + 160	3.3, 4.5, 5.2	12 weeks	53	McLaughlin et al., 1994
Loblolly/ Shortleaf[a]	CF, NF, 1.7×, 2.0×, 2.5×	—	4 months	2	Elsik et al., 1992
Loblolly	CF, NF, 1.5×, 2.25×, 3.0×	3.5, 5.2	20 months	3	Kress and Allen, 1991
Loblolly	CF, NF, 1.3×, 1.7×, 2.0×, 3.0×	3.3, 4.3, 5.3	32 months	4	Kress et al., 1992
Loblolly	CF, NF, 1.7×, 2.5×	3.3, 4.3, 5.3	30 months	2	Chappelka et al., 1990
Shortleaf	CF, NF, 1.7×, 2.5×	3.3, 4.3, 5.3	31 months	4	Flagler et al., 1992a
Shortleaf	CF, NF, 1.7×, 2.5×	3.3, 4.3, 5.3	26 months	4	Reardon et al., 1992
Slash	CF, NF, 2.0×, 3.0×	3.3, 4.3, 5.3	25 months	4	Dean and Johnson, 1992b

[a] Seedlings grown in containers.
[b] CF = charcoal-filtered air; A = ambient air; A + 40, A + 80, A + 160 = ambient air to which 40, 80, and 160 ppb ozone is added; NF = nonfiltered air; 1.3×, 1.5×, 1.7×, 2.0×, 2.25×, 2.5×, 3.0× = proportions of ambient air.

families; however, a common group of 10 half-sib families was exposed at all sites for cross-site comparisons.

All seedlings were grown in a common medium in the same type of container. They were watered and fertilized according to standardized protocols. A common set of growth measurements was taken at each study location. The exposure systems and methodologies varied from site to site; however, ozone exposure concentrations were standardized.

The types of exposure systems used included laminar flow growth chambers within a climate-controlled laboratory (Wiselogel et al., 1991) and continuous-stirred tank reactors (CSTRs) within either a climate-controlled laboratory (McLaughlin et al., 1988) or a greenhouse (Reinert et al., 1988). The advantages of these types of systems include precise control over exposure concentrations, ease of access for measurements and improved environmental control. Some disadvantages include reduced and artificial irradiance, restricted plant size and/or numbers, and unnatural pollutant exposure concentration distributions (Manning and Krupa, 1992). Exposure conditions for these studies are summarized in Table 10.1.

Controlled Field Studies

To maximize the usefulness of the information obtained from field studies, the SCFRC utilized multiple Intensive Field Sites. These were coordinated field studies providing a detailed examination of the response of pine seedlings and saplings to acidic precipitation and ozone at different locations across the South (Introduction, Fig. 2). Five sites were selected to reflect the regional distribution of southern pines, with emphasis on the three major commercial species (loblolly, shortleaf, and slash pine) and to represent the major physiographic regions.

The Intensive Field Sites conducted exposure–response research and mechanistic studies under field conditions. The five sites followed common protocols and used similar experimental designs. Each site employed the same range of rain acidity levels and chemical composition and range of ozone exposures (Table 10.2; see also Flagler et al., 1992b). Pollutant exposures were applied within either small (3-m diameter) or large (4.6-m diameter) open-top chambers (Heagle et al., 1973, 1989). Rain exclusion caps were used to prevent ambient rainfall from entering the chambers, and a standardized rain distribution system was used for applying simulated rain at each site.

Open-top chamber technology is a bridge between controlled environment chambers and ambient open-air systems. In most SCFRC studies, trees were grown in the ground and exposed to a range of pollutant concentrations that both span and track the natural variations in ambient levels. There are some drawbacks that must be recognized when using this technology, however. For example, ozone additions (i.e., above-ambient treatments) are generally applied only when a certain ambient threshold concentration has been reached (nominally 20 ppb); this may result in a bimodal distribution of ozone concentrations. While this method is uncertain regarding the effects of a bimodal exposure concentration distribution on plants (Lefohn et al., 1992), it is not a naturally occurring distribution. Other limitations of this technology involve tree size limitations, slightly elevated temperatures, reduced light, and the possible alteration of the chemical environment within the chambers (Manning and Krupa, 1992). Despite these limitations, the open-top chamber system is widely accepted for use in these types of studies.

Mature Tree Responses

The methodologies used by the SCFRC to measure mature tree responses were studies with clonal material (Kossuth and Biggs, 1989) and use of branch chambers (Teskey et al., 1991). The research with clonal material was designed to test the feasibility of using clonal or grafted material in air pollution research, since this material is genetically identical and grafted mature tissue may serve as a surrogate for mature trees. This method was tested using slash pine.

A second method used branch chambers on established mature loblolly pines. This technology has been developed to evaluate potential pollutant effects on mature trees, maintain a tree in a natural environment, and eliminate genetic variability or water and nutrient differences as confounding factors. Some disadvantages of this system are the assumption of branch autonomy, small sample size per tree, and inability to address carbon allocation questions and whole-tree responses (Manning and Krupa, 1992). To compare branch chamber treatments, two-year-old loblolly pines were planted at one site and growth patterns were compared with the mature trees.

A third method—enclosing the crown of mature loblolly pines with modified open-top chambers mounted on elevated platforms—was tried (Albaugh et al., 1992). This approach worked reasonably well on the small scale tested, and the method was recommended for use in future designed studies.

Quality Assurance

The goals for data quality assessment in the SCFRC were to provide procedures that reduce random and systematic error in treatment application and response measurements and to evaluate data accuracy and precision. Data quality objectives and measurement methodologies were established prior to research implementation for laboratory and field research parameters. Project Work Plans and Quality Assurance Plans were developed for environmental measurements from field sampling, laboratory analysis, and data reduction and reporting. Performance audits and on-site reviews of field and laboratory methodologies were conducted to quantify data quality. Quality assurance protocols were implemented within a program of coordinated research activities to facilitate the evaluation of environmental measurements for use as criteria for reporting of basic research findings and for wide-ranging environmental policy decisions (Peterson and Mickler, 1994).

When evaluating and comparing the various SCFRC studies, attention must be given to several factors. Among the major issues are length of exposure, use of potted versus planted material, tissue age, differences in location (environmental and edaphic factors), differences in exposure systems (CSTR, open-top chambers, laminar-flow growth chambers, branch chambers), tree species, family variation, exposure dynamics, environmental conditions within exposure facilities, preconditioning, and cumulative and carry-over effects. All of these factors can have an effect on the results of a particular study and may cause what appear to be contradictions between studies.

Foliar Symptoms

While foliar symptoms caused by ozone are well documented for many herbaceous and tree species (Hill et al., 1970; Skelly et al., 1989) there are no documented reports of visible injury symptoms in any tree species attributable to ambient acidic rain events. In pine species, ozone symptoms appear as chlorotic mottling, necrotic banding, or tip necrosis. The type of symptom generally reflects the ozone concentration–duration relationship. At low to moderate ambient ozone concentrations (typically <60 ppb), chlorotic mottling is the most common symptom (Fig. 10.1). Both recently expanded and more mature needles might be affected. At higher ambient ozone exposures (typically 80 to >100 ppb), necrotic banding (Fig. 10.2) and tip necrosis (Fig. 10.3) may become apparent. These necroses generally occur in leaf tissue that has just reached full expansion, as leaf cells are the most ozone sensitive at this stage.

Several factors are known to affect the expression of foliar symptoms due to ozone. In greenhouse and growth chamber experiments, factors that typically affect symptom expression are water deficit, relative humidity, nutrients, and irradiance (Davis and Wood, 1973; Heck et al., 1965). High humidity, balanced and adequate nutrients, and high levels of irradiance are usually conducive to symptom development. In the field, any factor that limits stomatal conductance, such as water deficit, will decrease the effective ozone dose and probably result in few if any foliar symptoms. Showman (1991) demonstrated this in field surveys conducted in 1988 and 1989. In 1988, there were considerably higher ozone levels in the survey area than in 1989; however, there were fewer foliar symptoms on known ozone-sensitive species in 1988. He attributed this to a major drought that occurred in late 1987 and continued in 1988. Other factors that affect expression of ozone injury include species, and even families within species (Kress et al., 1982).

In controlled exposure experiments, symptoms are often prevalent due to the exposure conditions being conducive to symptom development. In all controlled environment SCFRC studies that reported the variable, foliar symptoms were found on all families of loblolly pine examined at ozone exposure concentrations of 160 ppb and higher (Horton et al., 1990; Reinert et al., 1988; Wiselogel et al., 1991), usually within three to six weeks of exposure. There was only one report of foliar symptom development at lower (80 ppb) ozone exposure concentrations (Reinert et al., 1988). The injury symptoms found most often were chlorotic mottling and tip necrosis. The primary needles were reported as being more susceptible to foliar symptom development than the secondary (fascicles) needles (Horton et al., 1990). Differences among loblolly pine families in their foliar symptom response to ozone were reported in only one study (Horton et al., 1990), in which one family developed more symptoms than the other two under investigation. The difference in injury in this study did not result in differences in growth measured at the end of the study. The other two studies (Reinert et al., 1988; Wiselogel et al., 1991), which used 12 and 30 half-sib families, resulted in no differences in foliar response among the selections.

Foliar ozone symptoms in field-grown trees are usually mild to absent under most ambient ozone conditions in the southern United States. Ozone symptoms in field-grown trees do show up frequently in open-top chamber studies, especially at ozone exposures greater than ambient. Foliar symptoms (primarily chlorotic mottle and some necrotic flecking) range from light to moderate at near ambient ozone exposures and increase to severe at 3.0× current ambient levels (Flagler et al., 1989; Kress et al., 1992; Shafer and Heagle, 1989; Shafer et al., 1987). No foliar symptoms in any of these studies were attributable to acidic rain treatments (Flagler et al., 1989; Kress et al., 1992).

In the few studies that reported foliar symptoms in open-top chambers, there were differences among families in some studies but not others. In loblolly pine, there were differences among families in ozone-induced foliar symptom response in one study (Shafer and Heagle, 1989), with three of four families investigated showing more injury than the fourth, but not in another that looked at three families (Kress et al., 1992). In shortleaf pine, there was variation in foliar symptom response among four families, with one family showing significantly less chlorotic mottle than the others (Flagler et al., 1989).

In summary, ozone can cause foliar symptoms to develop in southern pines, even at current ambient exposure levels. At levels above current ambient, there is every indication that more severe and widespread symptoms would develop. Symptom development in controlled environ-

Figure 10.1. Chlorotic mottling on shortleaf pine exposed to ambient levels of ozone in east Texas. This is the most common type of foliar symptom that develops in response to ozone damage to needles. (Photo by R.B. Flagler.)

Figure 10.2. Necrotic banding on shortleaf pine. This type of foliar symptom is rare at ambient levels of exposure. It usually requires an acute exposure at a sensitive developmental stage in the needle. (Photo by R.B. Flagler.)

Figure 10.3. Tip necrosis on shortleaf pine foliage. Tip necrosis occurs when high concentrations of ozone are present when the needle is elongating. (Photo by R.B. Flagler.)

Figure 10.7. Effects of a range of ozone treatments on shortleaf pine after three growing seasons. Treatments are CF = charcoal-filtered air; NF = nonfiltered air; 1.7× and 2.5× = 1.7 and 2.5 times the ambient ozone concentration at the study site, respectively. Note the decrease in diameter and the loss of foliage as treatments progress from below ambient (CF) to the above-ambient exposures. (Photo by B. Toups.)

ment studies was more consistent than development of symptoms in field-grown trees. This is likely due to the relatively high-exposure concentrations used and the relative receptivity of the trees in greenhouse studies compared to the field. When greenhouse ozone exposure concentrations were 80 ppb, there were differences among families that did not show up at higher ozone exposure levels. This indicates that if screening is to be done for foliar injury symptoms, lower ozone concentrations should be used. The foliar symptom data frequently do not correlate well with growth data (see subsequent sections) and as such may not be a good indicator of relative ozone susceptibility among species or families within species.

Growth Responses

Height and Diameter

Tree height and diameter are two standard measurements used industry-wide in the evaluation of southern pine growth. Whether it is root collar diameter for a seedling or diameter at breast height for a mature tree, these measures are used to evaluate the performance of the parents in progeny tests and as indicators of performance in a wide range of silvicultural evaluations. These two responses were the most common variables collected within the SCFRC, with nearly every study represented.

Controlled Laboratory and Greenhouse Studies

All studies discussed in this section were conducted using loblolly pine. These studies involved square-wave ozone exposures of from 3 to 30 half-sib families of containerized seedlings. The results of five SCFRC controlled environment studies are summarized in Table 10.3. Height growth responses varied among the studies, but a majority showed decreased height growth as ozone exposure concentration increased. Decreases in height growth (averaged across families) ranged from 8 to 30% in response to ozone. Linear decreases in height growth with increasing ozone were clearly evident in two studies (Reinert et al., 1988; Wiselogel et al., 1991), with decreased height growth of 8% occurring at the relatively low ozone exposure concentration of 80 ppb in one study (Reinert et al., 1988). Increased height growth (averaged across families) was evident in two studies (Horton et al., 1990; McLaughlin et al., 1988); however, the increase in one study was only 1%, and probably not meaningful biologically (Horton et al., 1990). In the other study, the increase was linear with increasing ozone, resulting in a 21% maximum average increase in height growth (McLaughlin et al., 1988).

Diameter growth response to ozone was more consistent than height growth response (Table 10.3). In all but one study, the average response of the families to ozone was negative, with decreases in diameter growth ranging from 3 to 36% depending on ozone exposure concentration and study (Horton et al., 1990; McLaughlin et al., 1988). In two studies, the decrease in diameter growth due to ozone was strongly linear as ozone exposure concentration increased (Horton et al., 1990; Reinert et al., 1988). There was one instance in which the average family response to ozone was positive at the intermediate ozone exposure level only. This resulted in a 7% increase in diameter at a 160 ppb ozone exposure compared to a charcoal-filtered control (McLaughlin et al., 1988).

While the aforementioned responses reflect average responses (across families), it is important to note that not all families within a given study followed the "average" response. Some families were affected negatively, some positively, and some not at all (McLaughlin et al., 1988). This is discussed further in the section on genetic variation and response, later in this chapter.

Controlled Field Studies

Studies discussed in this section were conducted in open-top chambers, using either container-grown seedlings or seedlings/saplings growing in the ground. Most studies had an acidic rain treatment as a partial or complete factorial with ozone treatments.

Of the two studies involving exposure of containerized seedlings, one study (Elsik et al., 1992) involved both loblolly and shortleaf pine exposed to ozone only; the other (McLaughlin et al., 1994) involved 53 half-sib families of loblolly pine exposed to both ozone and simulated acidic rain. The ozone responses for these studies are summarized in Table 10.4. Ozone responses were similar to those obtained from the controlled environment studies. Height growth was unaffected by ozone in the one shortleaf pine family but had a variable response in loblolly pine. The average height growth response ranged from +4% to −26% (McLaughlin et al., 1994) or was unaffected (Elsik et al., 1992). There were differences in ozone response for the 53 half-sib families that varied with both the ozone exposure level and the families (McLaughlin et al., 1994). The vast majority of the families responded negatively at ambient ozone levels, but at slightly above ambient, the trend reversed itself. Diameter growth was more consistent in response to ozone treatments, with negative effects at all but one ozone treatment level (Table 10.4). The range of response was from a 4% increase to an 8% decrease in diameter, with both species responding similarly. Acidic rain at pH 4.5 resulted in greater height growth than either pH 5.2 or pH 3.3 simulated acid rain in 79% of the 53 families examined, but diameter growth was largely unaffected (McLaughlin et al., 1994).

Table 10.3. Ranges of Ozone Exposure and Average Height and Diameter Growth Responses for Container-Grown Loblolly Pine Seedlings Exposed to Square-Wave Ozone Fumigation Under Either Controlled Laboratory or Greenhouse Conditions

Exposure Conditions[a]				# Families	Growth Response[c]		Reference
System Type[b]	Duration (weeks)	Concentration (ppb)	Exposure (ppm · h)		Height (% change)	Diameter (% change)	
LF	9	160	46.1	30	−10[d]	−12[d]	Wiselogel et al., 1991
		320	92.2		−33[d]	−9[d]	
CSTR	22	160	46.1	8	+14	+7	McLaughlin et al., 1988
		320	92.2		+21	−3	
CSTR	12[e]	80	23.0	12	−8	−5	Reinert et al., 1988
		160	46.1		−17	−13	
		240	69.1		−24	−17	
		320	92.2		−35	−34	
CSTR	8	160	30.7	3	+1	−15	Horton et al., 1990
		320	61.4		−20	−36	
CSTR	12	240	161.3	21	−20	−25	Flagler and McKinley, 1992

[a] For details of exposure duration, see Table 10.1. Ozone exposure units (ppm.h) are 12-hour cumulative SUM0 values (sum of all 1-hour average ozone concentrations between 0800 and 1959 LST).
[b] LF is laminar flow growth chamber exposure system; CSTR is a continuous-stirred tank reactor chamber system.
[c] Range of response compared to the 0 ppb ozone (addition) treatment. The symbols (−) and (+) indicate decrease and increase, respectively; values are rounded to the nearest whole number.
[d] Relative height and relative diameter growth (response estimated).
[e] Study repeated second year. Growth responses reflect average of all family–year combinations.

Table 10.4. Ranges of Ozone Exposure and Average Height and Diameter Growth Responses for Container-Grown Loblolly and Shortleaf Pine Seedlings Exposed to Ozone in Open-Top Chambers.

Species	Exposure Conditions			# Families	Growth Response[c]		Reference
	Duration (weeks)	Treatment[a]	Exposure (ppm · h)[b]		Height (% change)	Diameter (% change)	
Loblolly	12	NF	37	44	−26	−5	McLaughlin et al., 1994
		NF + 40	41	44	+4	−8	
		NF + 80	59	53	−21	−8	
		NF + 160	79	53	−13	−6	
Loblolly	18	NF	58	1	ns[d]	−3	Elsik et al., 1992
		1.7×	118		ns	+2	
		2.0×	149		ns	−7	
		2.5×	171		ns	−7	
					ns		
Shortleaf	18	NF	58	1	ns	−5	Elsik et al., 1992
		1.7×	118		ns	−4	
		2.0×	149		ns	−6	
		2.5×	171		ns	−8	

[a]NF = nonfiltered treatment. NF + 40, NF + 80, and NF + 160 are square-wave additions of ozone to NF air at 40 ppb, 80 ppb, and 160 ppb, respectively. 1.7×, 2.0×, and 2.5× are proportional additions of ozone (based on study site ambient ozone) added to NF air.

[b]Daytime exposure period; see individual paper for details. Ozone exposure units (ppm.h) are 12-hour cumulative SUM0 values (sum of all 1-hour average ozone concentrations between 0800 and 1959 LST).

[c]Range of response compared to the charcoal-filtered (subambient) treatment. The symbols (−) and (+) indicate decrease and increase, respectively. Values are rounded to the nearest whole number.

[d]Data not reported, but no statistically significant differences compared to charcoal-filtered treatment stated.

Ozone effects on height and diameter growth of three southern pine species grown in the ground in open-top chambers were different from those exposed in the controlled environment studies (Table 10.5). At all but one ozone exposure level, height growth of loblolly pine was decreased by ozone (Chappelka et al., 1992; Kress et al., 1988). Decreases in height growth ranged from 1 to 26% across the two locations. Diameter growth response of loblolly pine to ozone was more variable than in the controlled environment studies. The positive response reported by Chappelka et al. (1992) was due largely to one of the two families (15-23) being very resistant and exhibiting increased diameter growth compared to the more susceptible family (15-91). Two of the four families grown in the North Carolina Piedmont were relatively susceptible to ozone compared to the others (Chappelka et al., 1992; Kress and Allen, 1991), including family 8-103, which was used by Kress et al. (1988) in an earlier study at the same location.

Acid rain treatments increased diameter growth at both locations as the acidity of the treatment increased. Height growth was similarly increased at the North Carolina location, but not in Alabama. Since the Alabama site had better soil fertility, these results were considered to be the result of additions of N (and to a lesser extent S) to the nutrient-poor soils and not any direct effect of the acidity. There were no significant interactions between ozone and acid rain at either site (Chappelka et al., 1992; Kress et al., 1988).

Shortleaf pine height and diameter response to ozone differed from those of loblolly pine (Table 10.5). The two locations that studied this species produced different results (Flagler et al., 1992a; Shelburne et al., 1991). Both height and diameter growth responded negatively to ozone at a site in the Western Gulf region of Texas, with decreases in height ranging from 3 to 13% and decreases in diameter ranging from 8 to 20% (Flagler et al., 1992a). There were three families that were relatively susceptible to ozone and one family, AR218, that was very resistant. At the site in the South Carolina Piedmont, height growth increased in all but the highest ozone exposure treatment, and diameter decreased at the two highest exposure levels, but there were no differences among the four families in their response to ozone (Shelburne et al., 1991).

The effects of acid rain on shortleaf pine also varied by location. At the South Carolina site, increasing acidity of simulated acidic rain resulted in increased height and diameter growth. Height and diameter growth increased 16 and 13%, respectively, as rain treatment acidity increased from pH 5.3 to pH 3.3 (Shelburne et al., 1991). In Texas, acidic rain treatments interacted with ozone treatments. The ozone × acid rain interaction for height growth resulted from a significantly steeper negative linear response to ozone in the pH 3.3 treatment than in the other rain treatments. The interaction for diameter growth was caused by less of a

Table 10.5. Ranges of Ozone Exposure and Average Height and Diameter Growth Responses for Field-Grown Loblolly and Shortleaf Pine Saplings Exposed to Ozone in Open-Top Chambers

Species	Exposure Conditions			# Families	Growth Response[c]		Reference
	Duration (months)	Treatment[a]	Exposure (ppm · h)[b]		Height (% change)	Diameter (% change)	
Loblolly	9	NF	137	3	−1	−3	Kress et al., 1988
		1.5×	201		+3	+3	
		2.3×	296		−9	−15	
		3.0×	393		−15	−24	
Loblolly	30	NF	609	4	−9	−3	Chappelka et al., 1992
		1.3×	726		−9	−5	
		1.7×	783		−2	−1	
		2.0×	945		−16	−13	
		3.0×	1250		−26	−30	
Loblolly	29	NF	702	2	−9	+4	Chappelka et al., 1992
		1.7×	974		−3	+2	
		2.5×	1249		−9	0	
Shortleaf	30	NF	367	4	−7	−8	Flagler et al., 1992a
		1.7×	674		−3	−9	
		2.5×	980		−13	−20	
Shortleaf	15	NF	167	4	+1	+2	Shelburne et al., 1991
		1.7×	282		+9	−1	
		2.5×	368		−6	−8	
Slash	28	3.0×	968	4	NR[d]	−8	Dean and Johnson, 1992

[a] NF = nonfiltered treatment. 1.3×, 1.5×, 1.7×, 2.0×, 2.3×, 2.5×, and 3.0× are proportional additions of ozone (based on study site ambient ozone) added to NF air.
[b] Daytime exposure period; see individual paper for details. Ozone exposure units (ppm.h) are 12-hour cumulative SUM0 values (sum of all 1-hour average ozone concentrations between 0800 and 1959 LST). In some cases, exposures were estimated from seasonal mean concentration data.
[c] Range of response compared to the charcoal-filtered (subambient) treatment. The symbols (−) and (+) indicate decrease and increase, respectively. Values are rounded to the nearest whole number.
[d] Data not reported.

response to ozone in the pH 5.3 rain treatment than in the other two rain treatments (Flagler et al., 1992a). These results may also be a function of overall site quality. The South Carolina site is on highly weathered and relatively infertile soil. This contrasts with the Texas site, which has relatively fertile, deep, and well-drained soils.

In slash pine, diameter growth was decreased in two separate studies, and height growth was decreased in the one study that reported it. Diameter increment growth in four half-sib families was decreased due to increasing ozone (Table 10.5) and increased as acidity of rain treatments increased from pH 5.3 to pH 3.3. There were no differences among the families and no interactions. Height growth was affected by ozone and an ozone × rain interaction (Dean and Johnson, 1992b). This compares with a study using small open-top chambers (Kossuth and Biggs, 1989), which compared grafted mature slash pine ramets from three clones with seedlings from the same parent trees. Trees were exposed to either subambient (charcoal-filtered) air or 150 ppb ozone for five days per week for nine months. The researchers found that height and diameter growth for both seedlings and ramets were decreased by ozone. Diameter growth was the more sensitive indicator of ozone stress in the ramets, and height growth was the more sensitive indicator in the seedlings. Percentage decreases in growth were similar between seedlings and ramets, indicating that, in this case, seedlings may be a good indicator of response for mature trees.

Data from six open-top chamber experiments at five locations, involving three southern pine species, were combined with the intent of making a regional assessment of the effects of ozone and acidic rain on southern pine sapling growth (Flagler et al., 1992b; Rawlings and Spruill, 1992). At the end of two growing seasons, ozone and acid rain had affected both height and diameter growth (Flagler et al., 1992b). As ozone concentrations increased, height and diameter decreased in a quadratic fashion (Figs. 10.4a and b). As acidity of rain treatments increased, growth increased. There were no ozone × acid rain interactions evident and the acid rain effects on the ozone-exposure growth responses were additive, with increased growth associated with the more acidic rain treatments. For diameter growth, there was a 1% decrease in growth and no effect on height growth responses due to ambient levels of ozone exposure (Table 10.6).

Summary

Height and diameter were extensively measured in SCFRC projects. Diameter growth was the most consistent and stable indicator of response. Most all of the studies found that diameter growth was either decreased or unchanged by ozone exposures. Decreases in diameter at ambient levels of ozone exposure were frequently found in the case of field studies

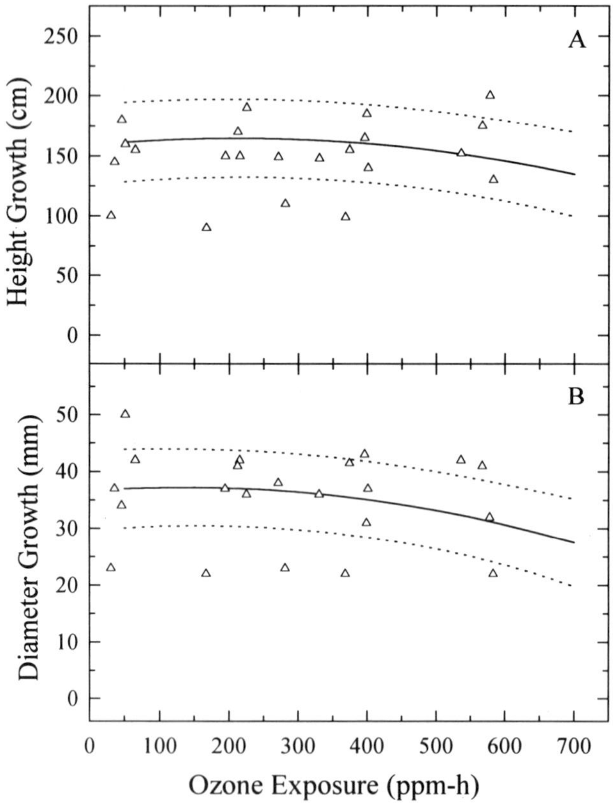

Figure 10.4. Fitted height (A) and diameter (B) response for the combined analyses of three species of southern pine exposed to ozone for two growing seasons. Ozone exposure is the 12-hour (SUM0) cumulative value. Data points are means from five locations and six separate studies (averaged across two to four families per location). Fitted response (—) with 95% confidence intervals (---) for pH 4.3 simulated acidic rain treatment (after Flagler et al., 1992b).

and at relatively low levels in the case of controlled and laboratory experiments. High-level ozone exposures decreased diameter growth in nearly all cases. Acidic rain tended to increase diameter growth with increasing acidity of the treatment. In all cases, these increases in growth were attributed to the addition of N and/or S to otherwise deficient soils and media rather than any direct effect of acidity. Differences in response to these pollutants would be expected due to the wide variety of exposure systems, species, and families used. These results are consistent with the findings of other studies with southern pines (Adams et al., 1988, 1990; Chappelka and Chevone, 1992; Edwards et al., 1992a; Kelly et al., 1993;

Table 10.6. Predicted Changes in Growth and Biomass of Southern Pines Exposed to Ozone for Two Growing Seasons[a]

Response Variable	Predicted Value at 150 ppm · h[b]	95% Confidence Interval	Percentage Change at 250 ppm · h[c]	Percentage Change at 400 ppm · h[d]
Height (cm)	164.34	±32.52	+0.08	−2.46
Diameter (mm)	37.21	±6.70	−1.00	−5.59
Stem biomass (g)	343.26	±95.55	−1.47	−7.71
Foliage biomass (g)	410.88	±149.14	−7.54	−18.85

Estimates are from pH 4.3 simulated rain treatments and represent three pine species and five locations (see table footnote a).

[a] Adapted from Flagler et al., 1992b. Ozone exposure units (ppm · h) are 12-hour cumulative SUM0 values (sum of all 1-hour average ozone concentrations between 0800 and 1959 LST).

[b] 150 ppm · h represents about half the ambient ozone exposure averaged across the five study sites.

[c] 250 ppm · h is the average ambient ozone from the five study sites.

[d] 400 ppm · h represents an approximately 50% decrease in current ozone air quality across the southern United States.

Shafer and Heagle, 1989). Height growth responses were variable. They increased, decreased, or remained unchanged in response to both ozone and acidic rain treatments. Height growth was apparently responding to factors other than intended treatments and should not be considered as reliable a response variable as diameter growth for seedlings.

Biomass Responses

Biomass measurements require the destructive harvesting of the seedling/sapling and are labor intensive. However, they are important measures in determining the effects of pollutant stress on trees. Determinations of ozone and acidic rain effects on seedling/sapling biomass and allocation of biomass among different tissues within the trees help in understanding the underlying mechanisms by which pollutants damage trees.

Controlled Environment and Greenhouse Studies

Results of six studies on the biomass responses of loblolly pine seedlings to ozone exposure are summarized in Table 10.7. The overwhelming response across families and studies was decreased biomass as ozone exposure concentration increased, with all plant parts affected. Many of the biomass responses were linearly related to ozone concentration. The seedlings involved in these studies are representative of the entire physiographic range of the species (Flagler and McKinley, 1992; Horton et

Table 10.7. Ranges of Ozone Exposure and Average Biomass Responses for Container-Grown Loblolly Pine Seedlings Exposed to Square-Wave Ozone Fumigation Under Either Controlled Laboratory or Greenhouse Conditions

Exposure Conditions				# Families	Biomass Response[c]			Reference
System Type[a]	Duration (weeks)	Concentration (ppb)	Exposure (ppm · h)[b]		Root (% change)	Stem (% change)	Foliage (% change)	
LF	9	160	46.1	30	−20	−8	−10	Wiselogel et al., 1991
		320	92.2		−20	−8	−12	
CSTR	22	160	46.1	8	NR	−2[d]	NR	McLaughlin et al., 1988
		320	92.2		NR	+1	NR	
CSTR	12[e]	80	23.0	12	NR	−4	−4	Reinert et al., 1988
		160	46.1		NR	−12	−13	
		240	69.1		NR	−22	−19	
		320	92.2		NR	−37	−32	
CSTR	8	160	30.9	3	−10	−4[d]	NR	Horton et al., 1990
		320	61.4		−28	−23[d]	NR	
CSTR	12	320	92.2	30	−14	−25[d]	NR	Shafer et al., 1993
CSTR	12	240	161.3	21	−40	−25	−49	Flagler and McKinley, 1992

[a] LF is laminar flow growth chamber exposure system; CSTR is a continuous-stirred tank reactor chamber system.
[b] For details of exposure durations, see Table 10.1. Ozone exposure units (ppm · h) are 12-hour cumulative SUM0 values (sum of all 1-hour average ozone concentrations between 0800 and 1959 LST).
[c] Range of response compared to the 0 ppb ozone (addition) treatment. The symbols (−) and (+) indicate decrease and increase, respectively; NR means response not reported. Values are rounded to the nearest whole number.
[d] Value reported is total weight.
[e] Study repeated second year. Growth responses reflect average of all family–year combinations.

al., 1990; McLaughlin et al., 1988; Reinert et al., 1988; Shafer et al., 1993; Wiselogel et al., 1991).

The plant tissue affected most by ozone exposure varied with the study and exposure concentration. Roots were most affected in two studies (Shafer et al., 1993; Wiselogel et al., 1991); the stem in one study (Reinert et al., 1988); and the foliage in another (Flagler and McKinley, 1992). Decreases in root biomass ranged from 10 to 40%, stem biomass from 0 to 37%, and foliage biomass from 4 to 49%. While there were differences in response among the families used in the various studies, there was only one report of an ozone × family interaction (Horton et al., 1990). Another factor that affected the response to ozone included the potting medium used (Horton et al., 1990). In this study, the authors found that root biomass was decreased more than shoot biomass, and the response was dependent on growth medium. Root biomass was decreased to a greater extent when the seedlings were grown in a peat–vermiculite mixture (37% at 320 ppb) than they were when grown in a mineral soil–peat mixture (19% at 320 ppb). The greater decrease in roots than shoots led to a decrease in the root-to-shoot ratio.

Controlled Field Studies

The results of field studies in which seedlings were grown in containers were very similar to the studies described in the last section. In three studies involving either loblolly and/or shortleaf pine, all responses were either negative or no effect (Table 10.8). Average decreases in total biomass ranged from 0 to 25%. Elsik et al. (1992) reported that ozone decreased total biomass, foliage biomass, and root biomass but not stem biomass for both species. There were decreases in biomass at ambient exposure levels (compared to the subambient treatment) for all biomass responses. The effect of ozone on roots was more pronounced than on shoots, leading to a decrease in root:shoot ratio in both species. This decrease in root:shoot ratio due to ozone was also observed by McLaughlin et al. (1994) in loblolly pine, but this effect was not observed with acid rain. Foliage was the only biomass component of shortleaf pine that responded to ozone, but only at 2.5 × the current ambient level; no biomass responses were affected by simulated acid rain; and no interactions were significant (Reardon et al., 1992). However, a significant interaction between ozone and acidity of rain was detected in loblolly pine (McLaughlin et al., 1994). Total biomass decreased as ozone increased at pH 5.2 and pH 4.5, but this trend reversed at pH 3.3.

When trees are grown in the ground, the responses to treatments become more variable. The response of biomass components of loblolly and shortleaf pine to ozone are summarized for four field studies in Table 10.9. When trees were exposed for at least two entire growing seasons, the biomass responses to ozone were largely negative (Chappelka et al.,

Table 10.8. Ranges of Ozone Exposure an Average Biomass Responses for Container-Grown Loblolly and Shortleaf Pine Seedlings Exposed to Ozone in Open-Top Chambers

Species	Exposure Conditions			# Families	Biomass Response[c]			Reference
	Duration (weeks)	Treatment[a]	Exposure (ppm · h)[b]		Root (% change)	Stem (% change)	Foliage (% change)	
Loblolly	12	NF + 80 ppb	59	9	NR	−6[d]	NR	McLaughlin et al., 1994
		NF + 160 ppb	79		NR	−6[d]	NR	
Loblolly	18	NF	58	1	−29	−17[d]	−17	Elsik et al., 1992
		1.7×	118		−28	−19[d]	−18	
		2.0×	149		−48	−29[d]	−29	
		2.5×	171		−44	−25[d]	−28	
Shortleaf	18	NF	58	1	−26	−15[d]	−4	Elsik et al., 1992
		1.7×	118		−12	−12[d]	−12	
		2.0×	149		−24	−10[d]	−12	
		2.5×	171		−38	−19[d]	−16	
Shortleaf	35	NF	340	1	ns[e]	ns	−1	Reardon et al., 1992
		1.7×	500		ns	ns	−2	
		2.5×	658		ns	ns	−31	

[a] NF = nonfiltered treatment. NF + 80 and NF + 160 are square-wave additions of ozone to NF air at 80 ppb and 160 ppb, respectively. 1.7×, 2.0×, and 2.5× are proportional additions of ozone (based on study site ambient ozone) added to NF air.
[b] Daytime exposure period; see individual paper for details. Ozone exposure units (ppm · h) are 12-hour cumulative SUM0 values (sum of all 1-hour average ozone concentrations between 0800 and 1959 LST).
[c] Range of response compared to the charcoal-filtered (subambient) treatment. The symbols (−) and (+) indicate decrease and increase, respectively. Values are rounded to the nearest whole number.
[d] Value reported is total weight.
[c] Data not reported, but no statistically significant differences compared to charcoal-filtered treatment stated.

Table 10.9. Ranges of Ozone Exposure and Average Biomass Responses for Field-Grown Southern Pine Seedlings Exposed to Ozone in Open-Top Chambers

Species	Exposure Conditions			# Families	Biomass Response[c]			Reference
	Duration (months)	Treatment[a]	Exposure (ppm · h)[b]		Total (% change)	Stem (% change)	Foliage (% change)	
Loblolly	9	NF	137	3	NR	−8	−9	Kress et al., 1988
		1.5×	201		NR	+20	+6	
		2.3×	296		NR	−22	−37	
		3.0×	393		NR	−15	−38	
Loblolly	19	NF	215	2	−6	+1	−12	Chappelka et al., 1990
		1.7×	374		−11	+1	−19	
		2.5×	537		−17	−10	−23	
Shortleaf	30	NF	367	4	−16	−14	−19	Flagler et al., 1992a
		1.7×	674		−29	−20	−49	
		2.5×	980		−52	−40	−77	
Shortleaf	15	NF	167	4	−13	−20	−10	Shelburne et al., 1993
		1.7×	282		+11	+12	+11	
		2.5×	368		−31	−32	−30	

[a] NF = nonfiltered treatment. 1.5×, 1.7×, 2.3×, 2.5×, and 3.0× are proportional additions of ozone (based on study site ambient ozone) added to NF air.
[b] Daytime exposure period; see individual paper for details. Ozone exposure units (ppm · h) are 12-hour cumulative SUM0 values (sum of all 1-hour average ozone concentrations between 0800 and 1959 LST). In some cases, exposures were estimated from seasonal mean concentration data.
[c] Range of response compared to the charcoal-filtered (subambient) treatment. The symbols (−) and (+) indicate decrease and increase, respectively. Values are rounded to the nearest whole number. NR indicates not reported.

1990; Flagler et al., 1992a). Responses were more variable with shorter term studies, regardless of species (Kress et al., 1988; Shelburne et al., 1993). Irrespective of exposure duration, foliage biomass was decreased more often and to a greater extent than stems and branches. When effects of ozone on foliage biomass were reported by growth flush (Kress and Allen, 1991), the earlier flushes were affected to a significantly greater extent than the later flushes. Root biomass data for loblolly pine were taken in one study, via root cores from two-year-old trees (Faulkner et al., 1991), and from ingrowth cores in another (Qiu et al., 1992). Faulkner et al. (1991) found no effects on root biomass in the intermediate (1.5× ambient) ozone treatment (relative to subambient) but found decreases of >50% in the 3.0× ozone treatment, with both coarse and fine root biomass affected in this high treatment. There were no significant changes in root biomass due to either ozone or acidic rain in the ingrowth core study (Qiu et al., 1992).

Ozone can alter the allometric relationships among tree tissues. Albaugh et al. (1991) showed that the relationship between components of biomass and volume (d^2h) [d^2 = diameter squared; h = height] for loblolly pine was affected by ozone. They found that increasing ozone exposure affected all relationships except stem biomass-d^2h. As ozone exposure increased from 50 ppm · h to 350 ppm · h, total biomass and foliage biomass decreased at a given d^2h value. Branch biomass increased as ozone exposure increased at a given d^2h value. However, Somerville et al. (1992), using the Alabama loblolly data, found that d^2h was not as sensitive an indicator of ozone effects as some of the biomass parameters.

Acidic rain effects on biomass components were much less frequent and not as great in magnitude. In loblolly pine, acid rain caused no changes in biomass components and did not interact with ozone to affect response (Chappelka et al., 1990; Kress and Allen, 1991; Kress et al., 1988). In shortleaf pine, there were no effects of acidic rain on biomass in one study (Shelburne et al., 1993), but in a second study, there was an interaction of level of acidity with ozone (Flagler et al., 1992a). Flagler et al. reported linear decreases in foliage and woody biomass due to ozone and increases in biomass as acidity of simulated acid rain increased. The ozone × pH interaction resulted from trees in the lower pH treatments being affected more severely by ozone than the trees in the higher pH treatments.

The effects of ozone and acidic rain on combined biomass data from five open-top chamber experiments involving three southern pine species were reported by Flagler et al. (1992b). At the end of two growing seasons, ozone and acid rain had affected total aboveground biomass and its components (Flagler et al., 1992b). As ozone exposures increased, woody biomass decreased quadratically and foliage biomass decreased linearly (Fig. 10.5a and b). As acidity of rain treatments increased, biomass increased. Acid rain effects were additive, and there were no

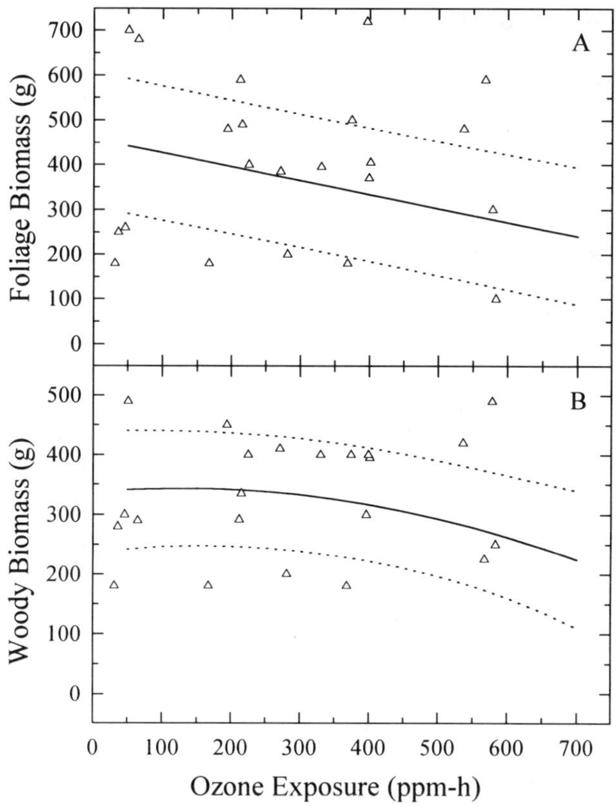

Figure 10.5. Fitted foliage biomass (A) and stem biomass (B) response for the combined analyses of three species of southern pine exposed to ozone for two growing seasons. Ozone exposure is the 12-hour (SUM0) cumulative value. Data points are means from five locations and six separate studies (averaged across two to four families per location). Fitted response (—) with 95% confidence intervals (---) for pH 4.3 simulated acidic rain treatment (after Flagler et al., 1992b).

ozone × acid rain interactions. Decreases in biomass occurred at ambient levels of ozone relative to subambient levels (Table 10.6), with foliage biomass decreased by 7.5% and stem biomass decreased about 1.5%.

Summary

Biomass responses varied somewhat among studies. Ozone had a much greater impact on biomass components than acidic rain. Acidic rain effects were sometimes stimulatory, and this was attributed to a fertilizer effect. Decreases due to ozone were most consistent for foliage biomass. This is consistent with findings of other investigators working with southern pines (Edwards et al., 1991; Shafer and Heagle, 1989). Root biomass in controlled exposure studies was often affected significantly, suggesting that allocation to roots is impacted negatively by ozone. While there has not been extensive work on ozone impacts on root systems in southern pines, changes in allocation of carbon in favor of shoots have been reported for crops and other tree species (Cooley and Manning, 1987; Edwards et al., 1991; Hogsett et al., 1985; Shafer and Heagle, 1989).

Foliage Dynamics

To understand decreases in foliage biomass (such as those reported in the preceding section), it is necessary to determine why biomass is decreasing. Decreases in foliage biomass can occur as a result of a host of different causes, such as decreases in needle number, decreases in mass per needle, accelerated needle loss, decreased needle length, or any combination of these factors. Perhaps more important than the biomass of foliage present is the surface area covered by the foliage and its duration on the tree. Decreases in foliage surface area may also be caused by several factors, including fewer needles, shorter needles, decreased specific needle areas, and accelerated senescence of needles.

Ozone can cause decreases in foliage area in all three of the southern pine species studied (Allen et al., 1992; Boutton and Flagler, 1990, 1991; Byres et al., 1992), but there are studies in which no effects were noted (Elsik et al., 1992) or effects were present only at very high exposure levels (Shelburne et al., 1993). Decreased leaf area due to ozone at near ambient levels also occurs (Boutton and Flagler, 1990; Byres et al., 1992). In all of these studies, leaf area was unaffected by acidic rain treatments.

Changes in needle characteristics can affect the total leaf area of trees. Ozone can affect leaf area ratio, and specific leaf area in loblolly and shortleaf pines (Elsik et al., 1992). Leaf area ratio (m^2 leaf area per g of total biomass) increased with increasing ozone exposure for both species. Specific leaf area (m^2 leaf area per g of foliage biomass) increased linearly with increasing ozone exposure in loblolly pine but was unaffected in shortleaf pine. Ozone effects on leaf area ratio reflect a compensatory response to the loss of photosynthetic capacity per unit leaf area. Specific leaf area is a component of leaf area ratio, and an increase in this factor as a result of ozone exposure indicates either less dense or thinner needles in response to the stress.

Needle length is another factor that relates to leaf area in pines. Kress et al. (1992) evaluated the foliage responses of three half-sib families of loblolly pine exposed to ozone and acid rain for two growing seasons in open-top chambers. They evaluated both branch and main-stem needle length and foliage retention in response to treatments. While there were differences among the three families in their foliage responses, none of the effects interacted with either ozone or acid rain treatments. Acidity of rain treatments affected only needle length (an increase) of the third flush branch foliage. Ozone effects on needle length was dependent on both the specific growth flush and the position of the foliage. In the early season flushes, increasing exposure to ozone decreased needle lengths; later season flushes increased needle lengths as ozone exposures increased. Stow et al. (1992) also reported that needle length was decreased by ozone, but only for the first growth flush. Foliage elongation rates for the first two flushes of the Kress et al. (1992) study were greater at subambient ozone levels compared to the 3.0× ambient ozone. Elongation rates of subsequent flushes were unaffected by ozone treatment. Expansion rates of foliage index, the product of fascicle number and fascicle length, were decreased in all flushes at the high ozone exposure relative to the subambient treatment (Mudano et al., 1992).

Ozone has a significant effect on foliage retention of loblolly pine, and acidity of rain treatments has no consistent effects (Kress et al., 1992). Foliage retention in all flushes and positions was decreased by increasing ozone exposure, and there were significant decreases in foliage retained in the near-ambient ozone exposures. Stow et al. (1992) found that ozone had no effects on foliage production of loblolly pine but caused a significant decrease in foliage retention as exposure level increased from subambient levels. Foliage abscised when cumulative ozone exposure exceeded 130 to 220 ppm · h. Abscission occurred irrespective of flush and/or position within the canopy (i.e., main stem vs. branch). Leaf area duration was decreased at elevated ozone exposures compared to subambient exposures, including a significant decrease at near-ambient exposures. Decreases in duration at elevated ozone exposures occurred for both stem and branch foliage.

Allen et al. (1992) reviewed the foliage dynamics of loblolly pine as affected by ozone for two seedling/sapling studies and one mature tree study. They concluded that ozone has little effect on the production of foliage but has a significant negative impact on the retention of foliage in young trees (Fig. 10.6). No abscission of foliage was noted on the mature loblolly pine trees (Fig. 10.7, p. 386), but they pointed out that the cumulative ozone exposure, even at the highest exposure level, was <160 ppm · h, less than what would be expected to induce premature senescence in loblolly pine (Stow et al., 1992). They reported that ambient levels of ozone induce premature senescence in young trees, but they cautioned that there are insufficient data to determine if this also applies to mature trees. Byres et al. (1992) determined that neither ozone nor acidic rain treatments had a significant effect on number of buds formed in slash pine and that fascicle survival was decreased by high (3.0× ambient) ozone levels.

The effects of ozone on foliage dynamics of the southern pine seem clear. Ozone does not appear to impede the production of foliage in young trees, but ozone exposure at or above current ambient levels causes the premature senescence of foliage. There are insufficient data to

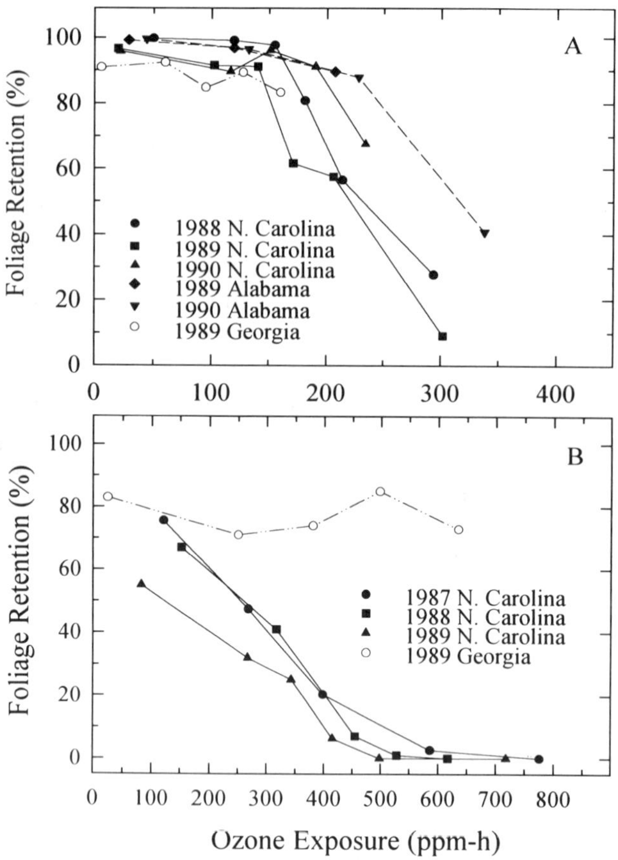

Figure 10.6. First flush foliage retention by year and site for loblolly pine after (A) first and (B) second growing seasons as related to 12-hour cumulative (SUM0) ozone exposures (after Allen et al., 1992).

determine whether the same process is occurring in mature trees. It may be that because of the differences in physiology between young and mature trees (e.g., differences in stomatal conductance), a higher exposure is necessary to effect the same internal dose. The effects of acidic rain on foliage dynamics are minimal and limited to the same effects one would expect from low-level fertilizer inputs. The results obtained from the SCFRC studies are in agreement with other studies with southern pines (Adams et al., 1988; Edwards et al., 1992b; Hogsett et al., 1985; Kelly et al., 1993) and other conifer species (Coyne and Bingham, 1982; Peterson et al., 1987; Skeffington and Roberts, 1985).

Root Growth and Mycorrhizae

Healthy roots are critical to the overall health of trees, as they provide the means by which water and nutrients are acquired. Mycorrhizae are fungi that form symbioses with roots of virtually all forest tree species in the southern United States. Simplistically, they derive carbohydrates from the host species and provide an increased absorbing surface area for water and nutrients for the host. This can be important in the many areas of the South that have phosphorus-poor soils.

Because roots are not readily accessible, growth measurements can be somewhat difficult. Measurement techniques are usually destructive in nature or require growing trees within a confined area. Qiu et al. (1992) grew seedlings in situ in root exclusion tubes within open-top chambers and exposed them to ozone and acidic rain (see Chappelka et al., 1990). Seedlings were harvested after one growing season, and root surface area was determined. Ozone exposure resulted in a significant decrease in root surface area in one (sensitive) family but not in the other (tolerant) family. No deleterious effects of acidic rain were found.

Another means of determining root growth is the use of minirhizotrons. These are clear tubes buried in the soil in proximity to the roots through which a camera or periscope is lowered. Root elongation, length, and/or surface area can then be estimated. This technique was used to evaluate root growth of slash pine seedlings to ozone and acidic rain in open-top chambers (Dean and Johnson, 1990, 1992a). After one growing season, no effects of ozone on root length were observed, but there was an increase in root length in trees exposed to pH 3.3 rain compared to either pH 4.3 or pH 5.3 rain for most of the growing season—March through August (Dean and Johnson, 1990). At the end of the second growing season, there were no treatment effects due to either ozone or simulated acidic rain (Dean and Johnson, 1992a).

Only a few SCFRC studies investigated the effects of pollutants on mycorrhizal status of seedlings for the same reason that roots were not widely studied. Ozone caused a decrease in both the percentage and total number of ectomycorrhizal tips per cm of long roots in loblolly pine in a CSTR study. There were no interactions between ozone treatments and either families or soil water status (Meier et al., 1990). The decrease in mycorrhizal tips due to ozone correlated well with a decrease in both starch and soluble sugar content of the roots. Field-grown trees exposed to ozone in open-top chambers had more variable results. Increasing exposure to ozone and increasing acidities of acidic rain caused an increase in the number of mycorrhizal short roots/cm of root in an ozone-sensitive family of loblolly pine, but not in an ozone-tolerant family. Ozone exposure also caused a shift in the morphotypes present on short roots (Qiu et al., 1993). Faulkner et al. (1991) evaluated root cores taken from two-year-old loblolly pine seedlings growing in a native soil and exposed to three ozone treatments ranging from subambient to 3.0× ambient ozone in open-top chambers. Relative to the subambient treatment, they reported no effects on mycorrhizae in either elevated ozone treatment.

The effects of ozone and acidic rain on roots and mycorrhizae of southern pines are variable among the few studies that have been published. There are too few data to draw any specific conclusions about root growth as affected by these pollutants. The same can be said for mycorrhizae: In some cases, ozone and acidic rain negatively affect mycorrhizae, and in other cases, they increase or have no effects on mycorrhizae. This is consistent with the findings of studies with other pollutant-stressed tree species, in which pollutants have decreased mycorrhizae (Adams and O'Neill, 1991; Reich et al., 1986), increased mycorrhizae (Reich et al., 1985; Stroo and Alexander, 1985), or not affected mycorrhizae (Keane and Manning, 1988; Mahoney et al., 1985; Meier et al., 1989; Simmons and Kelly, 1989). The effects of ozone on mycorrhizae are likely due to decreased carbon allocation to the roots; while the effects of acidic rain on mycorrhizae are probably the result of soil-mediated effects of treatments rather than direct effects on the foliage. Shafer (1992) reported that microbial populations in the rhizosphere increased as a function of increasing acidity of acidic rain only when the acidic solutions were applied to the soil. Increases in mycorrhizae may be the result of more favorable nutrient status of the soil rather than an effect of acidity.

Litter Decomposition

Air pollutants can have direct effects on plants (i.e., reductions in growth and biomass productivity). Air pollutants may also have subtle, indirect effects on forests, such as effects on litter decomposition. Ozone is known to increase senescence of foliage (Allen et al., 1992), producing more litter. It is also known to affect fungal growth (Chappelka and Chevone, 1992). Acidic precipitation has been shown to increase N in living foliage (Johnson et al., 1992) and accelerate the leaching of Ca, Mg, and K in litter (Reich et al., 1988). Therefore, these pollutants may affect litter decomposition.

Only one SCFRC study was found that addresses this issue (Meier and Flagler, 1990, 1991). The influence of acidic precipitation and ozone on litter decomposition/mineralization was observed over a 20-month period in open-top chambers in Texas. There were no effects due to ozone. Acidic rain had no effects on either mass loss or capitals of N and S. Nutrient capitals of the remaining elements in the decomposing shortleaf pine needles—K, P, Ca, and Mg—were affected by the acidic rain treatments. Litter K content decreased exponentially throughout the study, with linear decreases in loss occurring as acidity of rain simulant increased. Litter P content increased with time but was increased to a greater extent in pH 5.3 rain (14%) than pH 3.3 rain (1%) at the end of the study. Ca content of litter remained relatively constant throughout the study in the pH 5.3 and 4.3 treatments but decreased linearly with time in the pH 3.3

rain treatment. Loss of Mg from the litter was linear over time, with loss increasing with the acidity of acidic rain.

The results of this single study indicate that ozone has no effects on decomposition of shortleaf pine litter but that acidic rain can affect decomposition processes and nutrient content of litter. The results with Ca, Mg, and S are consistent with a study by Hagvar and Kjondal (1981), who worked with the effects of simulated acidic rain on decomposition in birch leaves. However, there a too few studies, especially with southern pines, to make generalizations.

Genetic Variation and Response

In the studies reviewed in this chapter, three pine species—loblolly, shortleaf, and slash pine—were examined. Further, in most studies, more than one family was used. In the controlled environment studies, up to 53 families were evaluated simultaneously (McLaughlin et al., 1994). In the field-oriented studies, usually only two to four families were used. The range of the southern pines is broad geographically and extends from eastern Texas east to Florida and north to New Jersey. Because of this broad range and the different physiographic regions that are spanned, these species have developed many adaptations to stress, both biotic and abiotic. While it is difficult to rationalize that the southern pines have developed specific genes for ozone resistance, there is ample evidence of a wide range of ozone response based on the differences in response among the many families examined (see Chappelka et al., 1990; Flagler et al., 1992a; McLaughlin et al., 1994; Reinert et al., 1988; Shafer and Heagle, 1989; Wiselogel et al., 1991). The purpose of this section is to underscore the importance of considering the genetic variability and the genotype–environment interactions that occur when interpreting the results of pollution stress studies.

For example, when experiments are repeated in time in the same facility and care is taken to keep as many factors constant as possible, genotypes that represent extremes in resistance (or susceptibility) will usually rank at that extreme. Shafer et al. (1993) and Reinert et al. (1988) demonstrated this with loblolly pine seedlings exposed to ozone. They each found that out of 12 half-sib families exposed to ozone in two different years, the three most resistant families from the first year ranked within the top three in the second year. However, when environments change, differences in results are not uncommon. This was seen in an evaluation of the relative susceptibility of the three loblolly pine half-sib families—8-130, 8-103, and 8-80—to ozone based on diameter growth. McLaughlin et al. (1988) determined that family 8-130 was the most resistant of the three, while Kress et al. (1988) found it to be the least resistant.

In a recent comprehensive review of genotype and ozone response in loblolly pine, Taylor (1994) concluded that foliar ozone symptoms do not relate well to growth; ozone resistance is inherited quantitatively; there are physiological and ecological costs associated with genetically determined ozone resistance; and species diversity has not been affected by the increases in ozone over the past century. He went on to state that the current levels of ozone in the Southeast are above the threshold for growth response of the average loblolly pine seedling. Exactly what this means for mature trees is unclear; however, if one infers that the ozone growth response (susceptibility) of seedlings is the same as for mature trees, then there are a large number of trees in both managed and natural stands that are being adversely affected by the current ambient ozone levels.

Conclusions

The studies reviewed in this chapter indicate that acidic rain, at current levels of acidity, is not likely to cause direct, detrimental effects on the growth of southern pine seedlings and saplings. Even if ambient rain acidity were to increase 10-fold, it is unlikely to produce any direct negative impacts. If in fact the acidity of rainfall was to increase, positive responses to the increased deposition of N and S into the forest ecosystem may cause short-term beneficial effects on the growth of young pines. However, these conclusions are for young trees exposed to acidic rain for short durations. There are no studies of the long-term effects of acidic deposition on mature southern pine trees. However, some soils in the southern United States are very susceptible to increased acidity. When these poorly buffered soils are acidified further, they may have decreased quantities of exchangeable nutrients. They may also release increased quantities of metals, such as aluminum, that are not only toxic but may interfere with the normal uptake of nutrients. Acid rain may affect the decomposition of pine needle litter, however, and result in further disturbance of the nutrient cycling processes within forests.

Ozone, however, has been demonstrated in several studies to cause adverse effects on growth and physiology of young southern pines, even at current ambient exposures. Ozone causes decreased height and diameter growth; it decreases biomass of seedlings, especially foliage biomass; it decreases net photosynthesis; and it adversely affects other gas exchange parameters. The decreased longevity and retention of foliage is probably one of the most striking results of ozone exposure and one that has caused some forest scientists in the South to rethink natural foliage longevity in some species. This premature senescence of foliage, coupled with decreased rates of carbon gain, is responsible for much of the decreased growth observed in field studies. This process is linked to internal and external nutrient cycling and may result in detrimental effects on tree nutrition.

Effects of ozone occur even at current ambient exposures. Although these are minor in many cases, they should not be dismissed as inconsequential. It is necessary to consider that pines are perennial species and, as such, small decreases in growth that occur in a single season, or after only two or three growing seasons, may compound over time. Southern pine plantations are grown in rotations from 35 to 45 years. Trees growing in natural ecosystems frequently grow for much longer periods of time. A decrease in growth of only a few percent per year can result in a significant decrease in productivity over the life of the tree. These decreases can occur without visible symptoms of injury; and in the absence of any region that has truly clean air, there is no means of assessing a change in growth on a large scale within the southern pine range. The argument is often made that nearly all the ozone exposure–response data for southern pines are from young trees and that there are many physiological differences between young and mature trees; while this is true, all mature trees started out as young trees. Even if any lost productivity is limited to only 10 to 15 years, it may still be substantial.

References

Adams MB, Kelly JM, Edwards NT (1988) Growth of *Pinus taeda* L. seedlings varies with family and ozone exposure level. Water, Air, and Soil Pollut 38:137–150.

Adams MB, Kelly JM, Taylor GE, Edwards NT (1990) Growth of five families of *Pinus taeda* L. during three years of ozone exposure. New Phytol 116:689–694.

Adams MB, O'Neill EG (1991) Effects of ozone and acidic deposition on carbon allocation and mycorrhizal colonization of *Pinus taeda* L. seedlings. For Sci 37:5–16.

Albaugh TJ, Allen HL, Kress LW (1991) Biomass D^2H relationships for young loblolly pine as affected by ozone. Biomass Bioenergy 1:143–148.

Albaugh TJ, Mowry FL, Kress LW (1992) A field chamber for testing air pollution effects on mature trees. J Environ Qual 21:476–485.

Allen HL, Stow TK, Chappelka AH, Kress LW, Teskey RO (1992) Ozone impacts on foliage dynamics of loblolly pine. In: Flagler RB (ed) *The Response of Southern Commercial Forests to Air Pollution*. TR-21. Air & Waste Management Association, Pittsburgh, PA, pp 149–162.

Binkley D, Driscoll CT, Allen HL, Schoeneberger P, McAvoy D (1989) *Acid Deposition and Forest Soils*. Springer-Verlag, New York.

Blank L (1985) A new type of decline in Germany. Nature 314:311–314.

Boutton TW, Flagler RB (1990) Growth and water-use efficiency of shortleaf pine as affected by ozone and acid rain. Proc Air & Waste Manage Assoc 90-187.7:1–16.

Boutton TW, Flagler RB (1991) $^{13}C/^{12}C$ ratios as indicators of plant physiological response to ozone and simulated acid rain. In: *Stable Isotopes in Plant Nutrition, Soil Fertility and Environmental Studies*: *Proceedings of an International Symposium*, October 1–5 1990. Vienna. International Atomic Energy Agency, Vienna, Austria, pp 627–631.

Byres DP, Dean TJ, Johnson JD (1992) Long-term effects of ozone and simulated acid rain on the foliage dynamics of slash pine (*Pinus elliottii* var. *elliottii* Engelm). New Phytol 120:61–67.

Chappelka AH, Chevone BI (1992) Tree responses to ozone. In: Lefohn AS (ed) *Surface Level Ozone Exposures and Their Effects on Vegetation*. Lewis Publishers, Chelsea, MI, pp 271–324.

Chappelka AH, Lockaby BG, Meldahl RS, Allen HL, Kress LW (1992) Effects of ozone and acidic precipitation after three years on growth of loblolly pine saplings in the piedmont of North Carolina and coastal plain of Alabama. In: Flagler RB (ed) *The Response of Southern Commercial Forests to Air Pollution*. TR-21. Air & Waste Management Association, Pittsburgh, PA, pp 163–172.

Chappelka AH, Lockaby BG, Mitchell RJ, Meldahl RS, Kush JS, Jordan DN (1990) Growth and physiological responses of loblolly pine exposed to ozone and simulated acidic rain in the field. Proc Air & Waste Manage Assoc 90-187.5:1–16.

Cooley DR, Manning WJ (1987) The impact of ozone on assimilate partitioning in plants: a review. Environ Pollut 47:95–113.

Coyne PI, Bingham GE (1982) Variation in photosynthesis and stomatal conductance in an ozone-stressed ponderosa pine stand: light response. For Science 28:257–273.

Davis DD, Wood FA (1973) The influence of environmental factors on the sensitivity of Virginia pine to ozone. Phytopathology 63:371–376.

Dean TJ, Johnson JD (1990) Seasonal root growth of slash pine (*Pinus elliottii* engelm. var. *elliottii*) seedlings exposed to acidic precipitation and ozone stress. In: Bongarten BC, Dougherty PM, Teskey RO (eds) *Program and Abstracts from the 11th North American Forest Biology Workshop*, June 13–15, University of Georgia, Athens, pp 8–9.

Dean TJ, Johnson JD (1992a) Acid rain and ozone effects of root-length density of *Pinus elliottii* grown in open-top chambers. Bull Ecol Soc Amer 73:156.

Dean TJ, Johnson JD (1992b) Growth response of young slash pine trees to simulated acid rain and ozone stress. Can J For Res 22:839–849.

Edwards GS, Edwards NT, Kelly JM, Mays PA (1991) Ozone, acidic precipitation and soil Mg effects on growth and nutrition of loblolly pine seedlings. Environ Exper Bot 31:67–78.

Edwards GS, Friend AL, O'Neill EG, Tomlinson PT (1992b) Seasonal patterns of biomass accumulation and carbon allocation in *Pinus taeda* seedlings exposed to ozone, acidic precipitation, and reduced soil Mg. Can J For Res 22:640–646.

Edwards GS, Kelly JM (1992) Ectomycorrhizal colonization of loblolly pine seedlings during three growing seasons in response to ozone, acidic precipitation, and soil Mg status. Environ Pollut 76:71–77.

Edwards NT, Edwards GL, Kelly JM, Taylor GE Jr (1992a) Three-year growth responses of *Pinus taeda* L. to simulated rain chemistry, soil magnesium status, and ozone. Water, Air, and Soil Pollut 63:105–118.

Elsik CG (1992) *Growth, physiology, and $\delta^{13}C$ of Loblolly and Shortleaf Pine as Affected by Ozone and Soil Moisture Deficit*. M.S. thesis, Texas A&M University, College Station, TX.

Elsik CG, Flagler RB, Boutton TW (1992) Effects of ozone and water deficit on growth and physiology of *Pinus taeda* and *Pinus echinata*. In: Flagler RB (ed) *The Response of Southern Commercial Forests to Air Pollution*. TR-21. Air & Waste Management Association, Pittsburgh, PA, pp 225–245.

Faulkner P, Schoeneberger MM, Kress LW (1991) Belowground changes in loblolly pine as indicators of ozone stress. In: Coleman SS, Neary DG (eds) *Proceedings of the Sixth Biennial Southern Silvicultural Research Conference*. USDA-Forest Service Gen Tech Rep SE-70, pp 332–341.

Flagler RB (ed) (1992) *The Response of Southern Commercial Forests to Air Pollution*. TR-21. Air & Waste Management Association, Pittsburgh, PA.

Flagler RB, Boutton TW (1991) Growth and physiological response of *Pinus echinata* exposed to ozone and acid rain. Bull Ecol Soc Amer 72:114.

Flagler RB, Lock JE, Toups BG (1992a) Growth and gas exchange characteristics of shortleaf pine exposed to ozone and acid rain. Paper no. IU-22A.06. *Proceedings of the 9th World Clean Air Congress*. Vol. 5. Montreal, Quebec, Canada, August 30–September 4.

Flagler RB, McKinley CR (1992) Response of two loblolly pine ecotypes to ozone. Bull Ecol Soc Amer 73:175.

Flagler RB, Spruill SE, Chappelka AH, Dean TJ, Kress LW, Reardon JC (1992b) Growth of three southern pine species as affected by acid rain and ozone: A combined analysis. In: Flagler RB (ed) *The Response of Southern Commercial Forests to Air Pollution*. TR-21. Air & Waste Management Association, Pittsburgh, PA, pp 207–224.

Flagler RB, Toups BG, Lock JE (1989) Response of shortleaf pine to acid rain and ozone. In: *NADP Technical Committee Meeting—Program and Abstracts*. October 23–26, Provincetown, MA. NADP/NTN, Ft. Collins, CO, p 5.

Fox S, Joyner KC, Bartuska AM (1992) Overview of the Southern Commercial Forest Research Cooperative. In: Flagler RB (ed) *The Response of Southern Commercial Forests to Air Pollution*. TR-21. Air & Waste Management Association, Pittsburgh, PA, pp 1–15.

Garner JHB, Pagano T, Cowling EB (1989) *An Evaluation of the Role of Ozone, Acid Deposition, and Other Airborne Pollutants in the Forests of Eastern North America*. General Technical Report SE-59. USDA Forest Service, Southeastern Forest Experiment Station, Asheville, NC.

Hagvar S, Kjondal BR (1981) Decomposition of birch leaves: Dry weight loss, chemical changes, and effects of artificial acid rain. Pedobiologia 22:232–245.

Heagle AS, Body DE, Heck WW (1973) An open-top field chamber to assess the impact of air pollution on plants. J Environ Qual 2:365–368.

Heagle AS, Philbeck RB, Ferrell RE, Heck WW (1989) Design and performance of a large field exposure chamber to measure effects of air quality on plants. J Environ Qual 18:361–367.

Heck WW, Dunning JA, Hindawi IJ (1965) Interactions of environmental factors on the sensitivity of plants to air pollution. J Air Pollut Control Assoc 15: 511–515.

Hill AC, Heggestad HE, Linzon SN (1970) Ozone. In: Jacobson JS, Hill AC (eds) *Recognition of Air Pollution Injury to Vegetation: A Pictorial Atlas*. Air Pollution Control Association, Pittsburgh, PA, pp B1–B21.

Hogsett WE, Plocher M, Wildman V, Tingey DT, Bennett JP (1985) Growth response of two varieties of slash pine seedlings to chronic ozone exposures. Can J Bot 63:2369–2376.

Horton SJ, Reinert RA, Heck WW (1990) Effects of ozone on three open-pollinated families of *Pinus taeda* L. grown in two substrates. Environ Pollut 65:279–292.

Johnson AH, Siccama TG (1983) Evaluating the relationship between acid deposition and forest decline. Environ Sci & Technol 17:294–305A.

Johnson JD, Byres D, Chappelka AH, Dean TJ (1992) The role of homeostatic growth in the response of slash and loblolly pine to ozone and acid rain. In: Flagler RB (ed) *The Response of Southern Commercial Forests to Air Pollution*. TR-21. Air & Waste Management Association, Pittsburgh, PA, pp 273–286.

Keane KD, Manning WJ (1988) Effects of ozone and simulated acid rain on birch seedling growth and formation of ectomycorrhizae. Environ Pollut 52: 55–65.

Kelly JM, Taylor GE, Edwards NT, Adams MB, Edwards GS, Friend AL (1993) Growth, physiology and nutrition of¨oblolly pine seedlings stressed by ozone and acidic precipitation: A summary of the ROPIS-South project. Water, Air Soil Pollut 69:363–391.

Kossuth SV, Biggs RH (1989) Response of seedlings and mature trees of the same genotype to ozone. In: Noble RD, Martin JL, Jensen KF (eds) *Air Pollution Effects on Vegetation Including Forest Ecosystems*. Proc. 2nd US–USSR Symp September 13–25, 1988, Corvallis, OR, Raleigh, NC, Gatlinburg, TN. USDA-Forest Service, Northeastern Forest Experiment Station, Broomall, PA, p 287.

Kress LW, Allen HL (1991) Impact of ozone and acidic precipitation on the growth of loblolly pine seedlings. Proc Air & Waste Manage Assoc 91-142.9: 1–10.

Kress LW, Allen HL, Mudano JE, Heck WW (1988) Response to loblolly pine to acidic precipitation and ozone. Proc Air Pollut Control Assoc 88-70.5:1–11.

Kress LW, Allen HL, Mudano JE, Stow TK (1992) Impact of ozone on loblolly pine seedling foliage production and retention. Environ Toxicol Chem 11: 1115–1128.

Kress LW, Skelly JM, Hinklemann JH (1982) Relative sensitivity of 18 full-sib families of *Pinus taeda* L. to ozone. Can J For Res 12:203–209.

Lefohn AS, Shadwick DS, Somerville MC, Chappelka AH, Lockaby BG, Meldahl. RS (1992) The characterization and comparison of ozone exposure indices used in assessing the response of loblolly pine to ozone. Atmospheric Environ 26A:287–298.

Mahoney MJ, Chevone BI, Skelly JM, Moore LD (1985) Influence of mycorrhizae on the growth of loblolly pine seedlings exposed to ozone and sulfur dioxide. Phytopathology 75:679–682.

Manning WJ, Krupa SV (1992) Experimental methodology for studying the effects of ozone on crops and trees. In: Lefohn AS (ed) *Surface Level Ozone Exposures and Their Effects on Vegetation*. Lewis Publishers, Chelsea, MI, pp 93–156.

McLaughlin SB (1985) Effects of air pollution on forests: A critical review. J Air Pollut Control Assoc 35:512–534.

McLaughlin SB, Layton PA, Adams MB, Edwards NT, Hanson PJ, O'Neill EG, Roy WK (1994) Growth responses of 53 open-pollinated loblolly pine famillies to ozone and acid rain. J Environ Qual 23:247–257.

McLaughlin SB, Adams MB, Edwards NT, Hanson PJ, Layton PA, O'Neill EG, Roy WK (1988) *Comparative Sensitivity, Mechanisms, and Whole Plant Physiological Implications of Responses of Loblolly Pine Genotypes to Ozone and Acid Deposition*. ORNL/TM-10777, Environmental Sciences Division publication no 3105. Oak Ridge National Laboratory, Oak Ridge, TN.

Medows JS, Hodges JD, Amundson RG, Hinckley TM, Houston DR, Edmonds RL, Switzer GL (1987) *Effects of Natural and Airborne Chemical Stresses on Growth and Development of Individual Trees and Forests*. USDA-Forest Service, SCFRC Report.

Meier CE, Flagler RB (1990) Decomposition of *Pinus echinata* Mill. litter as affected by ozone and acid rain. Bull Ecol Soc Amer 71:250.

Meier CE, Flagler RB (1991) Decomposition of *Pinus echinata* (Mill.) needle litter as affected by acid precipitation. Proc Air & Waste Manage Assoc 91-142.5:1–20.

Meier S, Grand LF, Schoeneberger MM, Reinert RA, Bruck RI (1990) Growth, ectomycorrhizae and nonstructural carbohydrates of loblolly pine seedlings exposed to ozone and soil water deficit. Environ Pollut 64:11–27.

Meier S, Robarge WP, Bruck RI, Grand LF (1989) Effects of simulated rain acidity on ectomycorrhizae of red spruce seedlings potted in natural soil. Environ Pollut 59:315–324.

Miller PR (1983) Ozone effects in the San Bernardino National Forest. In: Davis DD, Millen AA, Dochinger LS (eds) *Air Pollution and Productivity of the Forest*. Izaak Walton League, Arlington, VA, pp 161–197.
Mudano JE, Allen HL, Kress LW (1992) Stem and foliage elongation of young loblolly pine as affected by ozone. For Sci 38:324–335.
Peterson CE, Mickler RA (1994) Considerations for evaluating controlled exposure studies of tree seedlings. J Environ Qual 23:257–267.
Peterson DL, Arbaugh MJ, Wakefield VA, Miller PR (1987) Evidence of growth reduction in ozone-injured Jeffrey pine (*Pinus jeffreyi* Grev. and Balf.) in Sequoia and Kings Canyon National Parks. J Air Pollut Control Assoc 37: 906–912.
Qiu Z, Chappelka AH, Somers GL, Lockaby BG, Meldahl RS (1993) Effects of ozone and simulated acidic precipitation on ectomycorrhizal formation on loblolly pine seedlings. Environ Exper Bot 33:423–431.
Qiu Z, Chappelka AH, Somers GL, Lockaby BG, Meldahl RS (1992) Effects of ozone and simulated acidic precipitation on above- and below-ground growth of loblolly pine (*Pinus taeda* L.). Can J For Res 22:582–587.
Rawlings JO, Spruill SE (1992) Response model synthesis of the Southern Commercial Forest Research data. In: Flagler RB (ed) *The Response of Southern Commercial Forests to Air Pollution*. TR-21. Air & Waste Management Association, Pittsburgh, PA, pp 187–205.
Reardon JC, Paynter, VA, Shelburne VB (1992) Biomass and carbohydrate allocation of shortleaf pine seedlings exposed to ozone and acid rain. In: Flagler RB (ed) *The Response of Southern Commercial Forests to Air Pollution*. TR-21. Air & Waste Management Association, Pittsburgh, PA, pp 173–186.
Reich PB, Schoettle AW, Stroo HF, Amundson RG (1986) Acid rain and ozone influence mycorrhizal infection in tree seedlings. J Air Pollut Control Assoc 36:724–726.
Reich PB, Schoettle AW, Stroo HF, Amundson RG (1988) Effects of ozone and acid rain on white pine (*Pinus strobus*) seedlings grown in five soils. III. Nutrient relations. Can J Bot 66:1517–1531.
Reich PB, Schoettle AW, Stroo HF, Troiano J, Amundson RG (1985) Effects of O_3, SO_2, and acidic rain on mycorrhizal infection in northern red oak seedlings. Can J Bot 63:2049–2055.
Reinert RA, Shafer SR, Eason G, Horton SJ, Schoeneberger MM, Wells C (1988) Responses of loblolly pine half-sib families to ozone. Proc Air Pollut Control Assoc 88-125.2:1–14.
Shafer SR (1992) Responses of microbial populations in the rhizosphere to deposition of simulated acidic rain onto foliage and/or soil. Environ Pollut 76:267–278.
Shafer SR, Heagle AS (1989) Response of loblolly pine to ozone over three growing seasons. Can J For Res 19:821–831.
Shafer SR, Heagle AS, Camberato DM (1987) Effects of chronic doses of ozone on field-grown loblolly pine: Seedling responses in the first year. J Air Pollut Control Assoc 37:1179–1184.
Shafer SR, Reinert RA, Eason G, Spruill SE (1993) Analysis of ozone concentration–biomass response relationships among open-pollinated families of loblolly pine. Can J For Res 23:706–715.
Sheffield RM, Cost ND (1987) Behind the decline. J For 85:29–33.
Shelburne VB, Reardon JC, Paynter VA (1993) The effects of acid rain and ozone on biomass and leaf area parameters of shortleaf pine (*Pinus echinata* Mill.). Tree Physiol 12:167–173.
Shelburne VB, Reardon JC, Paynter VA (1991) The effect of acid rain and ozone exposure on growth parameters of shortleaf pine. In: Coleman SS, Neary NG (eds) *Proceedings of the Sixth Biennial Southern Silvicultural Research*

Conference. October 30–November 1, 1990, Memphis, TN. General Technical Report SE-70. USDA-Forest Service, Southeastern Forest Experiment Station, Asheville, NC.
Showman RE (1991) A comparison of ozone injury to vegetation during moist and drought years. J Air & Waste Manage Assoc 41:63–64.
Simmons GL, Kelly JM (1989) Influence of O_3, rainfall acidity, and soil Mg status on growth and ectomycorrhizal colonization of loblolly pine roots. Water, Air, and Soil Pollut 44:159–171.
Skeffington RA, Roberts TM (1985) The effects of ozone and acid mist on Scots pine saplings. Oecologia 65:201–206.
Skelly JM, Davis DD, Merrill W, Cameron AE, Brown HD, Drummond DB, Dochinger LS (1989) *Diagnosing Injury to Eastern Forest Trees*. The Pennsylvania State University, State College, PA.
Somerville MC, Shadwick DS, Meldahl RS, Chappelka AH, Lockaby BG (1992) Use of a non-linear model in examining growth responses of loblolly pine to ozone and acid precipitation. Atmos Environ 26A:279–286.
Stow TK, Allen HL, Kress LW (1992) Ozone impacts on seasonal foliage dynamics of young loblolly pine. For Sci 38(1):102–119.
Stroo HF, Alexander M (1985) Effect of simulated acidic rain on mycorrhizal colonization of *Pinus strobus* L. Water, Air, and Soil Pollut 25:107–114.
Taylor GE (1994) Role of genotype in the response of loblolly pine to tropospheric ozone: Effects at the whole-tree, stand, and regional level. J Environ Qual 23:63–82.
Teskey RO, Dougherty PM, Wiselogel AE (1991) Design and performance of branch chambers suitable for long-term ozone fumigation of foliage in large trees. J Environ Qual 20:591–595.
US Environmental Protection Agency (1986) *Air Quality Criteria for Ozone and Other Photochemical Oxidants*. EPA-600/8-84/020cF. Environmental Criteria and Assessment Office, Research Triangle Park, NC.
Wiselogel AE, Bailey JK, Newton RJ, Fong F (1991) Growth response of loblolly pine (*Pinus taeda* L.) seedlings to ozone fumigation. Environ Pollut 71:43–56.
Zahner R, Saucier JR, Myers RK (1989) Tree-ring model interprets growth decline in natural stands of loblolly pine in the southeastern United States. Can J For Res 19:612–621.

11. Physiological and Biochemical Effects of Air Pollutants on Southern Pines

Thomas W. Sasek and Richard B. Flagler

Concerns about the susceptibility of the commercially important southern pine forests arose from observations of conspicuous forest decline in central and eastern Europe, in coniferous forests at high elevations in eastern North America, and in southern California (Skärby and Selldén, 1983). In the most general terms, decline in certain European forests has been linked to impacts of air pollutants and acidic deposition on tree physiology and soil biogeochemistry, especially on a local scale (Schultze et al., 1989); whereas recent declines of high-elevation red spruce (*Picea rubens* Sarg.), balsam fir [*Abies balsamea* (L.) Mill.], and Fraser fir [*Abies fraseri* (Pursh.) Poir.] seem to be a result in part of natural stresses exacerbated by acidic mists that decrease cold tolerance and increase winter injury (Adams et al., 1985; Johnson et al., 1992a; Sheppard et al., 1989; Woodman, 1987). In contrast, forests in the San Bernardino and southern Sierra Nevada mountains in southern California have been directly affected by high ozone concentrations originating in the Los Angeles basin (Miller et al., 1989; Williams and Williams, 1986).

There are several natural processes that have been shown to lead to significant forest injury and even stand-level dieback (Mueller–Dombois, 1987). However, of greater concern in the southeastern United States are the significant increases in acidic precipitation and air pollution during the past several decades, which could also affect tree growth rates and forest productivity (McLaughlin, 1985). The long-term impacts of chronic stresses on pine forests include effects that are both direct and cumulative. Numerous studies have documented the physiological effects of acute exposures to high concentrations of pollutants, like ozone, that result in rapid foliar injury and internal damage (Heck et al., 1982). However, it is important to consider physiological responses to chronic exposure at relatively low pollutant concentrations that are more characteristic of current regional pollutants. Under such conditions, cumulative impacts of exposure may have increased importance compared to transient peaks of high concentration (Reich, 1987; Richardson et al., 1992a).

As described in the previous chapter, a priority of the SCFRC research program was to identify physiological mechanisms directly related to growth, with lower priority placed on mechanisms involving indirect physiological responses mediated through soils or soil microorganisms. Winter injury and the direct and indirect effects of nitrogen deposition on southern pines were not as likely to have significant detectable impacts because of the short experimental time scales proposed. While the growth-response studies at the various sites were conducted mostly in parallel so that important policy questions could be addressed, physiological studies varied among the sites depending on the resources available. Different sites emphasized different physiological aspects of pollution stress, such as gas exchange characteristics, water-use efficiency, nutrition, and biochemistry.

Methodologies and Experimental Design

Detailed descriptions of the controlled laboratory studies, controlled field site studies, and comparative studies of mature trees with seedlings at the various SCFRC research sites is described in the previous chapter. Most of the physiological studies considered subsets of the trees within the larger experimental designs because of the constraints of time and equipment. Many ad hoc studies were also conducted, especially by visiting scientists. These studies often suggested avenues for future research to understand the physiological mechanisms of response more fully.

Gas Exchange and Carbon Assimilation

Methodology

The effects of acid precipitation and ozone on the gas exchange of southern pines were not characterized in parallel at the various research sites. Initial studies in the laboratory/greenhouse and controlled field sites concentrated on the response of photosynthetic capacity—that is, maximum photosynthetic rates under nonstressful environmental conditions and saturating irradiance. This approach was also used to develop exposure–response relationships of photosynthesis versus acid rain or ozone treatments, and it utilized large, replicated designs to improve statistical power. Repeated measurements with portable photosynthesis systems were made to determine response to cumulative exposure. Seedlings were grown in pots with artificial substrates or with field soils, while seedlings in the controlled field sites were grown directly in the soil. Other studies addressed differential responses of mature trees and branches compared to seedlings (Georgia) but necessarily involved smaller sample sizes.

Other methodologies were employed to gain a more complete understanding of physiological responses. Dougherty et al. (1992) measured carbon fixation of branches continuously, providing a record of the diurnal effects of pollutant exposure and an estimate of daily and seasonal net carbon fixation. Measurements of the response of photosynthesis to irradiance, carbon dioxide concentration, temperature, and so on were utilized to provide a more mechanistic understanding of the effects of pollutants on carbon fixation (Hanson and McLaughlin, 1988; Richardson et al., 1989, 1990).

Results of Short-Term Studies

Wiselogel et al. (1991) measured needle photosynthetic rates of loblolly pine seedlings exposed to ozone treatments in growth chambers. After nine weeks of fumigation at 320 ppb ozone (8 h/day, 4 days/week), photosynthesis rates were decreased up to 15% compared to charcoal-filtered controls. Hanson and McLauglin (1988) compared whole-shoot photosynthesis of first-year seedlings of loblolly pine grown in open-top field chambers or in greenhouse continuous stirred tank reactors (CSTRs). Acid rain treatments of pH 3.3 or 4.5 enhanced light-saturated whole-shoot photosynthesis compared to the pH 5.2 treatment. Ozone-treated seedlings in the field had reductions in light-saturated photosynthesis of up to 25% after 12 weeks. Responses in the CSTRs were mixed, perhaps because of reduced light availability in the CSTRs, the interrupted exposure regimes (4 days/week), and the short duration (12 weeks) of the experiments.

Results of Multiple Exposure Seasons

In the open-top chamber field studies, seedlings were exposed to acid precipitation and ozone treatments for multiple exposure seasons. Rates of photosynthesis of individual fascicles were measured in the various age classes of flushes. Thus, long-term cumulative effects of low ozone doses could be studied throughout the functional life span of a fascicle.

Seasonal Patterns

There were few consistent or statistically significant effects of acid rain treatments (pH 3.3, 4.3, or 5.3) on photosynthesis in loblolly pine (Sasek et al., 1991), shortleaf pine (Flagler et al., 1994), or slash pine (Byres et al., 1992) after three exposure seasons. However, for all three species, there was a general trend for slightly higher photosynthetic rates at pH 3.3 compared to the other acid rain treatments after one season of exposure. Most likely, the enhanced nitrogen input from the most acidic treatment stimulated photosynthesis. Comparable studies with loblolly pine (Edwards et al., 1990b), white pine (Reich et al., 1987), Scots pine (Skeffington and Roberts, 1985), Norway spruce (Barnes et al., 1990a), jeffrey pine, and giant sequoia (Temple, 1988) have shown little direct effect of acidic rain on photosynthesis except at extreme acidities.

Elevated ozone concentrations consistently resulted in statistically significant decreases in photosynthesis in all three pine species (Byres et al., 1992; Flagler et al., 1993; Sasek et al., 1991); however, there were no consistent ozone × rain interactions in the three species. Analysis of periodic measurements of loblolly pine showed that there was a significant time × ozone effect in a repeated measures statistical analysis but no time × rain effects or time × ozone × rain interactions (Sasek et al., 1991).

There was a progressive divergence of photosynthesis rates among the ozone treatments each growing season in all three pine species (Byres et al., 1992; Flagler et al., 1993; Sasek et al., 1991) except between the charcoal-filtered and nonfiltered treatments. In loblolly pine and shortleaf pine, different age classes of flushes were measured separately, as exemplified by the 1989 first-flush of loblolly pine shown in Fig. 11.1. In slash pine, the most recent fully expanded flush was measured at each sampling date, as shown in Fig. 11.2.

In loblolly pine, the intermediate ozone treatments did not significantly affect first-flush needles compared to the charcoal-filtered treatment until the second growing season. Net photosynthesis in September of the second growing season was decreased 25% at 1.33× and 50% at 1.67× ambient ozone compared to charcoal-filtered air (Fig. 11.1). Photosynthesis at 2.0× and 3.0× was decreased approximately 40 and 60%, respectively, compared to charcoal-filtered in September of the first growing season. During three years of treatments, exposure to 3.0× resulted in premature senescence and abscission of most first-flush needles by the end of their first growing season (Stow et al., 1992). For slash pine in Florida, photosynthesis rates were not affected as strongly by the ozone treatments (Fig. 11.2); however, only the youngest flush was measured at each sampling date, precluding detection of long-term, cumulative effects on older needles, as seen for loblolly pine. In September of the second

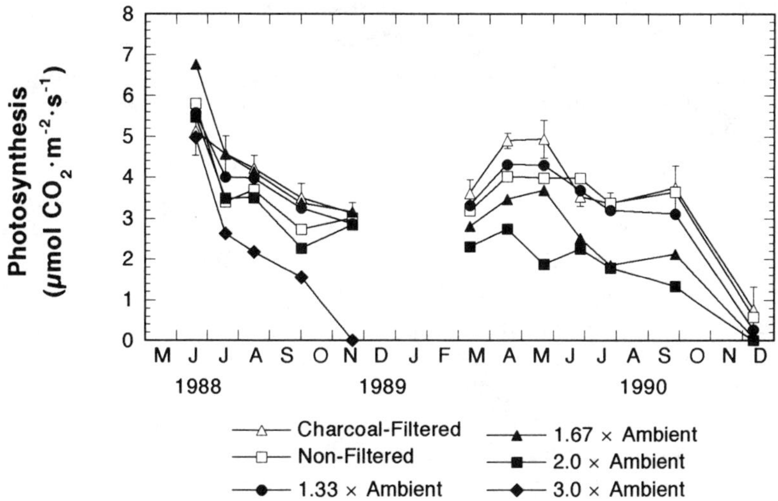

Figure 11.1. Photosynthesis of the first flushes of loblolly pine (North Carolina) produced in 1989 as affected by ozone treatment. Each point represents the mean of nine chambers. Error bars (±SE) on the charcoal-filtered treatment are representative of the other treatments.

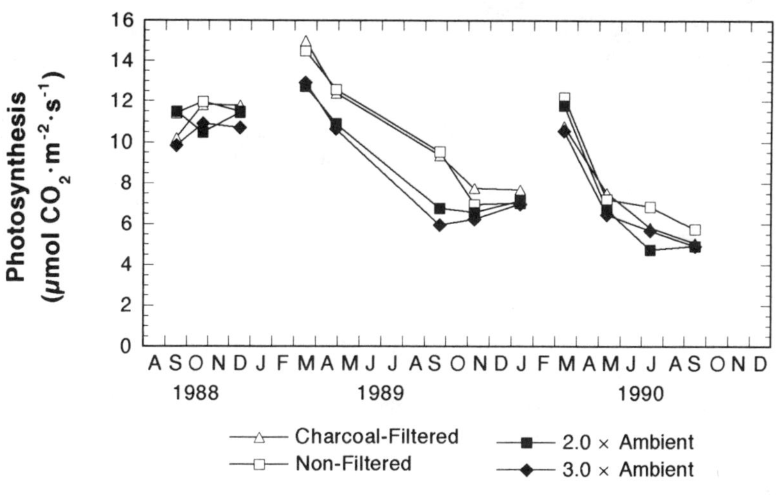

Figure 11.2. Photosynthesis of slash pine (Florida) as affected by ozone treatment. In 1988, the third flush of 1988 was measured. In 1989, the first flush of 1989 was measured. In 1990, the first flush of 1990 was measured (adapted from Byres et al., 1992).

growing season, photosynthesis at 2.0× and 3.0× decreased approximately 25 and 35%, respectively, compared to the control.

Photosynthesis rates of loblolly pine in North Carolina in winter were low, though some carbon was fixed during favorable temperature conditions (Sasek et al., 1991). Winter photosynthesis rates of slash pine in Florida and shortleaf pine in Texas remained higher given the warmer climates but were still less than half of the peak rates (Byres et al., 1992; Flagler et al., 1994). After overwintering, photosynthesis rates of both species increased to a peak in April or May of a flush's second growing season and then declined less sharply through summer than during the flush's first growing season. At the end of a flush's second growing season, rates dropped sharply as temperature decreased and needle senescence progressed. In loblolly pine, some charcoal-filtered needles were retained through a third growing season (Stow et al., 1992), but photosynthesis rates were not measured.

Although whole-tree gas exchange characteristics were not determined in these studies, it was possible to estimate whole-canopy photosynthesis by combining leaf area data with instantaneous photosynthesis measurements for each flush ageclass. For example, for shortleaf pine in Texas, Flagler et al. (1994) found that the significant impacts of ozone on needle retention and photosynthesis rates resulted in even more extreme differences at the whole-canopy level. Total carbon fixation per tree was estimated to be decreased by 34, 51, and 77% in the nonfiltered (NF), 1.7×, and 2.5× ozone treatments, respectively, compared to the charcoal-filtered treatment.

Although the sample sizes for all of the experiments were large, the relatively high variability in photosynthesis rates among trees decreased the statistical power of the analysis tests (Rotenberry and Wiens, 1985). For example, for Type I error (α) of 0.05 and Type II error (β) of 0.10, the minimum detectable differences between particular ozone treatments averaged about 25% in North Carolina (Table 11.1).

Cumulative Effects

The progressive divergence of photosynthesis rates among the treatments through time suggested a cumulative impact of ozone. Exposure–response relationships for photosynthesis were developed by relating the treatment "dose" to the measured photosynthetic potential, using the Weibull function to describe the relationship (Rawlings and Cure, 1985; Rawlings et al., 1988). For ozone, which directly impacts physiological processes, responses were related to the cumulative exposure received by each needle flush (based on flush emergence data), rather than the total cumulative exposure of the tree.

For shortleaf pine in Texas measured in the early spring (Flagler et al., 1994), photosynthesis rates of the previous season's flushes were well

Table 11.1. Statistical Power Tests Based on a Repeated Measures Analysis for Photosynthesis in Three Flushes of Loblolly Pine (North Carolina)

				Detectable Difference, R^2 (%)		
Source of Variation	Flush	n	Control Mean	$\beta = 0.20$	$\beta = 0.10$	$\beta = 0.05$
Between Subject						
Ozone	88-1	85	3.25	25	28	31
	89-1	99	3.82	22	24	28
	90-1	35	4.33	19	22	24
Rain	88-1	174	2.20	25	29	32
	89-1	204	2.87	20	23	25
	90-1	60	4.37	13	15	17
Within Subject						
Ozone	88-1	85	3.25	10	12	13
	89-1	99	3.82	10	11	13
	90-1	30	4.33	17	18	21
Rain	88-1	174	2.20	11	12	14
	89-1	204	2.87	9	11	12
	90-1	60	4.37	11	13	14

The minimum detectable differences (%) necessary to reject a null hypothesis (H_0) were calculated for three levels of Type II error (β) with Type I error at $\alpha = 0.05$. The reference control for ozone was the charcoal-filtered treatment at all pH levels, while the pH control was the pH 5.3 treatment at all ozone levels.

correlated to the ozone exposure they had received the previous growing season. Although older flushes had lower photosynthesis rates even in the charcoal-filtered control treatment, elevated ozone exposure further significantly decreased photosynthesis (Fig. 11.3). For loblolly pine in North Carolina (Richardson et al., 1992b), photosynthesis rates of the previous season's flushes were negatively correlated with their prior cumulative ozone exposure, once cumulative exposure exceeded approximately 150 to 200 ppm·h (Fig. 11.4). Thus, only the first and second flushes at 2.0× and 3.0× were significantly different from the other treatments since they had accumulated the greatest cumulative exposures the previous growing season. For the third and fourth flushes, there were no significant differences in photosynthesis among the ozone treatments because they had accumulated only a small cumulative exposure in the previous exposure season. In fact, about 50 ppm·h of the cumulative exposure occurred during winter, when ozone uptake rates into the leaves would be very low because stomatal conductance was low (Sasek and Richardson, 1992). Therefore, a cumulative ozone exposure of 100 to 150 ppm·h represents the true threshold in the exposure–response relationship for loblolly pine (Richardson et al., 1992b; Sasek and Richardson, 1992).

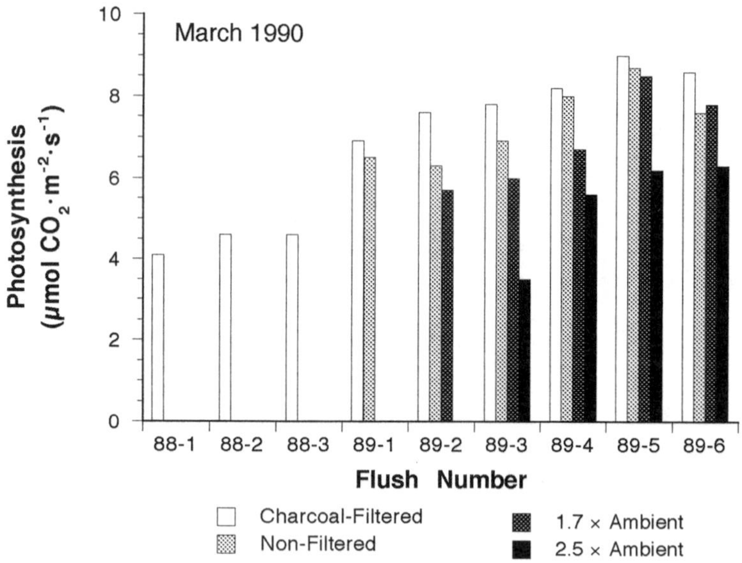

Figure 11.3. Photosynthesis of shortleaf pine (Texas) for each main-stem flush present in March 1990 (e.g., 88-1 represents the first flush produced in 1988). Seedlings had been exposed to the various ozone treatments for the previous two growing seasons (adapted from Flagler et al., 1994).

By expressing photosynthesis rates relative to the charcoal-filtered control rates at each sampling date, photosynthesis data could be combined into an overall exposure–response relationship for a particular flush based on the cumulative ozone exposure (12-h summation) at each sampling date (Fig. 11.5). This approach greatly reduces the influence of seasonal and aging effects on the exposure–response relationship (Sasek and Richardson, 1992). At each sampling date, photosynthesis was negatively correlated to cumulative ozone exposure during the life span of the needles, once a threshold exposure was exceeded. For loblolly pine first flushes, photosynthesis rates did not decrease until the cumulative exposure had exceeded 100 ppm·h. Relative photosynthesis rates were more than 80% inhibited at cumulative exposures greater than 400 ppm·h (Sasek et al., 1991). For slash pine in Florida, cumulative exposure–response relationships for photosynthesis were constructed without using relative rates and by measuring only the youngest flush at each sampling date (Byres et al., 1992). Photosynthesis rates declined with increasing cumulative exposure in the second and third exposure season but leveled off after 150 ppm·h (Fig. 11.6). Because relative rates were not used, the decrease was due to a combination of ozone effects, seasonal changes, tissue aging, and cumulative exposure, and is not directly comparable to the loblolly pine relationship (Fig. 11.5).

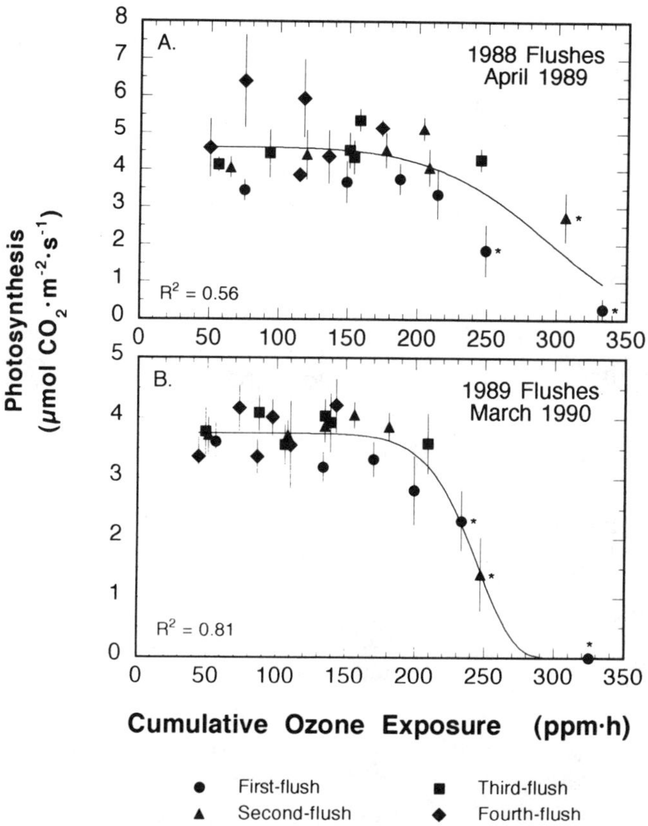

Figure 11.4. Photosynthesis versus the cumulative ozone exposure of the previous season's flushes for loblolly pine (North Carolina) measured in the spring of 1989 or 1990. Cumulative exposure was calculated from the time of needle elongation for each flush. Each point represents the mean ±SE of nine chambers. Within each flush, asteriks (*) indicate significant differences ($P < .05$). The data were fit with a Weibull function.

A dose–response relationship for a physiological process constructed from multiple sampling periods, like the aforementioned ones, can only be significant if the physiological process responds predominantly to cumulative exposure. For example, cumulative exposure indexes (concentration × time) often are not adequate because the absolute pollutant concentration may have a greater impact than the duration of exposure. However, in this study, measurements of loblolly pine made within the same growing season resulted in exposure–response relationships for photosynthesis that were highly significant (Sasek and Richardson, 1992). That is, for a given cumulative exposure, photosynthesis rates were inhibited to the same degree whether the dose was derived from shorter

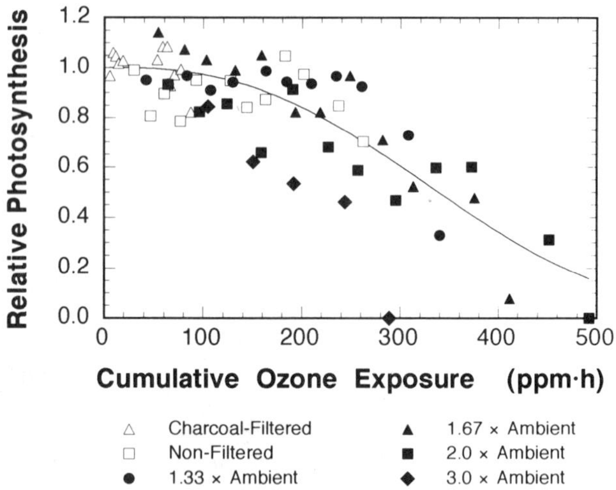

Figure 11.5. Photosynthesis versus the cumulative ozone exposure of the 1989 first flush for loblolly pine (North Carolina) measured during 1989 and 1990. At each sampling date, photosynthesis rates were expressed relative to the charcoal-filtered control. Cumulative exposure was calculated from the time of needle elongation for each flush. Each point represents the mean of nine chambers. The data were fit with a Weibull function (adapted from Sasek and Richardson, 1992).

exposures to higher concentrations of ozone or from longer exposures to lower concentrations of ozone (Richardson et al., 1992a and b). This approach was used successfully to demonstrate significant differences in the exposure–response relationships for photosynthesis among three half-sibling families of loblolly pine (Richardson et al., 1992a). The relative sensitivities of photosynthesis to ozone exposure among the three families matched the overall sensitivities of the families judged by previous screening trials.

However, when the sampling periods from multiple growing seasons were combined, the scatter between periods and among treatments increased. For example, photosynthesis rates at 2.0× and 3.0× were inhibited to a greater degree at a given cumulative exposure than were the other treatments at the same cumulative exposure (see Fig. 11.5). Thus, when many sampling periods are combined, the simple cumulative exposure index of concentration × time is less appropriate. Several previous studies have shown that peaks of concentration or the patterns of pollutant exposures can be most important in determining physiological response (Lee et al., 1988). Accordingly, weighted exposure indexes have been proposed to reflect the differences in concentration impacts (Pinkerton and Lefohn, 1987).

When multiple sampling periods extended over multiple growing seasons, other problems arose. There appeared to be a period during the

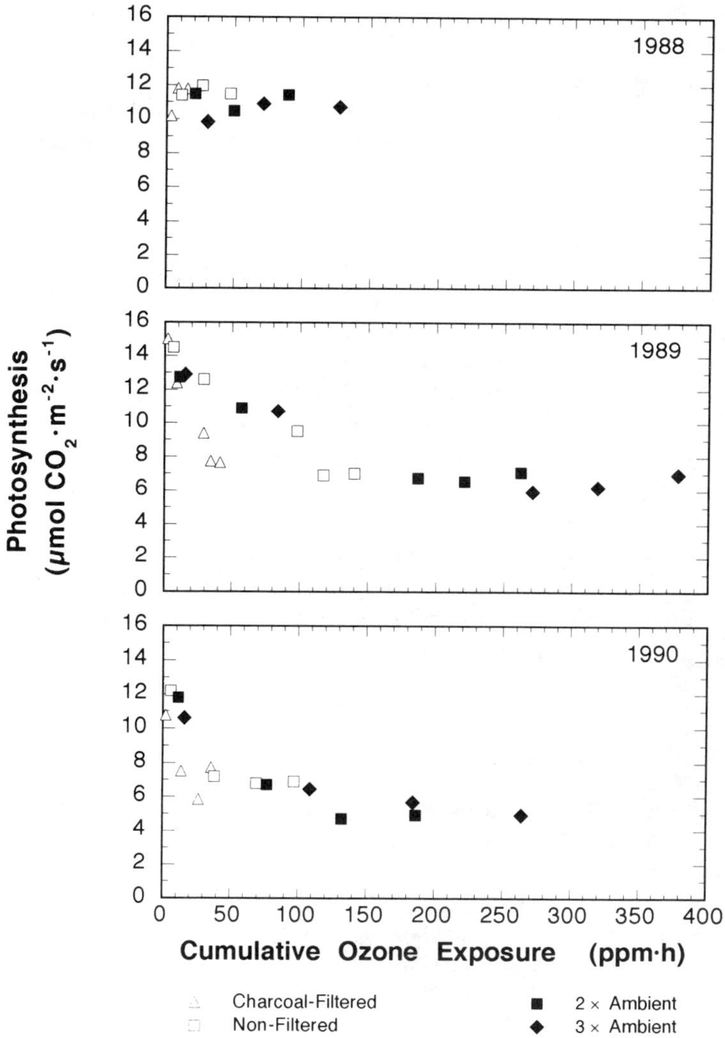

Figure 11.6. Photosynthesis of slash pine (Florida) in the youngest, fully expanded flush at each sampling date versus the cumulative ozone exposure of that flush measured in three successive exposure seasons (adapted from Byres et al., 1992).

winter in which cumulative ozone exposures had little physiological impact. During the winter (December through February), each treatment accumulated between 30 to 50 ppm·h in the Piedmont of North Carolina. There was no difference in the cumulative ozone exposure among the treatments during this period because there were no elevated ozone treatments during the winter. For the plants, this period corresponds to low physiological activity and low stomatal conductance, resulting in little ozone uptake into the leaves. Thus, cumulative exposure indexes should be modified to reflect seasonal patterns of physiological activity (Sasek and Richardson, 1992). A similar winter period with little physiological impact from ozone was observed for the patterns of foliage retention and abscission from these same trees (Stow et al., 1992). The reduced impact of ozone during the winter was also seen for slash pine in Florida (Byres et al., 1992).

The method of choice for developing a physiologically based dose–response relationship involves calculation of actual ozone uptake rather than external ozone exposure. A good approximation of ozone uptake can be made by using stomatal conductance data to estimate ozone flux into the leaf (Winner, 1994). The quality of the estimate is based on the amount of diurnal and seasonal conductance data that are available. For loblolly pine, a crude approximation of ozone uptake was calculated using the seasonal patterns of stomatal conductance measured monthly. The results showed a stronger relationship between photosynthesis and cumulative ozone uptake than cumulative ozone exposure (Sasek and Richardson, 1992). Thus, a dose–response relationship based on ozone uptake would reflect physiological processes more accurately. Estimations of ozone uptake should be a central goal of future physiological research in this area.

Carry-Over Effects

In a flush's second exposure season, there was a carry-over effect of ozone on photosynthesis from the previous season; plants in the elevated ozone treatments had rates still significantly lower than in charcoal-filtered air (see Figs. 11.1 to 11.3). That is, there was not complete recovery from ozone injury despite the winter period, when ozone exposures were low (Florida) or not elevated (Texas and North Carolina). During the remainder of a flush's second season, photosynthesis declined steadily in all treatments.

During successive years of chronic ozone exposures, there was some evidence of carry-over effects of ozone on the initial photosynthetic potential of new loblolly pine flushes produced on the trees. When measured for the first time, a young flush in the elevated ozone treatments exposed to only a small ozone dose had decreased photosynthesis rates that were related to the tree's prior ozone exposure (Sasek et al., 1991). However, in our second study, when the needles were measured closer in time to their emergence, the magnitude of initial carry-over effects was not significant (Fig. 11.7), similar to results found in slash pine (Byres et al., 1992). Thus, only the highest ozone treatment (3.0×) tended to decrease the initial photosynthetic capacities of new flushes significantly.

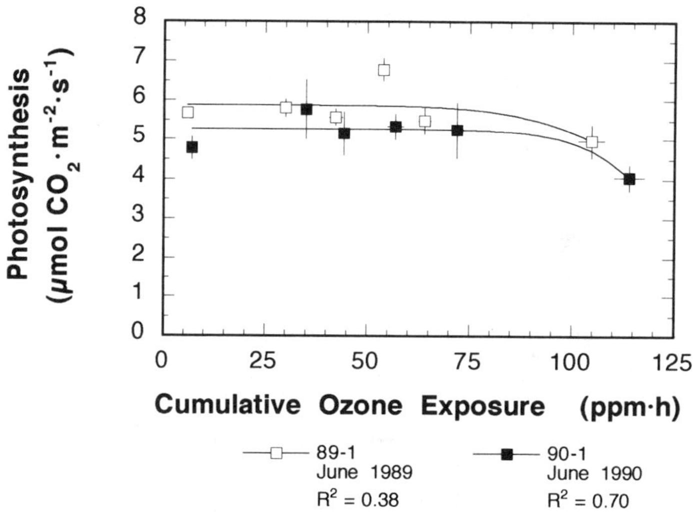

Figure 11.7. Photosynthesis in relation to the cumulative ozone exposure of loblolly pine (North Carolina) from a season's first flush when measured for the first time, just after elongation. Cumulative exposure was calculated from the time of needle elongation for each flush. The X- and Y-error bars represent the SE of nine chambers. The data were fit with a Weibull function.

Stomatal Conductance

The effects of ozone on stomatal conductance to CO_2 were generally similar to those on photosynthesis. Conductance decreased with increasing ozone exposure relative to that found in charcoal-filtered seedlings. Stomatal conductance also differed among different loblolly and shortleaf half-sib families (Flagler et al., 1994; Richardson et al., 1992a). For example, early in each exposure season, the stomatal conductance of the most resistant North Carolina family (8-80) was up to 30% less than in the other two families in most of the treatments. This difference diminished later in each exposure season for each flush. Lower stomatal conductance results in less ozone uptake into the leaves and may explain in part the apparent differences in sensitivity among the families. This is an important area for future research.

Stomatal responses to environmental factors are often closely linked to photosynthetic responses because of their impact on CO_2 flux into the leaf. This is often assessed by calculating the concentration of CO_2 inside the leaf based on the leaf-to-air concentration gradient created by the photosynthetic demand and the stomatal resistance to CO_2 entry. If stomatal resistance limits CO_2 uptake, intercellular CO_2 concentration (C_i) will decrease and thereby slow photosynthesis. In the elevated ozone treatments, stomatal conductances to CO_2 were lower than in the control treatment; however, C_i concentrations were actually higher than in the control treatment (Flagler et al., 1994; Sasek and Richardson, 1989). Therefore, decreased stomatal conductance did not significantly limit the supply of CO_2 for photosynthesis. Rather, the decreased photosynthesis rates that were observed in the elevated ozone treatments were a result of biochemical and physiological responses to ozone somewhere during photosynthesis (Sasek and Richardson, 1989). The decreased stomatal conductances in the elevated ozone treatments were more likely a response to the increases in C_i that resulted from decreased CO_2 fixation (Sharkey, 1985; von Caemmerer and Farquhar, 1981) rather than a direct response to ozone stress.

Seedlings Versus Mature Trees

The vast majority of experimental research on the gas exchange responses of forest trees to ozone has involved seedling or sapling studies. However, the relative ease of seedling studies is offset by apparent differences in mature tree response to similar ozone exposure. Consequently, it is difficult to scale the responses of seedlings and saplings to mature trees and forests. Differences in ozone uptake characteristics between seedlings and mature trees need to be clarified. Furthermore, given equivalent ozone uptake, there may be differences in the sensitivity to ozone from physiological differences such as antioxidant biochemical defenses or repair of oxidant damage.

It is generally assumed that seedlings are more sensitive to a given external ozone exposure than mature trees because stomatal conductance of seedlings is greater and therefore ozone uptake is greater. However, in red oak (*Quercus rubra*), seedlings had lower stomatal conductance than mature trees and consequently seedling photosynthesis was less sensitive to 2× ambient ozone fumigation (Samuelson and Edwards, 1993). Pine seedlings and mature trees may differ in stomatal conductance per unit leaf area either due to differences in stomatal density (Higginbotham, 1974; Knauf and Bilan, 1974; Thames, 1963) or due to differences in stomatal behavior (Cregg et al., 1989; Halpin, 1990). At the North Carolina site, a small study was conducted to compare the open-top chambered seedlings with nearby 20-year-old mature loblolly pines in which each tree's canopy was enclosed in an open-top chamber and exposed to various ozone treatments (Albaugh et al., 1992). Monthly measurements of photosynthesis at saturating irradiance suggested relatively small differences in seasonal patterns between seedlings and mature trees grown in charcoal-filtered air (Fig. 11.8). Photosynthesis rates of seedlings were inhibited more strongly through time than mature trees when exposed to 2× ambient ozone. Nevertheless, the 1989 first-flush needles of trees and seedlings both prematurely abscised at approximately

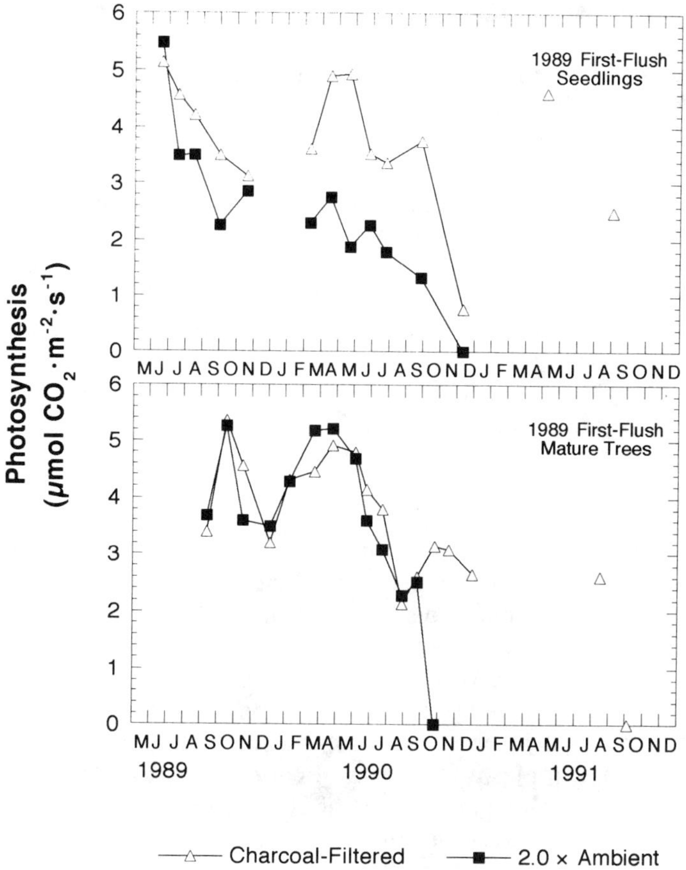

Figure 11.8. Photosynthesis of the 1989 first-flush lobolly pine (North Carolina) seedlings in charcoal-filtered or 2.0× ambient ozone treatment compared to the same flush in 20-year-old mature trees exposed to the same treatments.

the same date, an entire growing season sooner than charcoal-filtered trees or seedlings.

However, instantaneous measurements of photosynthesis do not reveal important diurnal patterns of response. Halpin (1990), in a more extensive comparative study in Georgia, found significant differences in the diurnal patterns of photosynthesis and stomatal conductance between seedling and mature loblolly pines. These comparisons were between branch chambers (mature trees) and open-top chambers (seedlings). Mature trees maintained relatively stable gas exchange rates through much of the day, whereas seedlings had higher gas exchange rates than mature trees early in the day; however, rates declined during the day to rates that were less than the mature trees by afternoon. When averaged for the entire day, mean daily stomatal conductance in mature trees was 20 to 40% less than in seedlings.

Comparisons of the gas exchange characteristics of seedlings and mature trees are also difficult because microenvironmental parameters may differ significantly. Mature trees with a well-developed canopy experience strong light gradients from the top of the canopy to the bottom or from one side to another. Physiologically distinct sun- and shade-adapted leaves develop (Kramer and Kozlowski, 1979). Leaf variation in seedlings is often much less because there is little self-shading within the canopy (Cregg et al., 1989). Sun- and shade-adapted foliage have different ozone uptake characteristics because stomatal behavior differs (Samuelson and Edwards, 1993). However, among leaves exposed to similar light levels, it is not clear to what degree seedlings and mature trees differ in gas exchange characteristics.

Carbohydrates

The carbon fixed during photosynthesis can be utilized directly for energy, exported for use in other, nonphotosynthetic tissues, or may be stored for future use. Carbohydrate pools are dynamic and respond to a variety of internal and external influences because starch synthesis and sucrose synthesis are competing biochemical pathways. Triose phosphates produced during the Krebs cycle of photosynthesis can be directed either toward starch synthesis in the chloroplast for storage or toward sucrose synthesis in the cytoplasm for storage or for export to other parts of the plant (Fig. 11.9). The partitioning of this carbon is controlled by the relative concentrations of orthophophate (P_i) and triose phosphate in the cytoplasm versus the chloroplasts.

The patterns of carbohydrate accumulation and distribution often reflect the diurnal and seasonal patterns of tree growth, photosynthesis, and respiration. Pollutant stresses have been hypothesized to have various effects on the carbon metabolism pathways (see the subsequent section on biochemical responses). In general, decreased photosynthesis results in less starch accumulation since there are fewer excess photosynthates for storage. In addition, increased respiratory demands for repair and defenses may decrease the flow of carbon toward starch synthesis and increase carbon flow toward sucrose synthesis and translocation. Therefore, changes in the relative concentrations of starch, sucrose, and other soluble sugars may be indicative of stress.

The quantification of specific carbohydrates in the various biomass compartments and the characterization of their diurnal and seasonal patterns are logistical challenges. In pines, biochemical methodology is further complicated by the presence of large quantities of interfering compounds such as phenolics. Consequently, there were only a few SCFRC projects that measured carbohydrates systematically.

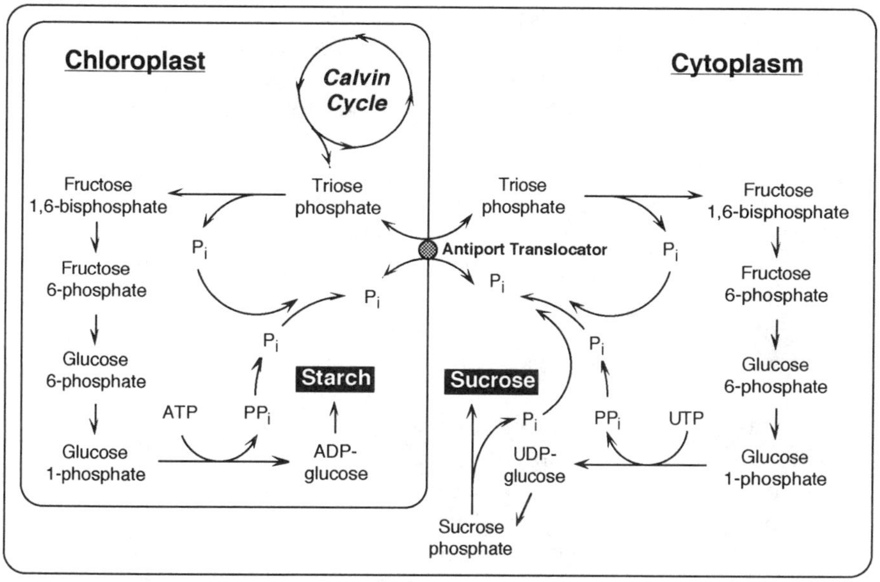

Figure 11.9. Carbon flow through the biosynthetic pathways within a typical plant cell toward either starch formation in the chloroplast or sucrose formation in the cytoplasm. The triose phosphate/orthophosphate (P_i) antiport translocator in the chloroplast membrane is an important control point in the two pathways.

Wiselogel et al. (1991) found no significant effects of ozone on root soluble sugars or starch content of 22-week-old loblolly pine seedlings exposed to either 160 or 320 ppb ozone for nine weeks in a controlled environment chamber study. In another study with loblolly pine growing in open-top chambers, Faulkner et al. (1991) reported that both soluble sugars and starch in fine roots were decreased after two growing seasons by 1.5× and 3.0× ambient ozone treatments when compared to a charcoal-filtered treatment. Carbohydrates in the foliage increased at the intermediate ozone level but decreased at the 3.0× ozone exposure.

In a comprehensive evaluation of pollutant effects on carbohydrates, Paynter et al. (1991, 1992) evaluated the effects of prolonged exposure to ozone and acid rain on carbohydrate pools in four families of shortleaf pine growing in open-top chambers in South Carolina. During 38 weeks of exposure, they reported no differences among the four families. There were generally no pollutant treatment effects until the thirty-second week of exposure. Subsequently, ozone caused a general decrease in foliar levels of starch and sucrose and increases in glucose, total reducing sugars, and total soluble sugars. These alterations in the carbohydrate pools balanced sufficiently that there was no net ozone effect on total nonstructural carbohydrates in the foliage (Paynter et al., 1991). At the thirty-eighth week of exposure, glucose and total reducing sugars in the foliage increased while sucrose levels decreased as acidity of rain increased.

Paynter et al. also determined the carbohydrate levels in new foliage developed during the third growing season in the same study (Paynter et al., 1992). No differences among the four families were evident during this phase of the study, and there were no ozone × acid rain interactions. Carbohydrate pools in two separate growth flushes, one current-year needles and one previous-year needles, were evaluated. Effects of the treatments varied considerably among the sampling dates, flushes, and the various carbohydrate pools. In current-year needles, ozone had no effect on starch, glucose, or sucrose, but midseason (April to June) total reducing sugars were increased at the highest ozone exposure level. Acid rain affected this flush only in the September sampling period, during which glucose increased slightly and starch and sucrose decreased as acidity of rain increased. In previous-season needles, ozone decreased midseason sucrose, starch, and nonstructural carbohydrates and increased glucose and total reducing sugars as exposure increased. Acid rain treatments affected only midseason (April) carbohydrate levels. Total soluble sugars and total reducing sugars increased and starch levels decreased as acidity of treatments increased.

Paynter and Reardon (1992) made diurnal measurements of several carbohydrates on two occasions (June and August) within the same study and evaluated the effects of ozone and acid rain treatment extremes on the various pool sizes. Acidic rain had no effects on any of the carbohydrates assayed. Ozone effects varied somewhat between the two sampling periods. Glucose levels were unaffected by ozone in June but were decreased at all times during the diurnal cycle in August. Sucrose levels were decreased by ozone only at midday during both months. Total reducing sugars were increased midday in June but decreased in the early morning hours in the August sampling. Starch content of needles was unaffected by ozone in the June sampling and increased in the early morning hours in the August sampling. Total soluble sugars were also unaffected in June but decreased in the early morning hours in the August sampling period.

Reardon et al. (1992), in a small study conducted within the aforementioned study, exposed containerized seedlings of one family of shortleaf pine to ozone and acid rain within the same open-top chambers. After 16 months of exposure, the seedlings were harvested and carbohydrate levels of needles, stems, and roots were determined. The authors found that acid rain had no effect on any of the carbohydrates assayed in any of the three tissues examined. Ozone effects on foliage were varied and generally were in agreement with the aforementioned results. There were no significant effects of ozone on either stem or root carbohydrate pools.

Faulkner et al. (1991) measured carbohydrates from foliage, taproot cores, and whole-tree root excavations of loblolly pines grown in open-top chambers for two growing seasons in North Carolina. Total available carbohydrates in foliage and fine roots were significantly decreased only at 3.0× ambient ozone, primarily due to decreased starch content.

Based on the various studies reviewed, it appears ozone has a more pronounced effect on carbohydrate content than acidic rain. From the work of Paynter et al. (1991, 1992), it is clear that it is important to sample intensively. Single point-in-time determinations are of limited value given the highly dynamic nature of the various carbohydrate pools. The work of Paynter et al. provides a comprehensive data set on foliar carbohydrate dynamics as affected by ozone and acid rain in loblolly pine. However, data are scarce regarding carbohydrate dynamics in other tissues, especially roots. Most studies indicated decreased carbohydrates in roots as a result of ozone exposure, especially after prolonged exposure. This is consistent with carbon allocation shifts due to ozone stress, which have been reported previously (Cooley and Manning, 1987; Edwards et al., 1992a). Allocation studies using ^{14}C-labeling (Adams et al., 1990) or studies using the short-lived ^{11}C isotope (Spence et al., 1990) also suggested a decreased allocation of carbon to roots of ozone-stressed southern pines.

Water-Use Efficiency

Water deficit is a major factor that limits tree growth in the southern United States (Kramer and Kozlowski, 1979; Teskey and Hinckley, 1986). The efficiency of water use by trees can determine where they grow, how much they grow, and if they survive during periods of drought. Plant water status affects stomatal behavior, which in turn can affect gas exchange rates. This includes not only the uptake of carbon dioxide for photosynthesis and the loss of water by transpiration but also the uptake of gaseous pollutants. While there have been several studies (mostly with crop and herbaceous species) that have reported effects of water status on pollutant response, there is only limited information concerning the effects of regional air pollutants on the water relations of southern pines.

Water-use efficiency (WUE), a measure of carbon gain per unit of water loss, can be determined by a number of techniques. The most direct method involves quantifying all water added to the tree, all water lost through evaporation and transpiration, and the amount of carbon gained by the tree over some measurement period. This time-consuming and intensive technique severely limits feasible sample sizes in most studies, especially for larger plants like trees.

A more convenient method that has become increasingly popular is based on portable photosynthesis systems that simultaneously measure net carbon gain and water loss. An "instantaneous" WUE can be calculated and has the advantage of allowing the operator to make many measurements in a relatively short time. A major disadvantage of the technique is that the instantaneous WUE accurately reflects only the time period measured (often less than one minute) and the environmental conditions within the cuvette.

Another important technique is the use of stable carbon isotopes to estimate average water-use efficiency integrated over long time periods (Rundel et al., 1989). This technique is based on the known atmospheric ratio of $^{12}CO_2/^{13}CO_2$ and the fact that plants discriminate against the heavier $^{13}CO_2$ both physically during uptake through the stomata and enzymatically during photosynthetic carbon fixation. Therefore, by comparing the $^{12}CO_2/^{13}CO_2$ ratio in plant tissues (determined by mass spectroscopy) with the atmospheric ratio, a discrimination value is calculated ($\delta^{13}C$). Over long time periods, the $\delta^{13}C$ value represents an average, integrated degree of stomatal opening once the enzymatic discrimination is factored out. The ratio of $^{12}CO_2/^{13}CO_2$ has been shown experimentally to be well correlated to WUE in many species of C_3 angiosperms. The major advantages of this technique are that it is relatively simple in terms of sampling and the long averaging periods reduce differences associated with short-term environmental influences. The major disadvantage is the high cost per sample for analysis on a mass spectrometer.

In an open-top chamber study, Boutton and Flagler (1990, 1991) evaluated both instantaneous and integrated WUE of shortleaf pine exposed to acidic rain and ozone treatments. After 15 months of exposure, they reported significant ozone effects on WUE but no effects of acidic rain. Ozone treatment caused a significant linear decrease in instantaneous WUE, due to the proportionately larger impact of ozone on photosynthesis compared to transpiration. Integrated WUE, as measured by $\delta^{13}C$ values, increased slightly, supporting the observation that ozone treatments decreased stomatal conductance.

Elsik et al. (1992, 1993) evaluated the effects of ozone and water deficit on instantaneous and integrated WUE of containerized loblolly and shortleaf pine seedlings grown in open-top chambers. At the end of four months of treatments, shortleaf pine had greater WUE than loblolly pine. Instantaneous WUE was not affected by ozone treatment in either species (Elsik et al., 1992), probably due to the short-term nature of the study. Integrated WUE, as measured by $\delta^{13}C$, was affected by ozone treatments (Elsik et al., 1993). Foliar $\delta^{13}C$ increased in both species as ozone exposure increased. Stem tissue $\delta^{13}C$ increased with increasing ozone exposure only in loblolly pine. These increases in $\delta^{13}C$ were attributed to decreases in stomatal aperture caused by ozone. These determinations were supported by instantaneous measurements of stomatal conductance.

While there have been numerous experiments that quantitatively relate $\delta^{13}C$ to WUE in crop plants, analogous experiments have not been conducted on conifers. Until this linkage has been tested, we must use caution in interpreting these results. In general, for southern pines, instantaneous WUE decreased as a function of increasing ozone exposure, due largely to effects on photosynthesis. Integrated WUE data suggest that elevated ozone exposures tend to increase WUE slightly. These data are supported by Taylor et al. (1989), who report that $\delta^{13}C$ of loblolly pine seedlings was increased by elevated ozone relative to ambient levels. The decrease in instantaneous WUE is misleading in some ways because with decreased stomatal conductance and less leaf area at elevated ozone concentrations, the trees may transpire less water per tree. Therefore, ozone-stressed trees may be more drought tolerant.

Tree Nutrition

Soil fertility can have a profound effect on plant growth, and plant response to air pollution can be influenced by nutritional status. Most often, visible injury of plants due to various pollutants was correlated to increasing or decreasing levels of soil fertility. Little research exists regarding the interactions of air pollutants, plant nutrition, and growth. A few studies, however, investigated the effects of acidic precipitation and ozone on the nutrition of southern pines.

Wright et al. (1990) observed the influence of acidic precipitation and ozone on nitrogen nutrition of young loblolly pine grown in open-top chambers for one growing season in Alabama. The N content (g/tree) and biomass of the trees increased with increasing rain acidity, indicating a possible fertilizer response. This hypothesis is plausible since the soil at the Alabama site has been characterized as N deficient (Crawford et al., 1989).

Concentrations of N increased with increasing ozone concentrations in stems but not in foliage (Wright et al., 1990). Total plant nitrogen content did not change with ozone concentrations, although the N content in the stems was increased in the high-ozone treatments. This increase in N in the stem tissues without a subsequent increase in total plant biomass in the higher ozone treatments is believed to result from an increase in internal translocation of N from prematurely senescing foliage. There was no evidence that ozone and acidic precipitation stimulated leaching of foliar N. The effect of ozone on internal translocation of N deserves further attention, since this mechanism is important in annual nitrogen requirements (Switzer and Nelson, 1972). Any alteration of internal translocation could have major implications for nutrient cycling in loblolly pine stands.

Johnson et al. (1992b) investigated the role of homeostatic growth in explaining the effects of ozone and acidic precipitation on loblolly and slash pine. The concept of homeostatic growth (Agren and Ingestad, 1987; Johnson et al., 1985) is based on the theory that resource allocations in trees are optimized in response to favorable and detrimental environmental conditions. For example, a factor that stimulates growth consequently dilutes N concentration, which in turn can slow the increased growth rate. Similarly, environmental stresses that decrease growth can actually increase nutrient concentration, thereby counteracting growth inhibition to some degree. In loblolly and slash pine, Johnson et al. (1992b) found adjustments in growth rates that correlated to shifts in foliar N and biomass. Growth rates generally decreased in response to elevated ozone concentrations and increased in response to acidic precipitation. Foliar N content was higher in elevated ozone treatments in both pine species and higher with increased acidity in loblolly pine but less with slash pine. Foliar biomass was decreased with elevated ozone for both species and greatest in the low-acidity treatments with slash pine but greatest in the high-acidity treatment with loblolly pine.

Manderscheid et al. (1992) investigated the effect of ozone on foliar nitrogen metabolism of loblolly pine in open-top chambers during two growing seasons in North Carolina. Soluble amino acids, glutamine synthetase, glutamate hydrogenase, soluble protein, and total chlorophyll were measured in second-flush foliage at the November 1989 harvest. The amino acid glutamate and the enzymes glutamine synthetase and glutamate dehydrogenase increased two- to threefold at 2× and 3× ambient ozone, respectively. Soluble protein content also increased with increasing ozone exposure. Since trees at the highest ozone concentrations exhibited the greatest amount of foliage senescence, the results indicate that the trees internally transported N from the senescing tissues, as suggested by Wright et al. (1990).

Elsik (1992) characterized the concentration and content of N, P, Ca, Mg, and K in containerized, one-year-old loblolly and shortleaf pines during one growing season in the Texas open-top chambers. The trees were exposed to five levels of ozone (0.2× to 2.5× ambient) and two drought stress regimes (well-watered and mild drought stress). Increasing ozone concentrations resulted in increases in N and P concentrations in the stem and increases in N, Mg, and Ca concentrations in the foliage. Ozone caused an increase in foliar and stem N, in foliar Mg, and in root K of shortleaf pine. Ozone exposures resulted in an increase in stem N content in shortleaf pine and a linear decrease in root N content for both loblolly and shortleaf pine.

Elsik (1992) also observed several ozone × drought stress interactions. Nitrogen concentration increased linearly with ozone in roots, stems, and foliage of loblolly pine and foliage of shortleaf pine in well-watered seedlings but not stressed seedlings. Phosphorus concentrations in loblolly pine stem and roots exhibited a linear increase with increasing ozone in well-watered but not drought-stressed seedlings. Nitrogen content decreased linearly with increasing ozone in drought-stressed but not well-watered seedlings. Nitrogen-use efficiency (NUE = seedling biomass ÷ seedling N content) decreased significantly with increasing ozone in well-watered but not drought-stressed loblolly and shortleaf pines. Increases in N and P concentrations resulted from increased internal translocation of these nutrients from prematurely senescing foliage, as shown in other studies (Manderscheid et al., 1992; Wright et al., 1990).

Edwards et al. (1992b) determined the effects of ozone and acidic precipitation on loblolly pine nutrient status after three growing seasons of ozone exposure in open-top chambers in Tennessee (a non-SCFRC-funded project). After three years of exposure, neither ozone nor acidic precipitation accelerated foliar leaching or affected foliar nutrient concentrations in loblolly pine.

In summary, based on limited research, results suggested that ozone exposure results in a shift or alteration in internal translocation of mobile nutrients such as N and P. Such responses would have important implications for pine stand nutrition. The effects of ozone on nutrient allocation and growth rates indicate the presence of homeostatic growth responses. In nutrient-poor soils, nutrient-rich acidic precipitation may be an important source of mineral nutrition input. However, after chronic exposure, H^+ deposition may override any "fertilizer effect" and could cause a reduction in growth.

Biochemical Responses

There is a strong concentration gradient driving ozone uptake from the atmosphere into the leaf, because ozone concentrations inside the leaf approach zero (Laisk et al., 1989). Ozone uptake is due almost entirely to stomatal flux rather than epidermal flux and is therefore strongly affected by stomatal conductance (Hanson and Taylor, 1990). Once inside the leaf, ozone can penetrate deep into the mesophyll tissue because it is not as highly reactive and highly water soluble as some other gases, like HNO_3 or SO_2 (Taylor et al., 1988). Approaching a cell, ozone passes through the apoplastic water sheath into the region of the cell wall. Only a negligible fraction of the ozone decomposes before reaching the plasmalemma (Chameides, 1989); however, ozone may be scavenged directly by the presence of ascorbate in the cell wall (Chameides, 1989; Heath, 1988). Thus, ascorbate and other antioxidant molecules in the intercellular fluid may be a primary defense against low-ozone fluxes.

Ozone first interacts with the living cell at the plasmalemma, reacting with unsaturated fatty acid residues of the polar lipids, aromatic residues and primary amines, or exposed sulfhydryl groups (Heath, 1980). As a result, altered passive membrane permeability may cause "leakiness," which disrupts membrane fluidity, chemical gradients, and osmotic potential. Changes in these characteristics can disrupt normal biochemical pathway activities, even those inside chloroplasts and mitochondria. For example, decreased photosynthesis rates may result in part from changes in osmotic potential and turgor pressure rather than from actual damage to the photosynthetic apparatus (Heath, 1988). Ozone passing through the plasmalemma similarly reacts with intracellular biochemicals and membranes and may induce further disturbance.

Once ozone enters the cell, a variety of biochemicals and metabolic pathways can be affected. These can result in acute effects or chronic, cumulative effects. However, there are many responses that are not specific to ozone stress; instead, certain responses are typical of stress in general. In Fig. 11.10, a generalized scheme of carbon biosynthesis and metabolic pathways is presented for a typical plant cell. Because most of the pathways are linked together and because carbon can flow in different directions through the pathways, pollutant effects in one part of the cell can have cascading effects through several pathways (Dizengremel et al., 1994). For example, as previously discussed, the biosynthetic pathways leading to starch or sucrose are competitive reactions. Increased sucrose synthesis necessitates a reduction in starch accumulation.

Ozone molecules diffusing through the cytoplasm react strongly with various molecules they encounter. In addition, this oxidant attack often leads to the propagation of free radicals inside the cell. Some of the most common free radicals generated from ozone include the hydroxyl radical, the superoxide radical, and hydrogen peroxide (Fig. 11.10). In turn, these highly reactive molecules can directly damage more cellular constituents. One indicator of oxidant damage from ozone decomposition products is lipid peroxidation, as evidenced by the presence of malondialdehyde (MDA). In loblolly pine seedlings, MDA content in current season's needles was doubled in the 3× ambient ozone treatment compared to the charcoal-filtered control, though MDA content at 1.5× ambient ozone was only slightly increased (Richardson et al., 1990). Fendick (1991) measured total sulfhydryl content (protein-bound and non-protein-bound) in loblolly seedlings as another possible indicator of oxidant damage to a variety of cellular molecules. There was some evidence for slightly decreased sulfhydryl content at 3× ambient ozone, perhaps indicating oxidant stress.

On the other hand, the direct attack of double bonds in membrane lipids by ozone itself ("ozonolysis") may be more important, especially at the plasmalemma (Chameides, 1989; Heath, 1980). Consistent with this prediction, elevated ozone treatments changed the degree of saturation of various 18-carbon fatty acids in loblolly pine needles (Fangmeier et al., 1990). With increasing ozone exposure up to 3×, the content of the more

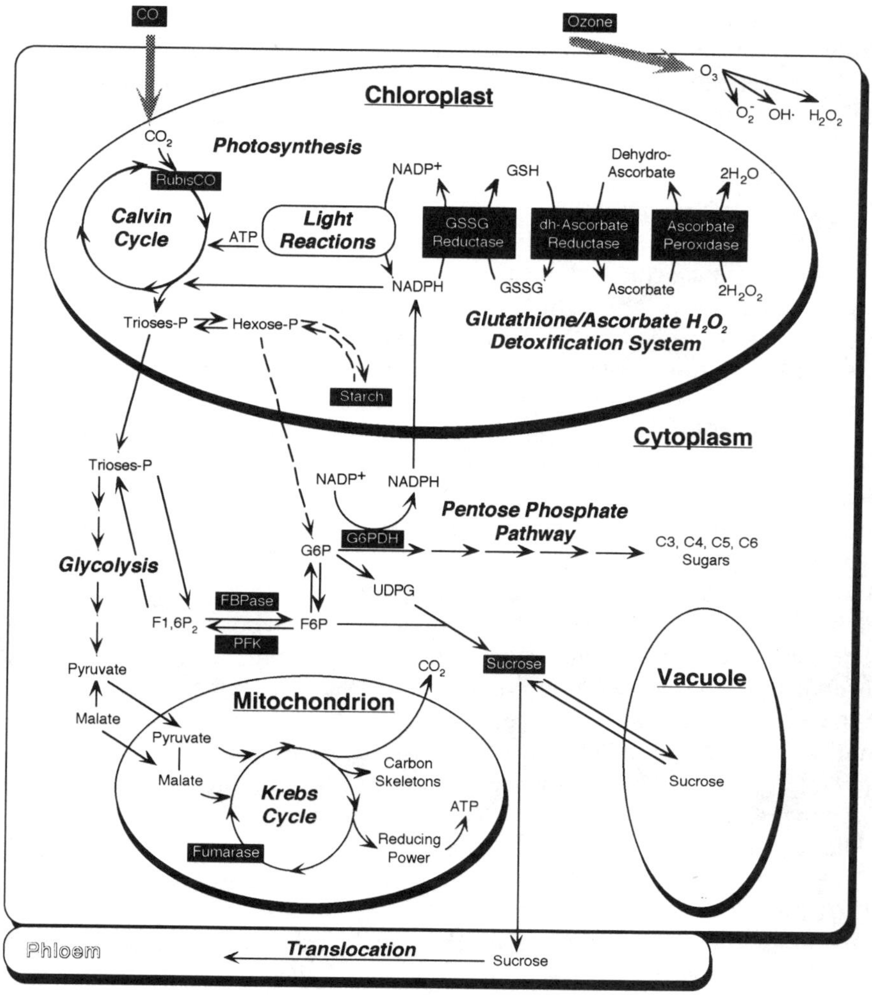

Figure 11.10. Carbon flow through the biochemical pathways within a typical plant cell indicating important processes affected by ozone exposure. Molecules and enzymes highlighted in black are some of the key control points discussed in the text. Key: F1,6P2 = Frutose 1,6-bisphosphate; FBPase = Fructose bisphosphatase; F6P = Fructose 6-phosphate; G6P = Glucose 6-phosphate; G6PDH = Glucose-6-phosphate dehydrogenase; GSSG = Oxidized glutathione; GSH = Reduced glutathione; PFK = Phosphofructokinase; RubisCO = Ribulose bisphosphate carboxylase oxygenase; UDGP = Uridine diphosphoglucose.

unsaturated linolenic acid (three double bonds) decreased by 30 to 40% compared to the charcoal-filtered controls, while the contents of the more saturated oleic acid (one double bond) and linoleic acid (two double bonds) increased by 15 to 25%. However, it was not clear whether this response was a result of ozonolysis of linolenic acid to oleic acid and linolenic acid or whether there was a long-term shift in lipid metabolism.

To scavenge free radicals and quench radical-generating chain reactions that are formed from a variety of oxidative processes (including ozone exposure), antioxidant defense mechanisms are present throughout the cell (Richardson et al., 1990). A variety of enzymatic and nonenzymatic biochemicals are considered to be antioxidants. Several of these antioxidants have been shown to be stimulated in pine seedlings in response to ozone stress. The activity of superoxide dismutase (SOD) increased up to 200% in a 3× ambient ozone treatment compared to charcoal-filtered air and increased up to 42% in a 1.5× ambient ozone treatment (Richardson et al., 1990; Sheu and Chappelka, 1993; Tandy et al., 1989). Peroxidase activity increased up to 116% at 3× ambient ozone compared to charcoal-filtered air and increased up to 50% at 1.5× ambient ozone. Manderscheid et al. (1991) found very similar increases in peroxidase activity at 2× and 3× ambient ozone in another study of loblolly pine. The glutathione/ascorbate H_2O_2 detoxification pathway produces scavengers of reactive hydrogen peroxide (Fig. 11.10). The pools of these reduced radical scavenging molecules, like glutathione and ascorbate, decreased as much as 50% at 3× ambient ozone compared to charcoal-filtered controls, suggesting radical scavenging activity under ozone stress (Fendick, 1991). Glutathione reductase and ascorbate peroxidase, which replenish the radical scavengers, more than doubled in activity at 3× ambient ozone compared to charcoal-filtered air. Manderscheid et al. (1991) found that pools of ascorbate increased in the elevated ozone treatments after several months of exposure, suggesting increased defenses.

Operation of the glutathione/ascorbate H_2O_2 detoxification pathway (Fig. 11.10) is a potential drain on the pool of reduced nicotinamide adenine dinucleotide phosphate (NADPH) produced during the light reactions of photosynthesis, which normally would be used in the Calvin cycle for carbon fixation (Alscher and Amthor, 1988; Fendick, 1991). Therefore, photosynthesis rates might decrease when detoxification was occurring. At low ozone fluxes, this diversion of photoreductant may be the most important factor affecting carbon fixation rates. However, with high ozone stress, other biochemical disruptions and cellular damage have a much greater relative impact on net photosynthesis (Fendick, 1991).

Few studies have been able to detail the effects of pollutants like ozone on enzyme activities and other metabolic responses in trees because of the difficulty of extracting active enzymes from woody plant tissues (Citerne et al., 1988; Gerant et al., 1989; Pitel and Cheliak, 1986). Nevertheless, the sensitivity of some key enzymes to gaseous pollutants was recently reviewed (Koziol et al., 1988). In North Carolina, Dizengremel et al. (1994) found that decreased needle photosynthetic rates were correlated to a parallel decline in RuBP carboxylase (RubisCO) activity and content (Fig. 11.10). In contrast, increases were observed in the activities of several enzymes linked to carbohydrate metabolism. There was a 75% increase in the activity of glucose-6-phosphate dehydrogenase (G6PDH) after 350 ppm·h (Fig. 11.10). The marked increase in the activity of G6PDH in the pentose phosphate pathway may be related to an increased need for secondary metabolites that can be used for repair of ozone-induced damage. Similarly, there was a smaller increase in phosphofructokinase (PFK) and fumarase activities (15 to 30% after 350 ppm·h). PFK is part of the degradative pathway for sugars, while fumarase is a key control point in the Krebs cycle (Fig. 11.10). Mitochondrial respiration provides a source of energy (adenosine triphosphate, ATP) for various repair processes as well as various carbon skeletons that may be needed. In previous studies, respiration increased in coniferous and broad-leaved trees that were fumigated with near-ambient concentrations of ozone (Guderian et al., 1989; Reich, 1983). Thus, current ambient ozone concentrations could lead progressively to biochemical disequilibrium within leaf cells, reduced photosynthate production, and increased respiration. Carbon flow through the various biochemical and biosynthetic processes may change direction or relative magnitude.

Implications of Physiological Responses

The inhibition of photosynthesis by ozone in older flushes and the premature abscission of the oldest flushes may result in altered whole-tree carbon and nutrient allocation patterns. Altered allocation patterns may then have various feedback effects that further affect carbon assimilation and metabolism. Initial results suggest that the effects of elevated ozone treatments may reduce over winter storage reserves, adversely affecting the photosynthetic capacity of new growth in the spring. On the other hand, during the growing season, reallocation of carbon and nutrients from ozone-damaged flushes may result in enhanced photosynthetic capacity in new flushes. However, subsequent exposure to ozone stress decreases photosynthesis in all flushes.

Direct Effects

The relative sensitivities of various plant species to external ozone exposure have been well correlated to leaf diffusive conductance, because physiological response to ozone depends on the actual ozone uptake into the leaves (Reich, 1983, 1987). Given an equivalent ozone uptake, most species have a similar response to ozone at the cellular level (Reich, 1987). Ozone and other oxidants are extremely reactive molecules and interact with many biochemicals in the cell wall and the plasmalemma

(Heath, 1980). These chemical interactions produce destructive free radicals that also oxidize organic molecules, including enzymes and membrane lipids deeper inside the cells. The primary defense against oxyradicals, the antioxidant defense system, includes components that are enzymatic (e.g., superoxide dismutase, peroxidases, and catalase) and nonenzymatic (e.g., α-tocopherol, β-carotenes, glutathione, and ascorbate). The presence of these antioxidant mechanisms suggests that there may be some buffering capacity for ozone, ameliorating the effects of low-ozone exposures. In fact, the synthesis of antioxidant biochemicals may be increased when oxidant stresses increase (Dizengremel et al., 1994; Hausladen et al., 1990; Richardson et al., 1990).

While Reich (1987) reported a generally linear relationship between ozone exposure and photosynthetic inhibition, the results for loblolly pine suggest that there is an initial plateau before initiation of photosynthetic inhibition. Exposure–response relationships indicated that photosynthesis was not affected by ozone until cumulative ozone exposures exceeded 100 ppm·h. The initial plateau in the ozone exposure–response relationship for photosynthesis may be related to antioxidant defenses or other threshold-related processes (Alscher and Amthor, 1988; Fendick, 1991). Phenotypic variation in physiological responses may also affect the apparent sensitivity among loblolly genotypes (Richardson et al., 1992a). Analogous thresholds of damage have been suggested for other pollutants, such as SO_2 (Roberts, 1984), and similar sigmoidal dose–response relationships between ozone and yield have been reported in agricultural crops (Heck et al., 1982).

The decrease in photosynthesis at higher cumulative ozone exposures reflects progressive damage to physiological processes in excess of biochemical defenses and repair capacities (Dizengremel and Citerne, 1988). In white pine clones, photosynthesis decreased during a daily 6-hour ozone fumigation of 50 ppb but recovered by the next day; however, photosynthesis decreased more rapidly each successive fumigation day, suggesting a progressive impairment of the photosynthetic apparatus (Boyer et al., 1986).

Photosynthesis rates of loblolly pine were inhibited more than 80% when cumulative exposure exceeded 400 ppm·h. These values correlate well with those reported for several pine species (Reich, 1987). In ponderosa pine, ecotypes with varying classes of ozone injury—slight, moderate, or severe—were deemed to have different tolerance to ozone exposure. As predicted, the most resistant ecotypes required 800 ppm·h (24-hour summation) of ozone exposure to result in a 90% reduction in net photosynthesis; whereas the most sensitive ecotypes required only 450 ppm·h of ozone exposure to be similarly inhibited (Coyne and Bingham, 1981). Considering the differences between a 12-hour cumulative dose used for southern pine studies and the 24-hour cumulative doses used by Reich (1987) and Coyne and Bingham (1981), photosynthesis in the three southern pines could be regarded as moderately sensitive to ozone.

Carry-Over Effects

The long-term, cumulative effects of chronic ozone exposure also resulted in some carry-over effects on photosynthesis, especially in the two highest elevated ozone treatments. During the overwintering period in which there were no elevated treatments, photosynthetic potential in the elevated ozone treatments did not recover to the control rates. Overwintering carry-over effects in loblolly pine may have resulted from incomplete repair of progressive ozone damage. In studies of Norway spruce, damage caused by ozone or SO_2 exposure was irreversible for at least 16 months after the exposure treatments (Barnes et al., 1990b; Dotzler and Schütt, 1990). Thus, elevated ozone exposures can lead to long-lasting effects on physiological processes and cell ultrastructure that persist even in the absence of elevated exposures. At the same time, ozone-induced damage may increase susceptibility to winter injury. In one study of loblolly pine, hardening and dehardening processes were affected by elevated ozone treatments due to changes in solute transport and osmoregulation mechanisms (Edwards et al., 1990a). Conversely, interactions between elevated ozone exposure and low temperatures in other loblolly pine seedlings were probably a result of photodestruction of chlorophyll pigments rather than desiccation or freeze injury (Chappelka et al., 1990). Red spruce exposed to elevated ozone the previous year also had lower pigment levels at the start of the second exposure season (Fincher et al., 1989).

New flushes did not show significant carry-over effects that decreased photosynthetic potential when measured for the first time and before accumulating a large ozone dose, except in the highest ozone treatments. Carry-over effects on the initial photosynthetic potential of new tissues are obviously an indirect response to elevated ozone concentrations. New needles in a study of loblolly pine had decreased esterase activity and decreased amounts of neutral lipids, proteins, and RNA compared to low-ozone treatments after the seedlings had been exposed to elevated ozone concentrations the previous year (Sigal et al., 1988). The first flush may be most highly dependent on nutrient and carbohydrate reserves accumulated by older needles during the previous season, as shown by ^{14}C tracer studies of white pine clones (McLaughlin et al., 1982; Wilkinson and Barnes, 1973). Decreased white pine photosynthesis rates and accelerated needle abscission of older needles in the elevated ozone treatments resulted in less carbohydrate reserves and storage capacity. In another study of white pine, both of these processes similarly decreased carbohydrate reserves, resulting in smaller new needles the following year (Mann et al., 1980). In our studies, needle production and length of loblolly pine first flushes were decreased in the elevated ozone treatments during the second and third exposure seasons (Stow et al., 1992).

Flushes produced later in the growing season had no carry-over effects on initial photosynthetic potential. Each successive flush may have been less affected by the tree's previous exposure because there was progressively less total biomass and leaf area per flush compared to the first flush (Stow et al., 1992), or there may have been a preferential reallocation of plant resources to subsequent flushes, to repair processes, or to antioxidant biochemical defenses (Richardson et al., 1990; Sasek et al., 1991). In fact, photosynthesis rates of the late-season flushes were often highest in the highest ozone treatments. The foliar N concentration of these flushes was also 40 to 60% higher in the 3.0× treatment compared to charcoal-filtered (Baker et al., 1993). Therefore, reallocation processes may partially compensate for the negative effects of ozone on photosynthesis in other flushes. Nevertheless, it is important to note that the contributions of late-season flushes in terms of leaf area display and total carbon fixation were relatively unimportant (Stow et al., 1992).

Mature loblolly pines produce fewer growth flushes per year than seedlings. Thus, the carry-over effects of ozone on first flushes may be of greater importance in mature trees. However, it is not yet known how important reallocation and compensation processes in late-season flushes may be in mature trees.

Near-Ambient Responses

Regional ozone concentrations in the Southeast currently average approximately 100 to 150 ppm·h (12-hour summation, growing season) in rural and remote areas greater than 50 km from urban areas (McLaughlin, 1985). The exposure–response relationships for loblolly pine photosynthesis suggested that these ambient exposures could decrease instantaneous photosynthesis rates compared to "clean" air, leading to less total carbon fixation during a needle's functional life span. Although reductions of only a few percent might seem insignificant, compounding effects during several decades of growth could result in significant growth decline. Ozone exposures in and around urban areas are significantly greater, with the potential for greater impacts on adjacent forests. More importantly, urban areas are characterized by a higher frequency of hourly ozone concentrations exceeding 120 ppb; such peaks are rare in rural and remote areas. Peaks of high ozone concentration can result in acute injury that differs significantly from chronic exposure to lower ozone concentrations. Forest response near urban areas may therefore be uncharacteristic of regional responses, which are more closely correlated to cumulative effects (Reich, 1987). Nevertheless, both rural and urban air quality should continue to be of concern for forest health in the southeastern United States of America.

Long-term cumulative effects and compounded responses are important when extrapolating the impact of current near-ambient pollution concentrations on photosynthesis to whole-tree and stand-level growth. There were few indications of long-term, compounding effects of the pollutants in our studies; that is, carry-over effects did not increase progressively in magnitude each season. Longer experiments will be necessary to detect the magnitude and significance of compounding effects.

Long-term, compounding effects of pollutant stresses may be difficult to detect during the rapid growth phases of young seedlings and saplings, in which there are exponential increases in foliage production from one year to the next. In older saplings and mature trees, the magnitude of annual leaf area production and biomass accumulation is more similar from one year to the next, perhaps resulting in greater functional dependence between the current season's needles and older tissues. Furthermore, the growth of mature trees in a forest may be controlled to a greater extent by light availability as well as nutrient and carbohydrate pools (Weinstein et al., 1991). More importantly, there may be long-term effects of pollutants on biomass allocation patterns, particularly between roots and shoots. Therefore, several physiological processes may be important in determining the compounding effects of pollutants through time and the increased susceptibility of forests to other environmental stresses.

Decreased growth at elevated ozone concentrations has been correlated to decreased photosynthesis in many tree species (Reich, 1987). While acute exposures to high concentrations of ozone result in rapid foliar injury and internal damage (Heck et al., 1982), lower concentrations of ozone can have serious long-term impacts due to cumulative effects (Reich, 1987; Richardson et al., 1992a). The sensitivity of a particular species to pollutants such as ozone depends on various morphological, physiological, and biochemical characteristics. Differential sensitivity to ozone has been well correlated to differential uptake of ozone into the plant (Reich, 1987). Therefore, differences in sensitivity to external ozone concentrations may be based on differential leaf diffusive conductance to ozone (Reich, 1987) or other morphological and physiological characteristics that affect overall ozone uptake (Evans and Miller, 1972; Sasek et al., 1991). For example, among four pine species (*Pinus ponderosa*, *P. jeffreyi*, *P. coulteri*, and *P. lambertiana*), the more sensitive species had more stomata per unit area, fewer mesophyll cells per stoma, fewer hypodermal layers, and thinner epidermal and hypodermal layers (Evans and Miller, 1972). Ozone sensitivity can also be related to a plant's ability to repair ozone-induced damage or to synthesize biochemicals involved in antioxidant defense mechanisms (McLaughlin and Shriner, 1980; Richardson et al., 1990).

Genetic variation within a species can lead to analogous differences in sensitivity to air pollutants like ozone (Coleman et al., 1990). Studies have shown a strong genetic component to sensitivity in clonal lines of white pine (Boyer et al., 1986; Yang et al., 1983) and among ecotypes of Scots pine (Oleksyn and Bialobok, 1986). Variation in sensitivity to ozone has been shown in both half- and full-sibling families of loblolly pine (Kress et al., 1982; Shafer and Heagle, 1989; Weir, 1977). Pollutant stress has been shown to lead to an increased coefficient of variation in plant populations (Coleman et al., 1990), and chronic air pollution has been shown to lead to selection of pollution-tolerant genotypes within populations of *Populus tremuloides* and *Acer rubrum* (Karnosky et al., 1989). Intraspecific genetic variation in sensitivity to ozone exposure has been documented in many species. However, mechanisms responsible for genotypic and phenotypic sensitivity and/or tolerance have seldom been suggested. More importantly, frequency distributions of sensitivity and tolerance to ozone stress are poorly understood. Mechanisms of tolerance and the range of genetic variability have important implications for understanding selection pressures in both managed and unmanaged vegetation.

Conclusions

Results from the various physiological studies failed to reveal any significant direct impacts of acidic precipitation. However, acidic precipitation most likely affects trees through altered soil nutrient status and other biogeochemical processes that affect whole-plant physiology. For example, deposition of N at the soil surface may result in a preferential allocation of root growth near the surface, predisposing the trees to drought stress in summer and winter desiccation when surface soils freeze (Deans et al., 1990). Red spruce seedlings exposed to pH 2.5 or 3.0 acid mists had smaller coarse root systems than seedlings exposed to pH 5.0. Few of the SCFRC studies considered the effects of belowground processes and biomass on physiological processes. Nevertheless, it is likely that the duration of the experiments was not sufficient to detect if there are any long-term effects of chronic acid deposition.

Ozone caused significant negative effects on photosynthesis and other physiological processes that directly impact on growth. Exposure–response relationships suggest that there are significant cumulative impacts of ozone on net carbon fixation. These become even more significant when coupled with the significant impacts of cumulative exposure on needle retention. However, the high variability of the measurements did not produce enough statistical power to resolve relatively small effects at near-ambient exposures compared to controls. Short-term studies showed long-lasting inhibition of photosynthesis from the previous exposure season through the following winter, when ozone concentrations were low. However, previous ozone exposure history had little effect on the initial photosynthetic potential of emerging needles. Longer studies will also be necessary to determine progressive, long-term compounding effects of ozone exposure.

To gain a better understanding of the effects of ozone on tree response, it is necessary to determine effects at the metabolic and biochemical levels and relate these to growth and productivity. The SCFRC made significant progress toward a more mechanistic understanding of the effects of ozone on southern pines, but there is still much that remains to be done. A major gap in our understanding is the determination of the mechanism(s) of ozone toxicity and host tolerance. Future research is needed to establish the prevalence of apoplastic antioxidant systems and other defensive compounds, such as phenolics. Most importantly, little of the research was able to consider actual ozone uptake in relation to the physiological responses. Only in this way can true dose–response relationships be developed.

We now have a better understanding of the extent to which ozone affects photosynthesis in southern pines, but we still know little about the relationship between carbon assimilation and subsequent transport of photoassimilates in trees. This factor becomes more complex when the aging process in trees is considered, since the relative amount of photosynthetic to nonphotosynthetic tissue changes with age.

References

Adams HS, Stephenson SL, Blasing TJ, Duvick DN (1985) Growth-trend declines of spruce and fir in mid-Appalachian subalpine forests. Environ Exper Bot 25:315–325.

Adams MB, Edwards NT, Taylor GE Jr, Skaggs BL (1990) Whole-plant ^{14}C-photosynthate allocation in *Pinus taeda*: seasonal patterns at ambient and elevated ozone levels. Can J For Res 20:152–158.

Agren GI, Ingestad T (1987) Root:shoot ratio as a balance between nitrogen productivity and photosynthesis. Plant, Cell and Environ 10:579–586.

Albaugh TJ, Mowry FL, Kress LW (1992) A field chamber for testing air pollution effects on mature trees. J Environ Qual 21:476–485.

Alscher RG, Amthor JS (1988) The physiology of free-radical scavenging: maintenance and repair processes. In: Schulte–Hostede S, Darrall NM, Blank LW, Wellburn AR (eds) *Air Pollution and Plant Metabolism*. Elsevier Applied Science, London, pp 94–115.

Baker TR, Allen HL, Schoeneberger MM, Kress LW (1993) Nutritional responses of loblolly pine exposed to ozone and simulated acid rain. Can J For Res 24:453–461.

Barnes JD, Eamus D, Brown KA (1990a) The influence of ozone, acid mist and soil nutrient status on Norway spruce [*Picea abies* (L.) Karst.] II. Photosynthesis, dark respiration and soluble carbohydrates of trees during late autumn. New Phytol 115:149–156.

Barnes JD, Eamus D, Davison AW, Ro–Poulsen H, Mortensen L (1990b) Persistent effects of ozone on needle water loss and wettability in Norway spruce. Environ Pollut 63:345–363.

Boutton TW, Flagler RB (1990) Growth and water-use efficiency of shortleaf pine as affected by ozone and acid rain. In: *83rd Annual Meeting of the Air & Waste*

Management Association, Air & Waste Management Association, Pittsburgh, PA, pp 1–16.
Boutton TW, Flagler RB (1991) $^{13}C/^{12}C$ ratios as indicators of plant physiological response to ozone and simualted acid rain. In: *Stable Isotopes in Plant Nutrition, Soil Fertility and Environmental Studies: Proceedings of an International Symposium*, International Atomic Energy Agency, Vienna, Austria, pp 627–631.
Boyer JN, Houston DB, Jensen KF (1986) Impacts of chronic SO_2, O_3, and $SO_2 + O_3$ exposures on photosynthesis of *Pinus strobus* clones. Eur J For Pathol 16:293–299.
Byres DP, Johnson JD, Dean TJ (1992) Seasonal response of slash pine (*Pinus elliottii* var. *elliottii* Engelm.) photosynthesis to long-term exposure to ozone and acidic precipitation. New Phytol 122:91–96.
Chameides WL (1989) The chemistry of ozone deposition to plant leaves: role of ascorbic acid. Environ Sci Technol 23:595–600.
Chappelka AH, Kush JS, Meldahl RS, Lockaby BG (1990) An ozone–low temperature interaction in loblolly pine (*Pinus taeda* L.). New Phytol 114: 721–726.
Citerne A, Banvoy J, Dizengremel P (1988) Pollution atmosphérique et métabolisme respiratoire de l'Epicéa. In: Mathy P (ed) *Air Pollution and Ecosystems*. D. Reidel Publishing, Dordrecht, pp 678–683.
Coleman JS, Mooney HA, Winner WA (1990) Anthropogenic stress and natural selection: variability in radish biomass accumulation increases with increasing SO_2 dose. Can J Bot 68:102–106.
Cooley DR, Manning WJ (1987) The impact of ozone on assimilate partitioning in plants: a review. Environ Pollut 447:95–113.
Coyne PI, Bingham GE (1981) Comparative ozone dose response of gas exchange in a ponderosa pine stand exposed to long-term fumigations. J Air Pollut Control Assoc 31:38–41.
Crawford DN, Lockaby BG, Somers GL (1989) Genetic × nutrition interactions in young loblolly pine. In: *American Society of Agronomy Abstracts*. American Society of Agronomy, Madison, WI, p 301.
Cregg BM, Halpin JE, Dougherty PM, Teskey RO (1989) Comparative physiology and morphology of seedling and mature forest trees. In: Noble RD, Martin JL, Jensen KF (eds) *Air Pollution Effects on Vegetation, Including Forest Ecosystems: Proceedings of the Second US–USSR Symposium*, Northeastern Forest Experiment Station, Broomall, PA, pp 111–118.
Deans JD, Leith ID, Sheppard LJ, Cape JN, Fowler D, Murray MB, Mason PA (1990) The influence of acid mists on growth, dry matter partitioning, nutrient concentrations and mycorrhizal fruiting bodies in red spruce seedlings. New Phytol 115:459–464.
Dizengremel P, Citerne A (1988) Air pollutant effects on mitochondria and respiration. In: Schulte–Hostede S, Darrall NM, Blank LW, Wellburn AR (eds) *Air Pollution and Plant Metabolism*. Elsevier Applied Science, London, pp 169–188.
Dizengremel P, Sasek TW, Brown KJ, Richardson CJ (1994) Ozone-induced changes in carbon metabolism enzymes of loblolly pine needles. J Plant Physiol 144:300–306.
Dotzler M, Schütt P (1990) Longlasting effects of O_3 and SO_2 on the sensitivity to drought in two clones of *Picea abies* (L.) Karst. Eur J For Pathol 20:59–63.
Dougherty PM, Teskey RO, Jarvis PG (1992) Development of MAESTRO, a process based model for assessing the impact of ozone on net carbon exchange of loblolly pine trees. In: Flagler RB (ed) *The Response of Southern Com-*

mercial Forests to Air Pollution. TR-21. Air & Waste Management Association, Pittsburgh, PA, pp 303–312.

Edwards GS, Friend AL, O'Neill EG, Tomlinson PT (1992a) Seasonal patterns of biomass accumulation and carbon allocation in *Pinus taeda* seedlings exposed to ozone, acidic precipitation, and reduced soil Mg. Can J For Res 22:640–646.

Edwards GS, Kelly JM, Mays PA (1992b) Ozone, acidic precipitation, and soil Mg impacts on soil and loblolly pine seedling nutrient status after three growing seasons. Water, Air, and Soil Pollut 63:281–294.

Edwards GS, Pier PA, Kelly JM (1990a) Influence of ozone and soil magnesium status on the cold hardiness of loblolly pine (*Pinus taeda* L.) seedlings. New Phytol 115:157–164.

Edwards NT, Taylor GE Jr, Adams MB, Simmons GL, Kelly JM (1990b) Ozone, acidic rain and soil magnesium effects on growth and foliar pigments of *Pinus taeda* L. Tree Physiol 6:95–104.

Elsik CG (1992) *Growth, Physiology, and $\delta^{13}C$ of Lobolly and Shortleaf Pine as Affected by Ozone and Soil Moisture Deficit*. M.S. thesis, Texas A&M University, College Station, TX.

Elsik CG, Flagler RB, Boutton TW (1992) Effects of ozone and water deficit on growth and physiology of *Pinus taeda* and *Pinus echinata*. In: Flagler RB (ed) *The Response of Southern Commercial Forests to Air Pollution*. TR-21. Air & Waste Management Association, Pittsburgh, PA, pp 225–245.

Elsik CG, Flagler RB, Boutton TW (1993) Carbon isotope composition and gas exchange of loblolly and shortleaf pine as affected by ozone and water stress. In: Ehleringer JR, Hall AE, Farquhar GD, Ting IP (eds) *Stable Isotopes and Plant Carbon–Water Relations*. Academic Press, New York, pp 227–244.

Evans LS, Miller PJ (1972) Comparative needle anatomy and relative ozone sensitivity of four pine species. Can J Bot 50:1067–1071.

Fangmeier A, Kress LW, Lepper P, Heck WW (1990) Ozone effects on the fatty acid composition of loblolly pine needles (*Pinus taeda* L.). New Phytol 115: 639–647.

Faulkner P, Schoeneberger MM, Kress LW (1991) Belowground changes in loblolly pine as indicators of ozone stress. In: Coleman SS, Neary DG (eds) *Proceedings of the Sixth Biennial Southern Silvicultural Research Conference*, USDA-Forest Service General Technical Report SE-70, pp 332–341.

Fendick EA (1991) *Antioxidant Defense and Physiological Gas-Exchange Response in Ozone-Exposed Pinus taeda L.* Ph.D. dissertation, Duke University, Durham, NC.

Fincher J, Cumming JR, Alscher RG, Rubin G, Weinstein L (1989) Long-term ozone exposure affects winter hardiness of red spruce seedlings. New Phytol 113:85–96.

Flagler RB, Lock JE, Elsik CG (1994) Leaf-level and whole-plant gas exchange characteristics of shortleaf pine exposed to ozone and simulated acid rain. Tree Physiol 14:361–374.

Flagler RB, Lock JE, Toups BG (1993) Growth and gas exchange characteristics of shortleaf pine exposed to ozone and acid rain. In: *Proceedings 9th World Clean Air Congress*, Vol 5. Air & Waste Management Association, Pittsburgh, PA.

Gerant D, Citerne A, Fabert C, Dizengremel P (1989) Extraction and study of enzymes linked to malate metabolism in tree leaves. Ann Sci For 46S:811–814.

Guderian R, Klumpp G, Klumpp A (1989) Effects of SO_2, O_3, and NO_x, singly and in combination, on forest species. In: Oztürk M (ed) *Proceedings International Symposium on Plants and Pollutants in Developed and Developing Countries*. Ege University, Izmir, Turkey, pp 231–268.

Halpin JE Jr (1990) *Phenology and Gas Exchange of Seedlings and Trees of Loblolly Pine (Pinus taeda* L.*) Under Ambient and Elevated Levels of Ozone*. M.S. thesis, University of Georgia.

Hanson PJ, McLaughlin SB (1988) *Net Carbon Dioxide Exchange Characteristics of Pinus taeda L. Shoots*. Oak Ridge National Laboratory, Environmental Sciences Division Publication No 3105 ORNL/TM-10777.

Hanson PJ, Taylor GE Jr (1990) Modeling pollutant gas uptake by leaves: an approach based on physicochemical properties. In: Dixon RK, Meldahl RS, Ruark GA, Warren WG (eds) *Process Modeling of Forest Growth Responses to Environmental Stress*. Timber Press, Portland, OR, pp 351–356.

Hausladen A, Madamanchi NR, Fellows S, Alscher G, Amundson RG (1990) Seasonal changes in antioxidants in red spruce as affected by ozone. New Phytol 115:447–458.

Heath RL (1980) Initial events in injury to plants by air pollutants. Ann Rev Plant Physiol 31:395–431.

Heath RL (1988) Biochemical mechanisms of pollutant stress. In: Heck WW, Taylor OC, Tingey DT (eds) *Assessment of Crop Loss from Air Pollutants*. Elsevier Applied Science, London, pp 259–286.

Heck WW, Taylor OC, Adams R, Bingham G, Miller H, Preston E, Weinstein L (1982) Assessment of crop loss from ozone. J Air Pollut Control Assoc 32:353–361.

Higginbotham KO (1974) *The Influence of Canopy Position and the Age of Leaf Tissue on Growth and Photosynthesis in Loblolly Pine*. Ph.D. dissertation, Duke University.

Johnson AH, McLaughlin SB, Adams MB, Cook ER, DeHayes DH, Eagar C, Fernandez IJ, Johnson DW, Kohut RJ, Mohnen VA, Nicholas NS, Peart DR, Schier GA, White PS (1992a) Synthesis and conclusions from epidemiological and mechanistic studies of red spruce decline. In: Eagar C, Adams MB (eds) *Ecology and Decline of Red Spruce in the Eastern United States*. Springer-Verlag, New York, pp 385–412.

Johnson JD, Byres D, Chappelka AH, Dean TJ (1992b) The role of homeostatic growth in the response of slash and loblolly pine to ozone and acid rain. In: Flagler RB (ed) *The Response of Southern Commercial Forests to Air Pollution*. TR-21. Air & Waste Management Association, Pittsburgh, PA, pp 273–286.

Johnson JD, Zedaker SM, Hairston AB (1985) Foliage, stem and root interrelations in young loblolly pine. For Sci 31:891–898.

Karnosky DF, Berrang PC, Scholz F, Bennett JP (1989) Variation in and natural selection for air pollution tolerances in trees. In: Scholz F, Gregorius H-R, Rudin D (eds) *Genetic Effects of Air Pollutants in Forest Tree Populations*. Springer-Verlag, Berlin, pp 29–37.

Knauf TA, Bilan MV (1974) Needle variation on loblolly pine from mesic and xeric seed sources. For Sci 20:88–90.

Koziol MJ, Whatley FR, Shelvey JD (1988) An integrated view of the effects of gaseous air pollutants on plant carbohydrate metabolism. In: Schulte–Hostede S, Darrall NM, Blank LW, Wellburn AR (eds) *Air Pollution and Plant Metabolism*. Elsevier Applied Science, London, pp 148–168.

Kramer PJ, Kozlowski TT (1979) *Physiology of Woody Plants*. Academic Press, Orlando, FL.

Kress LW, Skelly JM, Hinklemann JH (1982) Relative sensitivity of 18 full-sib families of *Pinus taeda* L. to ozone. Can J For Res 12:203–209.

Laisk A, Kull O, Moldau H (1989) Ozone concentration in leaf intercellular air spaces is close to zero. Plant Physiol 90:1163–1167.

Lee EH, Tingey DT, Hogsett WH (1988) Evaluation of ozone exposure indices in exposure–response modeling. Environ Pollut 53:43–62.

Manderscheid R, Jäger H-J, Kress LW (1992) Effects of ozone on foliar nitrogen metabolism of *Pinus taeda* L. and implications for carbohydrate metabolism. New Phytol 121:623–633.

Manderscheid R, Jager HJ, Schoeneberger MM (1991) Dose–response relationships of ozone effects on foliar levels of antioxidants, soluble polyamines and peroxidase activity of *Pinus taeda* (L.): assessment of the usefulness as early ozone indicators. Angewandte Botanik 65:373–386.

Mann LK, McLaughlin SB, Shriner DS (1980) Seasonal physiological responses of white pine under chronic air pollution stress. Environ Exper Bot 20:99–105.

McLaughlin SB (1985) Effects of air pollution on forests: a critical review. J Air Pollut Control Assoc 35:512–534.

McLaughlin SB, McConathy RK, Duvick D, Mann LK (1982) Effects of chronic air pollution stress on photosynthesis, carbon allocation, and growth of white pine trees. For Sci 28:60–70.

McLaughlin SB, Shriner DS (1980) Allocation of resources to defense and repair. In: Horsfall JB, Cowling EB (eds) *Plant Disease*. Academic Press, New York, pp 407–431.

Miller PR, McBride JR, Schilling SL, Gomez AP (1989) Trend of ozone damage to conifer forests between 1974 and 1988 in the San Bernardino Mountains of southern California. In: Olson RK, Lefohn AS (eds) *Effects of Air Pollution on Western Forests*. Air & Waste Management Association, Pittsburgh, PA, pp 309–323.

Mueller–Dombois D (1987) Natural dieback in forests. Bioscience 37:575–583.

Oleksyn J, Bialobok S (1986) Net photosynthesis, dark respiration and susceptibility to air pollution of 20 European provenances of Scots pine *Pinus sylvestris* L. Environ Pollut 40:287–302.

Paynter VA, Reardon JC (1992) The effects of ozone and acid rain on diurnal carbohydrate flux in shortleaf pine (*Pinus echinata* Mill.) needles. In: Flagler RB (ed) *The Response of Southern Commercial Forests to Air Pollution*. TR-21. Air & Waste Management Association, Pittsburgh, PA, pp 247–256.

Paynter VA, Reardon JC, Shelburne VB (1991) Carbohydrate changes in shortleaf pine (*Pinus echinata*) needles exposed to acid rain and ozone. Can J For Res 21:666–671.

Paynter VA, Reardon JC, Shelburne VB (1992) Changing carbohydrate profiles in shortleaf pine (*Pinus echinata*) after prolonged exposure to acid rain and ozone. Can J For Res 22:1556–1561.

Pinkerton JE, Lefohn AS (1987) The characterization of ozone data for sites located in the forested areas of the eastern U.S. J Air Pollut Control Assoc 37:1005–1010.

Pitel JA, Cheliak WM (1986) Effectiveness of protective agents for increasing the activity of five enzymes from vegetative tissues of white spruce. Can J Bot 64:39–44.

Rawlings JO, Cure WW (1985) The Weibull function as a dose-response model to describe ozone effects on crop yields. Crop Sci 25:807–814.

Rawlings JO, Lesser VM, Dassel KA (1988) Statistical approaches to assessing crop losses. In: Heck WW, Taylor OC, Tingey DT (eds) *Assessment of Crop Loss from Air Pollutants*. Elsevier Applied Science, London, pp 389–416.

Reardon JC, Paynter VA, Shelburne VB (1992) Biomass and carbohydrate allocation of shortleaf pine seedlings exposed to ozone and acid rain. In: Flagler RB (ed) *The Response of Southern Commercial Forests to Air Pollution*. TR-21. Air & Waste Management Association, Pittsburgh, PA, pp 173–186.

Reich PB (1983) Effects of low concentrations of O_3 on net photosynthesis, dark respiration, and chlorophyll contents in aging hybrid Poplar leaves. Plant Physiol 73:291–296.

Reich PB (1987) Quantifying plant response to ozone: a unifying theory. Tree Physiol 3:63–91.

Reich PB, Schoettle AW, Stroo HF, Troiano J, Amundson RG (1987) Effects of ozone and acid rain on white pine (*Pinus strobus*) seedlings grown in five soils. I. Net photosynthesis and growth. Can J Bot 65:977–987.

Richardson CJ, Di Giulio RT, Tandy NE (1989) Free-radical mediated processes as markers of air pollution stress in trees. In: Smythe RB, Polincansky D (eds) *Biological Markers of Air-Pollution Stress and Damage in Forests*. National Academy Press, Washington, DC, pp 251–260.

Richardson CJ, Sasek TW, Di Giulio RT (1990) Use of physiological and biochemical markers for assessing air pollution stress in trees. In: Wang W, Gorsuch JW, Lower WR (eds) *Plants for Toxicity Assessment*, ASTM STP 1091. American Society for Testing and Materials, Philadelphia, PA, pp 143–155.

Richardson CJ, Sasek TW, Fendick EA (1992a) Implications of physiological responses to chronic air pollution for forest decline in the southeastern U.S.A. Environ Toxicol Chem 11:1105–1114.

Richardson CJ, Sasek TW, Fendick EA, Kress LW (1992b) Ozone exposure–response relationships for photosynthesis in genetic strains of loblolly pine seedlings. For Ecol Manage 51:163–178.

Roberts TM (1984) Effects of air pollution on agriculture and forestry. Atmos Environ 18:629.

Rotenberry JT, Wiens JA (1985) Statistical power analysis and community-wide patterns. Amer Naturalist 125:164–168.

Rundel PW, Ehleringer JR, Nagy KA (1989) *Stable Isotopes in Ecological Research*. Springer-Verlag, New York.

Samuelson LJ, Edwards GS (1993) A comparison of sensitivity to ozone in seedlings and trees of *Quercus rubra* L. New Phytol 125:373–379.

Sasek TW, Richardson CJ (1989) Effects of chronic doses of ozone on loblolly pine: Photosynthetic characteristics in the third year. For Sci 35:745–755.

Sasek TW, Richardson CJ (1992) The dose–response approach for characterizing the effects of near-ambient ozone concentrations on photosynthesis. In: Flagler RB (ed) *The Response of Southern Commercial Forests to Air Pollution*. TR-21. Transactions of the Air & Waste Management Association, Pittsburgh, PA, pp 257–272.

Sasek TW, Richardson CJ, Fendick EA, Bevington SR, Kress LW (1991) Carryover effects of acid rain and ozone on the physiology of multiple flushes of loblolly pine seedlings. For Sci 37:1078–1098.

Schultze E-D, Oren R, Lange OL (1989) Processes leading to forest decline: a synthesis. In: Schultze E-D, Lange OL, Oren R (eds) *Forest Decline and Air Pollution*. Springer-Verlag, Berlin, pp 459–468.

Shafer SR, Heagle AS (1989) Growth responses of field-grown loblolly pine to chronic doses of ozone during multiple growing seasons. Can J For Res 19:821–831.

Sharkey TD (1985) Photosynthesis in intact leaves of C3 plants: physics, physiology and rate limitations. Bot Rev 51:53–105.

Sheppard LJ, Smith RI, Cannell, MGR (1989) Frost hardiness of *Picea rubens* growing in spruce-decline regions of the Appalachians. Tree Physiol 5:25–37.

Sheu B-H, Chappelka AH (1993) Effects of long-term ozone exposure on superoxide dismutase and peroxidase activities in loblolly pine (*Pinus taeda* L.). In: *86th Annual Meeting of the Air & Waste Management Association*, Denver, CO. Air & Waste Management Association, Pittsburgh, PA, pp 1–13.

Sigal L, Eversman S, Berglund DL (1989) Isolation of protoplasts from loblolly pine needles and their flow-cytometric analysis for air pollution effects. Environ Exper Bot 28:151–161.

Skärby L, Selldén G (1983) The effects of ozone on crops and forests. Ambio 13:68–72.

Skeffington RA, Roberts TM (1985) The effects of ozone and acid mist on Scots pine saplings. Oecologia 65:201–206.

Spence RD, Rykiel EJ Jr, Sharpe PJH (1990) Ozone alters carbon allocation in loblolly pine: assessment with carbon-11 labeling. Environ Pollut 64:93–106.

Stow TK, Allen HL, Kress LW (1992) Ozone impacts on seasonal foliage dynamics ofyoung loblolly pine. For Sci 38:102–119.

Switzer GL, Nelson LE (1972) Nutrient accumulation and cycling in loblolly pine (*Pinus taeda* L.) plantation ecosystems: the first twenty years. Proc Soil Sci Soc Amer 36:143–147.

Tandy NE, Di Giulio RT, Richardson CJ (1989) Assay and electrophoresis of superoxide dismutase from red spruce (*Picea rubens* Sarg.), loblolly pine (*Pinus taeda* L.), and scotch pine (*Pinus sylvestris* L.). Plant Physiol 90:742–748.

Taylor GE Jr, Gunderson CA, Edwards NT (1989) Patterns in stable carbon isotope discrimination in *Pinus taeda* as a function of ozone stress. Plant Physiol 89(suppl):129.

Taylor GE Jr, Hanson PJ, Baldocchi DD (1988) Pollutant deposition to individual leaves and plant canopies: Sites of regulation and relationship to injury. In: Heck, WW, Taylor OC, Tingey DT (eds) *Assessment of Crop Loss from Air Pollutants*. Elsevier Applied Science, London, pp 227–258.

Temple PJ (1988) Injury and growth of jeffrey pine and giant sequoia in response to ozone and acidic mist. Environ Exper Bot 28:323–333.

Teskey RO, Hinckley TM (1986) Moisture: effects of water stress on trees. In: Hennessey TC, Dougherty PM, Kossuth SV, Johnson JD (eds) *Stress Physiology and Forest Productivity*. Marinus Nijhoff, The Netherlands, pp 9–33.

Thames JL (1963) Needle variation on loblolly pine from four geographic seed sources. Ecology 44:168–169.

von Caemmerer S, Farquhar GD (1981) Some relationships between the biochemistry of photosynthesis and the gas exchange of leaves. Planta 153:376–387.

Weinstein DA, Beloin RM, Yanai RD (1991) Modeling changes in red spruce carbon balance and allocation in response to interacting ozone and nutrient stresses. Tree Physiol 9:127–146.

Weir RJ (1977) *Genetic Variation in Loblolly Pine (Pinus taeda* L.*) Response to Ozone*. Ph.D. dissertation, North Carolina State University.

Wilkinson TG, Barnes RL (1973) Effects of ozone on $^{14}CO_2$ fixation patterns in pine. Can J Bot 51:1573–1578.

Williams WT, Williams JA (1986) Effects of oxidant air-pollution on needle health and annual-ring width in a Ponderosa pine forest. Environ Conserv 13:229–234.

Winner WE (1994) Mechanistic analysis of plant responses to air pollution. Ecolog Applic 4:651–661.

Wiselogel AE, Bailey JK, Newton RJ, Fong F (1991) Growth response of loblolly pine (*Pinus taeda* L.) seedlings to ozone fumigation. Environ Pollut 71:43–56.

Woodman JN (1987) Pollution-induced injury in North American forests: facts and suspicions. Tree Physiol 3:1–15.

Wright LM, Lockaby BG, Meldahl R, Thornton F, Chappelka AH (1990) The influences of acid precipitation and ozone on nitrogen nutrition of young loblolly pine. Water, Air, and Soil Pollut 54:135–142.

Yang Y-S, Skelly JM, Chevone BI, Birch JB (1983) Effects of long-term ozone exposure on photosynthesis and dark respiration of eastern white pine. Environ Sci Technol 17:371–373.

4. Impacts of Atmospheric Deposition and Ozone on Southern Commercial Pines: A Synthesis of Epidemiological and Mechanistic Studies

12. Synthesis and Conclusions from Studies of Southern Commercial Pines

Robert O. Teskey

Identification of the possible is an intellectual, abstract activity called "logic," identification of the probable is an empirical, applied activity termed "science."
R.H. Peters, 1991, p 18.

In 1985, when the Southern Commercial Forest Research Cooperative was initiated, regional, national, and international concerns about the vitality of forests were emerging. Beginning in the late 1970's in Europe there had been reports of unhealthy forests, and by the early 1980s, 20 to 25% of these forests were classified as moderately or severely damaged from unknown causes (Schulze, 1989). Perhaps no European forests were so severely impacted as in Germany, where *Waldsterben* or "forest decline" became a focal point of international concern. In 1986, visible symptoms of damage were identified in over one half of the West German forests. These symptoms were primarily in older trees and consisted mostly of chlorotic leaves and premature leaf senescence (Krzak et al., 1988). As the intensity of forest surveys increased, damage appeared even more widespread and varied (Blank et al., 1988). Eventually in Germany and elsewhere in Europe, the amount of reported damage would stabilize (Schulze, 1989), but at the time of the initiation of the Southern Commercial Forest Research Cooperative, the health of European forests appeared to be declining rapidly. The European scientific community focused its attention on the possibility that air pollutants were the causal agents responsible for the damage, which by then was being called *neuartige Waldschaden* (i.e., new types of forest damage). Most hypotheses focused on the direct effects of gaseous pollutants NO_x, SO_2, and ozone and the direct and indirect effects of acidic precipitation (Prinz et al., 1987). The effect of acidic precipitation on soil chemical properties and nutrient availability was linked to decline symptoms, at least on sites with sensitive soils (Ulrich, 1987), while the direct effects of the gaseous pollutants NO_x, SO_2, and ozone appeared to be of secondary importance, except in localized situations such as near large pollution sources (Schulze, 1989).

Similar to the *Waldsterben* in Europe, disturbing symptoms of forest decline had appeared in many areas in the United States. The most dramatic examples were in southern California, where stands of Ponderosa pine (*Pinus ponderosa* Dougl. ex Laws.) had developed pronounced visual symptoms of stress, apparently from oxidant air pollution (Miller and Millecan, 1971; Muir and Armentano, 1988; Williams and Williams, 1986). Other tree species, including giant sequoia [*Sequoiadendron giganteum* (Lindl.) Buchholz] and Jeffrey pine (*Pinus jeffreyi* Grev. and Balf.), also appeared to be affected (Evans and Leonard, 1991; Miller et al., 1989; Peterson et al., 1987). In the Midwest, chlorotic mottling and tip burn of eastern white pine (*Pinus strobus* L.) were reported over a large area of eastern Wisconsin and were linked with the concentrations of ozone in the region (Rezabek et al., 1989). In the East, a correlation between ambient ozone concentrations and tip burn in eastern white pine had been recognized for many years (Berry and Ripperton, 1963). But in the 1980s, visual symptoms of apparent ozone damage were also found on other species in the region, including black locust (*Robinia pseudoacacia* L.) and yellow poplar (*Liriodendron tulipifera* L.) (Winner et al., 1989). At higher elevations in the Appalachian Mountains, red spruce (*Picea rubens* Sarg.) and, in the southern portion of the mountain range, Fraser fir [*Abies fraseri* (Pursh) Poir.] were showing symptoms of lost vitality and accelerated death (Peart et al., 1992). Although many biotic and abiotic factors could have been responsible for some of the visual symptoms that were reported (Skelly, 1989), air pollution was almost always identified as the probable cause of these phenomena, connecting it with the concept of forest decline.

In the southern United States, with the exception of the Appalachian Mountain region, there had been no reports of widespread visible damage in the forests. However, an analysis of long-term growth trends in pine stands in the region indicated that there had been a significant regional decline in growth (Sheffield et al., 1985). These authors only suggested that air pollution might be one of several possible causes for the apparent decline, but the regional nature of the growth decline, its correlation with the increase in air pollution in the region over the past 40 years, and the focus on air pollution as the cause of forest decline in other regions and countries provided strong support and justification for the research efforts of the Southern Commercial Forest Research Cooperative. The conclusion by Sheffield et al. (1985) that there had been a decline in growth rates in the region has been criticized (Cleveland et al., 1992, Lucier and Barnard, 1992), and different analyses made on the same data sets used by Sheffield et al., have led to conclusions that contradict those made by Sheffield et al. (Zeide, 1992). On the other hand, additional evidence for a regional decline in pine growth during that period has been reported (Bechtold et al., 1991; Ruark et al., 1991). Whether or not there had been a growth decline during the 1970s and early 1980s remains unresolved (e.g., Discussion Paper Comments, *Forest Science* 37:718–732). However, there is no doubt that the Sheffield et al. report provided an impetus for investigating the effects of air pollution on the forests of the southern United States, even though there had been no visible symptoms of widespread damage in the region.

Both ozone and acidic precipitation were identified as possible agents that could cause a regional decline in growth in the southern states. It became evident that acid precipitation was having no immediate negative impacts on the growth of the southern pine species (Chapters 10 and 11). In fact, the nitrogen and sulfur amendments to the soil were found to stimulate growth. Height growth and diameter growth were increased in acid rain treatments of pH 4.5 and 3.3 (compared with treatments of pH 5.2) in loblolly pine (Chappelka et al., 1992; Kress and Allen, 1991) and shortleaf pine (Flagler et al., 1992; Shelburne et al., 1993). Diameter growth increased with increasing rainfall acidity from pH 5.2 to 3.3 in slash pine (Dean and Johnson, 1992). These findings are consistent with reports of studies on other species (see Chapter 5) and consistent with observations of increased growth rates in European forests subjected to acid precipitation (Kauppi et al., 1992). These results are compelling and led to the conclusion stated in Chapter 10 that nitric and sulfuric acids present in the rain at current, or even 10 times greater, concentrations were unlikely to cause direct negative impacts on growth.

Although at present acid precipitation does not appear to be harmful to the growth of southern pines, there is cause for concern that in the future this may not be the case. There have been indications that acid precipitation is leaching cations from certain sensitive soil types in the region at a significant rate (Binkley et al., 1988, Chapter 8). Sensitive soil types have been identified using a variety of criteria but, in general, are those with a low cation exchange capacity and a low buffering capacity against pH change. Approximately 20 to 26% of the soils in the South have been identified as potentially sensitive to accelerated leaching of basic cations caused by acidic precipitation (Binkley et al., 1989). As much as 51% of the soils currently supporting forests in the region may be susceptible to acidification (see Chapter 8; see also Turner et al., 1986).

At present, there is a lack of compelling evidence linking acidic precipitation to decreases in growth or negative changes in physiological functions in the southern pines, so the remainder of this chapter will focus on ozone as a causal agent of a regional decline in pine growth. Ozone was targeted for intensive study in the Southern Commercial Research Forest Cooperative for many reasons. It had already been demonstrated that ambient levels of ozone were reducing the yields of a variety of crop species in the South and elsewhere (Heck et al., 1982, 1984) and was causing significant economic losses to farmers (Adams et al., 1989). Compared with other forested areas in the eastern United States, ozone concentrations in the South were relatively high (Lefohn and Shadwick, 1991). There had also been evidence of recent increases in regional (Pinkerton and Lefohn, 1987) and global (Fishman, 1991) tropospheric ozone concentrations. Of all the gaseous air pollutants, including SO_2 and NO_x, ozone was the only one that was detected across the region in high enough concentrations to, at least potentially, be a significant stress on tree growth and forest productivity. Other possible causes for a widespread decline in growth were considered, such as drought or long-term changes in weather (Chapter 4) or changes in silvicultural practices, but none appeared as likely as ozone to explain this regional phenomenon.

It is well known that, in sufficient quantities, ozone is harmful to plants. In a review of the effects of ozone on tree growth, Pye (1988) clearly demonstrated that there was already ample evidence on which to base the conclusion that ozone can reduce the growth of trees. However, the key question that remained unanswered was if the current ambient levels of ozone were sufficient to cause a decrease in the growth of southern pine species. At a more mechanistic level, could a link be demonstrated between ambient concentrations of ozone and a reduction in rates of plant processes known to affect growth? Alternatively, if ozone could not be shown to affect the southern pines at the present, at what level of exposure would the southern pines become susceptible to damage by ozone? These questions will be addressed in the following sections. As Allen and Hoekstra (1992) succinctly stated, "For an adequate understanding leading to robust prediction, it is necessary to consider at least three levels at once: 1) the level in question; 2) the level below that gives mechanisms; and 3) the level above that gives context, role or significance" (p. 9). The Southern Commercial Forest Research Cooperative operated with just such a structure. Its role was to evaluate the effects that air pollution might have on the health and productivity of pine forests. The level in question was growth; and the level providing mechanistic understanding was that of physiology (i.e., the processes that could be directly or indirectly related to growth, such as photosynthesis).

The level above was the effect of air pollution on forest productivity and an important economic resource in the region.

Effects of Ozone on Growth

The effects of ambient and above-ambient concentrations of ozone were evaluated in three commercial pine species: slash pine (*Pinus elliottii* Engelm.), shortleaf pine (*P. echinata* Mill.), and loblolly pine (*P. taeda* L.). Loblolly pine was the recipient of the most intensive study because of its greater economic importance and wider regional distribution, relative to the other pines. In comparison with other species, loblolly pine appears to be more sensitive to ozone than most conifers but less sensitive than most hardwoods (Table 12.1). Exposure of loblolly pine to doses of ozone above $130\,\mu l\,l^{-1}\cdot h$ accelerated leaf abscision, and doses above $160\,\mu l\,l^{-1}\cdot h$ consistently produced significant growth reductions, in the range of 5 to 12%. There was one report of reduced growth at a dose of $99\,\mu l\,l^{-1}\cdot h$ (Table 12.1), which is somewhat lower than that detected in the other studies, but experimental differences such as sample sizes and the genetic selection of the individuals in the study may account for what appears to be a more sensitive response in this case. Ambient ozone concentrations and total dose are site-specific factors and vary on a daily, seasonal, and yearly basis, so there is little consistency among the field studies with respect to the ozone doses received by the plants. Field studies conducted in open-top chambers make up a large majority of the studies summarized in Table 12.1. For studies conducted on loblolly pine, the accumulated dose of ozone in the ambient air for one year, or the part of the year when treatments were imposed, ranged from 82 to $139\,\mu l\,l^{-1}\cdot h$. For studies involving shortleaf and slash pines, the yearly mean dose at ambient was 137 and $119\,\mu l\,l^{-1}\cdot h$, respectively. Ambient doses for the other species shown in Table 12.1 varied widely, from as little as $44\,\mu l\,l^{-1}\cdot h$ to as much as $282\,\mu l\,l^{-1}\cdot h$.

At the Southern Commercial Forest Research Cooperative sites, after one year of treatment, there was no evidence that ambient ozone concentrations were reducing the growth of slash, shortleaf, or loblolly pine (Chapter 10). This is consistent with the findings from other studies using these species as well as studies on other conifer species. Of the 19 studies on conifers summarized in Table 12.1, conducted under field conditions and incorporating an ambient ozone treatment, none reported finding a significant growth effect due to ozone in the first year. After a second year of exposure to ambient ozone, Edwards et al. (1991) reported that there had been a significant reduction in height and diameter growth in loblolly pine seedlings. However, their twice-ambient treatment was not significantly different than the control, making the interpretation of the ambient-level response problematic. The ambient ozone effect was not

Table 12.1. Growth Effects of Ambient and Elevated Ozone, Relative to a Subambient Ozone Treatment

Species	Effect on Growth (+, 0, −)	Ozone Dose	Comments	Reference
Conifers				
Abies alba	0	184 $\mu l\, l^{-1} \cdot h$ (amb)	Subamb. dose = 47 $\mu l\, l^{-1} \cdot h$, 4 year study.	Billen et al., 1990
Picea rubens	0	dose? (amb)	Growth in ambient air compared to that in charcoal-filtered air.	Vann et al., 1992
Picea rubens	−	12 $\mu l\, l^{-1} \cdot h$	Subamb. dose = 3 $\mu l\, l^{-1} \cdot h$. Exposed for 10 wks, 3 d/wk, 4h/d. Total dry wt. was reduced.	Lee et al., 1990
Picea rubens	0	60 $\mu l\, l^{-1} \cdot h$		Patton et al., 1991
Picea rubens	0 0	135 $\mu l\, l^{-1} \cdot h$ (amb) 263 $\mu l\, l^{-1} \cdot h$	Subamb. dose = 71 $\mu l\, l^{-1} \cdot h$. Dry wt. decreased linearly with increasing O_3 conc., from subamb to 3× amb.	Amundson et al., 1991
Picea rubens	0 0	96 $\mu l\, l^{-1} \cdot h$ (amb) 188 $\mu l\, l^{-1} \cdot h$	Subamb. dose = 51 $\mu l\, l^{-1} \cdot h$.	Alscher et al., 1989
Picea rubens	0 0	43 $\mu l\, l^{-1} \cdot h$ (amb) 82 $\mu l\, l^{-1} \cdot h$	Subamb. dose = 22 $\mu l\, l^{-1} \cdot h$, 3 mo. study.	Laurence et al., 1989
Picea rubens	0 0	114 $\mu l\, l^{-1} \cdot h$ (amb) 254 $\mu l\, l^{-1} \cdot h$	Subamb. dose = 68 $\mu l\, l^{-1} \cdot h$, 2 yr. study.	
Picea abies	−	184 $\mu l\, l^{-1} \cdot h$ (amb)	Subamb. dose = 47 $\mu l\, l^{-1} \cdot h$, 4 year study.	Billen et al., 1990
Picea sitchensis	0 0	44 $\mu l\, l^{-1} \cdot h$ (amb) 108 $\mu l\, l^{-1} \cdot h$	Subamb. dose = 3 $\mu l\, l^{-1} \cdot h$, 3 mo. study.	Lucas et al., 1988
Pinus coulteri	0	282 $\mu l\, l^{-1} \cdot h$ (amb)	Ambient air versus clean air comparison.	Bytnerowicz et al., 1989

Pinus echinata	0 0 –	410 $\mu l\,l^{-1}\cdot h$ (amb) 620 $\mu l\,l^{-1}\cdot h$ 850 $\mu l\,l^{-1}\cdot h$	Subamb. dose = 100 $\mu l\,l^{-1}\cdot h$, 3 yr study. Lower leaf wt., no other effects on biomass.	Shelburne et al., 1993
Pinus elliottii	0 – –	365 $\mu l\,l^{-1}\cdot h$ (amb) 675 $\mu l\,l^{-1}\cdot h$ 966 $\mu l\,l^{-1}\cdot h$	Subamb. dose = 119 $\mu l\,l^{-1}\cdot h$, 3 yr study. Increased leaf abscission at 2× and 3× amb., but not at ambient.	Byres et al., 1992
Pinus elliottii	–	966 $\mu l\,l^{-1}\cdot h$	Subamb. dose = 119 $\mu l\,l^{-1}\cdot h$, 3 yr study. Stem vol. and leaf area reduced.	Dean and Johnson, 1992
Pinus jeffreyi	0 0	33 $\mu l\,l^{-1}\cdot h$ 66 $\mu l\,l^{-1}\cdot h$	Exposed 4 h/d, 3 d/wk for 41 d over 2 yrs.	Temple, 1988
Pinus jeffreyi	0	282 $\mu l\,l^{-1}\cdot h$ (amb)	Ambient air versus clean air comparison.	Bytnerowicz et al., 1989
Pinus ponderosa	–		O_3 conc. from 0.038 to 0.09 $\mu l\,l^{-1}$ in summer. Radial growth reduced (11%) in trees with visible symptoms.	Peterson et al., 1987
Pinus ponderosa	0	282 $\mu l\,l^{-1}\cdot h$ (amb)	Ambient air versus clean air comparison.	Bytnerowicz et al., 1989
Pinus ponderosa	0 –	574 $\mu l\,l^{-1}\cdot h$ (amb) 914 $\mu l\,l^{-1}\cdot h$	Subamb. dose = 117 $\mu l\,l^{-1}\cdot h$, 3 yr study. Total wt and stem dia. reduced.	Temple et al., 1993
Pinus rigida	0 – –	36 $\mu l\,l^{-1}\cdot h$ (amb) 73 $\mu l\,l^{-1}\cdot h$ 145 $\mu l\,l^{-1}\cdot h$	13 wk experiment. Reduced leaf, stem, and root wt.	Schier et al., 1990
Pinus rigida	–	2.4 $\mu l\,l^{-1}\cdot h$	Two 4 h exposures to 0.3 $\mu l\,l^{-1}$ O_3 caused reduced root growth.	Scherzer and McClenahen, 1989

(continued)

Table 12.1. *Continued*

Species	Effect on Growth (+, 0, −)	Ozone Dose	Comments	Reference
Pinus strobus	0 0 0	20 µl l^{-1}·h 34 µl l^{-1}·h 47 µl l^{-1}·h	Subamb. dose = 6.7 µl l^{-1}·h. Exposed for 7 h/d, 3 d/wk for 4 months.	Reich et al., 1987
Pinus sylvestris	0	282 µl l^{-1}·h (amb)	Ambient air versus clean air comparison.	Bytnerowicz et al., 1989
Pinus taeda	0 −	82 µl l^{-1}·h (amb) 160 µl l^{-1}·h	Subamb. dose = 41 µl l^{-1}·h. Shoot dry wt. reduced 12%.	Shafer et al., 1987
Pinus taeda	0 −	275 µl l^{-1}·h 413 µl l^{-1}·h	Subamb. dose = 138 µl l^{-1}·h, 3 yr. study. Biomass reduced 8% at higher dose.	Edwards et al., 1992
Pinus taeda	0 − − −	136 µl l^{-1}·h (amb) 257 µl l^{-1}·h 299 µl l^{-1}·h (amb) 564 µl l^{-1}·h	Subamb. dose = 70 µl l^{-1}·h yr^{-1}. ** After 2 yrs., ambient O_3 conc. reduced needle (−7%) and stem and branch biomass (−5%).	Edwards et al., 1991
Pinus taeda	−	326 µl l^{-1}·h	Subamb. dose = 96 µl l^{-1}·h. Increased leaf abscission caused reduced leaf biomass. Shoot growth decreased more than root growth.	Tjoelker and Luxmoore, 1991
Pinus taeda	0 −	139 µl l^{-1}·h (amb) 253 µl l^{-1}·h	Subamb. dose = 70 µl l^{-1}·h. Reduced total biomass and stem diameter.	Edwards and Taylov, 1990
Pinus taeda	−	397 µl l^{-1}·h	Subamb. dose = 60 µl l^{-1}·h, 2 yr study. Shoot growth and number of fascicles decreased.	Mudano et al., 1992

Pinus taeda	– –	$130\,\mu l\,l^{-1}\cdot h$ $220\,\mu l\,l^{-1}\cdot h$	Increased leaf abscission.	Stow et al., 1992
Pinus taeda	–	$99\,\mu l\,l^{-1}\cdot h$	Up to 25% reduction in total dry wt. for sensitive family.	Shafer et al., 1993
Sequoiadendron giganteum	0 0	$32\,\mu l\,l^{-1}\cdot h$ $66\,\mu l\,l^{-1}\cdot h$	Exposed 4 h/d, 3 d/wk for 41 d over 2 yrs.	Temple, 1988
Hardwoods				
Acer rubrum	0	$231\,\mu l\,l^{-1}\cdot h$ (amb)	Subamb. dose = $110\,\mu l\,l^{-1}\cdot h$, 2 yr. study. Exposed to ambient air or charcoal-filtered air.	Simini et al., 1992
Acer rubrum	–	$38\,\mu l\,l^{-1}\cdot h$	Subamb. dose = $19\,\mu l\,l^{-1}\cdot h$ (conc. $0.04\,\mu l\,l^{-1}$), 10 wk. study. Stem dia. and dry wt. reduced.	Davis and Skelly, 1992
Alnus incana	0 –	$12\,\mu l\,l^{-1}\cdot h$ (amb) $19\,\mu l\,l^{-1}\cdot h$	Subamb. dose = $9\,\mu l\,l^{-1}\cdot h$. Dose slightly above amb. reduced shoot, root, and leaf dry wt.	Mortensen and Skre, 1990
Betula pendula	–	$360\,\mu l\,l^{-1}\cdot h$	Subamb. dose = $0\,\mu l\,l^{-1}\cdot h$. Linear reduction in growth with intermediate, O_3 doses.	Matyssek et al., 1992
Betula pubescens	0 –	$12\,\mu l\,l^{-1}\cdot h$ (amb) $19\,\mu l\,l^{-1}\cdot h$	Subamb. dose = $9\,\mu l\,l^{-1}\cdot h$. Dose slightly above amb. reduced shoot, root, and leaf dry wt.	Mortensen and Skre, 1990
Betula verrucosa	0 –	$12\,\mu l\,l^{-1}\cdot h$ (amb) $19\,\mu l\,l^{-1}\cdot h$	Subamb. dose = $9\,\mu l\,l^{-1}\cdot h$. Dose slightly above amb. reduced shoot, root, and leaf dry wt.	Mortensen and Skre, 1990
Fagus sylvatica	0	$184\,\mu l\,l^{-1}\cdot h$ (amb)	Subamb. dose = $47\,\mu l\,l^{-1}\cdot h$, 4 year study.	Billen et al., 1990
Fagus sylvatica	–	$4\,\mu l\,l^{-1}\cdot h$	Root biomass reduced.	Davidson et al., 1992

(continued)

Table 12.1. *Continued*

Species	Effect on Growth (+, 0, −)	Ozone Dose	Comments	Reference
Fraxinus americana	−	12 $\mu l\,l^{-1}\cdot h$	Exposed 4 h/d, 5 d/wk for 6 wk. Control received 0 O_3 exposure. Leaf dry wt. reduced 18%, total biomass reduced 10%.	Chappelka et al., 1988
Fraxinus americana	0	67 $\mu l\,l^{-1}\cdot h$ (amb)	Seedlings exposed to ambient air compared with those in charcoal-filtered air or treated with antioxidant. Repeated 3 yrs.	Elliott et al., 1987
Fraxinus pennsylvanica	0	67 $\mu l\,l^{-1}\cdot h$ (amb)	Seedlings exposed to ambient air compared with those in charcoal-filtered air or treated with antioxidant. Repeated 3 yrs.	Elliott et al., 1987
Fraxinus pennsylvanica	−	12 $\mu l\,l^{-1}\cdot h$	Exposed 4 h/d, 5 d/wk for 6 wk. Control received 0 O_3 exposure. Height growth reduced.	Chappelka et al., 1988
Liriodendron tulipifera	0	231 $\mu l\,l^{-1}\cdot h$ (amb)	Subamb. dose = 110 $\mu l\,l^{-1}\cdot h$, 2 yr. study. Exposed to ambient air or charcoal-filtered air.	Simini et al., 1992
Liriodendron tulipifera	0	326 $\mu l\,l^{-1}\cdot h$	Subamb. dose = 96 $\mu l\,l^{-1}\cdot h$.	Tjoelker and Luxmoore, 1991
Liriodendron tulipifera	−	38 $\mu l\,l^{-1}\cdot h$	Subamb. dose = 19 $\mu l\,l^{-1}\cdot h$ (conc. 0.04 $\mu l\,l^{-1}$), 10 wk. study. Leaf dry wt. reduced.	Davis and Skelly, 1992
Populus deltiodes	0	58 $\mu l\,l^{-1}\cdot h$	Subamb. dose = 19 $\mu l\,l^{-1}\cdot h$. Exposed to ambient air or charcoal-filtered air for 4 summer months over 3 yrs.	Wang et al., 1986

Populus deltiodes x P. nigra	–	12 $\mu l\, l^{-1} \cdot h$	Leaf senescence accelerated. Number of leaves reduced 50%.	Taylor and Frost, 1992
Populus masimowiczii x P. trihocarpa	–	58 $\mu l\, l^{-1} \cdot h$	Subamb. dose = 19 $\mu l\, l^{-1} \cdot h$. Exposed to ambient air or charcoal filtered air for 4 summer months over 3 yrs. Height, leaf, and stem biomass reduced.	Wang et al., 1986
Populus tremuloides	0 0	Dose? (amb) amb + 20 $\mu l\, l^{-1} \cdot h$	Seedlings exposed to ambient air compared with those exposed to charcoal-filtered air.	Karnosky et al., 1992
Prunus serotina	0	231 $\mu l\, l^{-1} \cdot h$ (amb)	Subamb. dose = 110 $\mu l\, l^{-1} \cdot h$, 2 yr. study. Exposed to ambient air or charcoal-filtered air.	Simini et al., 1992
Prunus serotina	–	38 $\mu l\, l^{-1} \cdot h$	Subamb. dose = 19 $\mu l\, l^{-1} \cdot h$ (conc. 0.04 $\mu l\, l^{-1}$), 10 wk. study. Ht., dia., and dry wt. reduced.	Davis and Skelly, 1992
Quercus rubra	0	231 $\mu l\, l^{-1} \cdot h$ (amb)	Subamb. dose = 110 $\mu l\, l^{-1} \cdot h$, 2 yr. study. Exposed to ambient air or charcoal-filtered air.	Simini et al., 1992
Quercus rubra	–	38 $\mu l\, l^{-1} \cdot h$	Subamb. dose = 19 $\mu l\, l^{-1} \cdot h$ (conc. 0.04 $\mu l\, l^{-1}$), 10 wk. study. Root dry wt. reduced.	Davis and Skelly, 1992
Robinia pseudoacacia	0	58 $\mu l\, l^{-1} \cdot h$	Subamb. dose = 19 $\mu l\, l^{-1} \cdot h$. Exposed to ambient air or charcoal-filtered air for 4 summer months over 3 yrs.	Wang et al., 1986

Ozone dose was calculated as (concentration × total length of exposure). For multiyear studies, the annual dose can be approximated by dividing by the total number of years of exposure. Experimental growth conditions were mostly well watered and fertilized. (Amb) refers to ozone exposure that was, or mimicked, the ambient ozone dose at the study site. All studies were conducted on seedlings except for Peterson et al. (1987). Effects of exposure to ozone on growth were characterized as causing an increase in growth (+), having no effect on growth (0), or causing a decrease in growth (–), compared with plants exposed to a subambient concentration of ozone for the same length of time. For a summary of studies conducted prior to 1987, see Pye (1988).

apparent after the third year of treatment. Edwards et al. (1991) was the only study on southern pines that reported a decrease in growth at the current ambient ozone dose. However, using regression analysis, others were able to demonstrate a linear decrease in specific growth parameters, such as biomass or stem diameter, with increasing ozone dose (Shafer et al., 1987, 1993). These relationships were interpreted to mean that ambient levels of ozone were reducing growth in loblolly pine seedlings. The conclusion has validity but does not answer with certainty the question of the effect of ambient levels of ozone on growth. In almost all studies using tree species, ozone doses higher than ambient, produced by exposing the plants to higher concentrations of ozone for equal periods of time, resulted in reductions in growth (Table 12.1). Growth reductions were evident at two times the current ambient dose in loblolly pine and shortleaf pine and at three times the current ambient dose in slash pine. It is apparent that above-ambient concentrations of ozone can reduce growth, but the effects at the current ambient concentration are less distinct.

Uncertainty about concluding that current ambient levels of ozone are or are not causing reductions in growth arises from not knowing if equivalent ozone doses, produced from either a low concentration over a long period of time or a high concentration for a short period of time, have the same effect. The strongest evidence supporting equivalency is that the ozone dose response ($\mu l\, l^{-1} \cdot h$) can be fitted to linear models (e.g., Shafer et al., 1993). However, data from most of the dose–response experiments can also be analyzed satisfactorily using curvilinear or threshold models (Chapter 10, Figs. 10.3, 10.4, and 10.5), so these data are not sufficient alone to resolve the question of whether there are growth reductions caused by ambient ozone concentrations in the southern United States.

An argument can be made that, at present ambient levels, ozone is causing a decrease in growth (e.g., see Chapter 10; see also Shafer et al., 1993; Taylor, 1994). The National Crop Loss Assessment Network demonstrated that ambient levels of ozone were significantly affecting growth of crop species in the region (Heck et al., 1982, 1984). One would expect ozone effects to be more obvious in crops than in trees after one growing season due to the faster growth rates and higher stomatal conductances of crop species (Reich, 1987). Trees are exposed to ozone for many years rather than a single growing season, so small effects each year may have cumulative impacts on growth. There is no evidence that crops differ in their fundamental physiological processes from tree species; hence it seems likely that ambient ozone is affecting the growth of southern pines to some degree. There is no doubt that concentrations higher than the current ambient concentration decrease growth. However, based on the experiments in the Southern Commercial Forest Research Cooperative that are summarized in Chapter 10 (e.g., Figs. 10.3, 10.4, and 10.5) and similar results from other studies on these species (Table 12.1), it is also clear that if there is a negative effect of ambient levels of ozone on the growth of southern pines in a single year or even in successive years, it cannot be large. Assuming that this argument is correct, the maximum potential effect that average ambient ozone concentrations could have on annual growth (biomass increment) of loblolly pine seedlings, averaged across the experimental population, is likely to be no more than a 2 to 5% reduction between pristine air and actual ambient air. A reduction this small is within the range of variation, or experimental error, of these studies and so could not be detected by statistical analyses.

Alternatively, in almost every study, when growth in the ambient ozone treatment was compared to that in the subambient ozone treatment, no statistically significant difference could be detected (Table 12.1), which is justification for the conclusion that ambient levels of ozone do not affect the growth of southern pines. Yet there are reasons to suspect that this conclusion is analogous to a Type II error in statistical analyses, which is to accept wrongly the null hypothesis (i.e., falsely conclude that there was no treatment effect). Most studies reported a decrease in growth at the ambient ozone treatment, compared with a subambient treatment, although the differences were not statistically significant (Table 12.1). These growth reductions were larger at higher ozone doses and were statistically different from subambient treatments (Chapter 10, Table 10.1). The inability to detect significant differences between the ambient and subambient treatments may have resulted from experimental variability and mean differences that were smaller than expected. It is understandable that a growth reduction of 2 to 5% would be very difficult to detect in these experiments due to the variance caused by (1) small sample sizes, (2) imposition of both acidic precipitation and ozone treatments, (3) genetic variation (Taylor, 1994), (4) significant site variation in some locations (Spruill et al., 1993), (5) year-to-year differences in ambient ozone concentrations (Chapter 3), and (6) within-chamber variation in ozone dose and microclimate, particularly as the seedlings increased in size and leaf area. It must also be realized that these experiments extended over no more than 5 to 15% of the life span of these species, assuming that they live for about 20 years under intensive management and for about 60 years in natural stands. While a growth reduction of less than 5% may seem insignificant, when sustained year after year it will compound itself, producing much larger effects over the life span of the trees (for an example of the compounding effect, see the section later in this chapter entitled "Implications for Productivity of Pine Forests").

The existence of genetic variation in the response to ozone by loblolly pine was clearly demonstrated in many studies (Adams et al., 1988, 1990; Qui et al., 1992; Shafer and Heagle, 1989; Shafer et al., 1993; Winner et al., 1987). Evidence of genetic variation in response to ozone also appeared in studies of slash and shortleaf pines, but fewer families were used in these experiments. For loblolly pine, the range of response to ozone was large, both within and between families (Taylor, 1994). In one experiment, family differences ranged from no effect to a 39% reduction in total dry weight for the same ozone exposure (Shafer et al., 1993). Genetic variation in response to ozone has been observed in many tree species, including silver fir (*Abies alba* Mill.) (Larsen et al., 1990), sugar maple (*Acer saccharum* Marsh.), eastern white pine (Karnosky et al., 1993), and quaking aspen (*Populus tremuloides* Michx.) (Berrang et al., 1986). It is not known if selective pressures due to ambient levels of ozone have favored the more resistant members in the population of southern pines. This phenomenon has been observed in quaking aspen (Berrang et al., 1989, 1991; Karnosky et al., 1993).

The large genetic variation in loblolly pine with respect to ozone is important for two reasons. First, this variation alone may have been enough to prevent the detection of significant differences in growth between ambient and subambient treatments. In the Southern Commercial Forest Research Cooperative program, a large number of loblolly pine families were screened for their sensitivity to ozone (McLaughlin et al., 1988; Reinert et al., 1988; Wiselogel et al., 1991) and subsets were used in subsequent field experiments. This provided important information on the population, but experimental variation was increased and may have been a key reason why small effects of ambient ozone concentrations on growth could not have been detected if they existed. Moreover, some families incurred little or no damage from ozone, and this demonstrated that there was a highly resistant subset of individuals within the population (Taylor, 1994). Whether this resistance is due to the activity of biochemical pathways that reduce ozone damage or simply due to lower rates of uptake through the stomata cannot be ascertained at this time. However, there are correlations between ozone resistance in loblolly pine and slow rates of growth (Winner et al., 1987), suggesting that these individuals either have inherently lower stomatal conductances and therefore take up less ozone but also gain less carbon, or have different carbon allocation patterns, perhaps using more carbohydrates for protective compounds (i.e., antioxidants) or for repair of ozone-damaged tissues. In any case, the lower growth rates of ozone-resistant individuals indicate that the productivity of the forest will not be enhanced by their increase in the population from either natural selection or as the result of breeding programs.

Gas Exchange and Carbon Gain

Studies of the effects of ozone on the photosynthetic capacity of loblolly pine and slash pine have been summarized in Chapter 11. For these species, the relationships between ozone dose and maximum rates of net photosynthesis are in agreement with those relating ozone dose to growth. Studies in the Southern Commercial Forest Research Cooperative have established that photosynthesis of loblolly pine was negatively correlated with ozone exposure, once a threshold dose of 100 to 150 $\mu l\, l^{-1} \cdot h$ was exceeded (Chapter 11, Figs. 11.4 and 11.5). This threshold is not likely to be exceeded in the first growing season at the current ambient ozone concentrations but almost certainly will be during the second growing season (Chapter 3). Photosynthetic inhibition progressed with additional dose, as cumulative exposures of up to 400 $\mu l\, l^{-1} \cdot h$, caused by treatments of twice-ambient ozone concentrations or higher, inhibited photosynthesis by as much as 80% relative to plants in charcoal-filtered air (Chapter 11). Not all conifer species appear to be as sensitive to ozone as the southern pines—for example, Norway spruce [*Picea abies* (L.) Karst.] (Eamus et al., 1990; Keller and Häsler, 1987; Schweizer and Arndt, 1990) and red spruce (Taylor et al., 1986). Evaluating changes in rates of photosynthesis in relation to ozone dose has been suggested as a useful diagnostic tool to indicate sensitivity to ozone when visible symptoms are lacking (Saxe and Murali, 1989). The similarity between the reduction in photosynthesis and reduction in growth in response to ozone in the southern pines supports this notion.

Since the length of foliage retention in slash, shortleaf, and loblolly pine is usually two growing seasons, the photosynthetic capacity of the current year's foliage will not be affected by ambient ozone concentrations, but the previous year's foliage will have somewhat reduced photosynthetic rates during the growing season. The effect of this reduction in photosynthetic capacity on the total annual carbon gain of the tree is difficult to estimate. The issue may best be addressed through the use of process models because there are so many factors that need to be considered simultaneously, such as the fluctuating light environment within the canopy and the effects that changes in leaf area have on it, environmental influences on photosynthesis, seasonal and age-related changes in photosynthetic capacity, and the cumulative effects of ozone. Using a process model, Dougherty et al. (1992) estimated that the annual carbon gain of loblolly pine trees, at the present ambient ozone concentration, was reduced by about 3% for an average tree, and by as much as 9% for sensitive individuals, relative to the carbon gain expected at one half the ambient ozone dose. The percentage reductions are similar to that expected for growth at the same level of ozone dose.

It is important to note that the estimated reductions in growth, photosynthesis, and simulated carbon gain discussed in this chapter probably represent the maximum reductions that would be expected, since they reflect responses under well-watered conditions. As noted in Chapter 11, the effective dose (i.e., the actual amount of ozone that enters the leaf) may be different under field conditions due to changes in stomatal conductance. The relationship between stomatal conductance and ozone uptake has been well demonstrated (Reich, 1987) and has been indicated as a primary reason for differences in response to ozone among genotypes (Havranek et al., 1990; Taylor, 1994; Wieser and Havranek, 1993). Of all of the factors that affect stomatal conductance, water stress has been most often linked to reduced ozone uptake (Dobson et al., 1990). A prolonged drought has even been correlated with reduced visible damage symptoms on foliage, even in a year with higher than average ozone concentrations (Showman, 1991). An additional factor important to limiting ozone uptake is vapor pressure deficit, which can cause partial stomatal closure even when the soil is moist. Vapor pressure deficits are usually highest in the afternoon, at about the same time that ozone concentrations reach their peak. On the other hand, low vapor pressure deficits allow stomata to remain open and have been correlated with an increase in ozone damage in trees (Jensen and Roberts, 1986). Stomatal conductance will increase or decrease in response to irradiance, soil water availability, vapor pressure deficit, air temperature, carbon dioxide concentration, leaf nutrient concentration, leaf age, and plant age. The large number of factors that can reduce stomatal conductance suggests that ozone uptake under field conditions will often be lower, for a given level of ozone exposure, than that observed under well-watered and fertilized experimental conditions. However, the same conditions that reduce ozone uptake also reduce photosynthesis to some degree and, in this way, may reduce the resources available to the plant to repair damage to the mesophyll caused by ozone.

Implications for Productivity of Pine Forests

It seems probable that ozone is reducing growth of pines in the South at the present ambient concentrations. The maximum potential annual reduction in total seedling biomass appears to be relatively small, within in the range of 2 to 5%, but could be significant to yield given the long life spans of these organisms. The accumulating effects of small reductions in growth are often cited as a reason to be concerned about the long-term effects of ozone in the region. To demonstrate the basis for this concern, consider the following analysis: Based on data and model of Dougherty et al. (1992) for an average loblolly pine tree growing in a plantation, the loss of gross primary production due to the actual ambient concentration of ozone near Athens, Georgia was 3% per year for an average tree. This estimate will be used because it was based on data collected on mature trees, rather than seedlings, that were growing in a plantation. This estimate is more conservative than derived from seedlings, and there is evidence to suggest that loblolly pine trees have lower stomatal conductances than seedlings growing in the same conditions and hence have lower rates of ozone uptake and suffer less damage at the same ambient concentration of ozone (Cregg et al., 1989; Halpin, 1990). To estimate gross primary production, a process-based model was developed to estimate total carbon gain for an individual tree in a plantation over an entire year. Summing the total crown photosynthesis and respiration was the basis for estimating a 3% loss in gross primary production (Dougherty et al., 1992). To calculate the effect that this would have on growth over a 10-year period, the following assumptions were made: (1) 60% of the gross primary production was lost to respiration from all sources (both growth and maintenance) (Dougherty, personal communication; Kinerson, 1975); (2) 35% of the residual carbon was allocated to branch, foliage, and root biomass, leaving 65% in the stem (Kinerson et al., 1977); (3) the pattern of carbon allocation remained constant year to year; (4) annual tree growth rate from year 5 to year 15 was constant at 10% per year; and, (5) the relative effect of ozone on carbon gain was the same every year. Given these assumptions, if the initial stem biomass was 100 kg, at the end of this 10-year period there would be an additional 159 kg of biomass accumulated under pristine conditions and 143 kg accumulated under current ambient ozone concentrations. The difference, 16 kg, amounts to a 10% difference in stemwood biomass.

Obviously, there is uncertainty associated with this estimate; this uncertainty relates to questions of extrapolation of these data to forests with different age structures, more genetic variation, different site and climatic factors, and different insects, diseases, pathogens, competition, and silvicultural regimes. Perhaps most importantly, the uncertainty pertains to the fact that even though these studies extended for as long as three or four years, they covered over only a small fraction of the life span of the trees.

There is no doubt, however, that additional increases in tropospheric ozone will have negative effects on pine growth. While ambient level effects were subtle, effects of above-ambient concentrations were not. Especially given the short duration of the experiments, the large reductions in growth when seedlings were subjected to double the current ambient ozone concentrations indicate that worsening air quality will have significant negative impacts on the pine resource in the region. Assuming a 2% per year increase in tropospheric ozone, an estimate based on remote sensing data by Fishman (1991), we will achieve a 50% increase in ambient ozone concentration in 21 years and a doubling in 35 years. Others have estimated an increase of 40% by the year 2020 (National Academy of Sciences, 1992). If these rates of increase are sustained, concentrations that unequivocally reduce growth could be reached within a single saw-timber rotation or less than two pulpwood rotations.

There are indications that at present, the growth of the southern pines is increasing rather than decreasing, a reversal of trends reported in the mid-1980s (Sheffield et al., 1985). For example, in South Carolina the net annual growth of merchantable pines increased by 18 to 19% between 1986 and 1993, reversing the 43% decline recorded between 1977 and 1986 (Brown, 1993; Koontz and Sheffield, 1993). Net annual growth increased even more substantially (29%) on land receiving more intensive management (Koontz and Sheffield, 1993). A similar pattern was reported for North Carolina (Brown, 1993) and in other states in the region. However, net annual growth is defined as the net change in merchantable volume for a specific year in the absence of cutting (gross growth minus mortality for a specified year) and may reflect a change in the amount of forest with merchantable timber as well as the actual growth rate of stands. But assuming that growth rate is increasing in the region, that does not negate the experimental evidence for a detrimental effect of ozone on growth. It is likely that the effects of ozone cannot be detected on a regional basis because they are masked by other growth-promoting factors. While ozone may be reducing growth, other pollutants in the atmosphere (i.e., carbon dioxide and nitrogen and sulfur compounds) are simultaneously increasing growth. In addition, improvements in genetic selection and silviculture continue to increase growth rates and productivity in managed stands. The effects of these factors are of the magnitude that they could, at least in the near term, mask the deleterious effects of ozone, in effect inadvertently masking the true effects of ozone.

References

Adams MB, Kelly JM, Taylor GE Jr, Edwards NT (1990) Growth of five families of *Pinus taeda* L. during three years of ozone exposure. New Phytol 116: 689–694.

Adams MB, Kelly JM, Edwards NT (1988) Growth of *Pinus taeda* L. seedlings varies with family and ozone exposure level. Water, Air, and Soil Pollut 38:137–150.

Adams RM, Glyer JD, Johnson SL, McCarl BA (1989) A reassessment of the economic effects of ozone on U.S. agriculture. JAPCA 39:960–968.

Allen TFH, Hoekstra TW (1992) *Toward a Unified Ecology*. Columbia University Press, New York.

Alscher RG, Amundson RG, Cumming JR, Fellows S, Fincher J, Rubin G, Leuken P van, Weinstein LH (1989) Seasonal changes in the pigments, carbohydrates and growth of red spruce as affected by ozone. New Phytol 113:211–223.

Amundson RG, Alscher RG, Fellows S, Rubin G, Fincher J, Leuken P van, Weinstein LH, Van Leuken P (1991) Seasonal changes in the pigments, carbohydrates and growth of red spruce as affected by exposure to ozone for two growing seasons. New Phytol 118:127–137.

Bechtold WA, Ruark GA, Lloyd FT (1991) Changing stand structure and regional growth reductions in Georgia's natural pine stands. For Sci 37:703–717.

Berrang P, Karnosky DF, Bennett JP (1989) Natural selection for ozone tolerance in *Populus tremuloides*: field verification. Can J For Res 19:519–522.

Berrang P, Karnosky DF, Mickler RA, Bennett JP (1986) Natural selection for ozone tolerance in *Populus tremuloides*. Can J For Res 16:1214–1216.

Berry CR, Ripperton LA (1963) Ozone, a possible cause of white pine emergence tipburn. Phytopath 53:552–557.

Billen N, Schatzle H, Seufert G, Arndt U (1990) Performance of some growth variables Environ Pollut 68:419–434.
Binkley D, Driscoll CT, Allen HL, Schoeneberger P, McAvoy D (1988) *Acidic Deposition and Forests Soils. Context and Case Studies in the Southeastern United States*. Springer-Verlag, New York.
Blank LW, Roberts TM, Skeffington RA (1988) New perspectives on forest decline. Nature 336:27–30.
Brown MJ (1993) Northern California's forests, 1990. United States Department of Agriculture Forest Service Southeastern Forest Experiment Station Resource Bulletin SE-142, p 101.
Byres DP, Dean TJ, Johnson JD (1992) Long-term effects of ozone and simulated acid rain on the foliage dynamics of slash pine (*Pinus elliottii* var. *elliottii* Engelm.). New Phytol 120:61–67.
Bytnerowicz A, Olszyk DM, Huttunen S, Takemoto B (1989) Effects of photochemical smog on growth injury, and gas exchange of pine seedlings. Can J Bot 67:2175–2181.
Chappelka AH, Chevone BI, Burk TE (1988) Growth response of green and white ash seedlings to ozone, sulfur dioxide, and simulated acid rain. For Sci 34:1016–1029.
Chappelka AH, Lockaby BG, Meldahl RS, Allen HL, Kress LW (1992) Effects of ozone and acidic precipitation after three years on growth of loblolly pine saplings in the piedmont of North Carolina and coastal plain of Alabama. In: Flagler RB (ed) *The Response of Southern Commercial Forests to Air Pollution*. Air and Waste Management Association, Pittsburgh, PA, pp 163–172.
Cleveland GD, Haines LW, Jahromi ST, Bryant RL (1992) No evidence of growth decline in International Paper's southern commercial loblolly and slash pine forests. In: Flagler RB (ed) *The Response of Southern Commercial Forests to Air Pollution*. Air and Waste Management Association, Pittsburgh, PA, pp 37–48.
Cregg BM, Halpin JE, Dougherty PM, Teskey RO (1989) Comparative physiology and morphology of seedling and mature forest trees In: Noble RD, Martin JL, Jensen KF (eds) *Proceedings of the Second US–USSR Symposium—Air Pollution Effects on Vegetation*. USDA Forest Service, Northeast Forest Experiment Station, Broomall, PA, pp 111–118.
Davidson SR, Ashmore MR, Garretty C (1992) Effects of ozone and water deficit on the growth and physiology of *Fagus sylvatica*. For Ecol Manage 51:187–193.
Davis DD, Skelly JM (1992) Growth response of four species of eastern hardwood tree seedlings exposed to ozone. J Air and Waste Manage Assoc 42:309–311.
Dean TJ, Johnson JD (1992) Growth response of young slash pine trees to simulated acid rain and ozone stress. Can J For Res 22:839–848.
Dobson MC, Taylor G, Freer–Smith PH (1990) The control of ozone uptake by *Picea abies* (L.) Karst. and *P. sitchensis* (Bong.) Carr. during drought and interacting effects on shoot water relations. New Phytol 116:465–474.
Dougherty PM, Teskey RO, Jarvis PG (1992) Development of MAESTRO, a process based model for assessing the impact of ozone on net carbon exchange of loblolly pine trees. In: Flagler RB (ed) *The Response of Southern Commercial Forests to Air Pollution*. Air and Waste Management Association, Pittsburg, PA, pp 37–48.
Eamus D, Barnes JD, Mortensen L, Ro–Poulsen H, Davison AW (1990) Persistent stimulation of CO_2 assimilation and stomatal conductance by summer ozone fumigation in Norway spruce. Environ Pollut 63:365–379.
Edwards GS, Edwards NT, Kelly JM, Mays PA (1991) Ozone, acidic precipitation, and soil Mg effects on growth and nutrition of loblolly pine seedlings. Environ Exper Bot 31:67–78.

Edwards NT, Taylor GE Jr (1990) Ozone, acidic rain and soil magnesium effects on growth and foliar pigments of *Pinus taeda* L. Tree Physiol 6:95–104.
Edwards NT, Edwards GL, Kelly JM, Taylor GE Jr (1992) Three-year growth responses of *Pinus taeda* L. to simulated rain chemistry, soil magnesium status, and ozone. Water, Air, and Soil Pollut 63:105–118.
Elliott CL, Eberhardt JC, Brennan EG (1987) The effect of ambient ozone pollution and acidic rain on the growth and chlorophyll content of green and white ash. Environ Pollut 44:61–70.
Evans LS, Leonard MR (1991) Histological determination of ozone injury symptoms of primary needles of giant sequoia (*Sequoiadendron giganteum* Bucch.). New Phytol 117:557–564.
Fishman J (1991) The consequences of increasing tropospheric ozone concentrations. Chemosphere 22:685–695.
Flagler RB, Spruill SE, Chappelka AH, Dean TJ, Kress LW, Reardon JC (1992) Growth of three southern pine species as affected by acid rain and ozone: a combined analysis. In: Flagler RB (ed) *The Response of Southern Commercial Forests to Air Pollution*. Air and Waste Management Association, Pittsburgh, PA, pp 37–48.
Halpin JE (1990) *Phenology and Gas Exchange of Seedlings and Trees of Loblolly Pine (Pinus taeda L.) Under Ambient and Elevated Levels of Ozone*. MS thesis, University of Georgia.
Havranek WM, Pfeifhofer H, Grill D (1990) Pigment contents and gas exchange of low- and high-altitude Norway spruce after chronic exposure to ozone. Forstwissenschaftliches Centralblatt 109:200–209.
Heck WW, Cure WW, Rawlings JO, Zaragoza LJ, Heagle AS, Heggestad HE, Kohut RJ, Kress LW, Temple PJ (1984) Assessing impacts of ozone on agricultural crops: I. crop yield functions and alternative exposure statistics. J Air Pollut Control Assoc 34:810–817.
Heck WW, Taylor OC, Adams R, Bingham G, Miller J, Preston E, Weinstein L (1982) Assessment of crop loss from ozone. J Air Pollut Control Assoc 32:353–361.
Jensen KF, Roberts BR (1986) Changes in yellow poplar stomatal resistance with SO_2 and O_3 fumigation. Environmental Pollut 41:235–245.
Karnosky DF, Gagnon ZE, Reed DD, Witter JA (1992) Effects of genotype on the response of *Populus tremuloides* Michx. to ozone and nitrogen deposition. Water, Air, and Soil Pollut 62:189–199.
Karnosky DF, Dickson RE, Gagnon ZE, Colman MD, Pechter P, Isebrands JG (1993) Genetic variability in ozone response of trees: indicators of sensitivity. Agricoltura Ricerca 15:16–17.
Kauppi PE, Mielikäinen K (1992) Biomass and carbon budget of European forests, 1971 to 1990. Science 256:70–74.
Keller T, Häsler R (1987) Some effects of long-term ozone fumigations on Norway spruce. Trees: Structure and Function 1:129–133.
Kinerson RS (1975) Relationships between plant surface area and respiration in loblolly pine. J Appl Ecol 12:965–971.
Kinerson RS, Ralston CW, Wells CG (1977) Carbon cycling in a loblolly pine plantation. Oecologia 29:1–10.
Koontz BJ, Sheffield RM (1993) Forest Statistics for the southern coastal plain of South Carolina. United States Department of Agriculture Forest Service Southeastern Forest Experiment Station Resource Bulletin SE-140, p 47.
Kress LW, Allen HL (1991) Impact of ozone and acidic precipitation on the growth of loblolly pine seedlings. Proc Air and Waste Manage Assoc 91-142.9:1–10.

Krzak J, Dong PH, Buttner G, Huttermann A, Kramer H, Ulrich B (1988) Photosynthesis, nutrient, growth and soil investigations of a declining Norway spruce (*Picea abies*) stand in the coastal region of northern Germany. For Ecol Manage 24:263–281.

Larsen JB, Yang W, Tiedemann AV (1990) Effects of ozone on gas exchange, frost resistance, flushing and growth of different provenances of European silver fir (*Abies alba* Mill.). Eur J For Pathol 20:211–218.

Laurence JA, Kohut RJ, Amundson RG (1989) Response of red spruce seedlings exposed to ozone and simulated acidic precipitation in the field. Arch Environ Contam Toxicol 18:285–290.

Lee WS, Chevone BI, Seiler JR (1990) Growth response and drought susceptibility of red spruce seedlings exposed to simulated acidic rain and ozone. For Sci 36:265–275.

Lefohn AS, Krupa SV, Winstanley D (1990) Surface ozone exposures measured at clean locations around the world. Environ Pollut 63:189–224.

Lefohn AS, Shadwick DS (1991) Ozone, sulfur dioxide, and nitrogen dioxide trends at rural sites located in the United States. Atmosph Environ 25A: 491–501.

Lucas PW, Cottam DA, Sheppard LJ, Francis BJ (1988) Growth responses and delayed winter hardening in Sitka spruce following summer exposure to ozone. New Phytol 108:495–504.

Lucier AA, Barnard JE (1992) Integration of epidemiological and mechanistic studies of southern pine health and air quality effects. In: Flagler RB (ed) *The Response of Southern Commercial Forests to Air Pollution*. Air and Waste Management Association, Pittsburgh, PA, pp 49–79.

McLaughlin SB, Adams MB, Edwards NT, Hanson PJ, Layton PA, O'Neill EG, Roy WK (1988) *Comparative Sensitivity, Mechanisms, and Whole-Plant Physiological Implications of Responses of Loblolly Pine Genotypes to Ozone and Acid Deposition*. Oak Ridge National Laboratory ORNL/TM-10999, DE 89-001956, pub No 3105. US Gov Printing Office, Washington DC.

Miller PR, McBride JR, Schilling SL, Gomez AP (1989) Trend in ozone damage to conifer forests between 1974 and 1988 in the San Bernardino mountains of southern California. In: Olson RK, Lefohn AS (eds) *Transactions Series No. 16: Effects of Air Pollution on Western Forests*. Air and Waste Management Association, Pittsburgh, PA, pp 309–323.

Miller PR, Millecan AA (1971) Extent of oxidant air pollution damage to some pines and other conifers in California. Plant Dis Rep 55:555–559.

Mortensen LM, Skre O (1990) Effects of low ozone concentrations on growth of *Betula pubescens* Ehrh., *Betula verrucosa* Ehrh. and *Alnus incana* (L.) Moench. New Phytol 115:165–170.

Mudano JE, Allen HL, Kress LW (1992) Stem and foliage elongation of young loblolly pine as affected by ozone. For Sci 38:324–335.

Muir PS, Armentano TV (1988) Evaluating oxidant injury to foliage of *Pinus ponderosa*: a comparison of methods. Can J For Res 18:498–505.

National Academy of Sciences (1992) *Rethinking the Ozone Problem in Urban and Regional Air Pollution*. National Academy Press, Washington, DC.

Patton RL, Jensen KF, Schier GA (1991) Responses of red spruce seedlings to ozone and acid deposition. Can J For Res 21:1354–1359.

Peart DR, Nicholas NS, Zedaker SM, Miller-Weeks MM, Siccama TG (1992) Condition and recent trends in high-elevation red spruce populations. In: Eagar C, Adams MB (eds) Ecology and Decline of Red Spruce in the Eastern United States, Springer-Verlag, New York, pp 125–191.

Peters RH (1991) *A Critique for Ecology*. Cambridge University Press, Cambridge.

Peterson DL, Arbaugh MJ, Wakefield VA, Miller PR (1987) Evidence of growth reduction in ozone-injured Jeffrey pine (*Pinus jeffreyi* Grev. and Balf.) in Sequoia and Kings Canyon National Parks. J Air Pollut Control Assoc 37: 906–912.

Pinkerton JE, Lefohn AS (1987) The characterization of ozone data for sites located in forested areas of the eastern United States. JAPCA 37:1005–1010.

Prinz B, Krause GHM, Jung KD (1987) Development and causes of novel forest decline in Germany. In: Hutchinson TC, Meema KM (eds) *Effects of Atmospheric Pollutants on Forests, Wetlands and Agricultural Ecosystems.* NATO ASI Series G (Ecological Sciences) 16, Springer-Verlag, Berlin, pp 1–24.

Pye JM (1988) Impact of ozone on the growth and yield of trees: A review. J Environ Qual 17:347–360.

Qui Z, Chappelka AH, Somers GL, Lockaby BG, Meldahl RS (1992) Effects of ozone and simulated acidic precipitation on above- and below-ground growth of loblolly pine (*Pinus taeda*). Can J For Res 22:582–587.

Reich PB (1987) Quantifying plant response to ozone: a unifying theory. Tree Physiol 3:63–91.

Reich PB, Schoettle AW, Stroo HF, Troiano J (1987) Effects of ozone and acid rain on white pine (*Pinus strobus*) seedlings grown in five soils. I. Net photosynthesis and growth. Can J Bot 65:977–987.

Reinert RA, Shafer SR, Eason G, Horton SJ, Schoeneberger MM, Wells C (1988) Responses of loblolly pine half-sib families to ozone. In: *81st Annual Meeting of Air Pollution Control Assoc*, Dallas, TX, June 19–24. Air and Waste Manage Assoc, Pittsburgh, PA, pp 1–14.

Rezabek CL, Morton JA, Mosher EC (1989) Regional effects of sulfur dioxide and ozone on eastern white pine (*Pinus strobus*) in eastern Wisconsin. Plant Dis 73:70–73.

Ruark GA, Thomas CE, Bechtold WA, May DM (1991) Growth reductions in naturally regenerated southern pine stands in Alabama and Georgia. South J Appl For 15:73–79.

Saxe H, Murali NS (1989) Diagnostic parameters for selecting against novel spruce (*Picea abies*) decline: III. Response of photosynthesis and transpiration to O_3 exposures. Physiolog Plantar 76:356–361.

Scherzer AJ, McClenahen JR (1989) Effects of ozone or sulfur dioxide on pitch pine seedlings. J Environ Qual 18:57–61.

Schier GA, McQuattie CJ, Jensen KF (1990) Effect of ozone and aluminum on pitch pine (*Pinus rigida*) seedlings: growth and nutrient relations. Can J For Res 20:1714–1719.

Schulze ED (1989) Air pollution and forest decline in a spruce (*Picea abies*) forest. Science 244:776–783.

Schweizer B, Arndt U (1990) CO_2/H_2O gas exchange parameters of one- and two-year-old needles of spruce and fir. Environ Pollut 68:275–292.

Shafer SR, Heagle AS (1989) Growth responses of field-grown loblolly pine to chronic doses of ozone during multiple growing seasons. Can J For Res 19:821–831.

Shafer SR, Heagle AS, Camberato DM (1987) Effects of chronic doses of ozone on field-grown loblolly pine: seedling responses in the first year. JAPCA 37:1179–1184.

Shafer SR, Reinert RA, Eason G, Spruill SE (1993) Analysis of ozone concentration–biomass response relationships among open-pollinated families of loblolly pine. Can J For Res 23:706–715.

Sheffield RM, Cost ND, Bechtold WA, McClure JP (1985) *Pine Growth Reductions in the Southeast*. USDA Forest Service Resource Bull SE-83, USDA Forest Service Southeastern Forest Exp Station, Asheville NC.
Shelburne VB, Reardon JC, Paynter VA (1993) The effects of acid rain and ozone on biomass and leaf area parameters of shortleaf pine (*Pinus echinata* Mill.). Tree Physiol 12:163–172.
Showman RE (1991) A comparison of ozone injury to vegetation during moist and drought years. J Air and Waste Manag Assoc 41:63–64.
Simini M, Skelly JM, Davis DD, Savage JE, Comrie AC (1992) Sensitivity of four hardwood species to ambient ozone in north central Pennsylvania. Can J For Res 22:1789–1799.
Skelly JM (1989) Forest decline versus tree decline—the pathological considerations. Environ Monitor Assess 12:23–27.
Spruill SE, Richter D, Gumpertz ML, Rawlings JO, Allen HL (1993) Use of edaphic variables to control experimental error: a case study on blocking and use of covariance. Soil Sci 156:156–162.
Stow TK, Allen HL, Kress LW (1992) Ozone impacts on seasonal foliage dynamics of young loblolly pine. For Sci 38:102–119.
Taylor G, Frost DL (1992) Impact of gaseous air pollution on leaf growth of hybrid poplar. For Ecol and Manage 51:151–162.
Taylor, GE Jr (1994) Plant and environment interactions. J Environ Qual 23: 63–82.
Taylor GE Jr, Norby RJ, McLaughlin SB, Johnson AH, Turner RS (1986) Carbon dioxide assimilation and growth of red spruce (*Picea rubens* Sarg.) seedlings in response to ozone, precipitation chemistry, and soil type. Oecologia 70:163–171.
Temple PJ (1988) Injury and growth of Jeffrey pine and giant sequoia in response to ozone and acidic mist. Environ Exper Bot 28:323–333.
Temple PJ, Riechers GH, Miller PR, Lennox RW (1993) Growth responses of ponderosa pine to long-term exposure to ozone, wet and dry acidic deposition, and drought. Can J For Res 23:59–66.
Tjoelker MG, Luxmoore RJ (1991) Soil nitrogen and chronic ozone stress influence physiology, growth and nutrient status of *Pinus taeda* L. and *Liriodendron tulipfera* L. seedlings. New Phytol 119:69–81.
Turner RS, Olson RJ, Brandt, CC (1986) Areas having soil characteristics that may indicate sensitivity to acidic deposition under alternative forest damage hypothese. Environmental Sciences Division Publications no 2720, Oak Ridge National Laboratory, Oak Ridge TN.
Ulrich B (1987) Stability, elasticity, and resilience of terrestrial ecosystems with respect to matter balance. In: Schulze ED, Zwolfer H (eds) *Ecological Studies*, Vol 61. Springer-Verlag, Berlin–Heidelberg, pp 11–49.
Vann DR, Strimbeck GR, Johnson AH (1992) Effects of ambient levels of airborne chemicals on freezing resistance of red spruce foliage. For Ecol Manage 51:69–79.
Wang D, Bormann FH, Karnosky DF (1986) Regional tree growth reductions due to ambient ozone: evidence from field experiments. Environ Sci Technol 20:1122–1125.
Wieser G, Havranek WM (1993) Ozone uptake in the sun and shade crown of spruce: quantifying the physiological effects of ozone exposure. Trees: Struct Func 7:227–232.
Williams WT, Williams JA (1986) Effects of oxidant air-pollution on needle health and annual-ring width in a ponderosa pine forest. Environ Conserv 13:229–234.

Winner WE, Cotter IS, Powers HR Jr, Skelly JM (1987) Screening loblolly pine seedling responses to SO_2 and O_3: analysis of families differing in resistance to fusiform rust disease. Environ Pollut 47:205–220.
Winner WE, Lefohn AS, Cotter IS, Greitner CS, Nellessen J, McEvoy LR Jr, Olson RL, Atkinson CJ, Moore LD (1989) Plant responses to elevational gradients of O_3 exposures in Virginia. Proc Natl Acad Sci USA 86:8828–8832.
Wiselogel AE, Bailey JK, Newton RJ, Fong F (1991) Growth of loblolly pine (*Pinus taeda*, L.) seedlings to ozone fumigation. Environ Pollut 71:43–56.
Zeide B (1992) Reevaluation of forest inventory data from loblolly pine stands in the Georgia piedmont and mountain areas. In: Flagler RB (ed) *The Response of Southern Commercial Forests to Air Pollution*. Air and Waste Management Association, Pittsburgh, PA, pp 17–36.

Index

O

Ecological Studies

Volume 68
Stable Isotopes in Ecological Research (1989)
P.W. Rundel, J.R. Ehleringer, and K.A. Nagy (Eds.)

Volume 69
Vertebrates in Complex Tropical Systems (1989)
M.L. Harmelin-Vivien and F. Bourlière (Eds.)

Volume 70
The Northern Forest Border in Canada and Alaska (1989)
J.A. Larsen

Volume 71
Tidal Flat Estuaries: Simulation and Analysis of the Ems Estuary (1988)
J. Baretta and P. Ruardij (Eds.)

Volume 72
Acidic Deposition and Forest Soils (1989)
D. Binkley, C.T. Driscoll, H.L. Allen, P. Schoeneberger, and D. McAvoy

Volume 73
Toxic Organic Chemicals in Porous Media (1989)
Z. Gerstl, Y. Chen, U. Mingelgrin, and B. Yaron (Eds.)

Volume 74
Inorganic Contaminants in the Vadose Zone (1989)
B. Bar-Yosef, N.J. Barrow, and J. Goldshmid (Eds.)

Volume 75
The Grazing Land Ecosystems of the African Sahel (1989)
H.N. Le Houérou

Volume 76
Vascular Plants as Epiphytes: Evolution and Ecophysiology (1989)
U. Lüttge (Ed.)

Volume 77
Air Pollution and Forest Decline: A Study of Spruce (*Picea abies*) on Acid Soils (1989)
E.-D. Schulze, O.L. Lange, and R. Oren (Eds.)

Volume 78
Agroecology: Researching the Ecological Basis for Sustainable Agriculture (1990)
S.R. Gliessman (Ed.)

Volume 79
Remote Sensing of Biosphere Functioning (1990)
R.J. Hobbs and H.A. Mooney (Eds.)

Volume 80
Plant Biology of the Basin and Range (1990)
C.B. Osmond, G.M. Hidy, and L.F. Pitelka (Eds.)

Volume 81
Nitrogen in Terrestrial Ecosystem: Questions of Productivity, Vegetational Changes, and Ecosystem Stability (1990)
C.O. Tamm

Volume 82
Quantitative Methods in Landscape Ecology: The Analysis and Interpretation of Landscape Heterogeneity (1990)
M.G. Turner and R.H. Gardner (Eds.)

Volume 83
The Rivers of Florida (1990)
R.J. Livingston (Ed.)

Volume 84
Fire in the Tropical Biota: Ecosystem Processes and Global Challenges (1990)
J.G. Goldammer (Ed.)

Volume 85
The Mosaic-Cycle Concept of Ecosystems (1991)
H. Remmert (Ed.)

Volume 86
Ecological Heterogeneity (1991)
J. Kolasa and S.T.A. Pickett (Eds.)

Volume 87
Horses and Grasses: The Nutritional Ecology of Equids and Their Impact on the Camargue (1992)
P. Duncan

Volume 88
Pinnipeds and El Niño: Responses to Environmental Stress (1992)
F. Trillmich and K.A. Ono (Eds.)